# Cisco Certified Design Expert CCDE 400-007 Official Cert Guide

## Companion Website and Pearson Test Prep Access Code

Access interactive study tools on this book's companion website, including practice test software, review exercises, Key Term flash card application, a study planner, and more!

To access the companion website, simply follow these steps:

1. Go to **www.ciscopress.com/register**.
2. Enter the **print book ISBN: 9780137601042**
3. Answer the security question to validate your purchase.
4. Go to your account page.
5. Click on the **Registered Products** tab.
6. Under the book listing, click on the **Access Bonus Content** link.

When you register your book, your Pearson Test Prep practice test access code will automatically be populated with the book listing under the Registered Products tab. You will need this code to access the practice test that comes with this book. You can redeem the code at PearsonTestPrep.com. Simply choose Pearson IT Certification as your product group and log into the site with the same credentials you used to register your book. Click the Activate New Product button and enter the access code. More detailed instructions on how to redeem your access code for both the online and desktop versions can be found on the companion website.

If you have any issues accessing the companion website or obtaining your Pearson Test Prep practice test access code, you can contact our support team by going to pearsonitp.echelp.org.

# Cisco Certified Design Expert

## CCDE 400-007

## Official Cert Guide

ZIG ZSIGA

Cisco Press

# Cisco Certified Design Expert CCDE 400-007 Official Cert Guide

Zig Zsiga

Marwan Alshawi's best-selling Cisco Press book, the *CCDE Study Guide*, served as the inspiration and content basis for this new Official Cert Guide which covers the latest CCDE exam blueprint.

ScoutAutomatedPrintCode

Library of Congress Control Number: 2023902257

ISBN-13: 978-0-13-760104-2

ISBN-10: 0-13-760104-2

## Warning and Disclaimer

This book is designed to provide information about the Cisco Certified Design Expert CCDE 400-007 exam. Every effort has been made to make this book as complete and as accurate as possible, but no warranty or fitness is implied.

## Trademark Acknowledgments

## Special Sales

For information about buying this title in bulk quantities, or for special sales opportunities (which may include electronic versions; custom cover designs; and content particular to your business, training goals, marketing focus, or branding interests), please contact our corporate sales department at corpsales@pearsoned.com or (800) 382-3419.

For government sales inquiries, please contact governmentsales@pearsoned.com.

For questions about sales outside the U.S., please contact intlcs@pearson.com.

## Feedback Information

At Cisco Press, our goal is to create in-depth technical books of the highest quality and value. Each book is crafted with care and precision, undergoing rigorous development that involves the unique expertise of members from the professional technical community.

Readers' feedback is a natural continuation of this process. If you have any comments regarding how we could improve the quality of this book, or otherwise alter it to better suit your needs, you can contact us through email at feedback@ciscopress.com. Please make sure to include the book title and ISBN in your message.

We greatly appreciate your assistance.

**Vice President, ITP Professional:** Mark Taub

**Alliances Manager, Cisco Press:** Arezou Gol

**Director, ITP Product Management:** Brett Bartow

**Executive Editor:** Nancy Davis

**Managing Editor:** Sandra Schroeder

**Development Editor:** Ellie Bru

**Senior Project Editor:** Mandie Frank

**Copy Editor:** Bill McManus

**Technical Editors:** Martin J. Duggan, Nicholas Russo

**Editorial Assistant:** Cindy Teeters

**Designer:** Chuti Prasertsith

**Composition:** codeMantra

**Indexer:** Ken Johnson

**Proofreader:** Barbara Mack

··|··|··
CISCO.

**Americas Headquarters**
Cisco Systems, Inc.
San Jose, CA

**Asia Pacific Headquarters**
Cisco Systems (USA) Pte. Ltd.
Singapore

**Europe Headquarters**
Cisco Systems International BV Amsterdam,
The Netherlands

Cisco has more than 200 offices worldwide. Addresses, phone numbers, and fax numbers are listed on the Cisco Website at www.cisco.com/go

Cisco and the Cisco logo are trademarks or registered trademarks of Cisco and/or its affiliates in the U.S. and other countries. To view a list of Cisco trad to this URL: www.cisco.com/go/trademarks. Third party trademarks mentioned are the property of their respective owners. The use of the word partner a partnership relationship between Cisco and any other company. (1110R)

# Figure Credits

| | |
|---|---|
| Figure 4-1 | Al-shawi, M, CCDE Study Guide, 1st Ed, ©2016 Reprinted by permission of Pearson Education, Inc |
| Figure 4-2 | Al-shawi, M, CCDE Study Guide, 1st Ed, ©2016 Reprinted by permission of Pearson Education, Inc |
| Figure 4-3 | Al-shawi, M, CCDE Study Guide, 1st Ed, ©2016 Reprinted by permission of Pearson Education, Inc |
| Figure 6-1 | Al-shawi, M, CCDE Study Guide, 1st Ed, ©2016 Reprinted by permission of Pearson Education, Inc |
| Figure 6-2 | Al-shawi, M, CCDE Study Guide, 1st Ed, ©2016 Reprinted by permission of Pearson Education, Inc |
| Figure 6-3 | Al-shawi, M, CCDE Study Guide, 1st Ed, ©2016 Reprinted by permission of Pearson Education, Inc |
| Figure 6-4 | Al-shawi, M, CCDE Study Guide, 1st Ed, ©2016 Reprinted by permission of Pearson Education, Inc |
| Figure 6-5 | Al-shawi, M, CCDE Study Guide, 1st Ed, ©2016 Reprinted by permission of Pearson Education, Inc |
| Figure 6-6 | Al-shawi, M, CCDE Study Guide, 1st Ed, ©2016 Reprinted by permission of Pearson Education, Inc |
| Figure 6-7 | Al-shawi, M, CCDE Study Guide, 1st Ed, ©2016 Reprinted by permission of Pearson Education, Inc |
| Figure 6-8 | Al-shawi, M, CCDE Study Guide, 1st Ed, ©2016 Reprinted by permission of Pearson Education, Inc |
| Figure 6-9 | Al-shawi, M, CCDE Study Guide, 1st Ed, ©2016 Reprinted by permission of Pearson Education, Inc |
| Figure 6-10 | Al-shawi, M, CCDE Study Guide, 1st Ed, ©2016 Reprinted by permission of Pearson Education, Inc |
| Figure 6-11 | Al-shawi, M, CCDE Study Guide, 1st Ed, ©2016 Reprinted by permission of Pearson Education, Inc |
| Figure 6-12 | Al-shawi, M, CCDE Study Guide, 1st Ed, ©2016 Reprinted by permission of Pearson Education, Inc |
| Figure 6-13 | Al-shawi, M, CCDE Study Guide, 1st Ed, ©2016 Reprinted by permission of Pearson Education, Inc |
| Figure 6-14 | Al-shawi, M, CCDE Study Guide, 1st Ed, ©2016 Reprinted by permission of Pearson Education, Inc |
| Figure 6-15 | Al-shawi, M, CCDE Study Guide, 1st Ed, ©2016 Reprinted by permission of Pearson Education, Inc |

Figure 9-24    Al-shawi, M, CCDE Study Guide, 1st Ed, ©2016 Reprinted by permission of Pearson Education, Inc

Figure 9-25    Al-shawi, M, CCDE Study Guide, 1st Ed, ©2016 Reprinted by permission of Pearson Education, Inc

Figure 9-26    Al-shawi, M, CCDE Study Guide, 1st Ed, ©2016 Reprinted by permission of Pearson Education, Inc

Figure 9-27    Al-shawi, M, CCDE Study Guide, 1st Ed, ©2016 Reprinted by permission of Pearson Education, Inc

Figure 9-28    Al-shawi, M, CCDE Study Guide, 1st Ed, ©2016 Reprinted by permission of Pearson Education, Inc

Figure 9-29    Al-shawi, M, CCDE Study Guide, 1st Ed, ©2016 Reprinted by permission of Pearson Education, Inc

Figure 9-30    Al-shawi, M, CCDE Study Guide, 1st Ed, ©2016 Reprinted by permission of Pearson Education, Inc

Figure 9-31    Al-shawi, M, CCDE Study Guide, 1st Ed, ©2016 Reprinted by permission of Pearson Education, Inc

Figure 9-32    Al-shawi, M, CCDE Study Guide, 1st Ed, ©2016 Reprinted by permission of Pearson Education, Inc

Figure 9-33    Al-shawi, M, CCDE Study Guide, 1st Ed, ©2016 Reprinted by permission of Pearson Education, Inc

Figure 9-34    Al-shawi, M, CCDE Study Guide, 1st Ed, ©2016 Reprinted by permission of Pearson Education, Inc

Figure 10-1    Al-shawi, M, CCDE Study Guide, 1st Ed, ©2016 Reprinted by permission of Pearson Education, Inc

Figure 10-2    Al-shawi, M, CCDE Study Guide, 1st Ed, ©2016 Reprinted by permission of Pearson Education, Inc

Figure 10-3    Al-shawi, M, CCDE Study Guide, 1st Ed, ©2016 Reprinted by permission of Pearson Education, Inc

Figure 10-4    Al-shawi, M, CCDE Study Guide, 1st Ed, ©2016 Reprinted by permission of Pearson Education, Inc

Figure 10-5    Al-shawi, M, CCDE Study Guide, 1st Ed, ©2016 Reprinted by permission of Pearson Education, Inc

Figure 10-6    Al-shawi, M, CCDE Study Guide, 1st Ed, ©2016 Reprinted by permission of Pearson Education, Inc

Figure 11-1    Henry, J, CCNP Wireless Design, ©2021 Reprinted by permission of Pearson Education, Inc

Figure 13-15  Al-shawi, M, CCDE Study Guide, 1st Ed, ©2016 Reprinted by permission of Pearson Education, Inc

Figure 13-16  Al-shawi, M, CCDE Study Guide, 1st Ed, ©2016 Reprinted by permission of Pearson Education, Inc

Figure 13-17  Al-shawi, M, CCDE Study Guide, 1st Ed, ©2016 Reprinted by permission of Pearson Education, Inc

Figure 13-18  Al-shawi, M, CCDE Study Guide, 1st Ed, ©2016 Reprinted by permission of Pearson Education, Inc

Figure 14-1  Al-shawi, M, CCDE Study Guide, 1st Ed, ©2016 Reprinted by permission of Pearson Education, Inc

Figure 14-2  Al-shawi, M, CCDE Study Guide, 1st Ed, ©2016 Reprinted by permission of Pearson Education, Inc

Figure 14-3  Al-shawi, M, CCDE Study Guide, 1st Ed, ©2016 Reprinted by permission of Pearson Education, Inc

Figure 14-4  Al-shawi, M, CCDE Study Guide, 1st Ed, ©2016 Reprinted by permission of Pearson Education, Inc

Figure 14-5  Al-shawi, M, CCDE Study Guide, 1st Ed, ©2016 Reprinted by permission of Pearson Education, Inc

Figure 14-6  Al-shawi, M, CCDE Study Guide, 1st Ed, ©2016 Reprinted by permission of Pearson Education, Inc

Figure 14-7  Al-shawi, M, CCDE Study Guide, 1st Ed, ©2016 Reprinted by permission of Pearson Education, Inc

Figure 14-8  Al-shawi, M, CCDE Study Guide, 1st Ed, ©2016 Reprinted by permission of Pearson Education, Inc

Figure 14-9  Al-shawi, M, CCDE Study Guide, 1st Ed, ©2016 Reprinted by permission of Pearson Education, Inc

Figure 14-10  Al-shawi, M, CCDE Study Guide, 1st Ed, ©2016 Reprinted by permission of Pearson Education, Inc

Figure 14-11  Al-shawi, M, CCDE Study Guide, 1st Ed, ©2016 Reprinted by permission of Pearson Education, Inc

Figure 14-12  Al-shawi, M, CCDE Study Guide, 1st Ed, ©2016 Reprinted by permission of Pearson Education, Inc

Figure 14-13  Al-shawi, M, CCDE Study Guide, 1st Ed, ©2016 Reprinted by permission of Pearson Education, Inc

Figure 14-14  Al-shawi, M, CCDE Study Guide, 1st Ed, ©2016 Reprinted by permission of Pearson Education, Inc

Figure 17-17  Al-shawi, M, CCDE Study Guide, 1st Ed, ©2016 Reprinted by permission of Pearson Education, Inc

Figure 17-18  Al-shawi, M, CCDE Study Guide, 1st Ed, ©2016 Reprinted by permission of Pearson Education, Inc

Figure 17-19  Al-shawi, M, CCDE Study Guide, 1st Ed, ©2016 Reprinted by permission of Pearson Education, Inc

Figure 17-20  Al-shawi, M, CCDE Study Guide, 1st Ed, ©2016 Reprinted by permission of Pearson Education, Inc

Figure 17-21  Al-shawi, M, CCDE Study Guide, 1st Ed, ©2016 Reprinted by permission of Pearson Education, Inc

Figure 17-22  Al-shawi, M, CCDE Study Guide, 1st Ed, ©2016 Reprinted by permission of Pearson Education, Inc

Figure 17-23  Al-shawi, M, CCDE Study Guide, 1st Ed, ©2016 Reprinted by permission of Pearson Education, Inc

Figure 17-24  Al-shawi, M, CCDE Study Guide, 1st Ed, ©2016 Reprinted by permission of Pearson Education, Inc

Figure 17-25  Al-shawi, M, CCDE Study Guide, 1st Ed, ©2016 Reprinted by permission of Pearson Education, Inc

Figure 17-26  Al-shawi, M, CCDE Study Guide, 1st Ed, ©2016 Reprinted by permission of Pearson Education, Inc

Figure 17-27  Al-shawi, M, CCDE Study Guide, 1st Ed, ©2016 Reprinted by permission of Pearson Education, Inc

Figure 17-28  Al-shawi, M, CCDE Study Guide, 1st Ed, ©2016 Reprinted by permission of Pearson Education, Inc

Figure 17-29  Al-shawi, M, CCDE Study Guide, 1st Ed, ©2016 Reprinted by permission of Pearson Education, Inc

Figure 17-30  Al-shawi, M, CCDE Study Guide, 1st Ed, ©2016 Reprinted by permission of Pearson Education, Inc

## About the Author

**Zig Zsiga**, CCDE 2016::32, CCIE #44883, has been in the networking industry for 20 years. He is currently a principal architect supporting the Cisco CX US public sector business and customers. Zig holds an active CCDE and two CCIE certifications, one in Routing and Switching and the second in Service Provider. He also holds a bachelor of science in computer science from Park University. He is a father, a husband, a United States Marine, a gamer, a nerd, a geek, and a big soccer fan. Zig loves all technology and can usually be found in the lab learning and teaching others. This is his second published book, and he is also the host of the *Zigbits Network Design Podcast* (ZNDP), where he interviews leading industry experts about network design. All of Zig's content is located at https://zigbits.tech. Zig lives in Upstate New York, USA, with his wife, Julie, and their son, Gunnar.

## About the *CCDE Study Guide* by Marwan Alshawi

Marwan Alshawi's best-selling Cisco Press book, the *CCDE Study Guide*, served as the inspiration and content basis for this new Official Cert Guide which covers the latest CCDE exam blueprint.

First published by Cisco Press in 2016, the *CCDE Study Guide* by Marwan Alshawi (CCDE 2013:66) has been the go-to resource for anyone studying for the CCDE exam. Marwan's work is well-respected in the industry for the way it coached networking professionals on how to prepare for the rigorous CCDE exam.

Marwan Alshawi is a technology and cloud solutions architect who has been in the IT industry for more than 16 years. He's been involved in architecting and designing various IT solutions with different systems integrators, technology vendors, and cloud service providers. Marwan has multiple publications focused on design and architecture including blog posts, whitepapers, and the following books from Cisco Press: *CCDE Study Guide and Designing for Cisco Network Service Architectures (ARCH) Foundation Learning Guide, Fourth Edition*, as well as a cloud networking design course on Udemy. He holds a Master of Science degree in internetworking from the University of Technology, Sydney.

## About the Technical Reviewers

**Martin J. Duggan**, CCDE #2016::6 and CCIE #7942, is a principal network architect designing network solutions for global financial accounts at Systal Technology Solutions. Martin gained his CCIE Routing and Switching certification in 2001 and has been passionate about Cisco qualifications and mentoring ever since. He wrote the *CCIE Routing and Switching Practice Labs* series and *CCDE v3 Practice Labs* titles for Cisco Press and provides content for multiple Cisco exam tracks. Martin resides in the UK and enjoys gliding, cycling, snowboarding, and karate when not designing networks. Follow Martin on Twitter @Martinccie7942.

**Nicholas (Nick) Russo**, CCDE #20160041 and CCIE #42518, is an internationally recognized expert in IP/MPLS networking and design. To grow his skillset, Nick has been focused on advancing network DevOps via automation for his clients. Recently, Nick has been sharing his knowledge through online video training and speaking at industry conferences. Nick also holds a bachelor of science in computer science from the Rochester Institute of Technology (RIT). Nick lives in Maryland, USA, with his wife, Carla, and daughters, Olivia and Josephine.

## Dedications

You, the reader, will never truly know the journey this book endured to reach completion. I would like to share with you the CliffsNotes version, so please bear with me. I had truly wanted to complete this book months ago...even a year ago. I wanted this book in your hands as it is now, to be used as a resource for making better network design decisions and for passing the CCDE exam. But good intentions are just that...intentions. As with anything, there is a path a book must take. This book most definitely took the path that was not paved. It had to persevere through a global pandemic, numerous family emergencies, including my own, and a number of other global events that we don't have the time to dive into here.

This book made it through it all. I promise you it is not perfect; I can only imagine what I missed or got wrong in this process, but it is here, and it is a resource for you, the network designer and CCDE candidate. I hope as you read through these chapters you feel the passion I have for this industry, network design, and the CCDE. This book, with all of the time and energy put into it, is for you.

I dedicate this book to you, the reader, the CCDE candidate, and the network design expert. May it help you on your network design journey!

# Acknowledgments

I had this great idea to write a book, and while I had limited experience in the area of writing, I figured it was going to be a piece of cake. I was far from the truth. Writing this book was much harder than I thought it would be. I didn't know what I didn't know and, frankly, this book would not have been possible without the help of many people.

Thank you, Marwan Al-shawi, for graciously letting me leverage the truly outstanding content in your *CCDE Study Guide*. Your content is truly remarkable, and it didn't make sense to re-create it when it still applies today. Thank you!

Thank you to my two technical editors, Martin Duggan and Nick Russo. You both kept me honest and humble throughout this experience. I could have not asked for a better technical team. I can only imagine the thoughts that went through your heads when reading some of the chapter drafts…"What is Zig doing now?" I thank you both for being great network designers, and better friends than I could ever ask for.

Thank you, Dave Lucas, for your contributions in Chapter 8. I sincerely appreciate you.

Thank you to my Cisco leadership team: Mike Solomita, Maurice DuPas, Jim Lien, and Fred Mooney. You all gave me the space, time, and overwhelming encouragement to make this project happen. Thank you for always supporting me and my career aspirations.

Thank you, Elaine Lopez and Mark Holm, as without the two of you…well, I can't even imagine where we would all be without the CCDE. You both have been great mentors, colleagues, and friends. Thank you for always being available for my random and crazy questions. Mark, we have some work to do, my friend!

Thank you, Nancy Davis and Ellie Bru. This entire process would have been truly unbearable without the two of you. You have been with me from the start. Thank you for being so understanding about all the life events that happened during the writing of this book. Thank you for always being willing to help and guide me throughout this journey. From helping me with writer's block to creating figures to author reviews, you two have been truly amazing. You are truly the A team!

Thank you to my wife, Julie. Julie, you have always supported me with everything I thrive to do in my life. You gave me critical advice throughout this journey when I truly needed it. You kept me honest and let me know when I was just being dumb, which happened a few times. You were my sounding board on all things as they happened in real time. You are my rock, my constant, and my muse. I am truly lucky to be able to journey through our life side by side. Thank you from the bottom of my heart and always remember I love you the Everest! Kilo!

To my son, Gunnar. This book, and the journey I took to write it, is a perfect example of how you can set a goal, attack that goal, and achieve it. Will you know the steps to take to achieve every goal in your life? Most definitely not. I didn't know half of the steps to complete this book, but here it is. Will there be roadblocks, pitfalls, and hurdles in the way? Of course. What matters is what you do when you encounter one. It's what you do when you find something standing in your way of achieving that goal. Will you have to sacrifice to make it happen? Most likely. That sacrifice might be time, energy, or sleep, but you will most likely have to endure it to achieve your goal. Once again, take this as an example that you can literally do anything you set your mind to in life; just set your mind to it and make it happen.

## Reader Services

Register your copy at www.ciscopress.com/title/9780137601042 for convenient access to downloads, updates, and corrections as they become available. To start the registration process, go to www.ciscopress.com/register and log in or create an account. Enter the product ISBN, 9780137601042, and click Submit. When the process is complete, you will find any available bonus content under Registered Products.

*Be sure to check the box that you would like to hear from us to receive exclusive discounts on future editions of this product.

# Contents at a Glance

**Online Elements:**

# Contents

## Icons Used in This Book

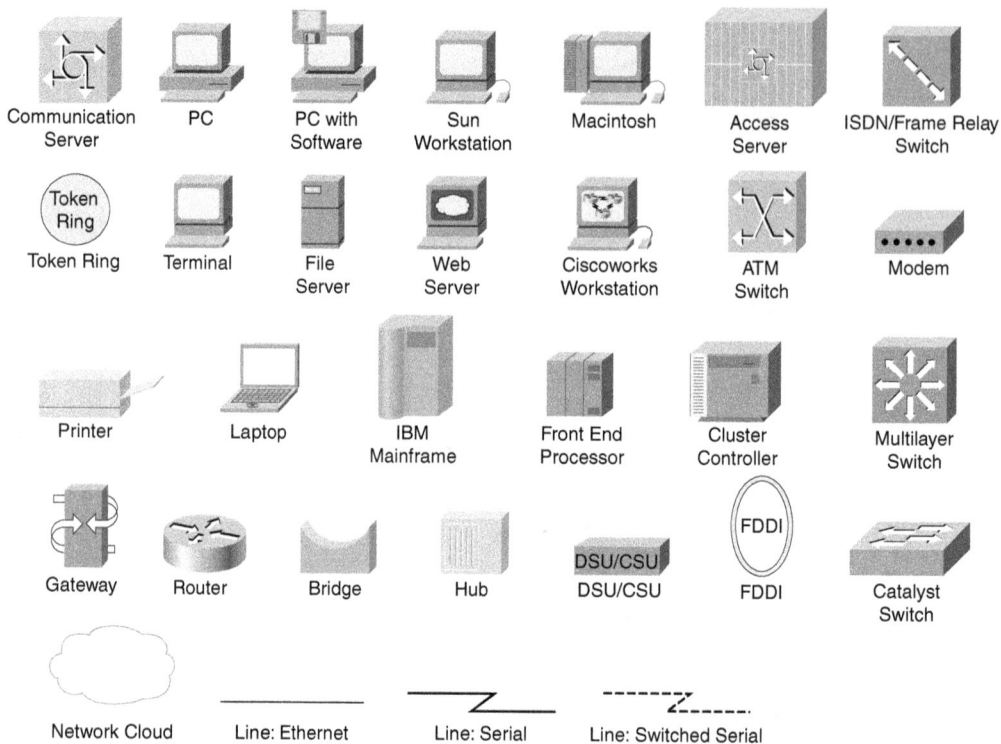

Communication Server   PC   PC with Software   Sun Workstation   Macintosh   Access Server   ISDN/Frame Relay Switch

Token Ring   Terminal   File Server   Web Server   Ciscoworks Workstation   ATM Switch   Modem

Printer   Laptop   IBM Mainframe   Front End Processor   Cluster Controller   Multilayer Switch

Gateway   Router   Bridge   Hub   DSU/CSU   FDDI   Catalyst Switch

Network Cloud   Line: Ethernet   Line: Serial   Line: Switched Serial

## Command Syntax Conventions

The conventions used to present command syntax in this book are the same conventions used in the IOS Command Reference. The Command Reference describes these conventions as follows:

- **Boldface** indicates commands and keywords that are entered literally as shown. In actual configuration examples and output (not general command syntax), boldface indicates commands that are manually input by the user (such as a **show** command).

- *Italic* indicates arguments for which you supply actual values.

- Vertical bars (|) separate alternative, mutually exclusive elements.

- Square brackets ([ ]) indicate an optional element.

- Braces ({ }) indicate a required choice.

- Braces within brackets ([{ }]) indicate a required choice within an optional element.

# Introduction

Congratulations! If you are reading this Introduction, then you likely have decided to pursue the Cisco Certified Design Expert (CCDE) certification, a mostly vendor-agnostic, expert-level network design certification. Network design is imperative to implementing a successful, secure network, yet networking professionals typically are not taught network design, including the fundamentals, principles, techniques, frameworks, and pitfalls, early enough in their careers.

One of the major difficulties of network design is that it can't be pre-scripted or templated, because each design decision, even if it's comparing the same design options, can have very different business drivers, outcomes, constraints, and overall circumstances that create a unique situation requiring a different design decision to be made. This is why this book is going to teach you how to answer the question "Why?" with regard to design decisions, and how to map those design decisions to the business. This book bridges the gap between the technology and business. Instead of attempting to teach, and require you to memorize, every design decision tree, the goal of this book is to teach you the required skills to make proper design decisions in any situation you are presented.

It's not enough to learn about the "why" to fully grasp the implications of our design decisions. We also need to see "how" something is done to help solidify our understanding of the "why" from a network design perspective. To accomplish this, this book (and its additional online resources) includes multiple mini-design scenarios, illustrations, lab topologies, and configurations to provide the required design perspective.

## Goals and Methods

The most important and somewhat obvious goal of this book is to help you pass the Cisco Certified Design Expert (CCDE) Written Exam (400-007). In fact, if the primary objective of this book were different, then the book's title would be misleading; however, the methods used in this book to help you pass the CCDE Written Exam are designed to also make you much more knowledgeable about how to make proper network design decisions that make businesses successful. Although this book and the companion website together have more than enough questions to help you prepare for the actual exam, the method in which they are used is not to simply make you memorize as many questions and answers as you possibly can.

One key methodology used in this book is to help you discover the exam topics that you need to review in more depth, to help you fully understand and remember those details, and to help you prove to yourself that you have retained your knowledge of those topics. So, this book does not try to help you pass by memorization, but rather helps you to truly learn and understand the topics. This book would do you a disservice if it didn't attempt to help you learn the material.

This is the first book to not only target the CCDE Written Exam (400-007) but also teach you the skillsets needed to make proper designs decisions in any situation. Technology is always changing and advancing, which makes it imperative that we learn how to make proper design decisions as these new technologies, solutions, and capabilities are developed and deployed. This book is the start of your "license to design" journey.

This book covers the different design fundamentals, principles, techniques, and topics using the following approach:

- It presents the different technologies, protocols, design fundamentals, design principles, design techniques, and the associated design decisions made with all of these items in mind.

- It identifies the impact to the business when adopting the different technologies and protocols.

- It addresses the implications of the addition or integration of an element to the overall design.

The network design topics covered in this book aim to prepare you to be able to

- Identify and analyze various network design requirements, constraints, and drivers that have an influence on the corresponding network design decisions.

- Understand the impact that the different network design fundamentals, principles, and techniques have on the business and the associated design decisions.

- Understand and compare the various network design architectures and the associated implications on various network design aspects.

- Identify and analyze network design limitations or issues, and how to optimize them, taking into consideration the business requirements and constraints.

- Identify and analyze the implications of adding new services or applications and how to accommodate the design or the design approach to meet the business outcome.

Whether you are preparing for the CCDE certification or just want to be a better network designer, you will benefit from the range of topics covered and the business success approach used to analyze, compare, and explain these topics to make proper design decisions.

## Who Should Read This Book?

This book is not designed to be a general networking topics book, although it can be used for that purpose. This book is intended to tremendously increase your chances of passing the CCDE Written Exam. Although other objectives can be achieved from using this book, the book is written with one goal in mind: to help you pass the exam.

In addition to those who are planning or studying for the CCDE certification, this book is for network engineers, network consultants, network architects, and solution architects who already have a foundational knowledge of the topics being covered and who would like to train themselves to think more like a network designer.

So, why should you want to pass the CCDE Written Exam (400-007)? Because it's one of the milestones toward getting the CCDE certification, and passing it is a prerequisite to being able to schedule the CCDE Practical Exam, which is no small feat. What would getting the CCDE certification mean to you? It might translate to a raise, a promotion, and recognition. It would certainly enhance your resume. It would demonstrate that you are serious about continuing the learning process and that you are not content to rest on your laurels. It might please your reseller-employer, who needs more certified employees.

## Your Path Towards the CCDE Certification

This section provides a quick high-level overview of the CCDE certification and how to obtain it. Figure I-1 shows the current path you will have to take to achieve the CCDE certification.

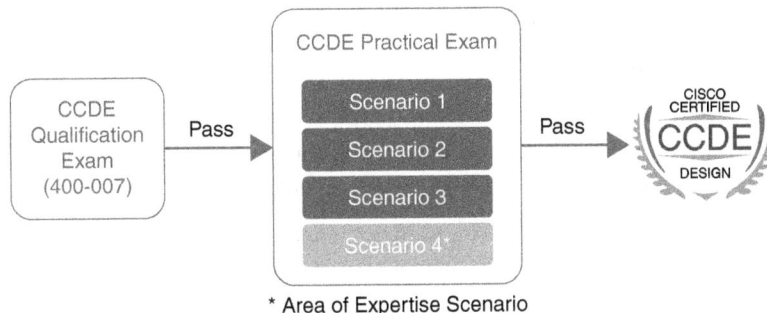

* Area of Expertise Scenario

**Figure I-1**    *CCDE Certification Path*

Both the CCDE Written and Practical Exams focus on high-level design (HLD) aspects as well as business requirements within the context of the enterprise network architectures.

Both exams share one single unified blueprint, "CCDE v3 Unified Exam Topics." The CCDE Written Exam will validate core enterprise network architecture HLD aspects. The CCDE Practical Exam is built to be modular and provides you the flexibility to focus on your area of expertise in addition to validating core enterprise architecture HLD aspects.

This flexibility is achieved by using multiple technology lists. The "CCDE v3 Core Technology List" contains all core enterprise architecture technologies, and multiple areas of expertise technology lists contain logically grouped technologies, focusing on your area of expertise. Figure I-2 shows the breakdown of the entire CCDE certification process, including where and when to leverage each technology list.

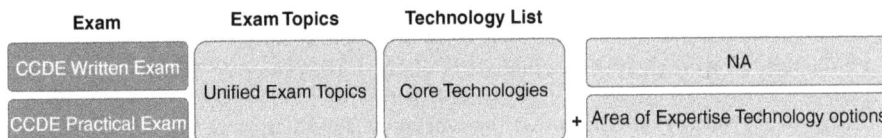

**Figure I-2**    *CCDE Certification Process Overview*

## CCDE Written Exam (400-007)

The CCDE Written Exam is a computer-based exam that's focused on technology and architecture operations, but not on the specific configurations to implement said technology or architecture. More importantly, the Written Exam is about determining what technology out of a list of technology options should be leveraged in a specific design situation. As a CCDE candidate, you have to know and understand which solution solves a specific problem. Then, you have to identify out of a group of solutions which one is the best option given the specific customer requirements, drivers, constraints, and circumstances, and why it is the best option. The why part is extremely important.

The Written Exam validates HLD aspects as well as business requirements within the context of enterprise network architecture. The exam is a two-hour, multiple-choice test with 90 to 110 questions that focus on core enterprise network architecture HLD aspects. The exam serves as a prerequisite for the CCDE Practical Exam. The Written Exam is closed book, meaning no outside reference materials are allowed.

For more information on what to expect on your CCDE Written Exam, refer to the following:

■ "CCDE v3 Unified Exam Topics" (https://learningcontent.cisco.com/documents/exam-topics/CCDEv3.0_UnifiedExamTopics_May2021.pdf)

■ "CCDE v3 Core Technology List" (https://learningcontent.cisco.com/documents/exam-topics/CCDEv30Practical_CoreTechnologyList_2.pdf)

Figure I-3 is an overview of the CCDE Written Exam process and how both the "CCDE v3 Unified Exam Topics" and "CCDE v3 Core Technology List" fit into it.

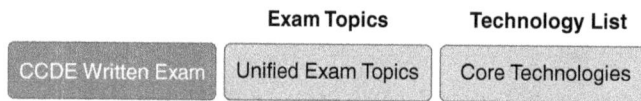

| | **Exam Topics** | **Technology List** |
|---|---|---|
| CCDE Written Exam | Unified Exam Topics | Core Technologies |

**Figure I-3**   *CCDE Written Exam (400-007) Process*

This book is the official certification guide for this exam, the CCDE Written Exam (400-007).

## CCDE Practical Exam

After successfully passing the CCDE Written Exam (400-007), you can schedule and take the CCDE Practical Exam. The CCDE Practical Exam is an 8-hour scenario-based exam, consisting of several scenarios, one of which you can select, providing the flexibility to include topics related to your area of expertise, in addition to validating core enterprise architecture technologies. While additional areas of expertise might be added in the future, the following are the options that are available at the time of writing:

■ **Large-Scale Networks:** Focuses on the aspects of designing large-scale networks, such as service provider networks or large enterprise networks and their associated technologies.

■ **On-Prem and Cloud Services:** Centered on design and integration of business-critical services from a networking perspective. These services may be placed in an on-premises data center, in the cloud, or a hybrid thereof. A deep understanding of applications and their requirements, as well as related networking technologies, is expected for this module.

■ **Workforce Mobility:** Focuses on the design of solutions that benefit users in their daily job and routines, allowing them to roam freely across campuses and buildings without losing access to the services they depend on. Achieving this requires a well-planned and well-designed wireless network.

Technologies relevant to these are grouped together in areas of expertise technology lists. Each area of expertise comes with its own corresponding list. Figure I-4 shows how these different technology lists are used within the CCDE Practical Exam.

■ CCDE v3 Technology Lists

  ■ "CCDE v3 Core Technology List"

    https://learningcontent.cisco.com/documents/CCDEv3_CoreTechnologyList.pdf

  ■ "Workforce Mobility Technology List"

    https://learningcontent.cisco.com/documents/CCDEv3_PracticalWorkforceMobilityTechnologyList.pdf

  ■ "On-Prem and Cloud Services Technology List"

    https://learningcontent.cisco.com/documents/CCDEv3_PracticalOnPrem_and_CloudServicesTechnologyList.pdf

  ■ "Large-Scale Networks Technology List"

    https://learningcontent.cisco.com/documents/CCDEv3_PracticalLargeScaleNetworksTechnologyList.pdf

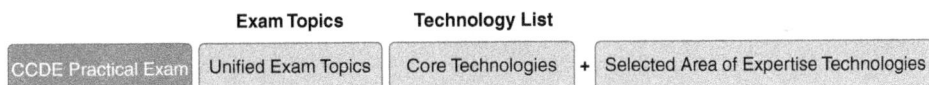

| | Exam Topics | Technology List | |
|---|---|---|---|
| CCDE Practical Exam | Unified Exam Topics | Core Technologies | + Selected Area of Expertise Technologies |

**Figure I-4**  *CCDE Practical Exam Process*

The CCDE v3 Practical exam is delivered at Cisco Certification Centers.

The Cisco Press book *CCDE v3 Practice Labs* (by Martin J. Duggan) is a great resource and guide for preparing to take the CCDE Practical Exam.

## Strategies for Exam Preparation

The strategy you use for the CCDE Written Exam (400-007) might be slightly different than strategies used by other readers, mainly based on the skills, knowledge, and experience you already have obtained.

Regardless of the strategy you use or the background you have, the book is designed to help you get to the point where you can pass the exam with the least amount of time required. For instance, there is no need for you to practice or read about enterprise Layer 2 design if you fully understand it already. However, many people like to make sure that they truly know a topic and thus read over material that they already know. Some readers might want to jump into new technologies or architectures, such as Cisco SD-Access, automation, Zero Trust Architecture, and DevOps. Several book features will help you gain the confidence that you need to be convinced that you know some material already, and to also help you know what topics you need to study more. As mentioned earlier, the CCDE Written Exam is not focused on configuration details, so there's no need to memorize how to configure each technology. However, some people learn better with a hands-on approach, so if you feel you get a better grasp of the technology by trying it out in a lab, there's absolutely nothing wrong with that.

## The Companion Website for Online Content Review

All the electronic review elements, as well as other electronic components of the book, are available on this book's companion website.

## How to Access the Companion Website

To access the companion website, which gives you access to the electronic content with this book, start by establishing a login at www.ciscopress.com and registering your book.

To do so, simply go to www.ciscopress.com/register and enter the ISBN of the print book: **9780137601042**. After you have registered your book, go to your account page and click the **Registered Products** tab. From there, click the **Access Bonus Content** link to get access to the book's companion website.

Note that if you buy the Premium Edition eBook and Practice Test version of this book from Cisco Press, your book will automatically be registered on your account page. Simply go to your account page, click the Registered Products tab, and select Access Bonus Content to access the book's companion website.

## How to Access the Pearson Test Prep (PTP) App

You have two options for installing and using the Pearson Test Prep application: a web app and a desktop app. To use the Pearson Test Prep application, start by finding the registration code that comes with the book. You can find the code in these ways:

- You can get your access code by registering the print ISBN (9780137601042) on ciscopress.com/register. Make sure to use the print book ISBN, regardless of whether you purchased an eBook or the print book. After you register the book, your access code will be populated on your account page under the Registered Products tab. Instructions for how to redeem the code are available on the book's companion website by clicking the Access Bonus Content link.

- If you purchase the Premium Edition eBook and Practice Test directly from the Cisco Press website, the code will be populated on your account page after purchase. Just log in at ciscopress.com, click Account to see details of your account, and click the digital purchases tab.

**NOTE**   After you register your book, your code can always be found in your account under the Registered Products tab.

Once you have the access code, to find instructions about both the PTP web app and the desktop app, follow these steps:

**Step 1**    Open this book's companion website, as described in the previous section.

**Step 2**    Click the Practice Exams button.

**Step 3**    Follow the instructions listed there both for installing the desktop app and for using the web app.

Note that if you want to use the web app only at this point, just navigate to www.pearsontestprep.com, establish a free login if you do not already have one, and register this book's practice tests using the registration code you just found. The process should take only a couple of minutes.

## How This Book Is Organized

Although this book could be read cover to cover, it is designed to be flexible and allow you to easily move between chapters and sections of chapters to cover just the material that you need more work with.

This book is organized into three distinct parts. Part 1, Chapters 1 through 5, covers network design generally. Then, we dive into the different technologies in Part 2, Chapters 6 through 14, and discuss how they work. Part 3, Chapters 15 to 17, covers enterprise network architectures in the real world today.

The following list describes what is covered in each chapter in this book:

■ **Chapter 1, "Network Design"**—This chapter focuses on the different network design fundamentals, principles, techniques, and pitfalls that all network designers need to know.

■ **Chapter 2, "Designing for Business Success"**—This chapter discusses the different aspects of the business and how to ensure that the network design decisions being made will make a business successful. In addition, project management methodologies are covered, as network designers should know how these methodologies impact a business.

■ **Chapter 3, "What's the Purpose of the Network?"**—This chapter covers the reason why there is a network in the first place. Coverage in this chapter includes applications, service models, cloud constructs, and data management.

- **Chapter 4, "Security Is Pervasive"**—This chapter explores the architecture-level security topics, including Zero Trust, the CIA triad, and regulatory compliance standards.

- **Chapter 5, "Reference Architecture Models and Frameworks"**—This chapter covers business architecture, enterprise architecture, and current architecture frameworks being leveraged at the time of this writing.

- **Chapter 6, "Transport Technologies"**—This chapter examines physical and logical transport topics, including Layer 2 media access, Virtual Private Wire Service (VPWS), Layer 2 VPN (L2VPN), and Ethernet VPN (EVPN).

- **Chapter 7, "Layer 2 Technologies"**—This chapter covers Layer 2 core technologies: Spanning Tree Protocol (STP), virtual local-area networks (VLANs), trunking, link aggregation, and First Hop Redundancy Protocol (FHRP).

- **Chapter 8, "Layer 3 Technologies"**—This chapter discusses Layer 3 control plane technologies from an enterprise perspective. This includes enterprise Layer 3 routing with different interior gateway protocols (IGPs), Border Gateway Protocol (BGP) routing, and associated design recommendations.

- **Chapter 9, "Network Virtualization"**—This chapter covers Multiprotocol Label Switching (MPLS) L2 and L3 VPNs and software-defined network solutions like SD-WAN and SD-LAN.

- **Chapter 10, "Security"**—This chapter explores infrastructure security topics, perimeter security and intrusion prevention, and network control and identity management.

- **Chapter 11, "Wireless"**—This chapter describes the most common IEEE 802.11 standards and protocols, and enterprise wireless network design use cases of high density, voice and video, and data.

- **Chapter 12, "Automation"**—This chapter covers automation concepts that impact a network design at the architecture level. These topics include zero-touch provisioning, Infrastructure as Code (IaC), and continuous integration/continuous delivery (CI/CD) pipelines.

- **Chapter 13, "Multicast Design"**—This chapter examines multicast switching and routing, along with the corresponding network design considerations.

- **Chapter 14, "Network Services and Management"**—This chapter covers critical IPv6 topics and network design elements for IPv6 that haven't been covered in other chapters. Additionally, this chapter covers Quality of Service and network management design.

- **Chapter 15, "Scalable Enterprise Campus Architecture Design"**—This chapter highlights campus hierarchical design models, campus Layer 3 routing design considerations, and campus network virtualization design considerations.

- Chapter 16, "Enterprise Internet Edge Architecture Design"—This chapter provides design recommendations and considerations from an enterprise Internet edge architecture standpoint. It explores how an enterprise gets to resources on the Internet and how customers get to locally hosted resources within the enterprise.

- Chapter 17, "Enterprise WAN Architecture Design"—This chapter provides design recommendations and considerations for the enterprise WAN module, WAN virtualization and overlay options, and enterprise WAN migration to MPLS VPN considerations.

- Chapter 18, "Final Preparation"—This chapter provides steps for you to take in your final phase of study before taking the CCDE Written Exam. This chapter also includes tools and resources that you can leverage to help in your journey.

## Certification Exam Topics and This Book

The questions for each Cisco certification exam are a closely guarded secret. However, Cisco does publish an exam blueprint so that you know which topics you must know to *successfully* complete this exam.

**NOTE**  This book covers only the "CCDE v3 Unified Exam Topics" blueprint and the "CCDE v3 Core Technology List," as they encompass all the knowledge areas required for the CCDE Written Exam.

Table I-1 lists each section in the "CCDE v3 Unified Exam Topics" blueprint along with a reference to the book chapter that covers the corresponding topic.

Table I-2 lists each section in the "CCDE v3 Core Technology List" along with a reference to the book chapter that covers the corresponding topic.

These are the same topics you should be proficient in when designing networks and making proper network design decisions in the real world.

**NOTE**  The two topic lists covered in Table I-1 and Table I-2 below are current as of the book's writing, but may be subject to updates, so always check the blueprint at cisco.com.

**Table I-1**  CCDE v3 Unified Exam Topics

| CCDE v3 Unified Exam Topics | Chapter(s) in Which Topic Is Covered |
| --- | --- |
| 1.0 Business Strategy Design (15%) | 2 |
| 1.1 Impact on network design, implementation, and optimization using various customer project management methodologies (for instance waterfall and agile) | 2 |
| 1.2 Solutions based on business continuity and operational sustainability (for instance RPO, ROI, CAPEX/OPEX cost analysis, and risk/reward) | 2 |

| CCDE v3 Unified Exam Topics | Chapter(s) in Which Topic Is Covered |
|---|---|
| 2.0 Control, data, management plane and operational design (25%) | 7, 8, 9, 12 |
| 2.1 End-to-end IP traffic flow in a feature-rich network | 8, 9 |
| 2.2 Data, control, and management plane techniques | 7, 8, 9 |
| 2.3 Centralized, decentralized, or hybrid control plane | 7, 8, 9 |
| 2.4 Automation/orchestration design, integration, and on-going support for networks (for instance interfacing with APIs, model-driven management, controller-based technologies, evolution of CI/CD framework) | 12 |
| 2.5 Software-defined architecture and controller-based solution design (SD-WAN, overlay, underlay, and fabric) | 9 |
| 3.0 Network Design (30%) | 1 |
| 3.1 Resilient, scalable, and secure modular networks, covering both traditional and software-defined architectures, considering: | 1 |
| 3.1.a Technical constraints and requirements | 1 |
| 3.1.b Operational constraints and requirements | 1 |
| 3.1.c Application behavior and needs | 1, 3 |
| 3.1.d Business requirements | 1 |
| 3.1.e Implementation Plans | 1 |
| 3.1.f Migration and transformation | 1 |
| 4.0 Service Design (15%) | 1, 3, 4 |
| 4.1 Resilient, scalable, and secure modular network design based on constraints (for instance technical, operational, application, and business constraints) to support applications on the IP network (for instance voice, video, backups, data center replication, IoT, and storage) | 1, 3 |
| 4.2 Cloud/hybrid solutions based on business-critical operations | 3, 4 |
| 4.2.a Regulatory compliance | 4 |
| 4.2.b Data governance (for instance sovereignty, ownership, and locale) | 4 |
| 4.2.c Service placement | 3 |
| 4.2.d SaaS, PaaS, and IaaS | 3 |
| 4.2.e Cloud connectivity (for instance direct connect, cloud on ramp, MPLS direct connect, and WAN integration) | 3 |
| 4.2.f Security | 4 |
| 5.0 Security Design (15%) | 4, 10 |

| CCDE v3 Unified Exam Topics | Chapter(s) in Which Topic Is Covered |
|---|---|
| *5.1 Network security design and integration* | 4 |
| *5.1.a Segmentation* | 4, 10 |
| *5.1.b Network access control* | 4 |
| *5.1.c Visibility* | 4 |
| *5.1.d Policy enforcement* | 4 |
| *5.1.e CIA triad* | 4 |
| *5.1.f Regulatory compliance (if provided the regulation)* | 4 |

**Table I-2**   CCDE v3 Core Technology List

| CCDE v3 Core Technology List | Chapter(s) in Which Topic Is Covered |
|---|---|
| 1.0 Transport Technologies | 6 |
| *1.1 Ethernet* | 6 |
| *1.2 CWDM/DWDM* | 6 |
| *1.3 Frame relay (migration only)* | 6 |
| *1.4 Cellular and broadband (as transport methods)* | 6 |
| *1.5 Wireless* | 11 |
| *1.6 Physical mediums, such as fiber and copper* | 6 |
| 2.0 Layer 2 Control Plane | 7, 15 |
| *2.1 Physical media considerations* | 7 |
| *2.1.a Down detection* | 7 |
| *2.1.b Interface convergence characteristics* | 7 |
| *2.2 Loop detection protocols and loop-free topology mechanisms* | 7, 15 |
| *2.2.a Spanning tree types* | 7 |
| *2.2.b Spanning tree tuning techniques* | 7 |
| *2.2.c Multipath* | 7 |
| *2.2.d Switch clustering* | 7 |
| *2.3 Loop detection and mitigation* | 7 |
| *2.4 Multicast switching* | 13 |
| *2.4.a IGMPv2, IGMPv3, MLDv1, MLDv2* | 13 |
| *2.4.b IGMP/MLD Snooping* | 13 |
| *2.4.c IGMP/MLD Querier* | 13 |
| *2.5 Fault isolation and resiliency* | 13 |
| *2.5.a Fate sharing* | 13 |
| *2.5.b Redundancy* | 13 |
| *2.5.c Virtualization* | 13 |

| CCDE v3 Core Technology List | Chapter(s) in Which Topic Is Covered |
|---|---|
| *2.5.d Segmentation* | 13 |
| **3.0 Layer 3 Control Plane** | 8, 9, 15, 16, 17 |
| *3.1 Network Hierarchy and topologies* | 8, 9 |
| *3.1.a Layers and their purposes in various environments* | 8, 9 |
| *3.1.b Network topology hiding* | 8, 9 |
| *3.2 Unicast routing protocol operation (OSPF, EIGRP, IS-IS, BGP, and RIP)* | 8 |
| *3.2.a Neighbor relationships* | 8 |
| *3.2.b Loop-free paths* | 8 |
| *3.2.c Flooding domains* | 8 |
| *3.2.d Scalability* | 8 |
| *3.2.e Routing policy* | 8 |
| *3.2.f Redistribution methods* | 8 |
| **3.3 Fast convergence techniques and mechanism** | 8, 9 |
| *3.3.a Protocols* | 8, 9 |
| *3.3.b Timers* | 8, 9 |
| *3.3.c Topologies* | 8, 9 |
| *3.3.d Loop-free convergence* | 8, 9 |
| **3.4 Factors affecting convergence** | 8 |
| *3.4.a Recursion* | 8 |
| *3.4.b Micro-loops* | 8 |
| **3.5 Route aggregation** | 8 |
| *3.5.a When to leak routes / avoid suboptimal routing* | 8 |
| *3.5.b When to include more specific routes (up to and including host routes)* | 8 |
| *3.5.c Aggregation location and techniques* | 8 |
| **3.6 Fault isolation and resiliency** | 8 |
| *3.6.a Fate sharing* | 8 |
| *3.6.b Redundancy* | 8 |
| **3.7 Metric-based traffic flow and modification** | 8 |
| *3.7.a Metrics to modify traffic flow* | 8 |
| *3.7.b Third-party next hop* | 8 |
| *3.8 Generic routing and addressing concepts* | 8 |
| *3.8.a Policy-based routing* | 8 |
| *3.8.b NAT* | 10 |
| *3.8.c Subnetting* | 8 |

| CCDE v3 Core Technology List | Chapter(s) in Which Topic Is Covered |
|---|---|
| *3.8.d RIB-FIB relationships* | 8 |
| *3.9 Multicast routing concepts* | 13 |
| *3.9.a General multicast concepts* | 13 |
| *3.9.b MSDP/anycast* | 13 |
| *3.9.c PIM* | 13 |
| **4.0 Network Virtualization** | 9 |
| **4.1 Multiprotocol Label Switching** | 9 |
| *4.1.a MPLS forwarding and control plane mechanisms* | 9 |
| *4.1.b MP-BGP and related address families* | 9 |
| *4.1.c LDP* | 9 |
| **4.2 Layer 2 and 3 VPN and tunneling technologies** | 9 |
| *4.2.a Tunneling technology selection (such as DMVPN, GETVPN, IPsec, MPLS, GRE)* | 9, 17 |
| *4.2.b Tunneling endpoint selection* | 9, 17 |
| *4.2.c Tunneling parameter optimization of end-user applications* | 9, 17 |
| *4.2.d Effects of tunneling on routing* | 9, 17 |
| *4.2.e Routing protocol selection and tuning for tunnels* | 9, 17 |
| *4.2.f Route path selection* | 9, 17 |
| *4.2.g MACsec (802.1.ae)* | 10 |
| *4.2.h Infrastructure segmentation methods* | 10 |
| *4.2.h (i) VLAN* | 10 |
| *4.2.h (ii) PVLAN* | 10 |
| *4.2.h (iii) VRF-Lite* | 10 |
| **4.3 SD-WAN** | 9 |
| *4.3.a Orchestration plane* | 9 |
| *4.3.b Management plane* | 9 |
| *4.3.c Control plane* | 9 |
| *4.3.d Data plane* | 9 |
| *4.3.e Segmentation* | 9 |
| *4.3.f Policy* | 9 |
| *4.3.f (i) Security* | 9 |
| *4.3.f (ii) Topologies* | 9 |
| *4.3.f (iii) Application-based routing* | 9 |
| **4.4 Migration techniques** | 9 |
| **4.5 Design considerations** | 9 |

| CCDE v3 Core Technology List | Chapter(s) in Which Topic Is Covered |
|---|---|
| *4.6 QoS techniques and strategies* | 14 |
| *4.6.a Application requirements* | 14 |
| *4.6.b Infrastructure requirements* | 14 |
| *4.7 Network management techniques* | 14 |
| *4.7.a Traditional (such as SNMP, SYSLOG)* | 14 |
| *4.7.b Model-driven (such as NETCONF, RESTCONF, gNMI, streaming telemetry)* | 14 |
| *4.8 Reference models and paradigms that are used in network management (such as FCAPS, ITIL, TOGAF, and DevOps)* | 5 |
| 5.0 Security | 4, 10 |
| *5.1 Infrastructure security* | 10 |
| *5.1.a Device hardening techniques and control plane protection methods* | 10 |
| *5.1.b Management plane protection techniques* | 10 |
| *5.1.b (i) CPU* | 10 |
| *5.1.b (ii) Memory thresholding* | 10 |
| *5.1.b (iii) Securing device access* | 10 |
| *5.1.c Data plane protection techniques* | 10 |
| *5.1.c (i) QoS* | 14 |
| *5.1.d Layer 2 security techniques* | 10 |
| *5.1.d (i) Dynamic ARP inspection* | 10 |
| *5.1.d (ii) IPDT* | 10 |
| *5.1.d (iii) STP security* | 10 |
| *5.1.d (iv) Port security* | 10 |
| *5.1.d (v) DHCP snooping* | 10 |
| *5.1.d (vi) IPv6-specific security mechanisms* | 10, 14 |
| *5.1.d (vii) VACL* | 14 |
| *5.1.e Wireless security technologies* | 11 |
| *5.1.e (i) WPA* | 11 |
| *5.1.e (ii) WPA2* | 11 |
| *5.1.e (iii) WPA3* | 11 |
| *5.1.e (iv) TKIP* | 11 |
| *5.1.e (v) AES* | 11 |
| *5.2 Protecting network services* | 10 |
| *5.2.a Deep packet inspection* | 10 |
| *5.2.b Data plane protection* | 10 |

| CCDE v3 Core Technology List | Chapter(s) in Which Topic Is Covered |
|---|---|
| *5.3 Perimeter security and intrusion prevention* | 10 |
| *5.3.a Firewall deployment modes* | 10 |
| *5.3.a (i) Routed* | 10 |
| *5.3.a (ii) Transparent* | 10 |
| *5.3.a (iii) Virtualization* | 10 |
| *5.3.a (iv) Clustering and high availability* | 10 |
| *5.3.b Firewall features* | 10 |
| *5.3.b (i) NAT* | 10 |
| *5.3.b (ii) Application inspection* | 10 |
| *5.3.b (iii) Traffic Zones* | 10 |
| *5.3.b (iv) Policy-based routing* | 10 |
| *5.3.b (v) TLS inspection* | 10 |
| *5.3.b (vi) User identity* | 10 |
| *5.3.b (vii) Geolocation* | 10 |
| *5.3.c IPS/IDS deployment modes* | 10 |
| *5.3.c (i) In-line* | 10 |
| *5.3.c (ii) Passive* | 10 |
| *5.3.c (iii) TAP* | 10 |
| *5.3.d Detect and mitigate common types of attacks* | 10 |
| *5.3.d (i) DoS/DDoS* | 10 |
| *5.3.d (ii) Evasion techniques* | 10 |
| *5.3.d (iii) Spoofing* | 10 |
| *5.3.d (iv) Man-in-the-middle* | 10 |
| *5.3.d (v) Botnet* | 10 |
| *5.4 Network control and identity management* | 10, 11 |
| *5.4.a Wired and wireless network access control* | 10, 11 |
| *5.4.b AAA for network access with 802.1X and MAB* | 10, 11 |
| *5.4.c Guest and BYOD considerations* | 10, 11 |
| *5.4.d Internal and external identity sources* | 10 |
| *5.4.e Certificate-based authentication* | 10 |
| *5.4.f EAP Chaining authentication method* | 10 |
| *5.4.g Integration with Multifactor authentication* | 10 |
| **6.0 Wireless** | 11 |
| *6.1 IEEE 802.11 Standards and Protocols* | 11 |
| *6.1.a Indoor and outdoor RF deployments* | 11 |
| *6.1.a (i) Coverage* | 11 |
| *6.1.a (ii) Throughput* | 11 |

| CCDE v3 Core Technology List | Chapter(s) in Which Topic Is Covered |
|---|---|
| *6.1.a (iii) Voice* | 11 |
| *6.1.a (iv) Location* | 11 |
| *6.1.a (v) High density / very high density* | 11 |
| **6.2 Enterprise wireless network** | 11 |
| *6.2.a High availability, redundancy, and resiliency* | 11 |
| *6.2.b Controller-based mobility and controller placement* | 11 |
| *6.2.c L2/L3 roaming* | 11 |
| *6.2.d Tunnel traffic optimization* | 11 |
| *6.2.e AP groups* | 11 |
| *6.2.f AP modes* | 11 |
| **7.0 Automation** | 12 |
| **7.1 Zero-touch provisioning** | 12 |
| **7.2 Infrastructure as Code (tools, awareness, and when to use)** | 12 |
| *7.2.a Automation tools (i.e., Ansible)* | 12 |
| *7.2.b Orchestration platforms* | 12 |
| *7.2.c Programming Language (e.g., Python)* | 12 |
| **7.3 CI/CD Pipeline** | 12 |

Each version of the exam can have topics that emphasize different functions or features, and some topics can be rather broad and generalized. The goal of this book is to provide the most comprehensive coverage to ensure that you are well prepared for the exam. Although some chapters might not address specific exam topics, they provide a foundation that is necessary for a clear understanding of important topics. Your short-term goal might be to pass this exam, but your long-term goal should be to become a qualified network designer that can make proper network design decisions that help to make businesses successful.

It is also important to understand that this book is a "static" reference, whereas the exam topics are dynamic. Cisco can and does change the topics covered on certification exams often.

This exam guide should not be your only reference when preparing for the certification exam. You can find a wealth of information available at Cisco.com that covers each topic in great detail. If you think that you need more detailed information on a specific topic, read the Cisco documentation that focuses on that topic.

Note that as technologies and architectures continue to develop, Cisco reserves the right to change the exam topics without notice. Although you can refer to the list of exam topics in Tables I-1 and I-2, always check Cisco.com to verify the actual list of topics to ensure that you are prepared before taking the exam. You can view the current exam topics on any current Cisco certification exam by visiting the Cisco.com website, choosing

Menu, and Training & Events, then selecting from the Certifications list. Note also that, if needed, Cisco Press might post additional preparatory content on the web page associated with this book at http://www.ciscopress.com/title/9780137601042. It's a good idea to check the website a couple of weeks before taking your exam to be sure that you have up-to-date content.

# CHAPTER 1

# Network Design

This chapter covers the following topics:

- **Network Design Fundamentals:** This section covers the foundation network design elements that all network designers should know.

- **Network Design Principles:** This section covers the short list of network design principles that will help every network designer make better design decisions.

- **Network Design Techniques:** This section covers the network design techniques and provides real-world examples of how they should be leveraged.

- **Network Design Pitfalls:** This section covers the most common design mistakes that a network designer can make and provides recommendations on how to ensure that they don't happen.

Before we jump into designing the different technologies, protocols, architectures—what most consider to be the fun stuff—we have to cover network design in its simplest form. What is it and why do we need it? No network design situation is the same, as each has unique customer requirements and demands that will dictate what network design option is best in that specific situation. With that said, there are foundational network design elements that all network designers should know. Knowing these elements will make it easier for network designers to make network design decisions.

This chapter covers the following "CCDE v3.0 Unified Exam Topics" section:

- 3.0 Network Design

## "Do I Know This Already?" Quiz

The "Do I Know This Already?" quiz allows you to assess whether you should read this entire chapter thoroughly or jump to the "Exam Preparation Tasks" section. If you are in doubt about your answers to these questions or your own assessment of your knowledge of the topics, read the entire chapter. Table 1-1 lists the major headings in this chapter and their corresponding "Do I Know This Already?" quiz questions. You can find the answers in Appendix A, "Answers to the 'Do I Know This Already?' Quizzes."

**Table 1-1**  "Do I Know This Already?" Section-to-Question Mapping

| Foundation Topics Section | Questions |
|---|---|
| Network Design Fundamentals | 1, 5, 10 |
| Network Design Principles | 2, 6, 9 |
| Network Design Techniques | 3, 7, 8 |
| Network Design Pitfalls | 4 |

1. Which of the following options are network design fundamentals? (Choose two.)
   a. Security
   b. Redundancy
   c. Constraints
   d. Scalability
   e. Mindset

2. Which of the following are security models? (Choose three.)
   a. Perimeter security
   b. Cybersecurity
   c. Zero Trust Architecture
   d. Session- and transaction-based security

3. Which network design technique allows for purpose-built building blocks to be leveraged as the business needs arise?
   a. Failure isolation
   b. Shared failure state
   c. Modularity
   d. Hierarchy

4. Overdesigning is one of the most common network design pitfalls; what's another phrase that means the same thing?
   a. Gold plating
   b. Preconceived notions
   c. Best practices
   d. Overthinking

5. From a network design perspective, what are the three categories of constraints?
   a. Application, security, and business
   b. Functional, technical, and application
   c. Compliance, technical, functional
   d. Business, application, and technology

6. The ability of the network to automatically fail over when an outage occurs is the definition of what availability component?
   a. Reliability
   b. Resilience
   c. Redundancy
   d. Routing failover

**7.** Which of the following network design techniques could help mitigate a Layer 2 broadcast storm from propagating from site to site?

   **a.** Hierarchy

   **b.** Shared failure state

   **c.** Modularity

   **d.** Failure isolation

**8.** Which of the following options leverages access, aggregation, distribution, and core layers to help structure a network design properly?

   **a.** Modularity

   **b.** Failure isolation

   **c.** Hierarchy

   **d.** Shared failure state

**9.** How much of the network data gets from source to destination locations in the right amount of time to properly be leveraged correctly is an example definition of what availability component?

   **a.** Reliability

   **b.** Resilience

   **c.** Redundancy

   **d.** Routing failover

**10.** Which network design use case focuses on when a company or business is split into two or more entities?

   **a.** Scaling

   **b.** Divest

   **c.** Design failure

   **d.** Merger

## Foundation Topics

Designing large-scale networks to meet today's dynamic business and IT needs is a complex assignment. This is especially true when the network was designed for technologies and requirements relevant years ago and the business decides to adopt new IT technologies and architectures to facilitate the achievement of its goals, but the business's existing network was not designed to address these new technologies' requirements. Therefore, to achieve the desired goal of a given design, the network designer must adopt an approach that tackles the design in a structured manner.

There are two common approaches to analyze and design networks:

- **Top-down approach:** The top-down design approach simplifies the design process by splitting the design tasks to make it more focused on the design scope and performed in a more controlled manner, which can ultimately help network designers to view network design solutions from a business perspective.

■ **Bottom-up approach:** In contrast, the bottom-up approach focuses on selecting network technologies and design models first. This can impose a high potential for design failures, because the network will not meet the requirements of the business or its applications.

To achieve a successful strategic design, there must be additional emphasis on the business. This implies a primary focus on business priorities, drives, and outcomes, technical objectives, and existing and future services and applications. In fact, in today's networks, business requirements are driving IT and network initiatives.

When it comes to network design, the elements all network designers must know and understand are network design fundamentals, network design principles, and network design techniques. Figure 1-1 and Figure 1-2 show how these network design elements work together to form the architecture basis of network design. As we go through this chapter, we will also examine the most common network design pitfalls that we can fall into as network designers.

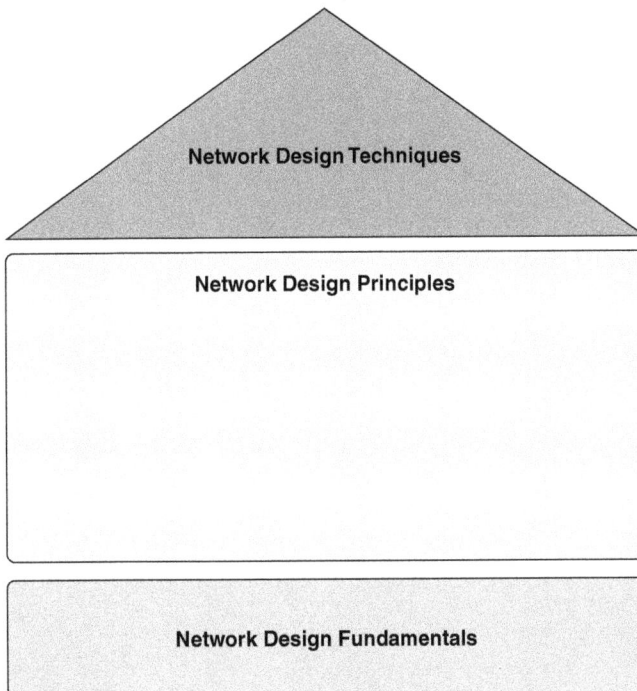

**Network Design Techniques**

**Network Design Principles**

**Network Design Fundamentals**

**Figure 1-1**  *Network Design Elements at a High Level*

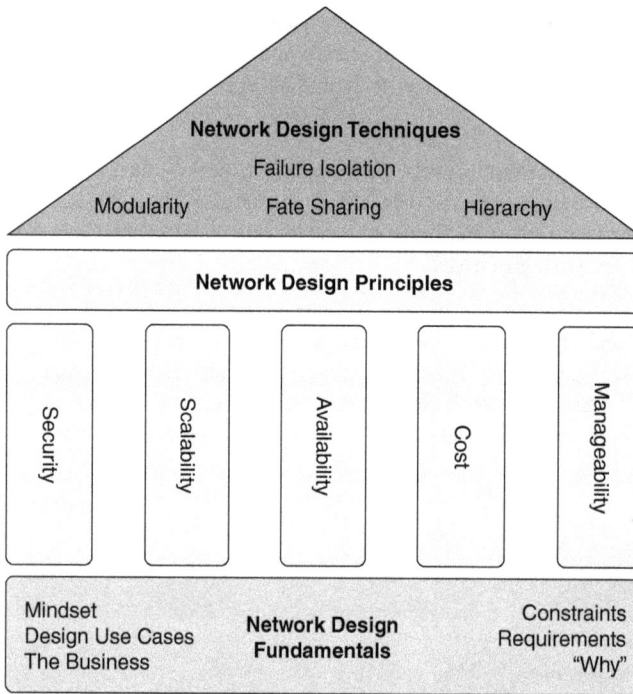

**Figure 1-2** *Network Design Elements in Detail*

# Network Design Fundamentals

The network design fundamentals are the foundational elements that all network designers should understand and know how to leverage.

- Mindset

- Requirements

- Design use cases

- The business

- Constraints

- "Why"

## Mindset

Above all else, your mindset is the most important factor for obtaining the CCDE certification. Knowing the technology is critical, but it is the relatively easy portion of this journey. We can put in the effort and time to learn what we don't know from a technology perspective. Many candidates that attempt the CCDE and Network Design do not have a proper *design mindset*. Most of us are not taught a proper design mindset until later in our careers. In this section we are going to highlight the different elements of a proper design mindset that can make you successful in all of your design situations, be it the CCDE or any network design situation. Mindset is one of six network design fundamentals.

An implementation mindset will not work for network design; you need to have a network design mindset to be successful both in network design situations and on the CCDE exam. This section starts to cover the items that need to be incorporated in a network designer's mindset.

## Functional Requirements

**Functional requirements** compose the foundation of any system design because they define system and technology functions. Specifically, functional requirements identify what these technologies or systems will deliver to the business from a technological point of view. For example, a Multiprotocol Label Switching (MPLS)-enabled service provider might explicitly specify a functional requirement in a statement like this: "The provider edge routers must send VoIP traffic over 10G fiber link while data traffic is to be sent over the OC-48 link." It is implied that this service provider network needs to have provider edge (PE) routers that support a mechanism capable of sending different types of traffic over different paths, such as MPLS Traffic Engineering (MPLS-TE). Therefore, the functional requirements are sometimes referred to as *behavioral requirements* because they address what a system does.

**NOTE** The design that does not address the business's functional requirements is considered a poor design: however, in real-world design, not all the functional requirements are provided to the designer directly. Sometimes they can be decided on indirectly, based on other factors. Most of the time, it is the responsibility of the network designer to find and document the functional requirements, in which case the network designer would also need to have proper sign-off on them before initiating the network design.

## Technical Requirements

The **technical requirements** of a network can be understood as the technical aspects that a network infrastructure must provide in terms of security, availability, and integration. These requirements are often called *nonfunctional requirements*. Technical requirements vary, and they must be used to justify a technology selection. In addition, technical requirements are considered the most dynamic type of requirements compared to other requirements such as business requirements because, based on technology changes, they change often. Technical requirements include the following:

- Heightened levels of network availability (for example, using First Hop Redundancy Protocol [FHRP])

- Support the integration with network tools and services (for example, NetFlow Collector, or authentication and authorization system "RADIUS servers")

- Cater for network infrastructure security techniques (for example, control plane protection mechanisms or infrastructure access control lists)

**NOTE** The technical requirements help network designers to specify the required technical specifications (features and protocols) and software version that supports these specifications and sometimes influence the hardware platform selection based on its technical requirements.

## Application Requirements

**Application requirements** are the driving factors that dictate and in most cases constrain a network design. If an application is created that requires Layer 2 connectivity, it limits the design to Layer 2 protocols like spanning VLANs between data centers. From a business point of view, user experience is one of the primary, if not the highest, priority that any IT and network design must satisfy. The term *end users* can be understood differently according to the type of business. The following are the most common categories of end users:

- **Customers:** Customers can be individuals, such as customers of a bank, or they can be a collective, such as the customers of an MPLS service provider. From a business point of view, customer satisfaction can directly impact the business's reputation and revenue.

- **Internal users:** In this category, the targeted users are internal users. Productivity of these users can translate to business performance efficiency, which has a direct relation to business success and revenue.

- **Business partners:** Partners represent those entities or organizations that work together to achieve certain goals. Therefore, efficient interaction between partners can enhance their business success in the service of strategic goal achievement.

Therefore, a network or a technology that cannot deliver the desired level of the users' expectation (also known as *quality of experience*) means a failure to achieve either one of the primary business goals or failure to satisfy a primary influencer of business success. Consequently, networks and IT technologies will be seen by the business as a cost center rather than a business enabler.

On this basis, network design must take into account how to deliver the desired level of experience. In fact, what influences users' experience is what they see in use. In other words, from a user's perspective, the quality of experience with regard to applications and services used by different types of users is a key deterministic factor.

Deploying network applications or services without considering the characteristics and network requirements of these applications will probably lead to a failure in meeting business outcomes. For example, a company providing modern financial services wants to distinguish itself from other competitors by enabling video communication with its customer service call center agents. If the network team did not properly consider video communication requirements as a network application, the application will probably fail to deliver the desired experience for the end users (customers in this example). Consequently, this will lead to a failure in achieving the company's primary business outcome. In other words, if any given application is not delivered with the desired quality of experience to the end users, the network can simply be considered as not doing its job (design failure).

Furthermore, in some situations, application requirements can drive functional requirements. For instance, if a service provider has a service-level agreement (SLA) with its clients to deliver Voice over IP (VoIP) traffic with less than 150 ms of one-way delay and less than 1 percent of packet loss, the VoIP requirements act as application requirements, which will drive the functional requirements of the PE devices to use a technology that can deliver the SLA. This technology may include, for example, a Class-Based Tunnel Selection (CBTS) MPLS-TE protected with fast reroute (FRR) to send VoIP traffic over high-speed links and provide network path protection in case of a failure within 50 ms. In addition, network designers should also consider the answers to the following fundamental questions when evaluating application requirements:

- How much network traffic does the application require?

- What is the level of criticality of this application and the service level requirement to the business?

- Does the application have any separation requirements to meet industry regulations and corporate security policies?

- What are the characteristics of the application?

- How long does the application need after losing connectivity to reset its state or session?

## Crafting the Design Requirements

This section demonstrates how different types of requirements collectively can lead to the achievement of the desired network design, which ultimately will facilitate achieving business goals. This demonstration is based on an example that goes through the flow of information gathering to build up the requirements starting from the business outcomes to the technical requirement level, such as the required features. Requirements are one of six network design fundamentals.

A national retail company is currently expanding their business to add several international sites within the next 12 months. However, with their current IT infrastructure, they face a high number of expenses in managing and maintaining their two current data and voice networks. In addition, the business wants to invest in technologies that offer enhancements to business lines of effort and increase employee productivity.

Leveraging this simple example scenario lets us identify what the design scope is and how it can impact the different network design options available in a given situation.

### Design Scope

It is important in any design project that network designers carefully analyze and evaluate the scope of the design before starting to gather information and plan network design. Therefore, it is critical to determine whether the design task is for a greenfield network or for a current production network. It is also vital to determine whether the design spans a single network module or multiple modules. In other words, the predetermination of design scope can influence the type of information required to be gathered, in addition to the time to produce the design. Table 1-2 shows an example of how identifying the design scope can help network designers determine the areas and functions a certain design must emphasize and address. As a result, the scope of the information to be obtained will be more focused on those areas.

**Table 1-2**   Design Scope

| Design Scope | Detailed Design Scope Example |
|---|---|
| Enterprise campus network and remote sites | Rollout of IP telephony across the enterprise, which may require a redesign of virtual LANs (VLANs), quality of service (QoS), and so on across the LAN, WAN, data center (DC), and remote-access edge |
| Campus only | Add multi-tenancy concept to the campus, which requires design of VLANs, IPs, and path isolation across the campus LAN only |
| Optimize enterprise edge availability | Add redundant link for remote access, which might require redesign of the WAN module and remote site designs and configuration such as overlay tunnels |

**NOTE**   Identifying the design scope in the CCDE exam is very important. For example, the candidate might have a large network to deal with, whereas the actual design focus is only on adding and integrating a new data center. Therefore, the candidate needs to focus on that part only. However, the design still needs to consider the network as a whole, a "holistic approach," when you add, remove, or change anything across the network.

**NOTE**   Identifying what is out of scope is equally important as well. This can protect situations of scope creep that can hinder the success of a project altogether. Furthermore, what's not in scope can also limit what is available to the network designer in their design decisions.

## Design Use Cases

To build a successful network design, you need to properly identify the specific network design use case(s) your situation involves. Design use cases are one of six network design fundamentals. The following are the primary use cases at the time of this writing. Keep in mind with this listing of design use cases, as the world and technology evolve, so will these design use cases.

### Greenfield

A **greenfield** network design use case is one of the best situations for network designers to encounter. It's a clean slate or a clean canvas for you to paint your picture on, but in this case, you are designing, architecting, and building an environment from scratch. Make sure what you are suggesting is actually needed by the business!

### Brownfield

Most of your network design situations will include a **brownfield** network design use case in some form. A brownfield use case is when there is already an environment with production traffic running through it. It is recommended you spend some time up front to discover the current state and properly assess it. This isn't just technical discovery with protocols and diagrams though. You need to discover the business and the associated lines of effort as

1

well. Your goal here is to understand what the business is trying to accomplish before you start making any design decisions.

Once you do make design decisions, make sure you prepare for the migration to the new design. Limit the potential failures, if any, and make sure to rely on the network design fundamentals covered earlier in this chapter and network design principles and network design techniques covered later in this chapter.

### Add Technology or Application

**Add technology**, or application, can be a very complicated design use case to solve for. The scope of this use case covers the impact of adding technology or an application to an existing network. Will anything break as a result of the new addition? In this use case, you must consider the application requirements in terms of traffic pattern, convergence time, delay, and so on across the network. By understanding these requirements (as a CCDE candidate and as a network designer), you should be able to make design decisions about fine-tuning the network (design and features such as QoS) to deliver the technology or application with the desired level of experience for the end users.

### Replace Technology

The **replace technology** use case covers a wide range of options to replace an existing technology to meet certain requirements. It might be a WAN technology, routing protocol, security mechanism, underlying network core technology, or some other technology. Also, the implications of this new technology on the current design, such as enhanced scalability or potential conflict with some of the existing application requirements, require network designers to tailor the network design so that these technologies work together rather than in isolation, so as to reach objectives, such as delivering business applications and services.

> **NOTE**   Make sure that when you are replacing a technology or adding new technology, you are doing it for the correct reasons. Deploying SD-WAN to replace your WAN architecture without a valid reason is not the way to go. Make sure there are direct business justifications for you to do what you are doing. When in this design use case, ensure you have a properly tested migration plan. If you were migrating from OSPF to EIGRP, you should have a plan listing each step along the way. Also, each step should have a validation task to ensure the migration is going as expected, and a proper backout plan in case something goes wrong.

### Merge or Divest

The merge or divest use case covers the implications and challenges (technical and non-technical) associated with merging or separating different networks. This can be one of the most challenging use cases for network designers because, most of the time, merging two different networks means integrating two different design philosophies, in which multiple conflicting design concepts can appear. Therefore, at a certain stage, network designers have to bring these two different networks together to work as one cohesive system, taking into consideration the various design constraints that might exist, such as goals for the merged network, security policy compliance, timeframe, cost, the merger constraints, the decision of which services to keep and which ones to divest (and how), how to keep services up and running after the divestiture, what the success criteria is for the merged network, and who is the decision maker. The following are some examples of each of these use cases and what you as a network designer could see:

■ **Merger:** A merger combines two independent businesses, creating one business with one end-to-end architecture. In such a scenario, make sure to watch for overlapping technologies. An easy example of this is subnetting. Most companies leverage RFC 1918 subnets (10.0.0.0/8, 172.16.0.0/12, or 192.168.0.0/16). When these issues do occur, because they will happen, have a short-term and long-term plan. A short-term plan might include using Network Address Translation (NAT) to present different IP addresses between organizations when overlapping addresses exist. A long-term plan might include renumbering one of the companies in conjunction with a network upgrade planned for the following year. Make sure you properly articulate both plans with senior leadership. Also, consolidate as much as possible. If between the two companies you have four data centers, consolidate them to two data centers (but keep in mind there might be a requirement to have four data centers still).

■ **Divestment:** This is where you as a network designer are splitting apart a business or a company and the corresponding architecture. This is the hardest of the design use cases. When you are forced to take apart an architecture, you are going to be left with two or more disjointed systems. It's your job as a network designer to ensure that all of these leftover businesses still function along their respective lines of effort. Each business needs to continue to operate, and in most cases that means making money. Network designers have to ensure that it is possible for each business to be successful with the independent architectures left over.

## Scaling a Network

The **scaling** use case covers different types of scalability aspects at different levels, such as physical topology, along with Layer 2 and Layer 3 scalability design considerations. In addition, the challenges associated with the design optimization of an existing network design to offer a higher level of scalability are important issues in this domain. For example, there might be some constraints or specific business requirements that might limit the available options for the network designer when trying to optimize the current design. Considerations with regard to this use case include the following: Is the growth planned or organic? Are there issues caused by the growth? Should a network designer stop and redesign the network to account for growth? What is the most scalable design model?

For example, you could be brought in to help solve a problem with a technology or architecture. It could be as simple as a single flat area 0 OSPF design that no longer scales to the business requirements. In this case, you could leverage multiple areas, multiple area types, and LSA filtering techniques if needed.

## Design Failure

Nine times out of ten, **design failure** is the design use case you will be brought in to fix. There is a problem, and you have to resolve it. An analogy is working at a hospital as an emergency room doctor, tasked with identifying the problems people have and resolving them as quickly as possible. This is the exact same situation for you as a network designer when you are dealing with a design failure use case. A simple technical example of this is not aligning the critical roles of Spanning Tree Protocol (STP) and First Hop Redundancy Protocol (FHRP). If your STP root bridge and your FHRP default gateways are not aligned to the correct devices, then you would have a suboptimal routing issue that would eventually lead to a design failure.

## The Business

Why do we make network design decisions?

Network engineers, network designers, and network architects tend to design networks without the correct purposes in mind. In most of these situations, there wasn't any reason for the design decisions made. When asked why they did what they did, they might state "just because," "it's how we have always done it," or "it's in the script, and we just copy it…"

This is not a path for success, not for you or your designs. The point of all of this is to answer the questions "Why do we design?" and "Why do we make design decisions?" The answers are actually pretty simple and straightforward but are not always clear.

As network designers, we do what we do for the sake of the businesses, companies, and organizations we support. Specifically, we make network design decisions for businesses so that those businesses can make money. This is not always the golden rule, but it is the situation more often than not. I remember having a conversation with a CIO at a company I worked at about 10 years ago, and at that time he was telling me I needed to care more about the business…I remember saying "The business doesn't matter; the technology is the only thing that matters!" I was totally and 100 percent naive.

Of course, there are cases where a business does not operate to make a profit. These not-for-profit organizations are more concerned with covering their expenses and reducing their day-to-day costs than making large profits on what they provide. These organizations have different goals, different outcomes that they are trying to achieve. In the public sector market, many organizations are focused not on making money but on providing a specific service or addressing a specific goal. These include, for example, local, state, and federal government agencies, public safety services (police, fire, ambulance), and environmental protection organizations.

We will discuss the business as a network design fundamental in more depth in Chapter 2, "Designing for Business Success."

**NOTE**   It has become more common to document all design decisions, and why they were made, in a network design binder of sorts. This allows all team members involved, past, present, and future, to be able to understand why a feature, design option, or functionality was implemented. Remembering why we made a design decision 6 months ago is much easier than remembering 2 years down the road why we added that routing adjustment at 2 a.m. during a maintenance window. Document everything to the best of your ability.

## Constraints

When we are designing an architecture for a customer, we start out clean with no requirements, no rules to follow. As we interview the customer, we learn about the different requirements and, more importantly, the specific constraints that box us in. Constraints are one of dix network design fundamentals and fall into three categories: Business, Application, and Technology constraints.

■ **Business constraints:** A business constraint can be as simple as a monetary budget limitation...the business just doesn't have the money. Other examples include a staffing constraint, such as a lack of skilled personnel with respect to a specific technology, or a contractual constraint, perhaps limited by the contract's scope or duration.

■ **Application constraints:** Application constraints are limitations to us as designers because an application was developed in a specific way. Maybe the application was created incorrectly, requiring Layer 2 connectivity between the different servers within the application. Now we have to instantiate some sort of Layer 2 architecture to allow this application to function. Maybe the application was created with hard-coded IP addresses, so now we have to design a solution that allows for those hard-coded IP addresses to live anywhere.

Some applications leverage cluster technologies that require multicast support. In these situations, we now have to design a multicast solution that is scalable and effective for this constraint.

■ **Technology constraints:** From a technology perspective, we can be constrained to specific vendor hardware or a specific vendor solution. Maybe we cannot leverage proprietary technology or, on the flip side, we are already leveraging a proprietary solution that we cannot move away from.

These constraints are hard rules and limitations network designers have to follow. Think of these rules as the scientific laws we learn in grade school, such as Newton's three laws of motion. These "laws of design" have the potential to change with each business and environment you encounter. No network design situation is exactly the same, as each one has its own business, application, and technology constraints.

Constraints are everywhere and come in a multitude of forms. There will always be constraints. It's up to you as the network designer to know which constraints are applicable in each situation. Don't assume constraints. Properly qualify each constraint with evidence.

The following are the most common constraints that network designers must consider:

■ **Cost:** Cost is one of the most common limiting factors when producing any design; however, for the purpose of the CCDE exam, cost should be considered as a design constraint only if it is mentioned in the scenario as a factor to be considered or a tiebreaker between two analogous solutions. The topic of "preconceived notions" is discussed in greater detail later, but for now, know that it's unwise to assume technology A costs more or less than technology B just because your own personal experience says so. MPLS circuits are not automatically more expensive than Internet circuits in the context of the CCDE exam, as an example.

■ **Time:** Time can also be a critical constraint when selecting a technology or architecture over another if there is a time frame to complete the project, for example. Given that most projects revolve around design failures, as mentioned earlier, designers will almost always encounter aggressive timelines for resolution.

■ **Location:** Location is one of the tricky types of constraints because it can introduce limitations that indirectly affect the design. For instance, a remote site may be located in an area where no fiber infrastructure is available, and the only available type of connectivity is over wireless. From a high-level architecture point of view, this might not be a serious issue. From a performance point of view, however, this might lead to a reduced link speed, which will affect some sensitive applications used by that site.

- **Infrastructure:** A good example here is that of legacy network devices. If a business has no plan to replace these devices, this can introduce limitations to the design, especially if new features or protocols not supported by these legacy platforms are required to optimize the design.

- **Staff expertise:** Sometimes network designers might propose the best design with the latest technologies in the market, which can help reduce the business's total cost of ownership (TCO). This can be an issue, however, if the staff of this company has no expertise in these technologies used to operate and maintain the network. In this case, you have two possible options:

  - **Train the staff on these new technologies:** This will be associated with a risk, because as a result of the staff's lack of experience, they may take a longer time to fix any issue that might occur, and at the same time, data center downtime can cost the business a large amount of money.

  - **Hire staff with experience in these technologies:** Normally, people with this level of expertise are expensive. Consequently, the increased operational cost might not justify the reduced TCO.

**NOTE**    In some situations, if the proposed solution and technologies will save the business a significant amount of money, you can justify the cost of hiring new staff.

## Identifying Requirements with "Why?"

You'll often hear customers say "That's a requirement," but when you ask "Why?" they won't be able to offer a valid reason. The concept of the five levels of "Why?" should be leveraged here. Envision a 5-year-old child always asking "Why?" They seem to never stop asking why until they receive an answer that they can understand and comprehend. As a network designer, you need to understand and comprehend your network design requirements. Children aren't the only people who do this, either. Toyota, the Japanese automotive manufacturer, uses the "Five Whys" system to investigate production problems by asking "Why?" five times to surface root causes of problems.

There are numerous evolving technologies (5G, Zero Trust, cloud, SASE, DevOps, etc.) that are revolutionizing how businesses solve problems. Customers have the impression that these evolving technologies are going to solve all of their business problems for them without valid justification as to why. They are simply the new shiny solutions. Network designers have to determine why these solutions are truly needed by the business.

Here are some example questions that network designers can ask that don't have why in them:

- What are you solving with this technology?

- What are your employees'/customers' workflows today?

- How will this technology change these workflows?

- How will your business (think governance, policy, processes, and culture) evolve because of this technology?

You want your customer to provide some specific problem information here. They are set on this "solution" or "technology" because they believe it will solve a number of problems/issues or that it will make their day-to-day activities easier. Maybe they will save money or reduce costs. Maybe they will increase security. Right or wrong, there is always a reason why, and it's the network designer's job to figure that out. "Why" is one of six of the network design fundamentals.

This is not an all-inclusive list of questions by any means. You are looking for the bottom level "Why?" You never want to come across as the 5-year-old child asking why over and over again. You need to find the "Why?" without asking why and fully understand the implication of it. This right here is where you start to design, architect, and engineer!

# Network Design Principles

There have been a number of changes over the last 10 years when it comes to how network designers should look at network design principles. The network design principles include the following:

- Security

- Scalability

- Availability

- Cost

- Manageability

Before we jump into the five network design principles, there is a new concept that we need to cover that doesn't exactly fit into any specific network design functional area, but it is imperative that all network designers know and understand it. This is called unstated requirements.

## Unstated Requirements

It has become more prevalent where customers do not come out and articulate their specific requirements. They assume requirements, and you as the network designer have to figure out what requirements are important to the network design. Your job is not done after you have identified the requirements, though; you also have to determine the level of each requirement. For example, does this network require no single points of failure or does this network require no dual points of failure?

You need to keep this concept of **unstated requirements** at the top of your mind, because every network design principle has become an unstated requirement.

## Pervasive Security

Historically, security hasn't been identified as a network design principle. It's been added here as a network design principle because of the impact it has to the overall business. Here is a list of questions to help set the stage on why security is a network design principle:

- What happens to your business if your network is compromised?

- What happens to your business if the integrity of your data is compromised?

These super-simple questions have very impactful possible answers:

- Business reputation suffers

- Customers lose trust in your business

- Business loses money and revenue

- Business is no longer meeting compliance standards, and can be fined or shut down

- In extreme cases, the business fails

The following are three security models that you should know as a network designer. The industry has been shifting between these models over the last 20-plus years.

**Key Topic**

- **Perimeter security (aka turtle shell):** This is the legacy way of doing security, with a firewall at the perimeter (the turtle shell). The firewall has a bunch of security capabilities that limit what traffic can get into your network and what can leave it. Inside the network, behind the firewall, there are no other security devices. In this model, there is full east–west (lateral movement) traffic between users and resources. There are no other security mechanisms in place to catch threats. Inside threats become prevalent.

- **Session- and transaction based security:** This model evolved from the Perimeter Security model. In this model, users and devices are locked down, including resources like printers, applications, security cameras, and so on. We are able to secure the east–west traffic dynamically based on what the device is, who is using it, where they are using it, and what they *need* to have access to, not what they *want* to have access to. This model leads to a 100 percent authentication model and then 100 percent authorization of each session and transaction.

- **Zero Trust Architecture:** The third security model is Zero Trust. This model brings an even larger shift from a security standpoint. Zero Trust adds real-time capture and analytics tools to the mix to allow for real-time artificial intelligence/machine learning (AI/ML) decision making. The other key concept of this model is that every device, user, application, server, service, and resource (even data itself) is assigned a trust score. This trust score changes based on what the analytics engine sees happening. This could be static changes like physical location or connectivity model. For example, a user connecting from a coffee shop over a VPN connection might get a lower trust score, and thus less access, than a user connected at a company location. There are a number of dynamic characteristics within this model that the Zero Trust Engine will leverage to increase or decrease a trust score on the fly. This could be as simple as the time of day and day of the week that a user is sending data, or it can be as complex as a user sending a different type of data than they've ever sent before or sending a larger amount of data than normal.

We will discuss Zero Trust in more depth in Chapter 4, "Security Is Pervasive." For now, understand that this model is the evolution of security.

Wherever possible, we want to include security capabilities to ensure we meet the confidentiality, integrity, and availability requirements of the business. Remember to think of *business assurance*. A business cannot fulfill its goals, outcome, or mission if the business or its data is compromised. This is also where you will start to find compliance requirements like the Health Insurance Portability and Accountability Act (HIPAA), National Institute of Standards and Technology (NIST), and Payment Card Industry Data Security Standard (PCI DSS). If a business is not compliant, then it will be disconnected or fully brought down (i.e., shut down and out of business).

## Shifting of Availability

Normally when we talk about network design principles, we talk about resiliency, reliability, fault tolerance...the list is never-ending. Availability includes all of these topics: redundancy, resiliency, reliability, and much more; this is why availability is one of the five network design principles. Here is where unstated requirements start to come into play.

- **Redundancy:** The concept of having multiple resources performing the same function/role so that if one of them fails the other takes over with limited to no impact on the production traffic

- **Resilience (aka resiliency):** The ability of the network to automatically fail over when an outage occurs

- **Reliability:** How much of the network data gets from source to destination locations in the right amount of time to be leveraged correctly

The need for availability is just assumed today. What level of availability is needed is the true question. Once that is identified, then a network designer must assess the complexity and cost, both monetary and non-monetary, of that level of availability.

As you increase the level of availability, the complexity and associated cost increase with it. A simple example, previously mentioned, is removing single points of failure versus removing dual points of failure. The latter option increases both the cost and complexity exponentially.

If you encounter a situation in which a customer or stakeholder says they need a highly available network that never allows an application outage, you should be ready to provide them with a couple estimates of the cost associated with making an environment with that level of availability. Make them realize what it truly is going to cost to have such a highly available environment. Show them the value and the cost.

A question arises when availability is in the forefront of the network design discussion: What level of redundancy, resiliency, and reliability is too much? The answer is, when the increased complexity, cost, and the associated return for availability are not worth it.

As an example, is it proper to design a solution that has eight Layer 2 or Layer 3 links between devices? It is normal to suggest redundant links for most devices in an architecture, with two links being the simplest option. In some designs and architectures, four links is actually preferred, and there are valid reasons for this. With that said, five or more links tends to add more complexity to the environment, with a higher level of cost, and has a diminishing return on the benefits of availability.

1

Network designers should apply this concept to all aspects of network design decisions: routing adjacencies, Layer 3 links, Layer 2 links, devices, pod architecture, aggregation, and core layers, and so on.

The large shift with availability is that the focus of network design is no longer network availability but rather application and service availability. Why does a network exist in the first place? The network is a service, getting data from point A to point B at the right time so that it can be properly leveraged. The network facilitates all of this, but it is seamless to the "resources." This is the shift in your design mindset that has to happen. As an analogy, the network is the plumbing to the running water in our house, and data is the water. Without the network, data cannot arrive.

This is a larger concept now than just data transporting the network. No one else understands what data is, or the bytes and bits on the network. Perspective matters here. The end user only cares if their application works when they go to leverage it.

For those of you who can remember the days when everyone had a landline at home, did you ever pick up a POTS (plain old telephone service) phone and not have a dial tone? I don't recall ever picking up a POTS phone and it not working. This is analogous to where the network sits now. If we pick up a VoIP phone and it doesn't work, what's the impact? If a user tries to access email but it doesn't load, what's the impact? If the cloud provider that hosts your company's Software as a Service (SaaS) application has an outage and your customers can no longer access your SaaS application, what's the impact?

As network designers, we have to identify the required level of availability for applications and services (again, in most cases requirements are unstated by the customer). We have to strive to maximize this identified level of availability while keeping the constraints in mind. We have to partner with the application owners to truly understand the requirements and interdependencies each application and service has. Not all customers know their applications and what they are supposed to be doing. From a network design perspective, each application will leverage different portions of the network. You will need to properly identify what the application is dependent on and make appropriate design decisions to ensure the required level of availability for that application is achieved.

## Limiting Complexity...Manageability

Probably one of the hardest tasks you will have as a network designer is to manage the complexity level of the design you are proposing. You have to keep it super simple (KISS). This is why manageability is one of the five network design principles. When comparing different design options that provide the same capability, choosing the simpler option is the way to go. A great question to ask yourself at this stage is, "Can the network design I'm proposing be managed by the team at hand?" For example, suppose you have a network design that meets the customer's needs. It is highly available, secure, redundant, and cost effective. However, your design has multiple CCIE-level design elements, but the customer doesn't have any CCIE skilled professionals on staff. How can your customer manage this design? How can your customer troubleshoot this design when there is a problem? This is an issue to consider. You as a network designer need to assess the team that will be owning this design and managing it day to day. They need to understand what is being done within the environment and why. The why here is actually more important than the how or what.

As a network designer, you cannot design a solution that is unmanageable by the staff who will operate the network. However, there are situations where no other choice exists but to leverage a more complex design. In this situation, if the local team does not have the skill sets to manage the design, you have to raise the issue with the business. This is where you have to assume the role of trusted advisor and tactfully explain to the business that they need higher-level skilled professionals to manage and maintain the network environment that the business requires. When you do this, you need to show the business why they need a complex solution in their terminology, not in technical terms. You need to show them the impact and the why.

Obfuscating the complexity of a solution still yields a complex solution. Leveraging other technology to hide the complexity of an environment does not make it a simple solution. It might make it a manageable solution, but it's still a complex solution. Oftentimes, the obfuscated solution becomes more complex. You have to understand not only the original complex environment, but also the technology being leveraged that is hiding that complex environment. Leveraging a GRE tunnel to form a routing adjacency over a complex OSPF multi-area design is a perfect example. Here you are hiding the "underlay" complex network by forming an "overlay" tunnel that you can then create a routing adjacency on top of. The original network is still complex, and just because the GRE tunnel makes it seem less complex does not mean it is.

## Making a Business Flexible with Scalability

Scalability has always been a network design principle and will most likely always be one. However, this design principle encompasses more than just making the network scalable. It's also about making the business flexible. You can provide flexibility to the business through your network design. You can allow the business to adapt on the fly as the business needs do. By doing this, the network becomes a business enabler and is no longer a cost center.

Designing the network to be scalable means providing different options to choose from as the business grows, which will allow the business to be flexible. This can be as simple as having different architecture options in your design. This can also be a more specific low-level design element, such as a routing architecture.

As an example, from an architecture perspective, designing a data center solution with purpose-built pods within a spine-leaf architecture provides extreme network scalability and business flexibility. If the business has a need to add a new stack of applications, this architecture offers the flexibility to accomplish it easily.

With a low-level design element like that of a routing architecture, leveraging routing boundaries, OSPF areas/area types, and LSA filtering techniques are all ways of increasing the scalability of the routing environment. This scalability allows for the business to grow as needed.

## Cost Constraints and What to Do

The final network design principle that we are going to cover is cost management.

As you design networks for businesses, there is always a cost constraint. You have to manage the cost of the network designs you are providing your customers. Cost isn't always monetary; it can also be a resource cost. Personnel, time, and technical costs like memory, CPU, storage, bandwidth, power, and cooling are all costs that you have to manage as a network designer.

One important rule to follow is to never be so disconnected from your customer that you don't realize the cost budget is drastically lower than the proposed design's cost. For example, if your customer has a budget of $100,000 but you propose a $5 million solution, that is a very large disconnect.

If you are in a situation like this, where the cost budget is drastically lower than the proposed design's cost, show the customer the pros and cons for each design option, including all associated costs. Let them make an educated decision as to what option works for them.

## Network Design Techniques

Bad network designs do happen. You might shun this thought, but it's true. We wouldn't have a role as a network designer if this weren't the case. To help address this concept of "bad network designs do happen," this section provides some of the most common network design techniques that all network designers should know not only to avoid creating bad network designs but also to remediate bad network designs others have created. To help explain the network design techniques in this section, we'll use as a case study a higher education campus. Figure 1-3 shows the current architecture topology diagram for this higher education campus case study.

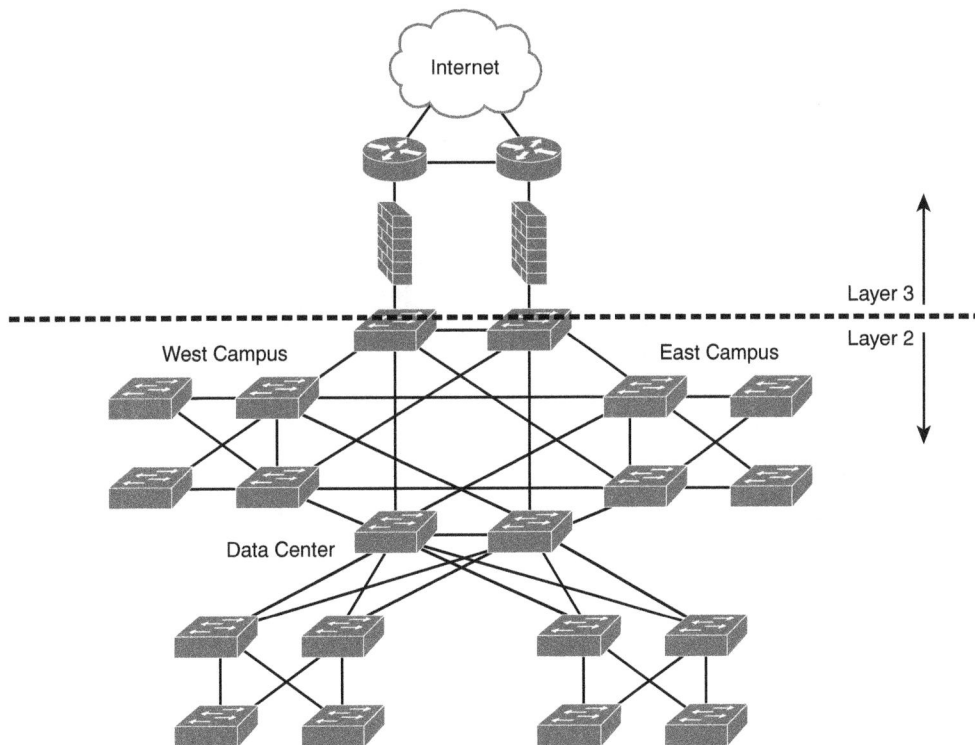

**Figure 1-3**   *Higher Education Campus Architecture Overview*

## Failure Isolation

Taking a closer look at the campus architecture depicted in Figure 1-3, there is a large failure domain that spans the data center, the west campus, and the east campus. A *failure domain* is an area in which an outage can propagate. Figure 1-4 shows an example of a failure situation.

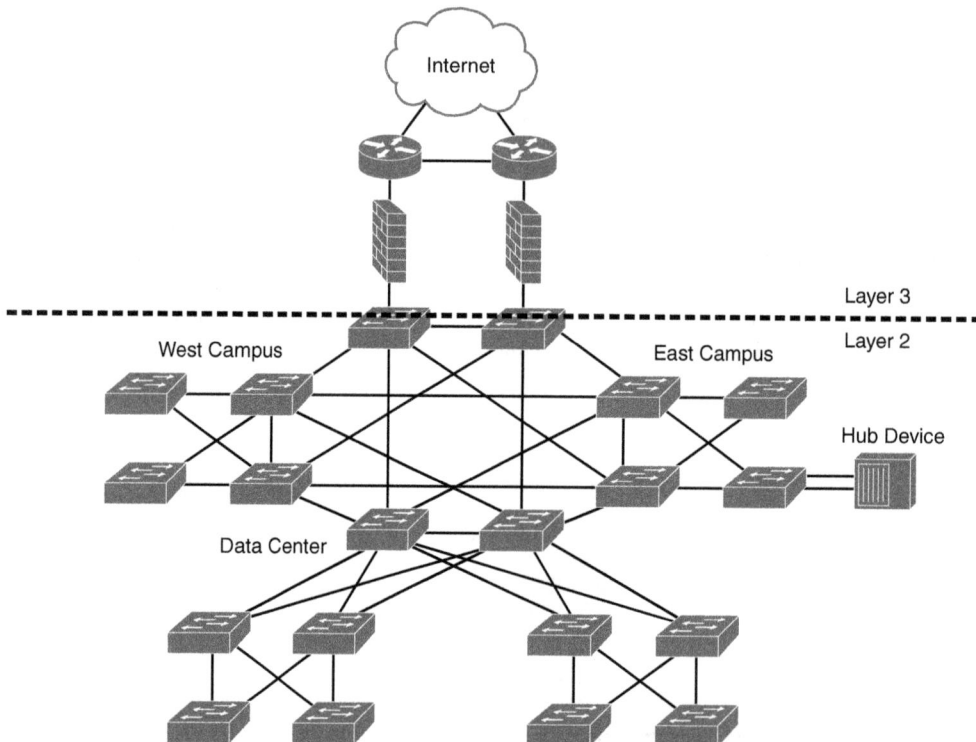

**Figure 1-4** *Higher Education Campus Architecture Failure Isolation*

In this campus architecture, there is a very large Layer 2 environment that spans multiple locations in the network. If a student plugged a hub device into one of the access switches twice, with two different ports and two unique cables, in the east campus location, a broadcast storm would be created. In this architecture and design, that broadcast storm would propagate to the main routers, the west campus, and to the data center. This one device and one student could very well bring down the entire campus in a matter of minutes. This is a design problem...specifically, this is a design failure. Figure 1-5 and Figure 1-6 show the impact of this failure situation on the rest of the higher education architecture.

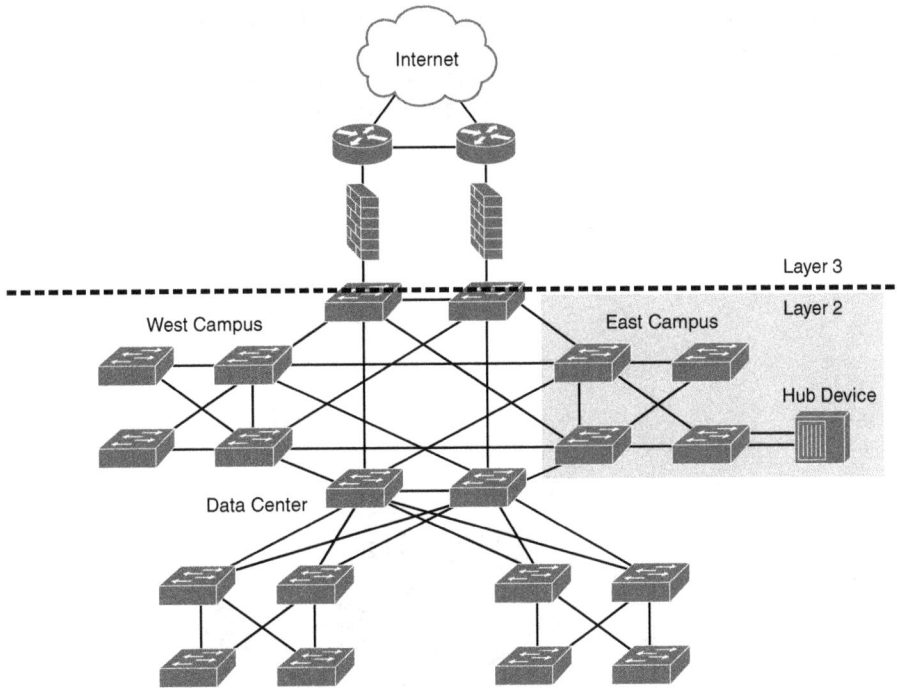

**Figure 1-5**  *Higher Education Campus Architecture Broadcast Storm*

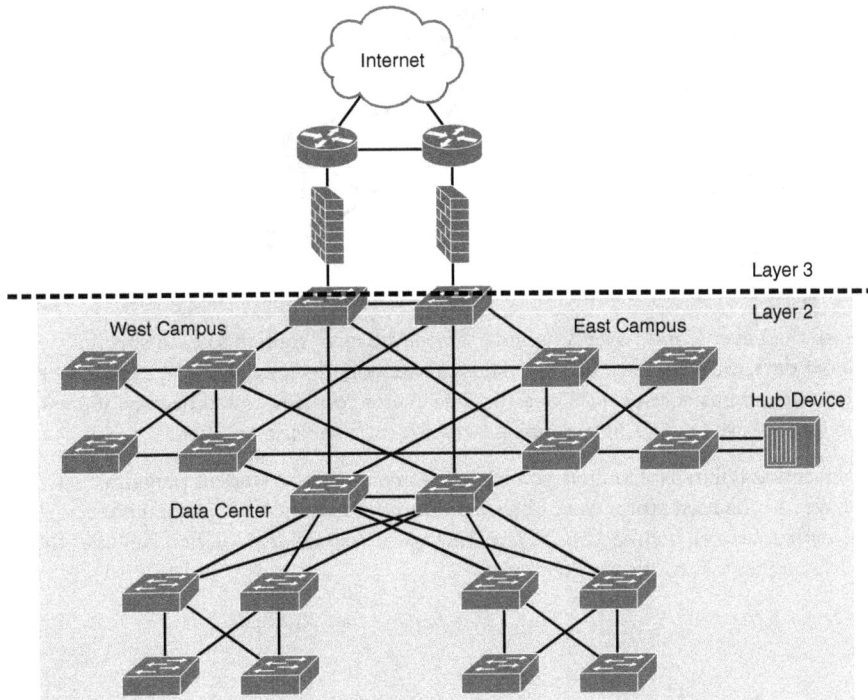

**Figure 1-6**  *Higher Education Campus Architecture Broadcast Storm Propagation*

How can you design this architecture and future architectures so something like this doesn't happen? You can leverage what's known as **failure isolation**, which involves creating logical boundaries to limit the propagation of failures. This is also referred to as failure radius, the impact domain, and the failure boundary. Figure 1-7 shows how we can leverage failure isolation to mitigate this failure situation, and many others like it.

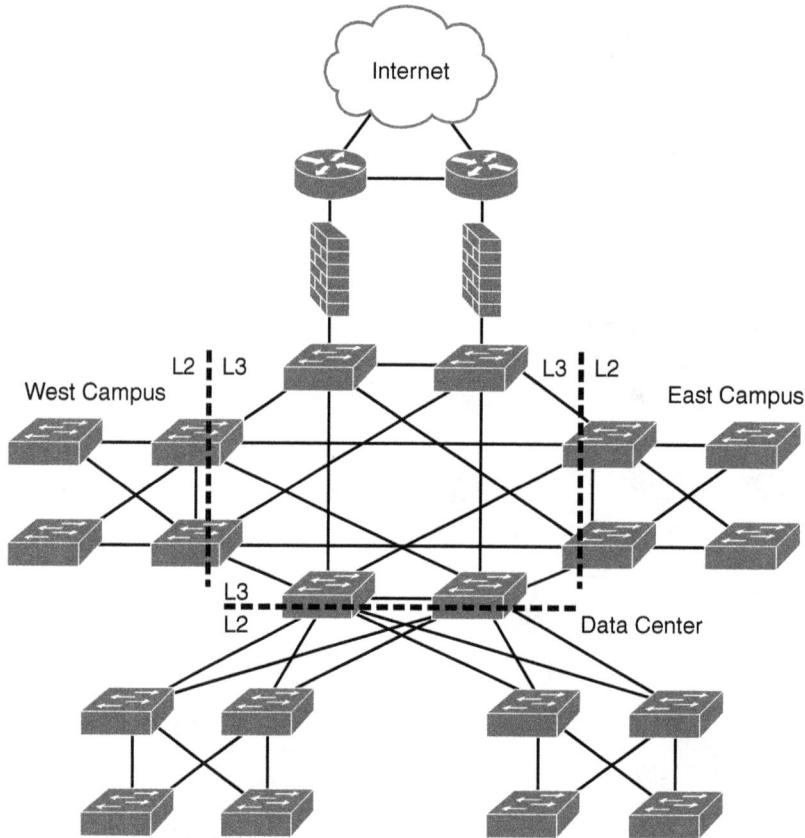

**Figure 1-7** *Higher Education Campus Architecture New Layer 2 and Layer 3 Boundary*

You can push the Layer 2 and Layer 3 boundary from the main routers to each specific location's core devices. When you do this, the local site core devices, the west campus core devices, the east campus core devices, and the data center core devices all become routers and you no longer have Layer 2 links running between each specific location.

After implementing failure isolation, if you had the same issue of a student plugging in a hub, that Layer 2 broadcast storm would be isolated to the specific location and not propagate to the entire network architecture. Figure 1-8 shows how failure isolation can limit the spread of a failure impact to the architecture.

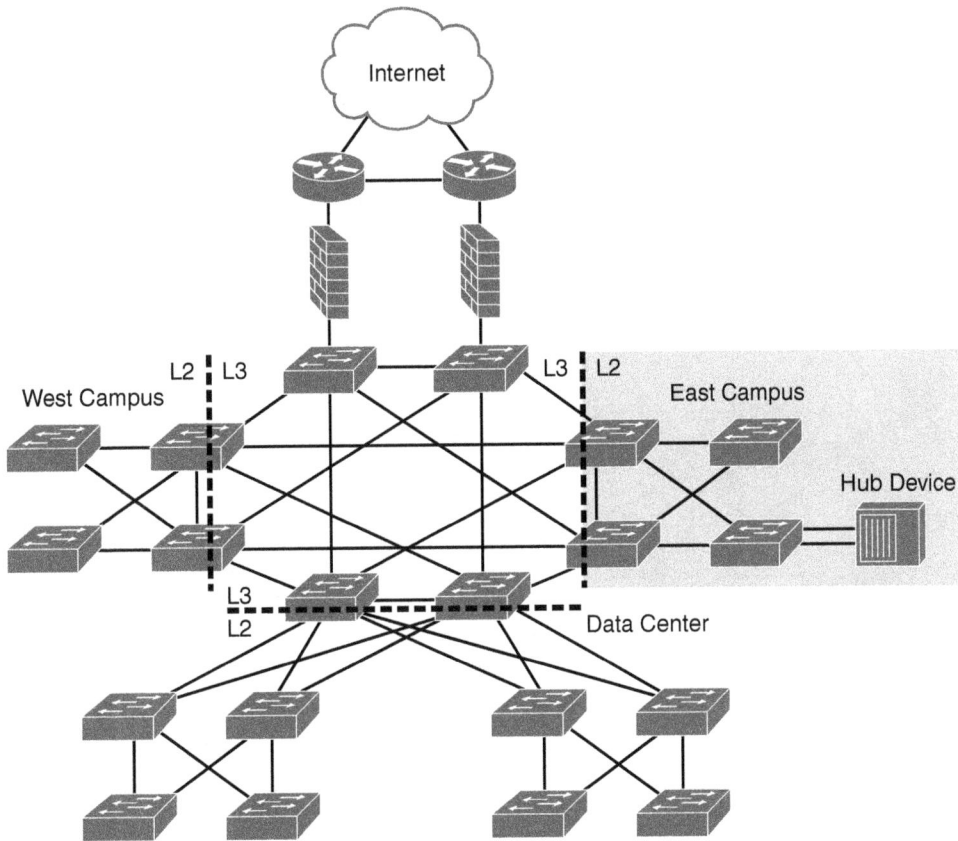

**Figure 1-8**  *Higher Education Campus Architecture New Failure Domain*

Failure isolation can be leveraged in almost every networking protocol.

This example shows the relationship between Layer 3 and Layer 2 technologies. You can also apply the failure isolation technique within Layer 3 technologies like OSPF. If you had a flat OSPF area 0 design for this new Layer 3 architected campus and you had to add 100 new satellite campus locations with routers hanging off of the main routers via VPN connections, all of these new routes would be in the same OSPF area 0. Every time a new router is added or removed, these routes would be added and removed in every router within the OSPF area 0 design. This causes route reconvergence churn that is very impactful to the devices and would degrade the network's performance and potentially cause a large outage. Figure 1-9 highlights how this flat OSPF design would be impacted by these additional satellite locations.

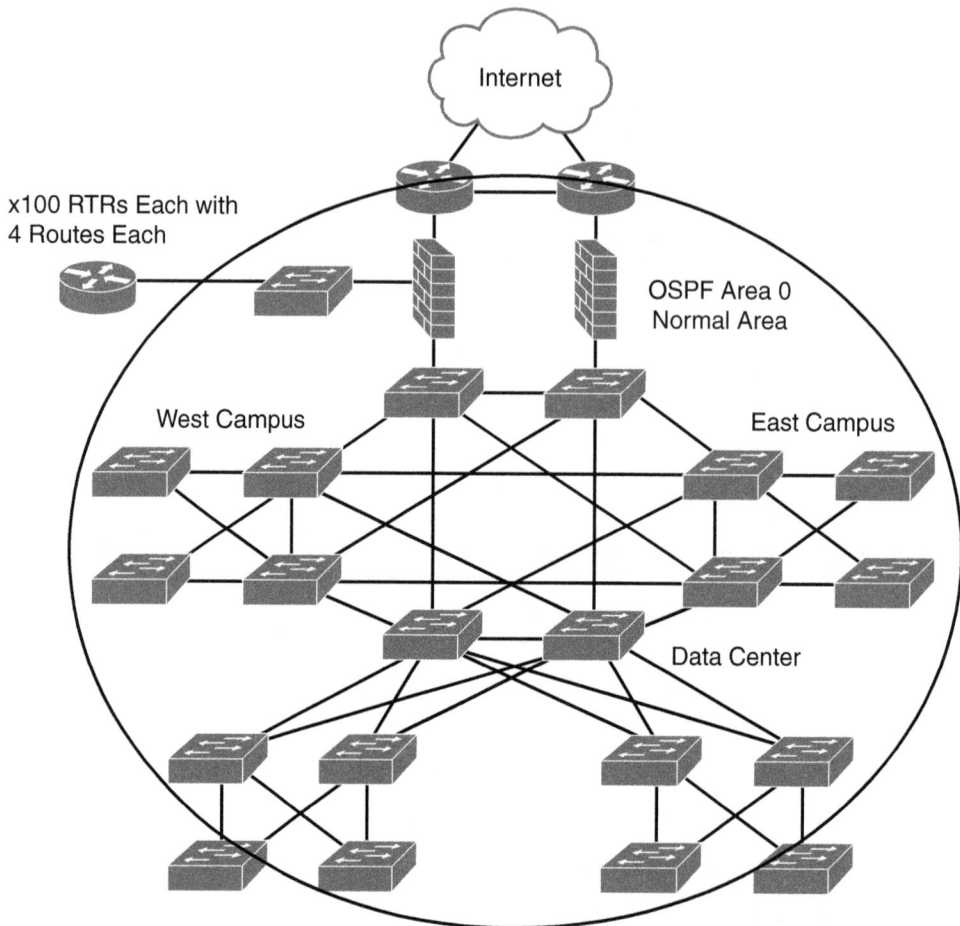

**Figure 1-9**  *Higher Education Campus Architecture Flat OSPF Area 0 Design*

Now if you leverage failure isolation, instead of designing a flat OSPF area 0 design, you could design a multi-area OSPF design with multiple area types. Figure 1-10 shows an example of how this could be achieved.

The main routers in the Internet section would be the area 0 demarcation, and each site's routers would be in their own specific site area. Each site would also have an OSPF area type of totally stubby area, which only allows a default route to be propagated into the area. The Internet site would be in its own area as well, area 20, but the OSPF area type would be different because we have a requirement to redistribute into OSPF here for the default route out to the Internet.

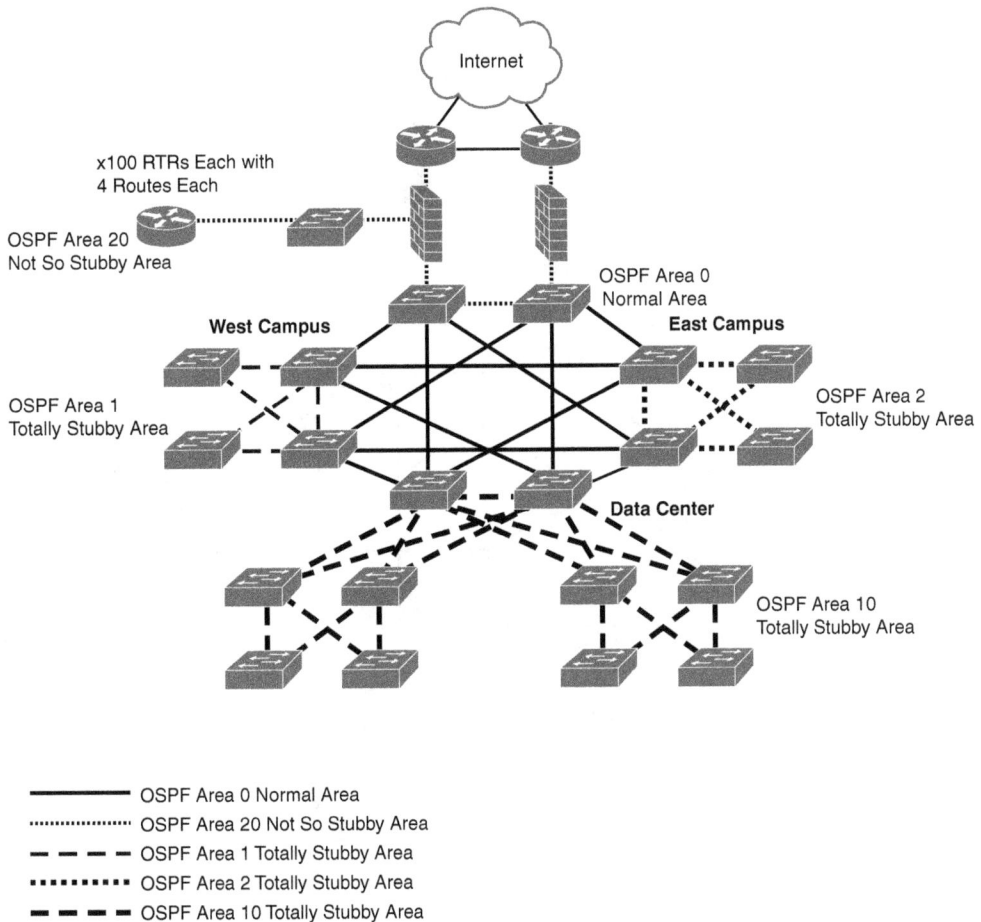

**Figure 1-10**  *Higher Education Campus Architecture Multi-Area OSPF Design*

## Shared Failure State

**Shared failure state (aka fate sharing)** is when a device, system, or portion of the network is filling multiple critical roles, services, and/or protocols.

Figure 1-11 provides an example of how you can solve issues with fate sharing, or at least design for it effectively.

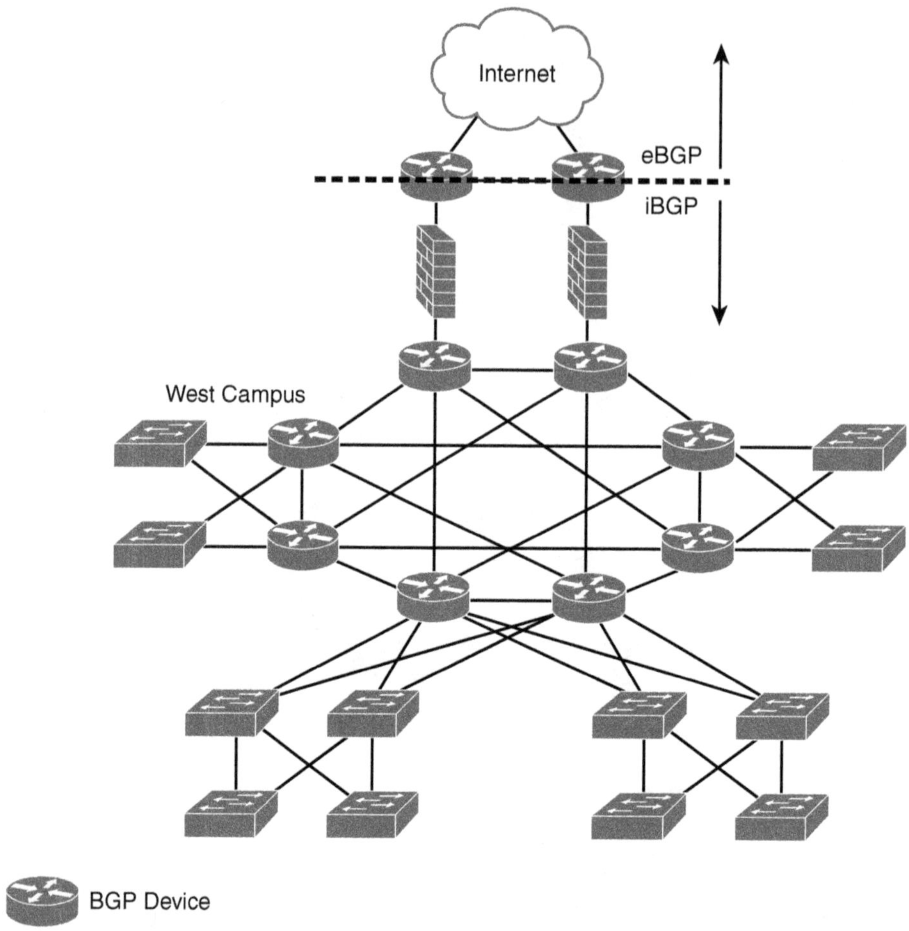

**Figure 1-11**  *Higher Education Campus Architecture BGP Design*

Here we have the higher education campus architecture, but this time we are focusing on Border Gateway Protocol (BGP). Each location has BGP routers running IPv4 and IPv6 routing tables. The Internet Point of Presence (POP) routers are configured for external BGP (eBGP) to the Internet providers to receive Internet-facing routes. These POP routers are also configured as internal BGP (iBGP) with each campus location's BGP routers. Figure 1-12 highlights this in detail.

For scalability purposes, we identify the routers in the Internet pod as BGP route reflectors that all campus locations' BGP routers would neighbor with. Doing this drastically increases the scale of iBGP and is preferred over a full mesh of iBGP neighborships in most design cases.

These BGP IPv4/IPv6 route reflectors represent a shared fate scenario. They are providing a critical role for both IPv4 and IPv6 routing. Let's assume here that there was an IPv4 vulnerability on both of these BGP route reflectors and that it eventually brought these routers down hard. Because of this shared fate scenario, this architecture also lost its IPv6 BGP route reflectors at the same time, even though there wasn't a vulnerability with IPv6.

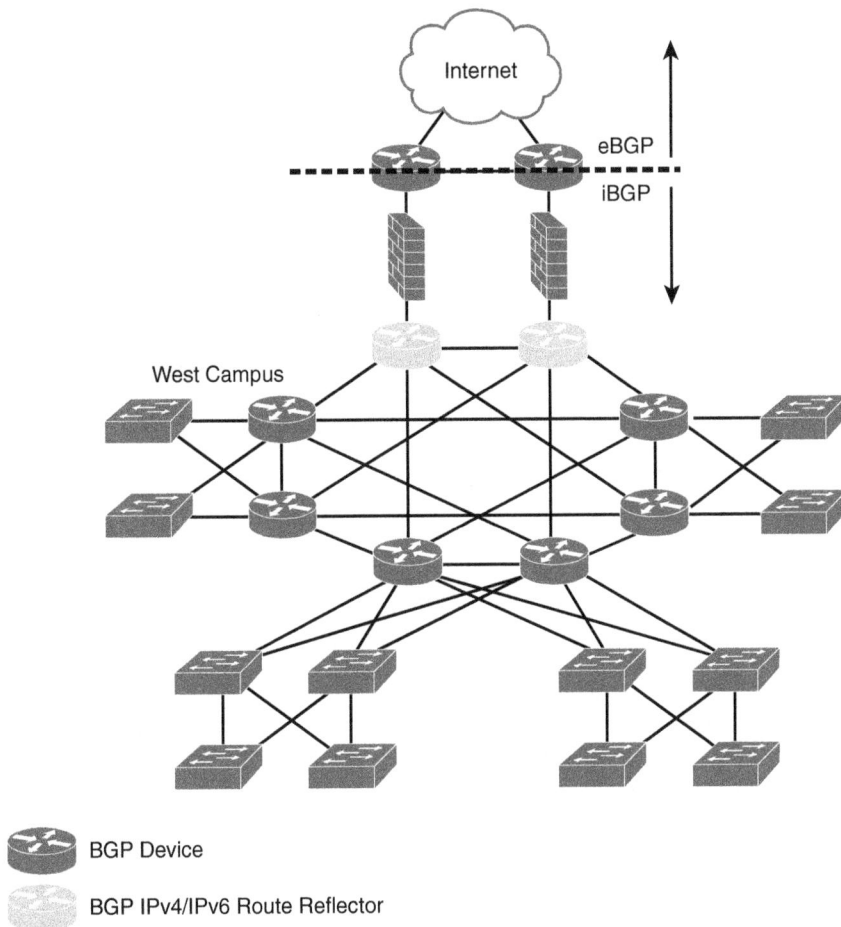

**Figure 1-12**  *Higher Education Campus Architecture BGP Fate Sharing*

To mitigate a shared fate scenario like this, you could break up the roles, IPv4 and IPv6 BGP route reflectors, on their own dedicated stacks. This is shown in Figure 1-13. While breaking up these roles mitigates the shared fate scenario, it increases the cost and complexity of the overall architecture. As a network designer you have to weigh the costs and benefits of leaving a shared fate scenario in place versus providing design options that limit the shared fate scenario but increase overall cost and complexity.

A shared fate scenario isn't always a bad situation. Shared fate scenarios can be leveraged as a design element to ensure when a specific function, capability, role, or device fails, the rest of the architecture affected fails over as well. Consider the IPv4 Internet block in Figure 1-14 for example. If the inside interface of one of the IPv4 BGP POP routers were to fail, we would want the entire IPv4 Internet flow to fail over from the A side of the stack to the B side of the stack. If we didn't design it to fail over that way, traffic would be black-holed, causing a significant outage to occur.

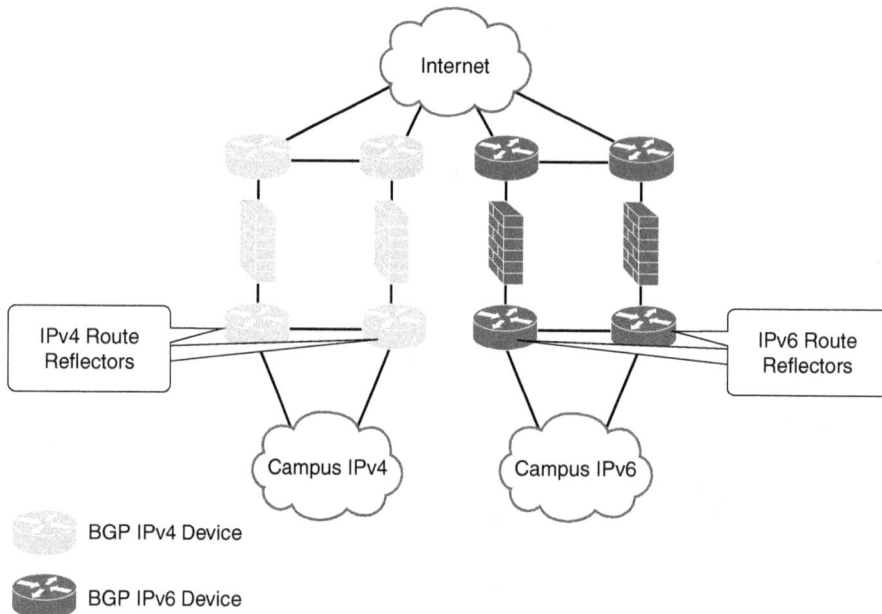

BGP IPv4 Device

BGP IPv6 Device

**Figure 1-13**  *Higher Education Campus Architecture BGP Dedicated Internet Stacks*

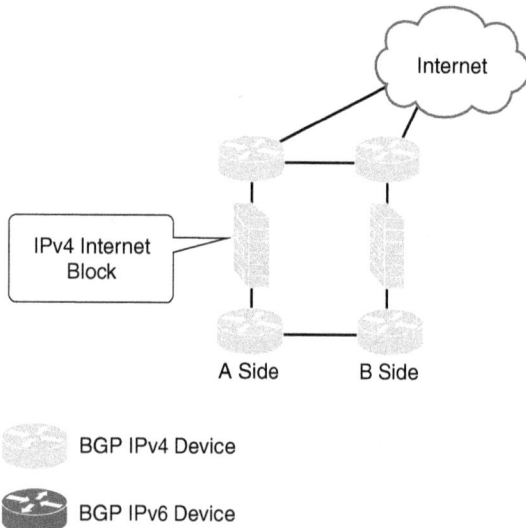

BGP IPv4 Device

BGP IPv6 Device

**Figure 1-14**  *Higher Education Campus Architecture BGP Dedicated IPv4 Internet Block*

## Modularity Building Blocks

**Modularity** is the concept of breaking design elements into functional blocks or pods. Another aspect of the modularity concept is to help isolate technologies and corresponding capabilities within that block. Think of purpose-built blocks. How do we put this into practice?

Let's use our campus example again. Figure 1-15 and Figure 1-16 show that we already have purpose-built pods created in this campus architecture. Currently we have the Internet pod, west campus pod, east campus pod, and data center pod.

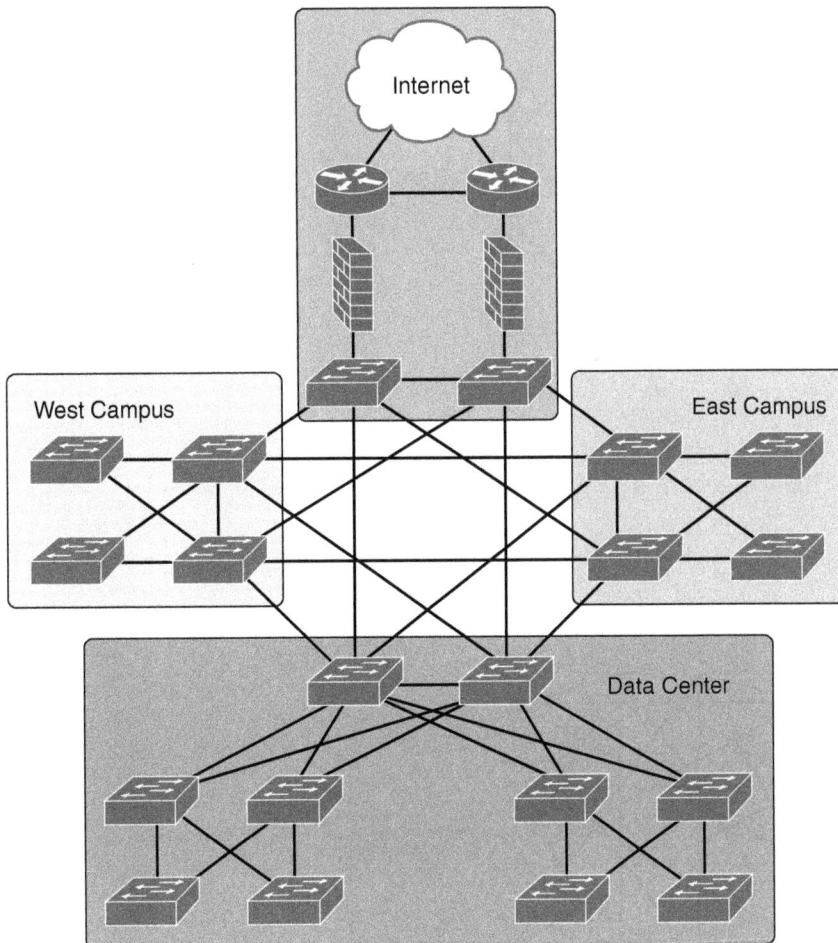

**Figure 1-15**  *Higher Education Campus Architecture Purpose-Built Pods*

Suppose that because of a merger with another higher education institute, these two architectures have to be merged. This new campus has two locations: north and south. If we were to continue the process of doing a full mesh of locations, this would be one very large, convoluted network, with an enormous amount of fiber running from location to location. The impact of this is shown in Figure 1-17.

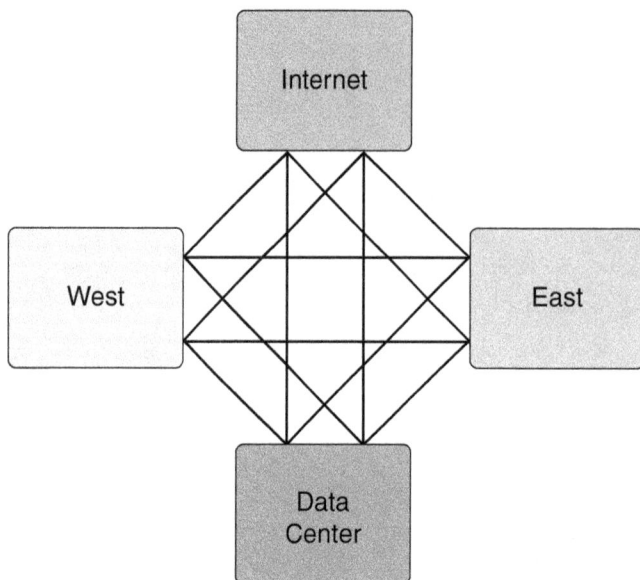

**Figure 1-16** *Higher Education Campus Architecture Modularity Full Mesh*

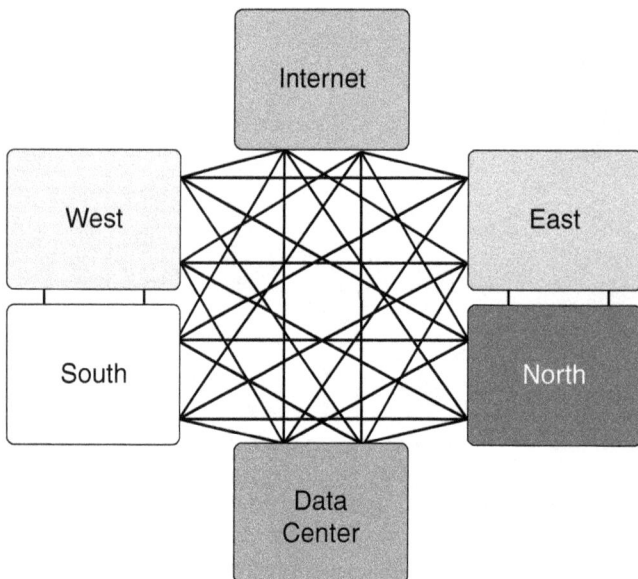

**Figure 1-17** *Higher Education Campus Architecture Modularity Merged Full Mesh*

Of course, this is an extreme situation.

Another potential design option would be to leverage the Internet location as the central connectivity point for the entire merged campus architecture. With this design option, the Internet "pod" is performing two roles. One role is the Internet edge connectivity, and the other role is that of a campus core. If the combined campus architectures were to have a new requirement to add another four sites to the mix, this would become a design problem.

To mitigate that, a second design would be a dedicated core pod that all other sites would connect into, including the Internet edge pod. This is shown in Figure 1-18.

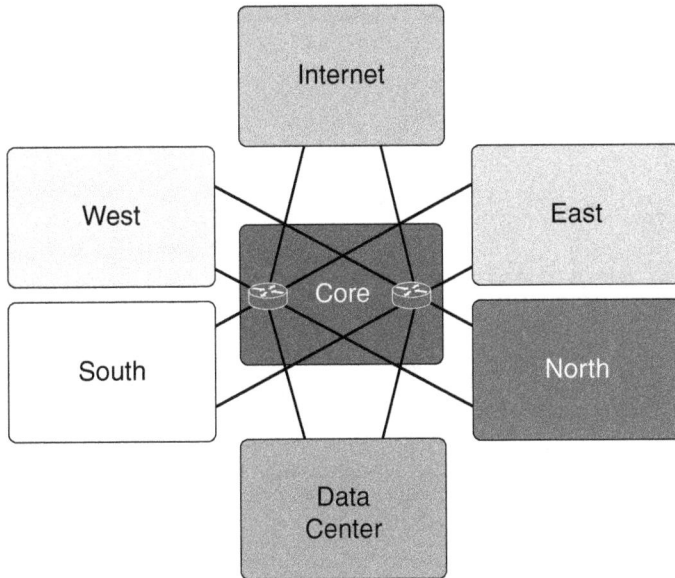

**Figure 1-18**   *Higher Education Campus Architecture Modularity Merged Core Pod*

From a scalability perspective, once the core pod hits its scale limitation, then you would simply add a second core pod interconnected with the original core pod. Within each pod construct, it would be its own Layer 2 isolated boundary, and Layer 3 mechanisms would be employed to limit the Layer 3 failure impact. In addition, each pod would have redundant devices and links, as a general rule, for availability requirements.

If we compare the design we started with, shown in Figure 1-3, to the design we have now, we can see how the modularity network design technique provides flexibility, scalability, and proper fault isolation that allows a network to function properly. Also, we now can easily identify problems in the network, because we have associated fault boundaries within each pod we've designed. When troubleshooting, we can literally rule out a pod in a matter of seconds to find the issue and resolve it.

With a modularity approach to network design, we can easily set performance factors and replicate each pod architecture as the network and business reach these performance guidelines. If a design needs a new core pod or block, we add it. If a design needs a new Internet edge block, we add it. If a design needs a new access block, we add it. This is repeatable, expandable, and manageable!

## Hierarchy of Design

**Hierarchy** of design is the idea of creating dedicated levels for different purposes within the architecture you are building. The traditional hierarchy model is access, distribution, aggregation, and core.

Let's use our higher education architecture again, but this time we are going to focus on the west campus. As shown in Figure 1-19, we have eight west campus locations connecting to the core pod. This is an extremely flat architecture that lacks hierarchy. By breaking this up into distribution and access layers, implementing a hierarchy of design, we can create a robust and scalable architecture.

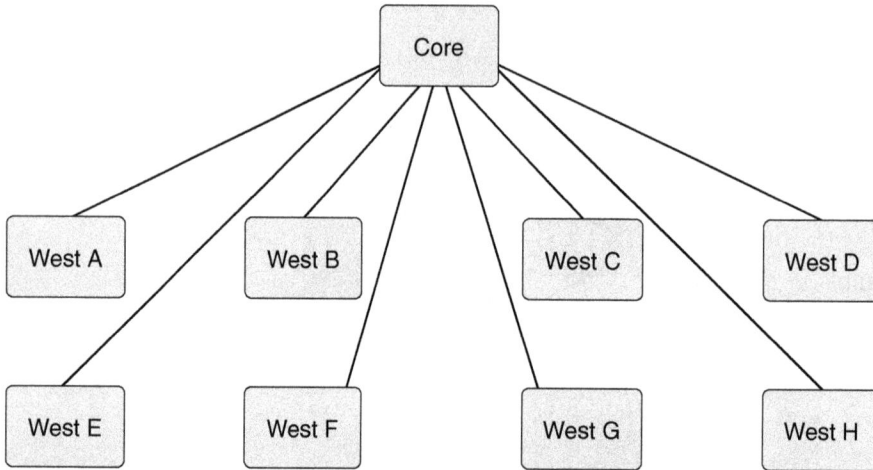

**Figure 1-19**  *Higher Education Campus Architecture Flat Hierarchy*

When we add a dedicated distribution layer, we can start to see how this changes the implications of the design. Figure 1-20 presents an example of how a distribution layer would change this design.

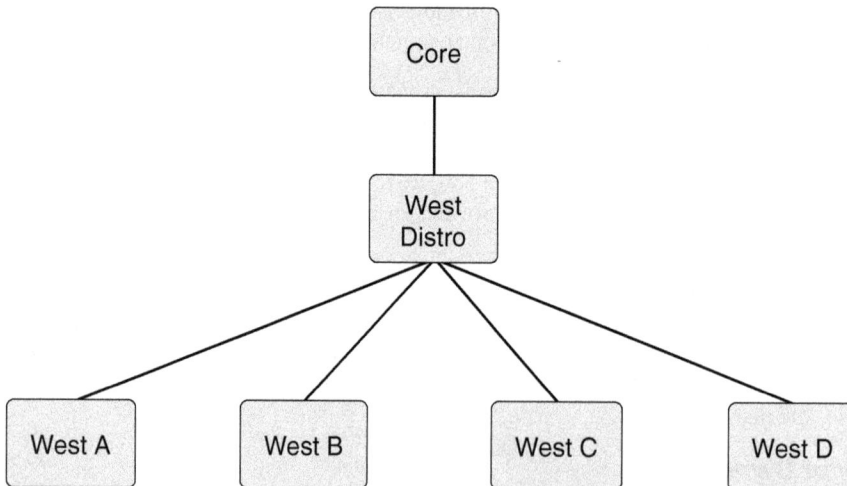

**Figure 1-20**  *Higher Education Campus Architecture Distribution Hierarchy*

Keep in mind, just adding a distribution layer may be enough for now. You do not need to add layers just because; you can easily add more layers as the requirements, business, and network need to. When the time comes to add a dedicated distribution, aggregation, and access layer to the overall design hierarchy, it might look very similar to that shown in Figure 1-21.

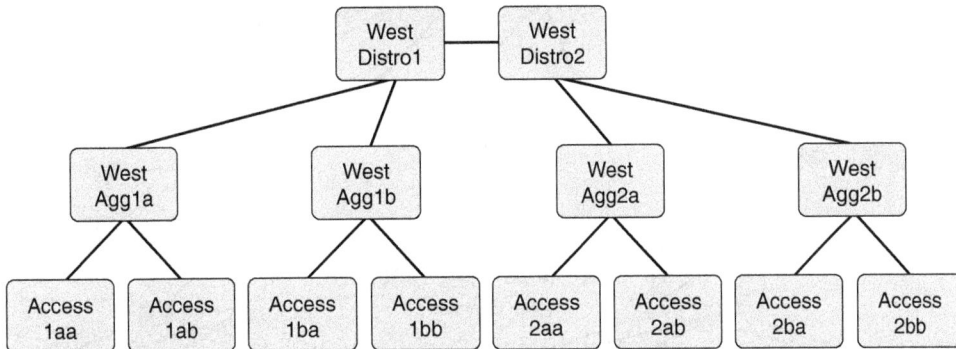

**Figure 1-21**  *Higher Education Campus Architecture Distribution, Aggregation, and Access Hierarchy*

## Putting It All Together

Now let's put together everything we've covered so far. Figure 1-22 shows the design we started with.

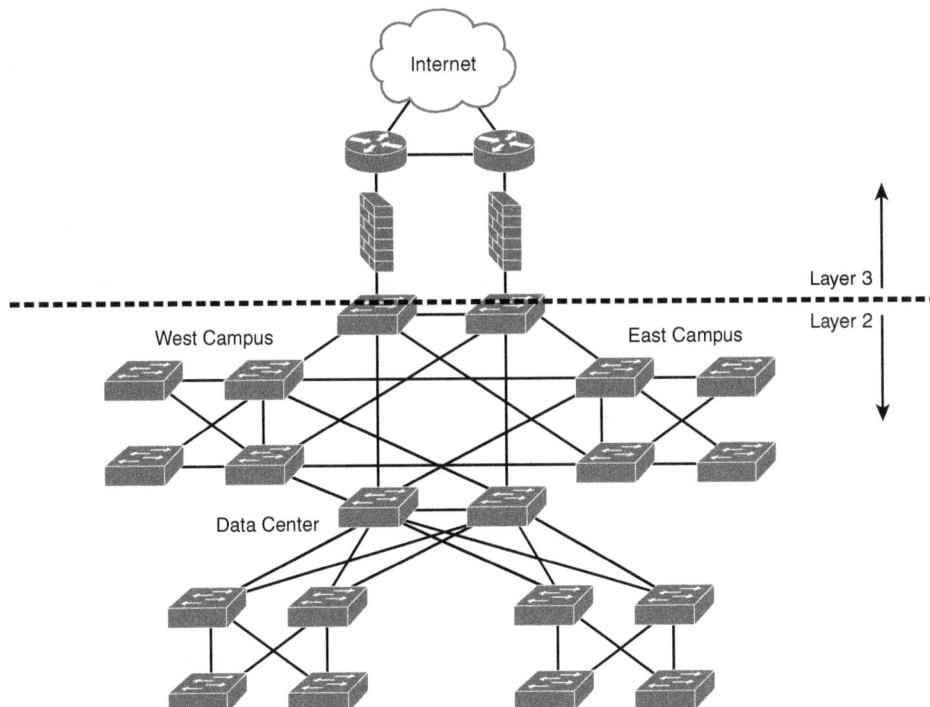

**Figure 1-22**  *Higher Education Campus Architecture Overview*

Figure 1-23 shows our new and improved design.

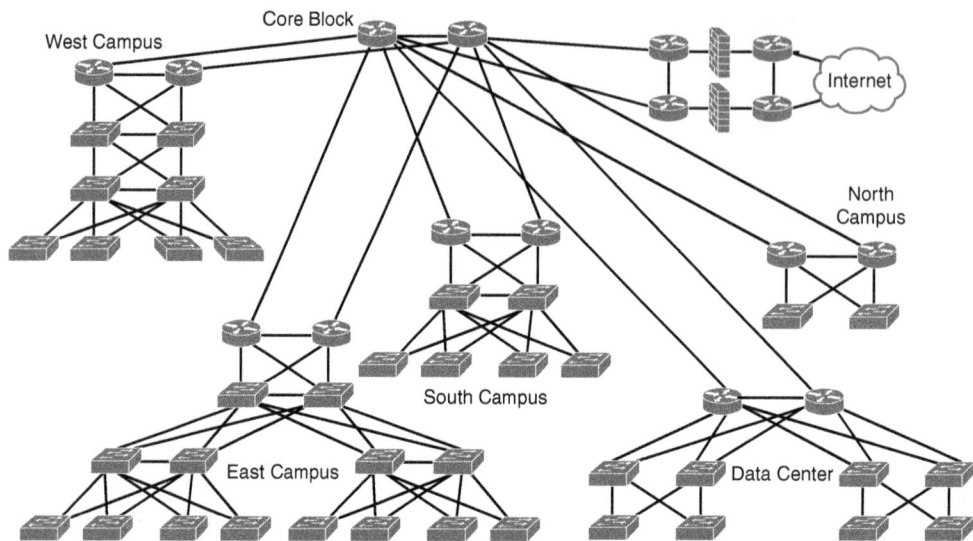

**Figure 1-23** *Higher Education Campus Architecture After Putting It All Together*

We now have a dedicated core block that interconnects our entire campus architecture. We still have our Internet and data center blocks, which haven't changed that much.

Our campus blocks are going to have different scale sizes based on their own needs and requirements. We start with simple, and we move up in complexity based on the need from the business.

When it's all said and done, we are left with a robust network architecture that is based on sound network design fundamentals, techniques, and principles.

These concepts, while simple and easy to understand, are extremely impactful from a network design perspective. This higher education example is a real-world example. Customer names were removed for anonymity, but otherwise everything else is the same.

# Network Design Pitfalls

Network design pitfalls are some of the most common network design mistakes that you can make as a network designer. These are the traps that can literally break any network design you create or break any network design decision you make.

## Making Assumptions

Here you are, all ready to help your customer, which is your company in this scenario, resolve a business problem. You step in and start leading the design discovery phase of the engagement or initiative. As you facilitate conversations with the different stakeholders you are working with, you start to think that complexity is bad. Maybe this was your thought from the beginning of the meeting—that this environment, this customer, your company, doesn't want a complex solution.

Here is the trap of making an assumption and not properly validating that assumption.

In any design situation, you never want to assume anything. Don't assume complexity is bad. Sure, you can have assumptions, but don't make any design decisions based on an assumption until you validate that the assumption you have is correct.

Validate your assumptions with questions to your customer, the business stakeholders, and your team. Leverage and lead workshops to drive customer dialog, but let them talk. Don't feel like you need to talk the entire time. Listen attentively, process what is being discussed, and take notes so you understand the customer and the situation better.

If you are assuming the customer doesn't want a complex solution, ask questions about how the customer is managing their network today. Could they manage a complex solution in this same way? Map everything back to the business and the requirements. If you do this properly, you might discover that the customer's needs require that you design a complex solution. This is out of your hands, and it is perfectly acceptable. Just don't assume complexity is bad. Don't assume that the customer has a limited budget. Don't assume the customer isn't open to new technologies or solutions that they've never used before. To reiterate, if you do assume, make sure you validate your assumptions before making any design decisions.

## Overdesigning Is Good, Right?

Overdesigning, also known as **gold plating**, is an easy trap to fall into (and a difficult one to escape) as a network designer. A common assumption is that added bells and whistles to a network design will impress the customer and enhance the job security. Other times network designers might feel like they are future proofing the network, but in reality, they are adding items that are not needed at all.

As an example of overdesigning a solution, suppose that you are designing a network for a customer and one of the customer's requirements is to have no single points of failure. You design a solution that meets that need, but then you have a brilliant idea that will also limit dual points of failure. The solution has an exponential amount of complexity and resource cost associated with it, but you're certain that your customer will appreciate your brilliance and readily approve the idea, so you include it in your design. This is a classic case of gold plating a solution and is more likely to aggravate your customer than to please them.

You are simply overthinking it.

If you find yourself going down this rabbit hole and ending up in this trap, bring it all back to the requirements. You need to refocus on what is truly needed versus what you are adding because you believed there was additional value in doing so.

The example of being tasked with eliminating single points of failure but deciding to eliminate dual points of failure on your own initiative is a fairly obvious overdesigning situation. Most scenarios involve a more subtle distinction between what is required and what is gold plating.

- What about receiving the full IPv4 Internet table on your Internet edge routers? In the past, some network designers would stick to this as the way to do things. They would assume and suggest this as a design because it provided so much value and flexibility to the environment. This was overdesigning the solution. There was no business need or requirement to do this.

- How about running an MPLS network at a service provider? Do you design MPLS-TE tunnels with sub-second failover, node protection, and link protection, or would the IGP timers be enough to meet the requirements at hand? The former solution, which would be considered gold plating, adds a ton of complexity and additional work to set up the solution, let alone the amount of time and resources that would now be needed to manage this solution on a daily basis.

- How about studying for your CCIE exam? Have you put a technology or solution into a protection network because you wanted to see how it worked? This has been done numerous times and there was no requirement for that solution in that environment at that time.

You have to catch yourself. The easiest way to mitigate this pitfall of overdesigning and gold plating is to hyperfocus on the requirements. Everything you do should have a direct business requirement that it maps to. This won't be a one-to-one mapping; it will be a one-to-many mapping, the business requirement mapped to many design decisions.

## Best Practices

Have you ever made a design decision and your justification was "Because it's best practice"? Did you fully understand the implications of what you were doing, or did you fill in the "best practice" variable with your own personal biased design, solution, or architecture?

A lot of us in the network design field make "best practice" calls without fully understanding the design implications of those decisions. Let's start with a simple example. Why do we enable an OSPF interface as a point-to-point interface? Is there a business requirement for it? Probably a better question to ask and answer is whether there is a business requirement that we are breaking because of this "best practice" decision.

Now, how about a more complex and realistic example. What about implementing sub-second failover for an IGP versus less than 5 seconds failover for an IGP? Are we implementing sub-second failover because it's "best practice" or are we correlating this choice to a business need?

As network designers, we cannot fall into the trap of "best practices." Instead, we take into consideration the best practices and we modify them, tweak them for each of our design decisions based on the business requirements. Just because something is best practice doesn't mean it's going to work that way in your design. Sometimes you have to build and design a network that is not preferred from a best practice standpoint, and you do this because of the requirements.

A perfect example of this would be spanning Layer 2 by leveraging STP between two data centers. This is something most network designers wouldn't want to do. Spanning Layer 2 like this creates a large failure domain between these two data centers. Sometimes we as network designers have no choice, especially when there are application requirements that force this design option to solve the application constraints. Nowadays there are better options to span Layer 2 between multiple sites than to leverage STP, but the implication is still the same.

## Preconceived Notions

Preconceived notions are pretty similar to assumptions, but they are defined by outside information from your experiences.

As network designers, we don't want to bring in our own preconceived notions or opinions. Just because a network designer likes EIGRP does not mean it's the correct IGP for every design situation. Just because MPLS L3VPN circuits are more expensive than MPLS L2VPN circuits in your experience does not mean you can make decisions based on that information in a design situation.

The design should always be tied back to the customer's business requirements!

## Summary

In this chapter, we focused on the network design elements that all network designers should know. We discussed the network design fundamentals, which are the foundation of all designs (like the foundation of a house). Then we added to this foundation with network design principles (like the framing of the house), which showed the give and take from a design perspective. The more availability a network architecture requires, the more it will cost. Once again, this cost can be both monetary and nonmonetary. We next covered the network design techniques (like a roof to the house), referencing a real-world use case to solidify how these techniques can be leveraged in every network design situation. Finally, we highlighted the mistakes that network designers tend to make. We talked about the pitfalls of assumptions, overdesigning, strict adherence to best practices, and preconceived notions.

If there are not any relevant business requirements for a specific design situation and you are not violating another business requirement, then best practice is probably the way to go, but *you need to understand the full picture* before making these decisions. In the end, it really boils down to doing what is right for the specific situation that you are presented with. A lot of people are looking for the "one fits all" solution or answer, but there isn't one. Especially for a network designer. People who take the "easy" way out end up doing a disservice to the networks they touch and customers they serve.

All of the items covered in this chapter are important and, yes, it's a different way of thinking altogether, which will not be easy. It will take some time and effort to instill these elements into your thought process. It's worth it to incorporate each element discussed in this chapter to ensure your continued success as you design a network and tackle the CCDE certification.

## Exam Preparation Tasks

As mentioned in the section "How to Use This Book" in the Introduction, you have a couple of choices for exam preparation: the exercises here, Chapter 18, "Final Preparation," and the exam simulation questions in the Pearson Test Prep Software Online.

## Review All Key Topics

Review the most important topics in this chapter, noted with the Key Topic icon in the outer margin of the page. Table 1-3 lists a reference of these key topics and the page numbers on which each is found.

**Key Topic**

**Table 1-3**   Key Topics for Chapter 1

| Key Topic Element | Description | Page Number |
|---|---|---|
| Section | Design Use Cases | 10 |
| List | Categories of customer design constraints | 14 |
| List | Common customer design constraints | 14 |
| List | Security models | 17 |

## Complete Tables and Lists from Memory

There are no Memory Tables or Lists in this chapter.

## Define Key Terms

Define the following key terms from this chapter and check your answers in the glossary:

top-down approach, bottom-up approach, functional requirements, technical requirements, application requirements, greenfield, brownfield, add technology, replace technology, merger, divestment, scaling, design failure, unstated requirements, Perimeter Security (aka Turtle Shell), Session- and Transaction-Based Security, Zero Trust Architecture, redundancy, resilience (aka resiliency), reliability, failure isolation, shared failure state (aka fate sharing), modularity, hierarchy, gold plating

## Reference

Al-shawi, Marwan, *CCDE Study Guide* (Cisco Press, 2015)

# CHAPTER 2

# Designing for Business Success

This chapter covers the following topics:

- **Business Success:** This section covers the business elements and terminology that all network designers should know.

- **Project Management Methodologies:** This section covers today's most common project management methodologies that will help every network designer make better design decisions.

As we start to uncover the ins and outs of designing networks, understanding the business side of the house is critical to ensuring that the design decisions being made will directly increase business success. We as network designers must bridge the gap between technology and the business. I challenge you to be the bridge. To help with this goal of being the bridge and ensuring business success, this chapter focuses on the different business elements, terminology, and project management styles that all network designers need to know.

This chapter covers the following "CCDE v3.0 Unified Exam Topics" section:

- 1.0 Business Strategy Design

## "Do I Know This Already?" Quiz

The "Do I Know This Already?" quiz allows you to assess whether you should read this entire chapter thoroughly or jump to the "Exam Preparation Tasks" section. If you are in doubt about your answers to these questions or your own assessment of your knowledge of the topics, read the entire chapter. Table 2-1 lists the major headings in this chapter and their corresponding "Do I Know This Already?" quiz questions. You can find the answers in Appendix A, "Answers to the 'Do I Know This Already?' Quizzes."

**Table 2-1** "Do I Know This Already?" Section-to-Question Mapping

| Foundation Topics Section | Questions |
|---|---|
| Business Success | 1–8 |
| Project Management | 9–10 |

**CAUTION** The goal of self-assessment is to gauge your mastery of the topics in this chapter. If you do not know the answer to a question or are only partially sure of the answer, you should mark that question as wrong for purposes of the self-assessment. Giving yourself credit for an answer you correctly guess skews your self-assessment results and might provide you with a false sense of security.

1. Which business term has the definition "typically based on strategies adopted for the achievement of goals"?

   a. Capability

   b. Outcome

   c. Driver

   d. Priority

2. Which business term is referenced as the "why" the business is doing something?

   a. Capability

   b. Outcome

   c. Driver

   d. Priority

3. Which business term equates to the end result, such as saving money, diversifying the business, increasing revenue, or filling a specific need?

   a. Capability

   b. Outcome

   c. Driver

   d. Priority

4. Which business term refers to what a solution provides, and most times you get multiple of them?

   a. Capability

   b. Outcome

   c. Driver

   d. Priority

5. Which design tool adds multiple dimensions to the design-making process?

   a. Decision tree

   b. Tactical planning approach

   c. Design matrix

   d. Strategic planning approach

6. What type of financial savings does a business achieve by consolidating its wholly owned data center locations from four to two?

   a. OPEX

   b. ROI

   c. CAPEX

   d. TCO

7. What type of financial savings does a business achieve by reducing its advertising and marketing budget?

   a. OPEX

   b. ROI

   c. CAPEX

   d. TCO

**8.** What business financial concept does spending $5,000 on a new application upgrade that includes new features and functionality that allow the business to charge a premium?

    **a.** OPEX

    **b.** ROI

    **c.** CAPEX

    **d.** TCO

**9.** Which of the following is *not* a characteristic of the waterfall project management framework?

    **a.** Linear

    **b.** Change adverse

    **c.** Structured

    **d.** Minimal documentation

**10.** Which of the following is a characteristic of the Agile project management framework?

    **a.** Change adverse

    **b.** End state is defined

    **c.** Faster delivery

    **d.** Feedback not welcome

## Foundation Topics

## Business Success

This section covers the primary aspects that pertain to the business needs and directions that (individually or collectively) can influence network design decisions either directly or indirectly. The best place to start understanding the business needs and requirements is to look at the big picture of a company or business and understand its associated business priorities, business drivers, and business outcomes. This enables network designers to steer the design to ensure business success. However, there can be various business goals and requirements based on the business type and many other variables. As outlined in Figure 2-1, with a top-down design approach, it is almost always the requirements, constraints, and drivers at higher layers, such as business and application requirements, that drive and set the requirements and directions for the lower layers. Therefore, network designers aiming to achieve a business-driven design must consider this when planning and producing a new network design or when evaluating and optimizing an existing one. The following sections discuss some of the business elements at the higher layers and how each can influence network design decisions at the lower layers. Remember, our goal as network designers is to ensure business success.

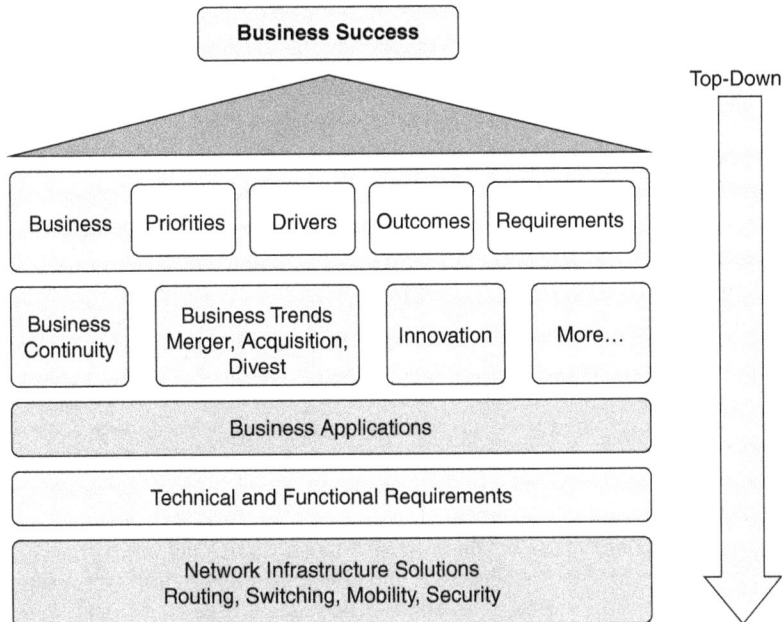

**Figure 2-1**  *Business Success Top-Down Approach*

## Business Priorities

Each business has a set of **business priorities** that are typically based on strategies adopted for the achievement of goals. These business priorities can influence the planning and design of IT network infrastructure. Therefore, network designers must be aware of these business priorities to align them with the design priorities, which ensures the success of the network they are designing by delivering business value. For example, suppose that a company's highest priority is to provide a more collaborative and interactive business communications solution, followed by the provision of mobile access for the end users. In this example, providing a collaborative and interactive communications solution must be satisfied first before providing or extending the solution over any mobility solution for the end users. Keep in mind, it is important to align the design with the business priorities, which are key to achieving business success and transforming IT into a business enabler.

An example business priority would be security. There are a number of other terms that can be used for this priority, such as Zero Trust Architecture, cybersecurity modernization, and risk management framework. No matter what the business priority is called, the intent is the same, to secure the network and maintain data integrity. If a business's data is compromised, that business is out of business.

## Business Drivers

Now that we know what the business priorities are, we need to start to identify the different constraints and challenges that apply to the business. What does this business have to do and why? This is what we call a **business driver**. Business drivers are what organizations must follow. A business driver is usually the reason a business must achieve a specific outcome. It is the "why" the business is doing something.

An example of a business driver would be a constraint on the business to follow a specific compliance standard like HIPAA or PCI DSS. This aligns perfectly with the business priority of security mentioned in the previous section. If the business does not adhere to this constraint, depending on the compliance standard in question, the business can be fined or even shut down. For this example, the business driver could be worded as "Required to follow HIPAA compliance standards."

## Business Outcomes

A **business outcome** equates to the end result, such as saving money, diversifying the business, increasing revenue, or filling a specific need. Essentially, a business outcome is an underlying goal a business is trying to achieve. A business outcome will specifically map to a business driver. Returning to our previous example of security as a business priority, the business driver could be phrased as "Required to follow HIPAA compliance standards" while the business outcome could be "Properly maintain HIPAA compliance to stay in business."

At a minimum, there will be one outcome per driver, but there can be multiple outcomes mapping to the same driver. This is perfectly fine. If, and when, an organization achieves its business outcomes, then its business drivers are met, which ensures the organization's business success.

## Business Capabilities

Before we go down the "solutionizing" path, which most of us engineers tend to do too early, we need to know what business capabilities are and how they apply to network design. Business capabilities are not solutions. **Business capabilities** are what you get from a solution. Most solutions provide multiple capabilities. Some solutions provide parts of multiple capabilities; when combined with other solutions, the business can get a number of capabilities that will make them successful.

Session-based security is a great example of a capability that a business might need to have to meet compliance standards. A vendor-agnostic solution that provides this capability would be a network access control (NAC) solution. There are many different vendor-specific NAC solutions; the point here is that all of them, no matter what vendor solution we highlight, provide the capability of session-based security. Moving forward along this thought process, Cisco Identify Services Engine (ISE) is an example of a vendor-specific solution that provides the capability of session-based security. Cisco ISE actually provides multiple capabilities in addition to session-based security.

Business capabilities map directly to business outcomes. As a network designer, you will find that multiple capabilities often map to the same outcome, and that multiple outcomes often map to the same capability. This is expected and perfectly fine. Table 2-2 shows the relationship between business priorities, drivers, outcomes, and capabilities.

**Table 2-2** Business Priorities, Drivers, Outcomes, and Capabilities Relationship Mapping

| Priorities | Drivers | Outcomes | Capabilities |
|---|---|---|---|
| Priority # 1 | Driver # 1 | Outcome # 1 | Capability # 1, Capability # 2 |
| Priority # 1 | Driver # 2 | Outcome # 2 | Capability # 1 |
| Priority # 2 | Driver # 3 | Outcome # 2 | Capability # 1 |
| Priority # 3 | Driver # 3 | Outcome # 3 | Capability # 3, Capability # 4 |
| Priority # 4 | Driver # 4 | Outcome # 4 | Capability # 4 |

## Business Continuity

*Business continuity (BC)* refers to the ability to continue business activities (business as usual) following an outage, which might result from a system outage or a natural disaster like a fire that damages a data center. Therefore, businesses need a mechanism or approach to build and improve the level of resiliency to react to and recover from unplanned outages.

The level of resiliency is not necessarily required to be the same across the entire network, however, because the drivers of BC for the different parts of the network can vary based on different levels of impact on the business. These business drivers may include compliance with regulations or the level of criticality to the business in case of any system or site connectivity outage. For instance, if a retail business has an outage in one of its remote stores, this is of less concern than an outage to the primary data center, from a business point of view. If the primary data center were to go offline for a certain period of time, this would affect all the other stores (higher risk) and could cost the business a larger loss in terms of money (tangible) and reputation (intangible). Therefore, the resiliency of the data center network is of greater consideration for this retailer than the resiliency of remote sites.

Similarly, the location of the outage sometimes influences the level of criticality and design consideration. Using the same example, an outage at one of the small stores in a remote area might not be as critical as an outage in one of the large stores in a large city. In other words, BC considerations based on risk assessment and its impact on the business can be considered one of the primary drivers for many businesses to adapt network technologies and design principles to meet their desired goals.

### Recovery Point Objective

The amount of data that can be lost during an outage at peak business demand before harm occurs to the business is called the **recovery point objective (RPO)**. The harm to the business can be monetary or industry reputation based. For example, if a cloud-based infrastructure company has a 30-minute outage worldwide, the impact to its reputation may very well be greater than the impact to its financial standing. On the flip side, if a large financial banking company has an outage at a single ATM location, this most likely has less of a reputation based impact to the overall business.

The amount of data that can be lost from an RPO perspective is given a specific time value, which is measured against the last backup that took place. In most cases the RPO is less than an hour, with the average value being 10 minutes. This means a company with an RPO of 10 minutes can withstand losing 10 minutes' worth of data from when the critical outage started to the most recent backup.

### Recovery Time Objective

The length of time an application, system, network, or resource can be offline without causing significant business damage as well as the time it takes to restore the service is called the **recovery time objective (RTO)**. The RTO is focused on the time to recover a failing system or network outage.

Returning to the example of a cloud-based infrastructure company, suppose it had a critical outage for 10 minutes that affected all online access. The company has an RPO of 15 minutes and an RTO of 20 minutes, so the company was within both metrics. Later that week, the company had another outage that lasted 2 hours during a period of peak business demand. With the RPO only allowing for 15 minutes' worth of data loss, and the RTO only allowing for 20 minutes of downtime, it means 1 hour and 40 minutes of the outage time was not accounted for. This example shows that the loss of data was exponential, as it was during peak business demand for the cloud-based infrastructure company.

## Elasticity to Support the Strategic Business Trends

*Elasticity* refers to the level of flexibility a certain design can provide in response to business changes. A change in this context refers to the direction the business is heading, which can take different forms. For example, this change may be a typical organic business growth, a decline in business, a merger, or an acquisition.

To illustrate the concept of elasticity in network design, let's return to the higher education campus architecture in Chapter 1, "Network Design." This architecture has four locations interconnected directly, as illustrated in Figure 2-2. Any organic growth in this network that requires the addition of a new location will introduce a lot of complexity in terms of cabling, control plane, and manageability. These complexities result from the inflexible design, which makes the design incapable of responding to the business's natural growth demand.

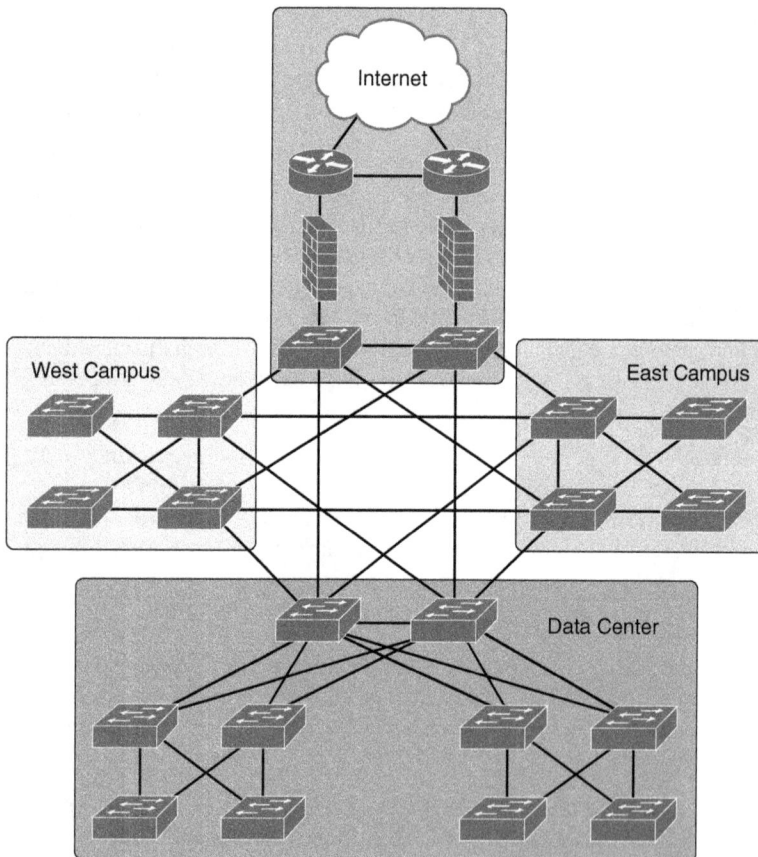

**Figure 2-2**  *Higher Education Campus Architecture Purpose-Built Pods*

To enhance the level of flexibility of this design, you can add a core module to optimize the overall design modularity to support business expansion requirements. As a result, adding or removing any module or location to or from this network will not affect other locations, and does not even require any change to the other modules, as illustrated in Figure 2-3. In other words, the design must be flexible enough to support the business priorities, drivers, and outcomes. If network designers understand business trends in directions in this area, such understanding may influence, to a large extent, design choices.

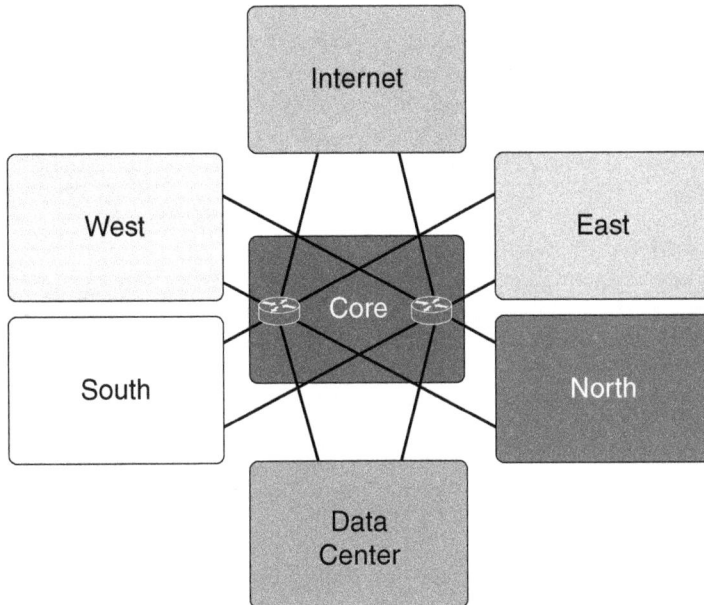

**Figure 2-3**  *Higher Education Campus Architecture Modularity Merged Core Pod*

Similarly, a flexible network design must support the capability to integrate with other networks (for example, when mergers and acquisitions occur). With mergers and acquisitions, however, the network can typically grow significantly in size within a short period of time, and the biggest challenge, in both scenarios (mergers and acquisitions), is that network designers have to deal with different design principles, the possibility of overlapping IP address space, different control plane protocols, different approaches, and so on.

## IT as a "Business Innovation" Enabler

In today's market, many businesses understand how IT technologies enhance their services and provide innovation to their customers. Therefore, when a certain technology can serve as a business enabler that can help the organization to compete in the market or increase its customer's satisfaction, the adoption of that technology will be supported by the business.

For example, cloud-based data centers have been opening new opportunities for hosting service providers to generate more revenue for businesses. To offer cloud-based services, a business requires a reliable, flexible, and high-performance data center infrastructure to deliver this service. Consequently, this forces the way forward, driving the business to build a high-performance, next-generation data center network. The network, by acting as a basis for cloud services, will be the enabler of the business's revenue-generation solution.

**Key Topic**

## IT as a "New" Utility

Over the last decade, IT, and networking specifically, has become more akin to a utility provider (such as an electricity or water provider). Identifying the network as a utility is important today because most businesses assume that the "Internet" is just going to work, that a business application will always be accessible for its end users. Everywhere we go—hotels, airports, restaurants—some form of wireless connection is available. We've come to expect it, just like we expect lights to come on when we flip a switch and water to pour out of the faucet when we move the handle.

This assumption of the network to be available and to properly provide the business whatever the business needs or requires at any specific time is the concept of *unstated requirements* introduced in Chapter 1. As a network designer, you will have to identify how far to take this expectation of the network being a utility and always being available. Table 2-3 shows how we can compare the associated business risk versus reward versus cost of the different availability options.

**Table 2-3**  Business Risk vs. Reward vs. Cost Analysis

| Design Decision | Associated Cost | Design Complexity | Business Risk | Business Reward |
|---|---|---|---|---|
| Single points of failure | Low; no additional cost for redundancy. | Low | Very high; outages are more likely to occur that would directly bring the business offline, which would make the business lose revenue and market reputation. | Low; initial cost savings. |
| No single points of failure | Medium; the solution cost increases to allow for redundant components to mitigate any single points of failure. | High | Low; the solution has been designed to allow for single failures to occur that would still allow the business to function and make money. | High; the initial cost is higher but the reward is substantially better because the business can function, and continue to make money, while a single failure occurs. With this design comes a level of complexity that needs to be properly managed. |

| Design Decision | Associated Cost | Design Complexity | Business Risk | Business Reward |
|---|---|---|---|---|
| No dual points of failure | High; a much higher initial cost is needed to create this design with no dual points of failure. | Very high | Very low; the solution has been designed to allow for dual failures to occur that would still allow the business to function and make money. | Very high; the initial cost is substantially higher than limiting single points of failure, but now the solution is more robust and can withstand a higher degree of failures and still allow the business to function. One of the drawbacks here besides the high cost is the very high complexity level. A business running solutions with no dual points of failure requires a highly skilled and technical team to manage and maintain it. |

We can see that no single points of failure would be the best design option to allow for a redundant solution, increasing overall business availability while also limiting the monetary cost to the business. As network designers, we should be able to present the information captured in Table 2-3 to the different business leaders within a company to allow them to make properly informed business decisions. The business leaders might assume the risk of the lower-cost option, or they may spend a ton of money to mitigate the risk to the business, thus increasing the business reward. Business risk versus reward and cost analysis are extremely important concepts to understand for network designers.

## Money, Money, Money

Approximately 90 percent of business organizations are for-profit companies. This means that the business is focused on making money through a combination of increasing revenues while concurrently decreasing costs. This could be as simple as a reduction in the company's **capital expenses (CAPEX)** by consolidating its wholly owned data center locations from four to two, which reduces the number of buildings (fixed assets) it owns. On the flip side, the company could focus on reducing its **operating expenses (OPEX)** by reducing its advertising and marketing budgets, for example.

Both of these examples of CAPEX and OPEX in the previous paragraph can also be in the form of investment. Completing that new application upgrade will provide a direct investment on either lowering the costs or increasing revenue being brought into the business because of new features and functionality. We call this **return on investment (ROI)**. ROI is specifically the concept of identifying what the perceived potential benefit is going to be for the business if the business does a specific action. This is a critical concept that is not always tied to money, i.e., a monetary return on the initial investment. A perfect example is the CCDE certification. Is there a positive ROI for you as a network designer to invest your time in this certification journey? That's the question you should be asking yourself as you embrace this journey. An overwhelming and profound YES is my answer for you. The CCDE and the associated journey that comes with it has an extremely high ROI, in the form of knowledge, mindset, and being a certified network designer.

## The Nature of the Business

Classifying the industry in which the business belongs or identifying the business's origins can aid in the understanding of "unstated requirements" previously introduced in the chapter and in Chapter 1. For example, information security is almost always a must for a banking business whenever traffic crosses any external link. So by default, when planning a design for a business base in the banking industry, the design must support or offer security capabilities to gain acceptance from the business. In addition, industry-specific standards that apply to IT infrastructure and services need to be considered (for instance, healthcare organizations may consider complying with the IEC-80001-1 standard).

## Planning

An enduring adage you've likely heard is "if you do not have a plan, you are planning to fail." This adage is accurate and applicable to network design. Many network designers focus on implementation after obtaining the requirements and putting them in design format. They sometimes rely on the best practices of network design rather than focusing on planning "what are the possible ways of getting from point A to point B?" This planning process can help the designer devise multiple approaches or paths (design options). At this point, the designer can ask the key question: Why? Asking *why* is vital to making a business-driven decision for the solution or design that optimally aligns with the business's short- or long-term strategy or objective. In fact, the best practices of network design are always recommended and should be followed whenever applicable and possible.

However, reliance on best practices is more common when designing a network from scratch (greenfield), which is not common with large enterprises and service provider networks. In fact, IT network architectures and building constructions and architectures are quite similar in the way they are approached and planned by designers or consultants.

For example, several years ago a Software as a Service (SaaS) company built a new headquarters location in a large city in the United States, which was architected and engineered based on the business priorities, drivers, outcomes, and requirements at that time. Recently, this SaaS company has acquired a number of other companies and is in the process of merging them all together. The stakeholders have requested the network designers to review the design and make suggestions for modification to address the increased number of people accessing the headquarters location, because this increase was not properly projected and planned for during the original design five years ago.

Typically, the architects and engineers will then evaluate the situation, identify current issues, and understand the stakeholders' goals. In other words, they gather the business priorities, drivers, and outcomes and identify the issues. Next, they work on optimizing the existing building (which may entail adding more parking space, expanding some areas, and so forth) rather than destroying the current building and rebuilding it from scratch. However, this time they need to have proper planning to provide a design that fits current and future means.

Similarly, with IT network infrastructure design, there are always new technologies or broken designs that were not planned well to scale or adapt to business and technology changes. Therefore, network designers must analyze business issues, requirements, and the current design to plan and develop a solution that could optimize the overall existing architecture. This optimization might involve the redesign of some parts of the network (for example, WAN), or it might involve adding a new data center to optimize BC plans. To select the right design option and technologies, network designers need to have a planning approach to connect the dots at this stage and make a design decision based on the information gathering and analysis stage. Ultimately, the planning approach leads to the linkage of design options and technologies to the gathered requirements and goals to ensure that the design will bring value and become a business enabler rather than a cost center to the business. Typical tools network designers use at this stage to facilitate and simplify the selection process are the decision tree and the decision matrix.

### Decision Tree

A **decision tree** is a helpful tool that a network designer can use to compare multiple design options, or perhaps protocols, based on specific criteria. For example, a designer might need to decide which routing protocol to use based on a certain topology or scenario, as illustrated in Figure 2-4.

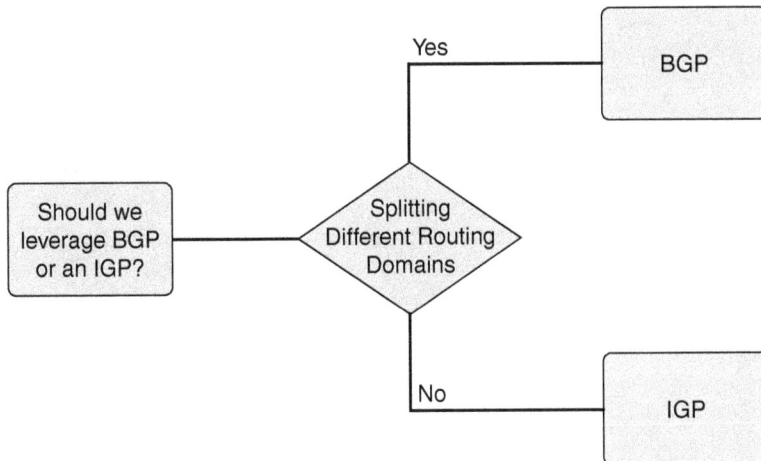

**Figure 2-4**   *Decision Tree*

### Decision Matrix

Decision matrices serve the same purpose as decision trees; however, with the **decision matrix**, network designers can add more dimensions to the decision-making process. Table 2-4 present two dimensions a network designer can use to select the most suitable design option. In these two dimensions, both business requirements and priorities can be taken into account to reach the final decision, which is based on a multidimensional approach.

**Key Topic**

**Table 2-4** Decision Matrix

| Business Requirements | Priority | Design Option 1 | Design Option 2 |
|---|---|---|---|
| Cost Savings | 4 | Moderate | Low |
| Scalable | 1 | High | High |
| Secure | 2 | Low | High |

When using the decision matrix as a tool in the preceding example, design option 2 is more suitable based on the business requirements and priorities. The decision matrix is not solely reliant on the business requirements to drive the design decision; however, priorities from the business point of view were considered as an additional dimension in the decision-making process, which makes it more relevant and focused.

## Planning Approaches

To develop a successful network design, a proper planning approach is required to build a coherent strategy for the overall design. Network designers can follow two common planning approaches to develop business-driven network designs and facilitate design decisions:

- **Strategic planning approach:** Typically targets planning to long-term business outcomes and strategies. For example, a service provider needs to migrate its core from legacy (ATM) to be based on MPLS instead. The design decision in this approach has to cater to long-term goals and plans.

- **Tactical planning approach:** Typically targets planning to overcome an issue or to achieve a short-term goal. For instance, an enterprise might need to provide remote access to some partners temporarily (for example, over the Internet) until dedicated and extranet connectivity is provisioned. Design decisions in this approach generally need to provide what is required within a limited period of time and are not necessarily required to consider long-term goals.

## Strategic Balance

Within any organization, there are typically multiple business units and departments, all with their own stakeholders. Each has its own strategy, some of which are financially driven, whereas others are more innovation-driven. For example, an IT department is more of an in-house service provider concerned with ensuring service delivery is possible and optimal, whereas the procurement department is cost-driven and always prefers cheaper options. The marketing department, in contrast, is almost always innovation-driven and wants the latest technology. Consequently, a good understanding of the overall business strategy and goals can lead to a compromise between the different aims of the different departments. In other words, the idea is that each business unit or entity within an organization must have its requirements met at a certain level so that all can collectively serve the overall business priorities, drivers, outcomes, and strategies.

As an example of achieving strategic balance, let's consider a case study of a retail business wanting to expand its geographic presence by adding more retail shops across the globe with low CAPEX. Based on this goal, the main point is to increase the number of remote sites with a minimal cost (expansion and cost):

- **IT point of view:**

  - The point of sales (PoS) application being used does not support offline selling or local data saving. Therefore, it requires connectivity to the data center to operate.

  - The required traffic volume from each remote site is small, but it is real-time application traffic requiring guaranteed treatment.

  - Many sites are to be added within a short period of time.

  - **Optimum solution:** IT suggested that the most scalable and reliable option is to use an MPLS VPN as a WAN.

- **Marketing point of view:** If any site cannot process purchased items due to a network outage, this will impact the business's reputation in the market.

  - **Optimum solution:** High-speed, redundant links should be used.

- **Financial point of view:** Cost savings.

  - **Optimum solution:** One cheap link, such as an Internet link, to meet basic connectivity requirements.

Based on the preceding list, it is obvious that the consideration for a WAN redundancy is required for the new remote sites; however, cost is a constraint that must be considered as well.

When applying the strategic balance (alignment) concept, each department strategy can be incorporated to collectively achieve the overall optimum business goal by using the suboptimal approach from each department's perspective.

In this particular example, the retail business can use two Internet links combined with a VPN overlay solution to achieve the business goal through a cost-effective solution that offers link redundancy to increase the availability level of the remote sites, meeting application bandwidth requirements while at the same time maintaining the brand reputation in the market at the desired level.

# Project Management

As a network designer, we need to understand how projects are managed. This doesn't mean we need to be a project manager, nor do we actually have to manage the projects in question. We need to understand the process each project will go through and the associated advantages and disadvantages of the methodology being used. This section covers the most common project management methodologies and their associated advantages and disadvantages.

## Waterfall Methodology

The waterfall project management framework follows a sequential, linear process and historically has been the most popular version of project management for software engineering and IT projects. The waterfall project management framework is sometimes planned using a Gantt chart that shows the start dates, end dates, assigned resources, dependencies, and overall status for each project task. Figure 2-5 shows an example of a waterfall project plan, and Figure 2-6 shows an example of the corresponding Gantt chart for this same waterfall project plan.

| Task Name | Duration | Predecessors | % Complete | Start | Finish | Assigned To | Status |
|---|---|---|---|---|---|---|---|
| − Super Large Project | 82d | | 27% | 09/13/21 | 01/04/22 | Zig Zsiga | In Progress |
| − Requirements | 19d | | 100% | 09/13/21 | 10/07/21 | Zig Zsiga | |
| Meeting # 1 | 5d | | 100% | 09/13/21 | 09/17/21 | Zig Zsiga | Complete |
| Task # 1 | 14d | 3 | 100% | 09/20/21 | 10/07/21 | Zig Zsiga | Complete |
| − Design | 10d | | 33% | 10/08/21 | 10/21/21 | Zig Zsiga | In Progress |
| Meeting # 2 | 3d | 2 | 50% | 10/08/21 | 10/12/21 | Zig Zsiga | In Progress |
| Task # 2 | 7d | 6 | 25% | 10/13/21 | 10/21/21 | Zig Zsiga | In Progress |
| − Implementation | 29d | | 0% | 10/22/21 | 12/01/21 | Zig Zsiga | Not Started |
| Meeting # 3 | 1d | 5 | 0% | 10/22/21 | 10/22/21 | Zig Zsiga | Not Started |
| Task # 3 | 28d | 9 | 0% | 10/25/21 | 12/01/21 | Zig Zsiga | Not Started |
| − Verification | 6d | | 0% | 12/02/21 | 12/09/21 | Zig Zsiga | Not Started |
| Meeting # 4 | 1d | 8 | 0% | 12/02/21 | 12/02/21 | Zig Zsiga | Not Started |
| Task # 4 | 5d | 12 | 0% | 12/03/21 | 12/09/21 | Zig Zsiga | Not Started |
| − Deployment | 18d | | 0% | 12/10/21 | 01/04/22 | Zig Zsiga | Not Started |
| Meeting # 5 | 4d | 11 | 0% | 12/10/21 | 12/15/21 | Zig Zsiga | Not Started |
| Task # 5 | 14d | 15 | 0% | 12/16/21 | 01/04/22 | Zig Zsiga | Not Started |

**Figure 2-5**  *Waterfall Project Plan Example*

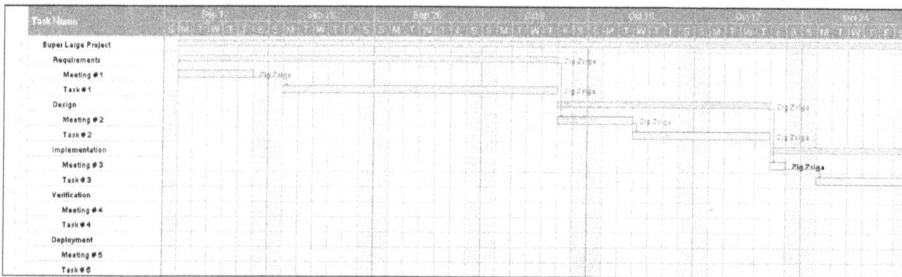

**Figure 2-6**  *Waterfall Project Plan Gantt Chart Example*

Once one of the stages is complete, the project team moves on to the next step. The team can't go back to a previous stage without starting the whole process from the beginning. And, before the team can move to the next stage, requirements may need to be reviewed and approved by the customer. This is why the waterfall project plan is a linear process.

## Advantages of Waterfall Project Management

The waterfall project management framework is best used for simple, unchanging projects. Its linear, rigid nature makes it easy to use and allows for in-depth documentation.

■ **Ease of manageability:** Because the waterfall methodology follows the same pattern for each project, it is easy to use and understand. The project team doesn't need any prior knowledge or training before working on a waterfall managed project. The waterfall project management model is very rigid; each phase has specific deliverables and a review process, so it's easy to manage and control.

- **Structure:** Every phase in waterfall has a start and endpoint, so sharing progress with stakeholders and customers is easy. By focusing on requirements and the overall network design before implementation, the team can reduce the risk of a missed deadline.

- **Documentation:** Waterfall requires documentation for every phase, resulting in a better understanding of the network design and the specific design decisions. Within this documentation, it is critical to highlight why a design decision is being made. This process also leaves a paper trail for any future projects or if stakeholders need to see more detail about a certain phase.

### Disadvantages of Waterfall

The biggest limitation of the waterfall project management model is its adversity to change throughout the project. Because waterfall is linear, you can't bounce between phases, even if unexpected changes occur. Once you're done with a phase, the project team moves on to the next phase and cannot go back to previous phases.

- **Change adverse:** Once the project team completes a phase, they can't go back. If they reach the testing and verification phase and realize that a specific business capability is missing, it is very difficult and expensive to go back and fix it.

- **Solution delivery is late:** The project has to complete multiple phases before the solution implementation can begin. As a result, business leaders won't see a working solution until late in the entire process.

- **Requirements gathering is difficult:** One of the first phases in a waterfall project is to complete requirements gathering with the business stakeholders. The problem with this is that it can be extremely difficult to properly identify what the stakeholders truly need and want this early in the process. In most cases, business leaders learn and identify requirements throughout the process as the project moves forward.

## Agile Methodologies

Agile methodologies are based on an incremental, iterative approach. Instead of in-depth planning at the beginning of the project like that of the waterfall model, the agile methodologies are open to changing requirements over time and encourage constant feedback from the different business leaders and stakeholders. Cross-functional teams work on iterations of a product over a period of time, and this work is organized into a backlog that is prioritized based on business or customer value. The goal of each iteration is to produce a working product.

Agile methodologies were built for software development, so why are we focusing on them here? With the industry shift to automation and orchestration within networking, there have been a number of development-focused adoptions within the networking industry, such as infrastructure as code, network as code, leveraging APIs to complete networking tasks at scale, and reworking manual workflows into a continuous integration/continuous delivery (CI/CD) pipeline process. Because of this market shift, network designers need to understand how an Agile methodology works to make proper design decisions when businesses are leveraging the different Agile methodologies. As we cover Agile throughout this section, you will see references to software development because that's what it was built for, but it can easily be leveraged from a network design perspective as well.

## 12 Principles of Agile Methodologies

The "Manifesto for Agile Software Development," commonly called the Agile Manifesto, lists the following 12 principles to guide teams on how to execute with agility (see https://agilemanifesto.org/principles.html). The 17 authors of this manifesto met at a ski resort called The Lodge at Snowbird in the Wasatch mountains in Utah, US between February 11th-13th 2001 with the intent to find common ground. During this time, the group created this manifesto and here are the 12 principles in it.

1. Our highest priority is to satisfy the customer through early and continuous delivery of valuable software.

2. Welcome changing requirements, even late in development. Agile processes harness change for the customer's competitive advantage.

3. Deliver working software frequently, from a couple of weeks to a couple of months, with preference to the shorter timescale.

4. Business people and developers must work together daily throughout the project.

5. Build projects around motivated individuals. Give them the environment and support they need, and trust them to get the job done.

6. The most efficient and effective method of conveying information to and within a development team is face-to-face conversation.

7. Working software is the primary measure of progress.

8. Agile processes promote sustainable development. The sponsors, developers, and users should be able to maintain a constant pace indefinitely.

9. Continuous attention to technical excellence and good design enhances agility.

10. Simplicity—the art of maximizing the amount of work not done—is essential.

11. The best architectures, requirements, and designs emerge from self-organizing teams.

12. At regular intervals, the team reflects on how to become more effective, then tunes and adjusts its behavior accordingly.

### Advantages of Agile

An Agile methodology is focused on flexibility, continuous improvement, and speed. Unlike the waterfall methodology, change is fully embraced and welcomed in an Agile methodology.

- **Change is welcomed:** With a shorter process and cycles of work, it's easier to add changes throughout the project. The goal is to always refine and reprioritize the work, including the recent changes that have been added to the backlog. This allows a project team to add changes to the project in a matter of weeks or days.

- **Unknown future state:** Agile is very beneficial for projects where the future state is not clearly defined. As the project progresses, the future state comes to light and network designers can easily adapt to these evolving requirements.

- **Faster delivery:** Breaking down the project into an iterative process allows the project team to focus on higher-quality work that is collaborative.

- **Feedback loop:** Agile projects encourage feedback from users and team members throughout the whole project, so lessons learned are used to improve future iterations.

### Disadvantages of Agile

An Agile methodology has some disadvantages and trade-offs. With all of the changes being added throughout the project, the project end date can be hard to predict as timelines are pushed because of these changes to the project. With an Agile methodology, documentation is not prioritized and can be forgotten altogether.

- **Lack of scheduling:** It can be hard to identify a completion date. Because Agile methodologies are based on time-bound delivery and tasks can be reprioritized based on the new additions to the project, it's very likely that some tasks may not be complete in the original time. Also, if there is a need for a change in a project, additional sprints (a short repeatable timeframe dedicated for a specific portion of work) may be added at any time, adding to the overall timeline.

- **Highly skilled teams:** The team makeup must include highly skilled self-starters who do not require micromanagement. They also must know and understand the specific Agile methodology being leveraged.

- **Time commitment:** Agile is most successful when all team members are completely dedicated to the project. Consistent active involvement and collaboration from all team members is required throughout the Agile process. It also means that the team members need to be allowed to commit to the entire duration of the project to ensure its success.

- **End solution creep:** In most cases, an Agile project does not have a definitive plan at the beginning, so the end product or solution can look very different from what was intended. Because Agile is meant to be flexible, new iterations will most likely be added based on customer feedback, which can lead to a very different end solution from the start. This can be a great situation, but as network designers we also need to be aware of solution creep that can introduce never-ending projects and a significant reduction in profitability for the parties involved.

## Scrum Methodology

Scrum is a subset of Agile and one of the most popular process frameworks for implementing Agile. Scrum was built to be an iterative model used to manage projects. With Scrum, there are roles, responsibilities, and meetings that never change. For example, Scrum leverages four ceremonies that provide a process structure to each sprint: sprint planning, daily stand-up, sprint demo, and sprint retrospective.

### Advantages of Scrum

Scrum is a prescriptive framework with specific roles, responsibilities, and meetings.

- **Increased transparency:** With consistent daily stand-up meetings, the entire project team knows who is doing what and when, so there is full transparency, which ensures there is no confusion. In addition, problems are found and discussed as a team, which leads to real-time resolutions before they become a larger issue in the future.

- **Increased team accountability:** With a Scrum team there is no micromanagement being done by a project manager, telling the team members what to do each step of the way. Instead, a team of highly skilled self-starters collectively decide what should be included in each sprint. The entire team works together to complete the sprint.

- **Easy to accommodate changes:** With short sprints and constant feedback, it's easier to cope with and accommodate changes. For example, if the team discovers a new user story (smallest portion of work within the project) during one sprint, they can easily add that feature to the next sprint during the backlog refinement meeting.

- **Increased cost savings:** Real-time consistent communication ensures the team is aware of all issues and changes as soon as they arise, helping to lower expenses, CAPEX and OPEX, and increase the overall quality of the solution. By testing functionality and completing verification steps in smaller chunks, there is continuous feedback and mistakes can be corrected early on, before they get too large to resolve.

### Disadvantages of Scrum

While Scrum offers some concrete benefits, it also has some downsides. Scrum requires a high level of experience and commitment from the team, and projects can be at risk of scope creep.

- **Scope creep:** Most Scrum projects experience scope creep at some point during the project. This is due to a lack of a specific project completion date. With no end date in sight, customer stakeholders and business leaders can, and often do, keep requesting additional functionality.

- **Requires Scrum experience:** With defined roles and responsibilities, the team needs to be familiar with Scrum principles to succeed. Because there are no defined roles in the Scrum team (everyone does everything), it requires team members with technical experience. The team also needs to commit to the daily Scrum meetings and to stay on the team for the duration of the project.

- **The wrong Scrum Master can ruin everything:** The Scrum Master is very different from a project manager. The Scrum Master does not have authority over the team; he or she needs to trust the team they are managing and never tell them what to do. If the Scrum Master tries to control the team, the project will fail.

- **Poorly defined tasks can lead to inaccuracies:** Project costs and timelines won't be accurate if tasks are not well defined. If the initial goals are unclear, planning becomes difficult and sprints can take more time than originally estimated.

## Kanban Methodology

Kanban is a Japanese term meaning "visual sign" or "card." The Kanban methodology is a visual framework used to implement Agile that shows what to produce, when to produce it, and how much to produce. It encourages small, incremental changes to your current system and does not require a certain setup or procedure. Kanban can be easily overlaid on top of the other project management methodologies discussed so far.

When looking at Kanban versus Agile, it's important to remember that Kanban is one flavor of Agile. It's one of many frameworks used to implement Agile software development.

A Kanban board is a tool to implement the Kanban method for projects. A Kanban board is made up of different "swim lanes" or columns. The simplest boards have three columns: To Do, Work In Progress (WIP), and Done.

I personally leverage a physical Kanban board. My columns include To Do, Working, and Completed. Within the Working column, I have two nested columns, In Progress and Pending. Oftentimes, when working on a task or project, you will get to a point where no more work can be done until someone else completes a task or something else happens. This is when I leverage the Pending column to place these cards until the corresponding work is completed so that I can jump right back into that card and complete my associated work on it.

I actually wrote this entire book leveraging Kanban. I broke down each chapter and section in this book into eight different Kanban cards, for a total of 160 cards. Figure 2-7 shows the cards for Chapter 2.

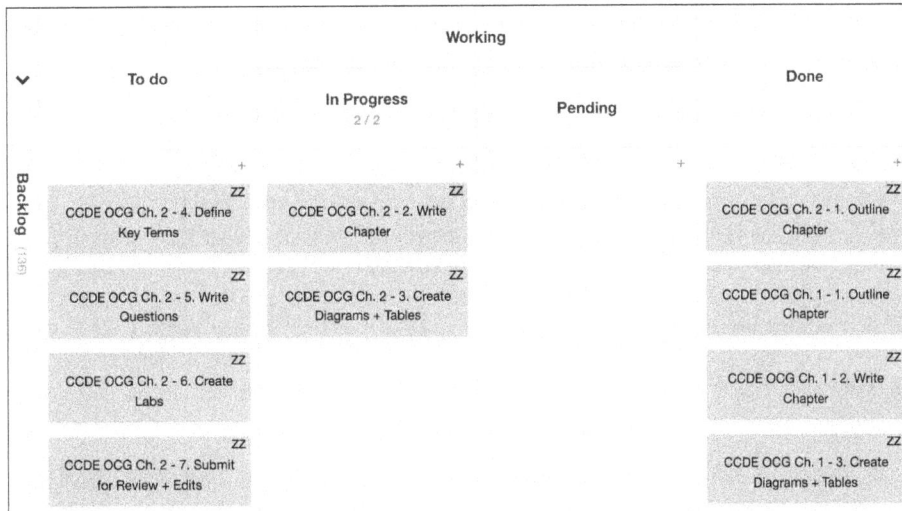

**Figure 2-7**   *Kanban Board Example*

Kanban cards represent the work, and each card is placed on the board in the swim lane that represents the status of that work. These cards communicate status at a glance. You could also use different color cards to represent different details. For example, in my Kanban board setup, green cards are for health, pink cards are for home-related items, orange cards are for Cisco-related work, and blue cards are for Zigbits-related work.

### Advantages of Kanban

Kanban's visual nature offers a unique advantage when implementing Agile. The Kanban board is easy to learn and understand, it improves the flow of work, and minimizes cycle time.

- **Increases flexibility:** Kanban is an evolving, fluid model. There are no set phase durations or times. As a new item of work is added to the board, it is prioritized along with the rest of the board items.

- **Reduces waste:** Kanban revolves around reducing waste, ensuring that teams don't spend time doing work that isn't needed or doing the wrong kind of work.

- **Easy to understand (KISS):** The visual nature of Kanban helps to make it incredibly intuitive and easy to learn, aka keep it super simple (KISS). The team doesn't need to learn a completely new methodology, and Kanban can be easily implemented on top of other systems in place.

## Disadvantages of Kanban

Many of the disadvantages associated with Kanban come with misuse or mishandling of the Kanban board. An outdated or overcomplicated board can lead to confusion, inaccuracies, or miscommunication.

- **Outdated board:** The team must be committed to keeping the Kanban board up to date; otherwise, they'll be working off inaccurate information. And once work is completed based on an out-of-date board, it can be hard to get things back on track.

- **Overcomplicate the board:** The Kanban board should remain clear and easy to read. Adding bells and whistles to the Kanban board just buries the important information. Sometimes too many options can be an issue that complicates the board. When I first started using a Kanban board, I kept adding columns after columns. At one point I had 10-plus columns, which was way too complicated. In essence, I was overdesigning the solution. Keep your Kanban boards simple, clear, and concise.

- **Lack of timing:** A frequent complaint about Kanban is that you don't know when things will be done. The columns on the Kanban board are only marked by phase (To Do, WIP, Done). There are no timeframes associated with each phase, so you really don't know how long the to-do phase could last. You can estimate how long a task will take and then track how long it actually took to complete it, but this only helps from a metric-tracking perspective.

## Kanban Rules

Every Kanban project should follow these rules:

- **Visualize:** Visually seeing the work on the board allows the team to fully understand the entire project and how it will move forward, or not move forward. By seeing it all, the team can find problems quicker in the process and resolve them before they become larger issues.

- **Limit work:** Work in progress limits (WIP limits) determine the minimum and maximum amount of work for each column on the board or for each workflow. Originally, I put my WIP limit at 10... this was way too much WIP. Today I leverage a WIP limit of 2. Keep in mind, a WIP limit of 10 might be perfectly fine for a large team working on a project together. By putting a limit on WIP, you can increase speed and flexibility, and reduce the need for prioritizing tasks.

- **Manage the card flow:** The movement of work (cards) throughout the Kanban board should be monitored and improved upon. Ideally, you want a fast, smooth flow, which shows that the team is creating value quickly.

- **Reserve the right to improve:** As the team leverages the Kanban method, the team should be able to identify and understand any issues that come up. The group should reserve the right to improve the process anytime that will help the flow of work, the overall cycle of time, and increase the work quality being delivered.

## Project Management Methodology Comparison

Table 2-5 compares the different project management methodologies covered in this section.

**Table 2-5**    Project Management Methodology Comparison

| Criteria | Waterfall | Agile | Scrum | Kanban |
|---|---|---|---|---|
| Scope creep | Not likely to happen. | Happens often in most Agile projects. | Happens often in most Scrum projects. | — |
| Manageability | Easy. | Hard. | Hard. | Easy. |
| Structure | Very structured. | Not structured. | Not structured. | Structured. |
| Documentation | Required/ included in each phase. | Not required. | Not required. | Not required. |
| Changes | Not welcome/ change adverse. | Change is welcomed. | Easy to include changes. | Easy to include changes. |
| Solution delivery | Late in process. | Faster delivery. | Faster delivery. | Fastest delivery. With WIP limits, the focus is on limiting the work in progress, to ensure a faster delivery of the work at hand. |
| Requirements gathering | Requirements are identified early. Hard to identify all requirements in the beginning phases. | Requirements are identified throughout the project. Hard to predict when the project will be completed by. | Requirements are identified throughout the project. Hard to predict when the project will be completed by. | Requirements are identified throughout the project. Hard to predict when the project will be completed by. |
| Future/end state | Clearly defined at beginning. | Future state is not defined at beginning of the project; it comes to light as the project moves forward. | The future state is not defined at beginning of the project; it comes to light as the project moves forward. | — |
| Stakeholder feedback | Feedback is not normally requested nor incorporated after the requirements gathering phase. | Constant feedback is requested and incorporated in future iterations. | Constant feedback is requested and incorporated in future iterations. | Constant feedback is requested and incorporated in future iterations. |

| Criteria | Waterfall | Agile | Scrum | Kanban |
|---|---|---|---|---|
| **Project completion date** | Easy to identify and schedule out. | Can slide because it's hard to identify a realistic completion date. As changes are incorporated, the end date slides into the future. | Can slide because it's hard to identify a realistic completion date. As changes are incorporated, the end date slides into the future. | Hard to identify because time is not tracked. |
| **Teams** | Team makeup can be a mix of highly skilled and lower-skilled members. No team accountability. | Team makeup requires highly skilled team members. Team members must also know and understand the Agile methodology. Limited team accountability. | Team makeup requires highly skilled team members. The team members must also know and understand the Scrum methodology. Increased team accountability. | No need to know about a methodology or process. Work needs to be visible and the Kanban board needs to be updated constantly. |
| **Time commitment** | Team resources can be shared with other projects at the same time. | All team resources should be completely dedicated to the project. | All team resources should be completely dedicated to the project. | Team/resources should be dedicated to the card until it's completed, or in a pending state, before work starts on another card. |
| **Work transparency/ visibility** | No transparency at all. | Semi-transparency. | Full transparency. | Full work visibility. |

## Summary

In this chapter, we focused on the business elements, terminology, and project management styles that all network designers need to know. We discussed business priorities, drivers, outcomes, and capabilities, and how each relates to the other to drive the requirements for the design. We highlighted how risk versus reward, ROI, CAPEX, OPEX, and business continuity all impact the design decisions we have to make as network designers. We labeled IT, including networking, as the new utility that everyone expects to be online and available all the time, just like running water and electricity. Understanding the business is critical to ensuring the design decisions will directly increase business success. Finally, we covered the most common project management methodologies of today. We talked about waterfall, Agile, Scrum, and Kanban, the advantages and disadvantages of each, and how they compare to each other based on various criteria.

Network designers must be the bridge between technology and the business. Be the bridge my friends!

## References

- Al-shawi, Marwan, *CCDE Study Guide* (Cisco Press, 2015)

- White, Russ and Donohue, Denise, *The Art of Network Architecture* (Cisco Press, 2014)

- Beck, Kent, et al., "Manifesto for Agile Software Development" (Agile Alliance, 2001)

## Exam Preparation Tasks

As mentioned in the section "How to Use This Book" in the Introduction, you have a couple of choices for exam preparation: the exercises here, Chapter 18, "Final Preparation," and the exam simulation questions in the Pearson Test Prep Software Online.

## Review All Key Topics

Review the most important topics in this chapter, noted with the Key Topic icon in the outer margin of the page. Table 2-6 lists a reference of these key topics and the page numbers on which each is found.

**Table 2-6**  Key Topics for Chapter 2

| Key Topic Element | Description | Page Number |
|---|---|---|
| Paragraph | Business priorities | 45 |
| Paragraph | Business continuity | 47 |
| Paragraph | IT as a "new" utility | 50 |
| Paragraph | Decision matrix | 53 |
| Table 2-4 | Decision Matrix | 54 |
| Table 2-5 | Project Management Methodology Comparison | 63 |

## Complete Tables and Lists from Memory

Print a copy of Appendix D, "Memory Tables" (found on the companion website), or at least the section for this chapter, and complete the tables and lists from memory. Appendix E, "Memory Tables Answer Key," also on the companion website, includes completed tables and lists to check your work.

## Define Key Terms

Define the following key terms from this chapter and check your answers in the glossary:

business priority, business driver, business outcome, business capability, recovery point objective (RPO), recovery time objective (RTO), capital expenses (CAPEX), operating expenses (OPEX), return on investment (ROI), decision tree, decision matrix, tactical planning approach, strategic planning approach

# CHAPTER 3

# What's the Purpose of the Network?

This chapter covers the following topics:

- **Business Applications:** This section covers business applications and the associated network design elements for them.

- **Service Models:** This section covers the different service models, how to leverage them, and what the network design characteristics are for each model.

- **The Cloud:** This section covers the cloud in all its forms and the associated network design elements.

- **Data Management:** This section covers data and the data management methodologies that will help every network designer make better design decisions.

What's the purpose of the network? Why do we need a network? Why is a network there in the first place? These questions should always be at the top of your mind as network designer because understanding the purpose and goal of the network is critical to properly designing it.

As we answer these questions, a technical (and potentially a tactical) answer is that the network's purpose is to get data from point A to point B in the right amount of time. This is a generally good answer to these questions, and usually network engineers, not network designers, can easily identify.

From a network design perspective, we need to take this answer a step further. Why is the network transferring data in the first place? To meet the business outcomes and objectives… to make the business successful!

Businesses have become so reliant on networks that the required availability of the network and its associated services has to be designed at an extremely high level. This is similar to what transpired with the plain old telephone service (POTS) network. It became so relied upon that the overall availability had to be designed at an extremely high level.

There is a shift in the network, as discussed in the previous chapters, toward a service-focused network. When we talk about services in this context, we are talking about applications, service models, the cloud in all its forms, and data.

This chapter covers the following "CCDE v3.0 Unified Exam Topics" section:

- 4.0 Service Design

# "Do I Know This Already?" Quiz

The "Do I Know This Already?" quiz allows you to assess whether you should read this entire chapter thoroughly or jump to the "Exam Preparation Tasks" section. If you are in doubt about your answers to these questions or your own assessment of your knowledge of the topics, read the entire chapter. Table 3-1 lists the major headings in this chapter and their corresponding "Do I Know This Already?" quiz questions. You can find the answers in Appendix A, "Answers to the 'Do I Know This Already?' Quizzes."

**Table 3-1** "Do I Know This Already?" Section-to-Question Mapping

| Foundation Topics Section | Questions |
| --- | --- |
| Business Applications | 1–3 |
| Service Models | 4–7 |
| The Cloud | 8–10 |
| Data Management | 11 |

**CAUTION** The goal of self-assessment is to gauge your mastery of the topics in this chapter. If you do not know the answer to a question or are only partially sure of the answer, you should mark that question as wrong for purposes of the self-assessment. Giving yourself credit for an answer you correctly guess skews your self-assessment results and might provide you with a false sense of security.

1. Which of the following options is the correct application model for the statement "This is the simplest application model and it is equivalent to running the application on a personal computer"?

   a. 3-tier model

   b. Single-server model

   c. 2-tier model

   d. SaaS

2. Which of the following application models is like the client/server architecture?

   a. 3-tier model

   b. Single-server model

   c. 2-tier model

   d. SaaS

3. Which of the following application models has web, application, and database layers?

   a. 3-tier model

   b. Single-server model

   c. 2-tier model

   d. SaaS

**4.** Which service model is best used when a business requires full control of all components within an application?

    **a.** PaaS

    **b.** IaaS

    **c.** SaaS

    **d.** On-premises

**5.** Which service model is best for application developers?

    **a.** PaaS

    **b.** IaaS

    **c.** SaaS

    **d.** On-premises

**6.** Which service model is best if a business wants complete control over its virtual infrastructure but also wants to operate on a pay-as-you-go basis?

    **a.** PaaS

    **b.** IaaS

    **c.** SaaS

    **d.** On-premises

**7.** Which service model is best if a business wants an application to run with ensured availability but doesn't want the headache of managing the application in any form?

    **a.** PaaS

    **b.** IaaS

    **c.** SaaS

    **d.** On-premises

**8.** Which cloud type has the highest cost?

    **a.** Hybrid cloud

    **b.** Private cloud

    **c.** Multi-cloud

    **d.** Public cloud

**9.** Which cloud type is best if a business wants the most control possible?

    **a.** Hybrid cloud

    **b.** Private cloud

    **c.** Multi-cloud

    **d.** Public cloud

**10.** Which cloud type is best if a business wants to ease into a cloud computing environment over a long period of time?

    **a.** Hybrid cloud

    **b.** Private cloud

    **c.** Multi-cloud

    **d.** Public cloud

**11.** Which option is the proper data management pillar for the definition "The planning of all aspects of data management"?

   **a.** Data quality

   **b.** Data governance

   **c.** Data architecture

   **d.** Data security

# Foundation Topics

## Business Applications

The network is the information highway for the business applications of today, and for the business to be successful, these applications must be able to properly communicate as required between users, devices, data, databases, and other application components.

### Application Models

Network designers need to understand how an application is built to properly design the network for that application. The following are the different application models being leveraged today and their associated design elements that you need to know as a network designer:

■ **Single-server model:** This is the simplest application model and it is equivalent to running the application on a personal computer. All of the required components for an application to run are on a single application or server.

■ **2-tier model:** This application model is like the client/server architecture, where communication takes place between client and server. In this model, the presentation layer or user interface layer runs on the client side while the dataset layer gets executed and stored on the server side.

   There is no intermediate layer (aka application layer) in between client and server.

■ **3-tier model:** This application model is the most common at the time of writing. This model has three tiers or layers:

   ■ **Presentation:** This is the front end of the application that all end users access. This is how an end user sees and interacts with the application. This is often called the web tier or GUI tier of the application. The main function of this layer is to translate tasks and results to something the end user can understand.

   ■ **Intermediate:** This is the layer where all of the application's functions and logic occur. This layer processes tasks, functions, and commands, makes decisions and evaluations, and performs calculations. It also is how data is moved between the presentation (web/GUI) and database layers. This is often referred to as the application or logic layer of the application.

   ■ **Database:** Here information is stored and retrieved from a database. The information is then passed back to the intermediate (application) layer and then eventually back to the end user.

Breaking elements of an application into different layers like the *n*-tier architectures allows network designers to properly design the network for each tier or layer individually. Each layer may need its own load balancing, applying source NAT, DNS, source routing, and traffic engineering design. This means more work from a network design perspective but a better purpose-built environment for each tier with all associated elements needed, which allows the different application layers to scale out as needed.

Table 3-2 shows the different network design elements for each layer of the 3-tier model and provides leading questions to ask to help elicit the information needed to make a proper design.

**Key Topic**

**Table 3-2**  3-Tier Application Model Network Design Elements

| Tier | Traffic Pattern | Network Design Elements | Questions to Ask |
|---|---|---|---|
| **Web tier** | End user and application layer access only. | No database layer access. The web tier needs to be globally accessible for the end users. Normally located in a DMZ. | How are end users accessing the web tier globally? How are the web tier–specific networks/IP addresses being routed? What's the web tier's high-availability architecture? (Active/active, active/standby, anycast, etc.) |
| **Application tier** | Web and database access only. No end user should ever access this tier directly. | This tier is internally accessed only, so no external addresses or routing are needed. Load balancing should be implemented, but how depends on the other tier's communication method with this tier (SNAT, NAT, Sticky, etc.). Normally located internally behind multiple security layers. | How does the web tier communicate with the application tier? How does the database tier communicate with the application tier? |
| **Database tier** | Application layer access only. No end user or web tier should have access. | This tier is internally accessed only, so no external addresses or routing are needed. Normally located internally behind multiple security layers. | How is replication being done between the different database member servers? How are the database changes synchronized? The answer to his question is especially critical when there are multiple data center locations. |

## Application Constraints and Requirements

As a network designer, there are a number of common application constraints and requirements that you should be aware of. These are topics that a network designer should be asking about as a network is being designed to support an application. This is by no means an all-inclusive list.

■ **Multicast:** Usually leveraged between a cluster of servers to keep data synchronized, such as a backend database replication architecture, or as a transport mechanism for data streaming applications like IPTV and real-time stock market updates for day traders. In these situations, not having multicast breaks the application in question.

- **Layer 2 extension:** Probably one of the most common network design requirements after an application has completed its development process. As the application is being deployed, it is quickly identified that the application servers, maybe in a 3-tier model, do not communicate outside of their Layer 2 segments. Now it's the network designer's job to provide Layer 2 extension options that allow the application to properly function. This leads to bad network designs with large Layer 2 fault domains that are generally unreliable. Even though these are bad network design options from a network design perspective, they do solve the application requirement and thus they make the business successful. If there is a requirement to extend Layer 2, some of these limitations can be mitigated with overlay technologies while still allowing for an expanded Layer 2 environment.

- **Hardcoded items:** Thankfully, we are seeing issues with hardcoded items in the code less frequently today. They do happen, and that's why network designers need to know about them. These hardcoded items bring a security element into the mix with compliance controls and overall security requirements for the application in question. From a network design perspective, though, how do we handle hardcoded IP addresses in the code? The simplest answer is to not allow hardcoded items in the code, but what do you do as a network designer when it does happen? This is where solutions like Network Address Translation (NAT), traffic engineering, and source routing can be leveraged to help mitigate this issue.

- **High availability:** How an application is designed for high availability has a large impact on the network supporting that application. Is the application in multiple locations, such as geographically separated data centers? If so, how is the application data synchronized between these locations? How do end users access the applications in each location? Is one location preferred over the other at all times (active/standby), or can the application be accessed from either location at any time (active/active)? What about the load-balancing options for the application? Is it leveraging DNS load balancing or a physical load balancer? Does the application require the use of source NAT (SNAT) between its different application layers? There are so many network design–related questions that we have to ask and answer to properly facilitate the creation of a network design that makes the application successful.

When creating a network design that's goal is to make an application successful, it really comes down to the applications, services, and so forth being created incorrectly. Network designers have been forced to provide band-aid solutions like Layer 2 extension options because of these problems. This is most definitely not solving the true issue of proper application development. We simply extend Layer 2 as a short-term solution that ends up becoming a permanent one. This is similar to hardcoding IP addresses and hostnames in an application's code. Network designers always have to provide bad network design options because of these application issues.

To solve these issues, a network designer and a security specialist should be part of the team that builds and reviews an application, to ensure network design and security controls are being properly considered in the application. It's not fair to expect an application developer to know and understand the details of network design or security; network designers have to help them, teach them, and show them.

If we want to change these situations with the plan to limit them from happening, we need to be a part of the creation process so we can explain the reasons to the business at those critical steps. The business doesn't know what a network designer knows. A network designer can tell the business *why* they shouldn't rely on a Layer 2 extension for the application, *why* they shouldn't hardcode IP addresses, hostnames, usernames, and passwords in code, and *why* they should ensure security controls are implemented during the creation process.

In the end, it all comes back to business decisions and the respective trade-offs.

## Service Models

We highlighted the different application models for how an application can be created earlier. This section takes that discussion a step further by covering the different service models that can be leveraged for the application. These service models determine where the application is located and what elements of the application are owned and managed by the business. The following are the most common service models:

> **NOTE** There are other service models that are not covered in this section, such as Database as a Service (DBaaS), Compliance as a Service (CaaS), and Security as a Service (SECaaS). What is covered in this section are the most common service models at the time of writing.

- **On-premises:** On-premises is the service model where a business owns and manages an application. A business will procure all of the infrastructure required to run the service and then fully manage, maintain, and operate it. In some situations, the management is outsourced but the infrastructure is procured and owned by the business.

- **Software as a Service (SaaS):** SaaS is where a vendor makes its software available to users, usually for a monthly or annual subscription service fee.

- **Platform as a Service (PaaS):** PaaS is where a vendor provides hardware and software tools, and people use these tools to develop applications. PaaS users tend to be application developers.

- **Infrastructure as a Service (IaaS):** IaaS is a pay-as-you-go service model for storage, networking, and virtualization—all of a business's infrastructure needs. IaaS gives users cloud-based alternatives to on-premises infrastructure, so businesses can avoid investing in expensive onsite resources.

Table 3-3 shows the comparison between these service models.

**Key Topic**

**Table 3-3** Service Model Comparison

| Service Model | Characteristics | Advantages | When to Use |
|---|---|---|---|
| On-premises | Business owned and managed. Available locally. Hosted within the business's server environment. | Full control over all components of the application. | When a business requires full control of all components within the application. This is most often seen with security compliance and data classification requirements. |

| Service Model | Characteristics | Advantages | When to Use |
|---|---|---|---|
| SaaS | Available over the Internet. Hosted on a remote server by a third-party provider. Scalable, with service offerings based on need. | No need to install and run software on any computer. Everything is available to the end user over the Internet. Access to software can be from any device, at any time, with Internet connectivity. | When a business wants an application to run with ensured availability but without the headache of maintaining that application at any level. |
| PaaS | Accessible by multiple users. Scalable. Built on virtualization technology. Easy to run without extensive IT knowledge. | Primarily used by developers to create software or applications. Developers don't need to start from scratch when creating applications. | When a business wants to create a unique application without spending a ton of money or taking on all the responsibility. |
| IaaS | Highly flexible. Highly scalable. Accessible by multiple users. Cost-effective. | On-premises IT infrastructure is expensive. The business maintains control over the infrastructure. | When a business requires complete control over its infrastructure and wants to operate on a pay-as-you-go basis. |

## The Cloud

When a business starts planning to leverage cloud in any form, there are three use cases that network designers should consider throughout the design process:

- **Securely extending a private network to a single or multiple public cloud environments:** Includes multiple clouds (for example, multiple Amazon Web Services [AWS] and Azure), multiple regions in a cloud, or multiple VPCs in a cloud; VPN; multi-cloud and multi-VPC connectivity; scaling; and performance optimization of transit VPC. Also supports extending data centers into the cloud and enabling direct branch-to-cloud connectivity.

- **Optimizing data center and branch connectivity performance to cloud IaaS and SaaS:** Includes best path to a destination, cloud segmentation, monitoring to assure the best performance, visibility into traffic going to applications, and traffic shaping/Quality of Service (QoS). Also supports extending data centers into the cloud and enabling direct branch-to-cloud connectivity.

- **Securing access to the Internet and SaaS from the branch:** Includes connecting and protecting branch office users directly to the multi-cloud environment using Direct Internet Access (DIA) and properly securing them.

## Cloud Connectivity Models

When businesses start to leverage cloud in any form, be it public, private, hybrid, or multi-cloud, how the business is going to connect to cloud environments is a topic for a network designer to address. There are multiple options, each with its own pros and cons.

### Direct Cloud Access

Direct cloud access (DCA) allows a remote site to access SaaS applications directly from the Internet and through dedicated private connections. The cloud permits only the designated application traffic to use the directly connected Internet transport securely, while all other Internet-bound traffic takes the usual path, which could be through a regional hub, a data center, or a **carrier-neutral facility (CNF)**. This feature allows the remote site to bypass the latency of tunneling Internet-bound traffic to a central site, subsequently improving the connectivity to the prioritized SaaS application; this feature is commonly referred to as Direct Internet Access (DIA). The edge router chooses the most optimal Internet path for access to these SaaS applications. Different applications could traverse different paths because the path selection is calculated on a per-application basis.

If any SaaS application path becomes unreachable or its performance score falls below an unacceptable level, the path is removed as a candidate path option. If all paths cannot be path candidates because of reachability or performance, then traffic to the SaaS application follows the normal, routed path. Figure 3-1 illustrates a remote site using DIA to access SaaS applications

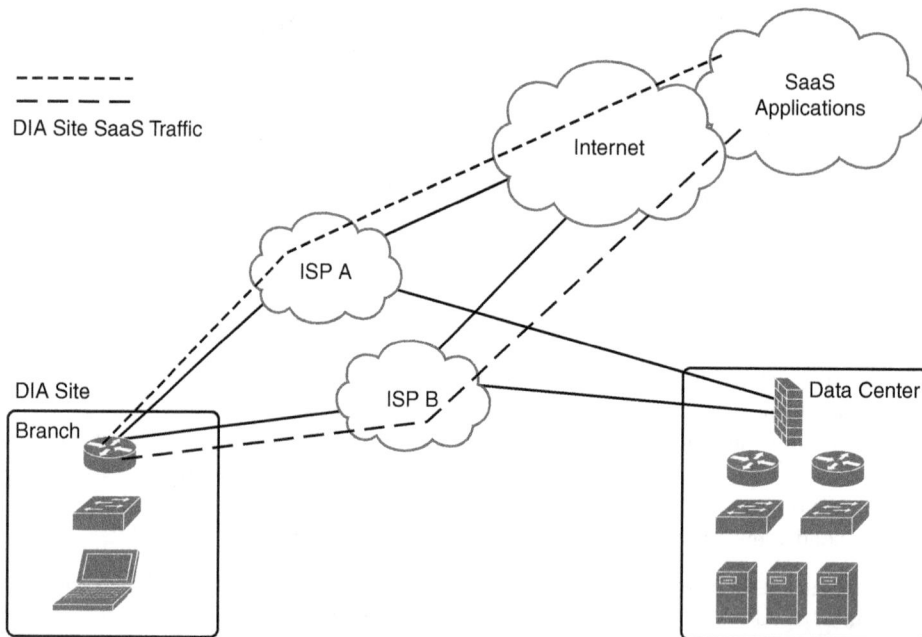

**Figure 3-1**  *DCA/DCI Remote Site with DIA to Access SaaS Applications*

## Cloud Access Through a Gateway (Cloud Access Point)

Many businesses do not use DIA at the branch office, because either their sites are connected only by private providers (MPLS, VPLS, etc.) or centralized policy or security requirements do not permit it. They may use data centers, regional hubs, or even CNFs to enable Internet connectivity. In this case, SaaS traffic is tunneled to the best-performing gateway site, where it is subsequently routed to the Internet to reach the requested SaaS application service.

**NOTE** Different remote sites and different applications may use different gateway sites and paths, depending on the application and measured application performance. Remote sites that use gateway sites for Internet access are referred to as *client sites*.

Figure 3-2 shows how cloud access can be achieved through a gateway in a data center or a **cloud access point (CAP)**. A branch office tunnels SaaS traffic to a gateway location and then uses the Internet at the gateway location to access the SaaS application.

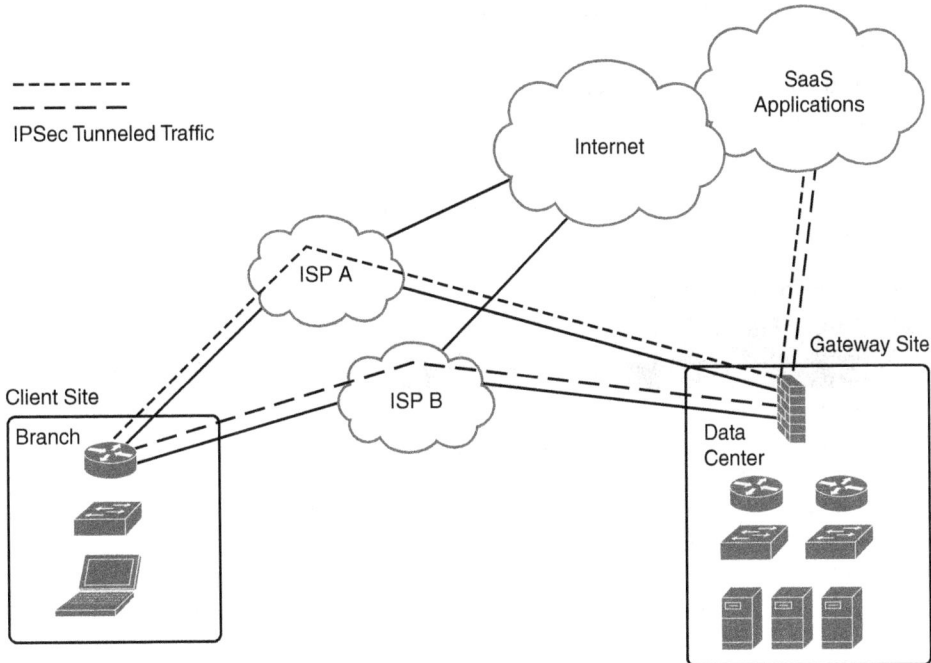

**Figure 3-2** *Cloud Access Through a Gateway*

## Hybrid Approach

It is possible to have a combination of DIA and client/gateway sites. When defining both DIA and gateway sites, SaaS applications can use either the DIA exits of the remote site or the gateway sites for any given application, depending on which path provides the best performance. DIA sites are, technically, a special case of a client site, but the Internet exits are local instead of remote.

## Cloud Types

When selecting a cloud solution, there are a number of different types to choose from, each with its own associated benefits and limitations:

- **Private cloud:** A private cloud consists of cloud computing resources used by one business. This cloud environment can be located within the business's data center footprint, or it can be hosted by a cloud service provider (CSP). In a private cloud, the resources, applications, services, data, and infrastructure are always maintained on a private network and all devices are dedicated to the business.

- **Public cloud:** A public cloud is the most common type of cloud computing. The cloud computing resources are owned and operated by a CSP. All infrastructure components are owned and maintained by the CSP. In a public cloud environment, a business shares the same hardware, storage, virtualization, and network devices with other businesses.

- **Hybrid cloud:** A hybrid cloud is the use of both private and public clouds together to allow for a business to receive the benefits of both cloud environments while limiting their negative impacts on the business.

- **Multi-cloud:** Multi-cloud is the use of two or more CSPs, with the ability to move workloads between the different cloud computing environments in real time as needed by the business.

Table 3-4 compares the different cloud types in relation to each other based on various characteristics.

**Key Topic**

**Table 3-4**  Cloud Types Detailed Comparison

| Cloud Type | Control | Maintenance | Flexibility | Scalability | Migration | Cost |
|---|---|---|---|---|---|---|
| Private cloud | Most control | High | Least flexibility | High scalability | Hard migration | High cost |
| Public cloud | Least control | None | Flexibility | High scalability | Hard migration | Lowest cost |
| Hybrid cloud | Mix of both | Medium | Flexible | Lowest scalability | Ease of migration | High cost |
| Multi-cloud | Least control | No maintenance for each CSP, but across the CSPs is high | Most flexibility | Highest scalability | Hardest migration | Highest cost |

## Cloud-Agnostic Architecture

A cloud-agnostic architecture is when there are no vendor specific features and functionality that are proprietary. It is focused on leveraging the same cloud capabilities across the different cloud providers no matter what vendor it is. When looking at cloud service providers and migrating applications to the cloud, there are three primary focus points that should be leveraged within a cloud-agnostic architecture:

- **Portability:** Moving to the cloud inherently provides a level of portability, but if not carefully architected, applications and services can lose their portability as they get locked into specific CSP services. Portability here specifically allows mobility between different CSPs with a proper abstraction layer.

- **Abstraction:** Leveraging an abstraction layer within the cloud architecture allows for a decoupling from the underlying cloud-specific platform functionality, which provides a direct cost reduction and an increase in flexibility. For example, using this abstraction layer to seamlessly invoke the same cloud capability between cloud provider one and cloud provider two, when there are different mechanisms and processes to do so. In addition, this same capability could be proprietary, but using this abstraction layer mitigates a potential hardcoded proprietary service call.

- **Interoperability:** Developing applications and services with cloud interoperability as a key priority will not be tied to a specific cloud feature set. This allows for these applications and services to leverage different cloud platforms without major redevelopment or changes. This specifically allows for a cloud-agnostic approach.

To achieve a cloud-agnostic architecture, network designers should consider adopting the following practices.

### Decoupling

There are two perspectives to think about for decoupling. First, all applications should be designed to be inherently decoupled from the underlying cloud platform they are on. This can be accomplished by leveraging service-oriented architecture (SOA), which is discussed in detail a bit later in this chapter. Second, all cloud components should be decoupled from the applications that leverage them.

### Containerization

All applications should follow a containerized architecture. This is critical for cloud applications as well as on-premises data center applications. Ensuring all applications are developed with containerization in mind allows for real cloud adoption and portability. Container technology helps decouple applications from the cloud-specific environment, which provides an abstraction layer away from any of the CSP dependencies. The goal is to ensure that it is relatively easy to migrate applications between different cloud vendors if the mission requires it. Cloud containerized architectures is a topic that is covered in detail in an upcoming section.

### Agnostic Versus Proprietary Cloud Services

Each cloud service provider is different and has unique services, with its own avenues to provision them to customers. There is a need to provide a mechanism to differentiate where these specific services interact with applications while also allowing for the standardization of agnostic services. Figure 3-3 shows how to delineate from an architecture perspective between cloud-agnostic services and cloud-proprietary services. This is how you should plan to migrate applications to a CSP.

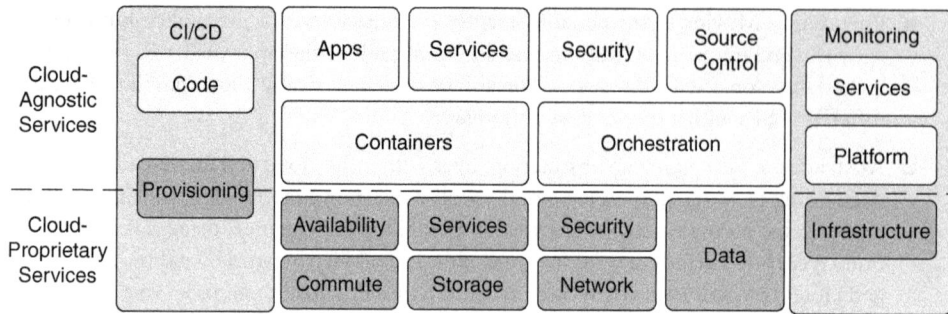

**Figure 3-3** *Agnostic vs. Proprietary Cloud Services*

## Service-Oriented Architecture

To ensure a successful cloud-agnostic architecture, incorporating the software design service-oriented architecture (SOA) is hyper-critical. SOA is a style of software design where services are provided to other parts of an application component themselves. This is accomplished through network communication protocols. The underlying principles are vendor and technology agnostic. In SOA, services communicate with other services in two ways. The first way is to simply pass data between the different services. The second way is to logistically coordinate an activity event between two or more services. There are many benefits to SOA:

- Code can be created so that it is reusable, which cuts down on time spent in the development process.

- Developers can leverage multiple coding languages with SOA because it uses a central interface, which allows for flexibility and scalability within the software development cycle.

- With SOA, a standard communication process is created that allows systems to function on their own and communicate effectively between them.

- SOA is much more scalable, limiting client-server interaction, which allows for a direct increase in efficiency.

## Cloud Containerized Architecture

Containerization is a large part of a cloud-agnostic architecture. Figure 3-4 shows the progression from a traditional on-premises deployment to a containerized cloud deployment.

### Traditional Deployment Architecture

Traditionally, organizations and companies ran applications on physical servers. You could deploy multiple applications on the same physical server, but there was no way to properly restrict resources or set up controls to govern application guidelines. Because of these issues, there were a number of allocation and performance issues. Most of the time, each physical server was dedicated to a single application because of these limitations. This increased cost and resources and limited overall scalability.

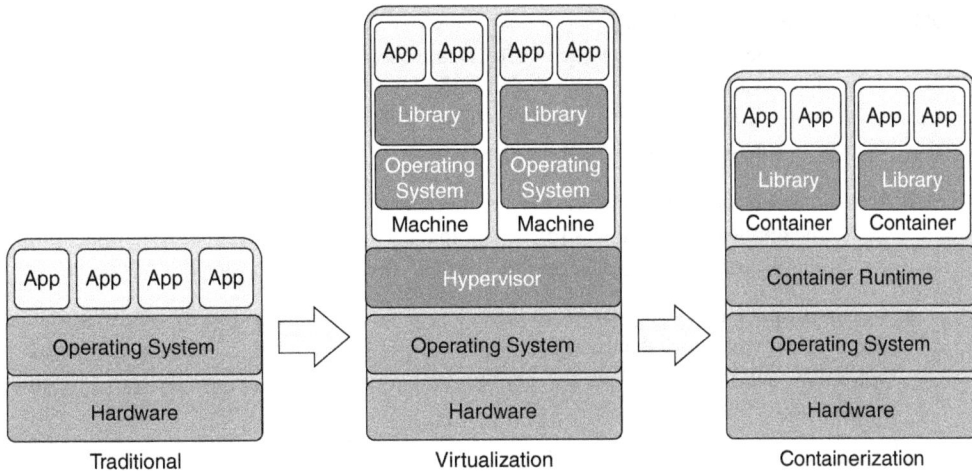

**Figure 3-4**    *Progression from Traditional to Containerized Deployment*

## Virtualization Deployment Architecture

Virtualization allows for multiple virtual machines (VMs) to be deployed on a single physical server. Each VM is isolated from other applications with its own resources and security controls allocated to it individually. Virtualization allows for better utilization of resources on a physical server and scales better because applications (VMs) can be added and removed as needed depending on the needed resources. Each VM runs all components that a physical server would run, such as the application and the operating system.

## Container Deployment Architecture

Containers are similar to VMs but have less stringent isolation controls that allows them to share the operating system among other applications. Because of this, containers are lightweight. A container has its own file system, shares CPU time, memory, process space, and more. Because containers are decoupled from the underlying infrastructure, they are moveable between different fabrics as needed by the underlying business requirements.

Containers provide a number of benefits:

- Agile application creation and deployment (CI/CD)

- Separation of responsibilities between development and operational tasks

- Real-time application-level health analytics

- Standardization and consistency across all environments and enclaves

- Real-time distribution with the capability to port into other operating systems and locations as required

- Increased overall predictability of application performance and requirements

- Increased resource efficiency

## Cloud Application Strategy

As a business readies its business lines of efforts and their respective applications for migration to the cloud, it is highly recommended that the business incorporate an application assessment process. As part of this process an application assessment team should be created with the following roles and purposes:

- **Line of business owner:** The business stakeholder for this application. They understand the application's business role and impact. They also understand the implications of this application and can appropriate business resources and priorities to this effort.

- **Security specialist, compliance auditor:** The security team member in charge of the security controls, compliance regulations, and auditing of code. These are all critical roles that will direct decisions and actions from a risk management perspective for this application.

- **Application owner, application developer:** The software engineering member responsible for this application. Creates code, modifies current code, and drives associated technical requirements for the application.

- **Network engineer, network designer, network architect:** Facilitates the network resources to properly service the application based on the different requirements from the line of business owner, security specialist, and application owner.

Each application will have different requirements as it's being reviewed in this process. The team will need to properly identify what the application is dependent on and make appropriate decisions to ensure the application is ready for the migration to the cloud environment.

The application assessment team will document in an *application binder* (or *run book*) everything that is discovered, decided on, and implemented for this application. The application binder should include all requirements and where they originated from; all security controls and regulatory standards that this application must comply with; and where the application is in the migration process and what is needed for it to be successful.

# Data Management

Data is the most critical resource that all other resources will be leveraging. We have to manage all data effectively, accurately, and securely so that these additional resources can properly leverage that data with ensured integrity, availability, and confidentiality. Data management in essence lays the foundation for data analytics. Without good data management, there will be no data analytics. Data management can be broken down into 11 pillars:

1. **Data governance:** The planning of all aspects of data management. This includes availability, usability, consistency, integrity, and security of all data within the organization.

2. **Data architecture:** The overall structure of an enterprise's data and how it fits into the enterprise architecture.

3. **Data modeling and design:** The data analytics and the corresponding analytics systems. This includes the designing, building, testing, and ongoing maintenance of these analytics systems.

4. **Data storage and operations:** The physical hardware used to store and manage the data within the enterprise.

5. **Data security:** Encompasses all security requirements, controls, and components to ensure the data is protected and accessed only by authorized users.

6. **Data integration and interoperability:** The transformation of data into a structured form to be leveraged by other systems and resources.

7. **Documents and content:** All forms of unstructured data and the work necessary to make it accessible to the structured databases.

8. **Reference and master data:** The process of managing data in a way that allows it to be redundant, and if there are any errors or mistakes that can be normalized by standard values.

9. **Data warehousing and business intelligence:** Involves the management and application of data for analytics and business decision making.

10. **Metadata:** Involves all elements of creating, collecting, organizing, and managing metadata (i.e., data that references other data).

11. **Data quality:** Involves the practices of data monitoring to ensure the integrity of the data being delivered is maintained.

For a true data management model, all of these pillars need to be included. Without one of these pillars, there is an area of data management that is not being addressed. For example, if there isn't a solution for metadata management, the business loses the ability to easily categorize data. Without data quality being ensured, all data is at risk and the analytics of that data becomes useless.

## Summary

What is the purpose of the network? To ensure business success! This chapter went into great detail on how a network designer can accomplish this.

This chapter covered how businesses rely heavily on the network and the corresponding services and applications riding on it. This chapter also covered application and service models, showing how the location and architecture of the application or service directly affect the required network design elements. In addition, this chapter highlighted the multitude of cloud options and the associated advantages of each option. This chapter highlighted the preference for agnostic cloud services over proprietary cloud services, to ensure a business doesn't lock itself into a specific cloud service provider, and how adopting a service-oriented architecture can be beneficial to the business. Last but not least, this chapter gave a quick overview of the importance of data and data management by highlighting the 11 data management pillars. Ensuring the confidentiality, integrity, and availability of a business's data is paramount to the business's success. If a business's data is compromised, it can no longer make valid decisions on that data, which handicaps the business until the data is fixed.

## Reference

Al-shawi, Marwan, *CCDE Study Guide* (Cisco Press, 2015)

## Exam Preparation Tasks

As mentioned in the section "How to Use This Book" in the Introduction, you have a couple of choices for exam preparation: the exercises here, Chapter 18, "Final Preparation," and the exam simulation questions in the Pearson Test Prep Software Online.

## Review All Key Topics

Review the most important topics in this chapter, noted with the Key Topic icon in the outer margin of the page. Table 3-5 lists a reference of these key topics and the page numbers on which each is found.

**Key Topic**

**Table 3-5**  Key Topics for Chapter 3

| Key Topic Element | Description | Page Number |
|---|---|---|
| Table 3-2 | 3-Tier Application Model Network Design Elements | 70 |
| Table 3-3 | Service Model Comparison | 72 |
| Table 3-4 | Cloud Types Detailed Comparison | 76 |

## Complete Tables and Lists from Memory

Print a copy of Appendix D, "Memory Tables" (found on the companion website), or at least the section for this chapter, and complete the tables and lists from memory. Appendix E, "Memory Tables Answer Key," also on the companion website, includes completed tables and lists to check your work.

## Define Key Terms

Define the following key terms from this chapter and check your answers in the glossary:

single-server model, 2-tier model, 3-tier model, web tier, application tier, database tier, on-premises, Software as a Service (SaaS), Platform as a Service (PaaS), Infrastructure as a Service (IaaS), carrier-neutral facility (CNF), cloud access point (CAP), private cloud, public cloud, hybrid cloud, multi-cloud

# CHAPTER 4

# Security Is Pervasive

This chapter covers the following topics:

- **Zero Trust Architecture:** This section covers the high-level pillars, capabilities, and components of Zero Trust Architecture that today's network designers should know.

- **Security CIA Triad:** This section covers confidentiality, integrity, and availability and their respective design elements.

- **Regulatory Compliance:** This section focuses on how a network designer should be able to take a compliance standard and design a solution that meets it appropriately.

This chapter discusses the primary design principles and considerations that network designers must evaluate when examining a network design from a network security point of view.

As discussed earlier in this book, to achieve a successful business-driven design, network architects and designers must always consider a top-down design approach to build the foundation or the roadmap of the design. With regard to security, it is pervasive and overlaid on top of every other component in an environment. Network designers should consider the following questions when designing a new network design or when evaluating an existing network design to help them draw the high-level picture of the design direction with regard to the security aspects:

- What are the business objectives and goals?

- What is the targeted industry (manufacturing, financial services, retail)? (This helps to identify industry-specific standards.)

- Is there any security policy or standard the design must comply with? (This is specific to the identified industry, such as Payment Card Industry Data Security Standard [PCI DSS] compliance.)

- What must be protected and what is desirable to be protected (priorities)?

- What is the impact of a failure to protect the intended systems/networks on the business (such as cost or reputation)?

- Will security services or appliances break any mission-critical applications? If yes, how can this risk be avoided or mitigated?

- Are there any capital expenditure (CAPEX) or operational expenditure (OPEX) concerns with regard to network security?

This chapter covers the following "CCDE v3.0 Unified Exam Topics" section:

- 5.0 Security Design

# "Do I Know This Already?" Quiz

The "Do I Know This Already?" quiz allows you to assess whether you should read this entire chapter thoroughly or jump to the "Exam Preparation Tasks" section. If you are in doubt about your answers to these questions or your own assessment of your knowledge of the topics, read the entire chapter. Table 4-1 lists the major headings in this chapter and their corresponding "Do I Know This Already?" quiz questions. You can find the answers in Appendix A, "Answers to the 'Do I Know This Already?' Quizzes."

**Table 4-1** "Do I Know This Already?" Section-to-Question Mapping

| Foundation Topics Section | Questions |
|---|---|
| Zero Trust Architecture | 1–6 |
| Security CIA Triad | 7, 8 |
| Regulatory Compliance | 9, 10 |

**CAUTION**   The goal of self-assessment is to gauge your mastery of the topics in this chapter. If you do not know the answer to a question or are only partially sure of the answer, you should mark that question as wrong for purposes of the self-assessment. Giving yourself credit for an answer you correctly guess skews your self-assessment results and might provide you with a false sense of security.

1. Which of the following are examples of static factors in regard to a Zero Trust Architecture? (Choose all that apply.)
   a. Credentials
   b. Threat intelligence
   c. Real-time data analytics
   d. Network
   e. GPS coordinates
   f. Biometrics

2. Which of the following are examples of dynamic factors in regard to a Zero Trust Architecture? (Choose all that apply.)
   a. Credentials
   b. Threat intelligence
   c. Real-time data analytics
   d. Network
   e. GPS coordinates
   f. Biometrics

**3.** Which of the following make up the trust score in regard to a Zero Trust Architecture? (Choose two.)

  **a.** Risk score

  **b.** Static factors

  **c.** Entitled to access

  **d.** Dynamic factors

**4.** Which of the following make up the authorization for access in regard to a Zero Trust Architecture? (Choose two.)

  **a.** Entitlement

  **b.** Static factors

  **c.** Dynamic factors

  **d.** Trust score

**5.** Which of the following is where policy is implemented, rules are matched, and associated access is pushed?

  **a.** Trust engine

  **b.** Endpoint device

  **c.** Inventory

  **d.** Policy engine

**6.** Which of the following evaluates overall trust by continuously analyzing the state of devices, users, workloads, and applications?

  **a.** Trust engine

  **b.** Endpoint device

  **c.** Inventory

  **d.** Policy engine

**7.** Which of the following are the correct elements of the CIA triad? (Choose three.)

  **a.** Compliance, integrity, availability

  **b.** Confidentiality, integrity, and availability

  **c.** Confidentiality, information, and availability

  **d.** Confidentiality, integrity, and assurance

**8.** Which of the following maintains accurate information end to end and is a key element of the CIA triad?

  **a.** Compliance

  **b.** Availability

  **c.** Integrity

  **d.** Confidentiality

**9.** Which of the following compliance standards would a retail clothing franchise that leverages point of sale systems for payment be required to follow?

  **a.** HIPAA

  **b.** PCI DSS

  **c.** Policy Enforcement Points

  **d.** Policy Engine

**10.** Which of the following compliance standards would a hospital that stores and transmits patient records be required to follow?

    **a.** HIPAA

    **b.** PCI DSS

    **c.** Policy Enforcement Points

    **d.** Policy Engine

# Foundation Topics

Today's converged networks carry much business-critical information across the network (whether it is voice, video, or data), which makes securing and protecting that information extremely important, for two primary reasons. The first reason is for information security and privacy purposes. The second reason is to maintain business continuity at the desired level, such as protecting against distributed denial of service (DDoS) attacks, regardless of whether this protection is within the internal network or between the different sites over an external network. Therefore, the design of network infrastructure and network security must not be performed in isolation. In other words, the holistic design approach discussed earlier in this book is vital when it comes to network security design considerations. No network should be designed independently from its security requirements. A successful network design must facilitate the application and integration of the security requirements by following the top-down approach, starting from the business goals and needs, to compliance with the organization's security policy standards, to a detailed design and integration of the various network technologies and components.

To secure any system or network, there must be predefined goals to achieve and specifications to comply with to ensure that the outcomes are measurable and always meet the organization's standards. Therefore, to achieve this, organizations almost always develop security standards, policies, and specifications that all collectively aim to achieve the desired goal with regard to information security. This is what is commonly known as a *security policy*. A security policy is a formal statement of the rules by which people who are given access to an organization's technology and information assets must abide. It should also specify the mechanisms through which these requirements can be met and audited for compliance with the policy. As a result, a good understanding of the organization's security policy and its standards is a crucial prerequisite before starting any network design or optimizing an existing design. This understanding ensures that any suggested solution will comply with the security policy standards and expectations of the business with regard to information security.

For instance, you may suggest redesigning an existing 1G dark fiber (owned by the organization) to a virtual leased line (VLL) solution (L2VPN based) that offers the same quality at a lower cost. However, the security policy may dictate that any traffic traversing any network that is not owned by the organization must be encrypted. By taking this point into consideration, the network architect or designer here can add IPsec or MACsec to the proposed solution to provide an encrypted VLL, to ensure the suggested design supports and complies with the organization's security policy standards.

The integration of network infrastructure and network security (including security components such as firewalls and configurations such as infrastructure access control lists ACL [iACLs]) can be seen as a double-edged sword. On the one hand, security will offer privacy, control, and stability to the network (for example, protect against DDoS attacks and unauthorized access). On the other hand, if both the network infrastructure and the security components are designed in isolation (siloed approach), then when they integrate together at some point, there will be a mix of the following issues:

- Complex integration

- Traffic drop

- Reduced performance

- Redesign or major design changes

- Design failure

For instance, in Figure 4-1, the network infrastructure of the Internet edge was designed to provide Internet access for the enterprise. Based on this, the network designer considered Enhanced Interior Gateway Routing Protocol (EIGRP) as the internal dynamic routing protocol and external Border Gateway Protocol (eBGP) to provide the edge routing to provide end-to-end connectivity.

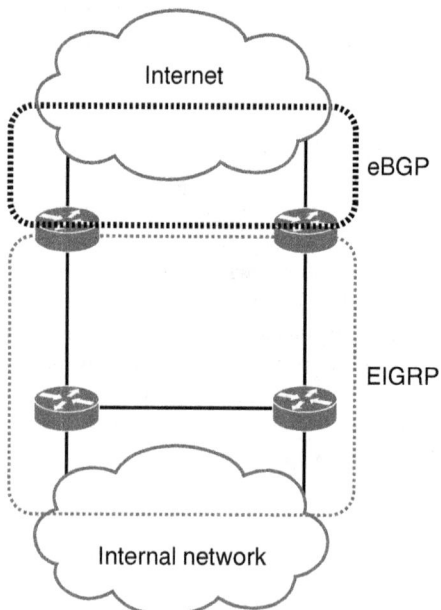

**Figure 4-1**  *Network Design Before Inserting Security Devices*

However, after this design was reviewed by the security team, they suggested that a pair of firewalls in routed mode has to be introduced into this design, and remote-access VPN sessions have to be terminated on these firewalls to comply with the enterprise security standards, as shown in Figure 4-2. In addition, the added firewalls support only static routing. Although it seems to be a simple change to introduce the firewalls into the original design, this simple change requires a review of the entire design (IGP, BGP, inbound and outbound traffic flow, policies to avoid asymmetrical routing, and NAT set up on the edge routers to allow users' remote-access VPN sessions to terminate at the firewalls), in addition to whether these firewalls have any impact on the traffic flows of mission-critical applications passing through them. Therefore, it is important to adopt a holistic approach to promote an optimal, integrated, and secure network design that is flexible enough to achieve the intended goals with minimal operational and design changes. Security cannot be an afterthought and should be at the forefront of each design decision.

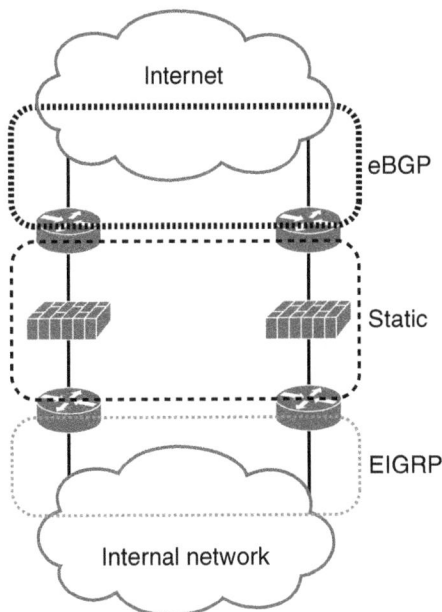

**Figure 4-2**   *Network Design After Inserting Security Devices*

To simplify the overall security design in a more controlled manner, a structured modular approach should be used by splitting the network into multiple domains. This approach helps to introduce chokepoints between the different domains to enforce structured control between these domains. Each domain should be given a level of trust based on the security requirements to facilitate controlling who and what is allowed to pass between the different domains by using various mechanisms such as packet filtering with iACLs or specialized security appliances (physical or virtual) such as firewalls and intrusion prevention systems (IPSs) to be positioned at each chokepoint (domain boundary point) between the different security domains. Furthermore, security requirements sometimes dictate that within each security domain there must be subdomains, each of which should have its own specific security policies and access restrictions among other subdomains. In this case, network designers can introduce what is commonly known as *zones*. For example, a data center can be placed under a single security domain. Normally, inside the data center, various services and applications require different levels of security enforcement between them, like web servers that can be placed in their own zone, database servers in different zones, and so on. With this approach, network designers can facilitate the security control to a large extent and optimize its manageability, as shown in Figure 4-3.

**NOTE** In Figure 4-3, the core block was not placed in its own domain to illustrate one of the recommended and proven design considerations that offloads any additional processing from the network core, such as security policies. However, some design requirements require the core to be treated as a separate security domain.

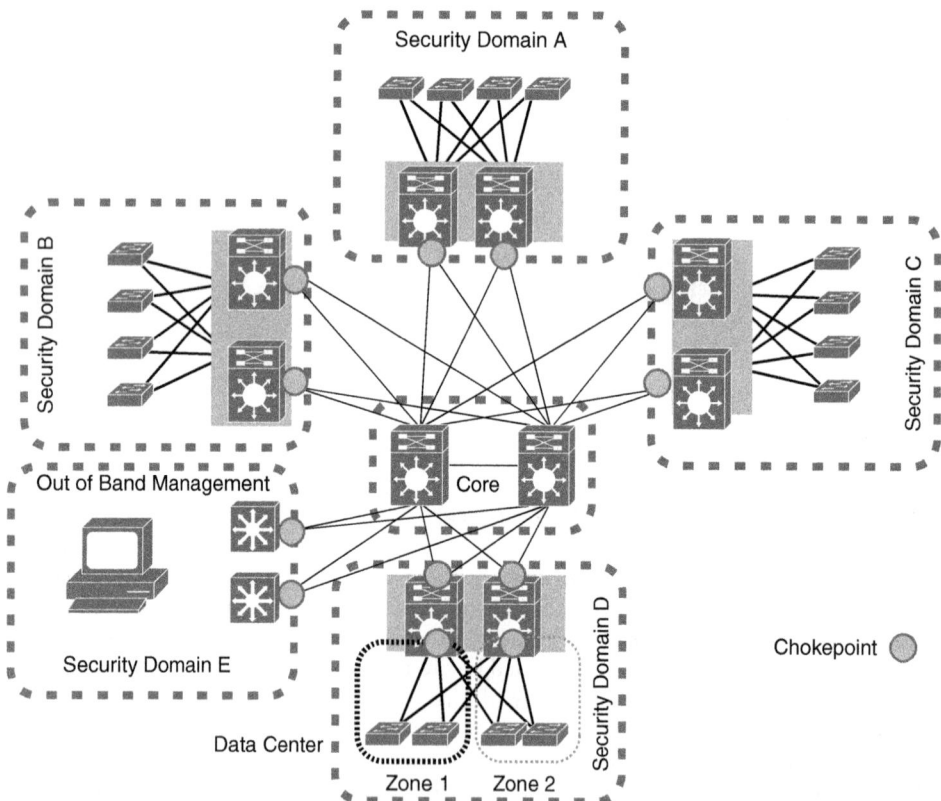

**Figure 4-3** *Security Domains and Zones*

Security devices are almost always placed at the chokepoints (domains or zone boundary points). However, the types of security devices and roles can vary based on the domain or zone to be protected and its location in the network. For example, in Figure 4-3, a firewall is placed at the network management zone boundary to fulfill packet filtering and inspection requirements for OAM traffic flows in both directions, while at the public demilitarized zone (DMZ), there are multiple specialized security nodes such as web application firewall data loss prevention and IPS.

Nevertheless, it is critical that network architects and designers consider the type of the targeted network and its high-level architecture. Each of these networks (irrespective of its detailed design) has different traffic flow characteristics. For instance, enterprise networks always define chokepoint boundaries such as the Internet edge or connections to extranet networks where traffic is always controlled by strict and confined rules.

# Zero Trust Architecture

Zero Trust Architecture, or Zero Trust Networking, is a paradigm shift for security. No longer is it the security model with a perimeter firewall allowing and restricting traffic in and out of an enterprise network. This section covers a short overview of what ZTA is, the pillars, the rules, and the capabilities that are involved to create it.

From a network design perspective, our focus is less on the implementation of a Zero Trust Architecture and more on the associated capabilities that are included in a Zero Trust Architecture and how those capabilities have an impact on the business and overall design of the enterprise network.

**Key Topic**

The following are the pillars of Zero Trust Architecture:

- The network is always assumed to be hostile.

- External and internal threats exist at all times and in all places.

- Every device, app, user, and network flow is authenticated and authorized.

- Automation systems are what allow a Zero Trust network to be built and operated.

- Policies must be dynamic and calculated from as many sources of data as possible.

- All activity is logged.

## Zero Trust Architecture Concepts

ZTA is a perimeterless architecture focused on denying all access until explicitly allowed. All resources (users, devices, infrastructure, applications, services, etc.) are authenticated and all access is authorized by a real-time policy engine. Contextual data on these resources is collected and, based on the correlation of that data and the trust level, the associated risk is evaluated. Depending on this evaluation of the correlated data, the provided access for a resource can change dynamically. Figures 4-4 and 4-5 show how the contextual evaluation and the correlated data in real time can impact the trust score of a resource's sessions.

**Static Factors  +  Dynamic Factors  =  Trust Score: 70%**

*Static Factors are assigned trust values and weights:*

- Credential
- Level of Confidence
- Device Trust
- Network
- Physical Location
- Biometrics
- Device Orientation and Peripherals

*Dynamic Factors are assessed and scored at time of access:*

- Threat intelligence
- Geovelocity
- GPS Coordinates

*Trust Score* is a combination of factors and are used to continually provide identity assurance. Trust Score determines level of access as required by the level of risk value of the resource being accessed.

**Trust Score  +  Entitlement  =  Access Authorization**

*Trust Score* is a combination of factors and are used to continually provide identity assurance. Trust Score determines level of access as required by the level of risk value of the resource being accessed.

Users have various roles and are *entitled to access* varies resources. Types of users can be Employees, Contractors, 3rd Party Partners, Affiliates, and Devices / IoT.

**Figure 4-4**  *Zero Trust Architecture Concept of Trust Score and Access Authorization*

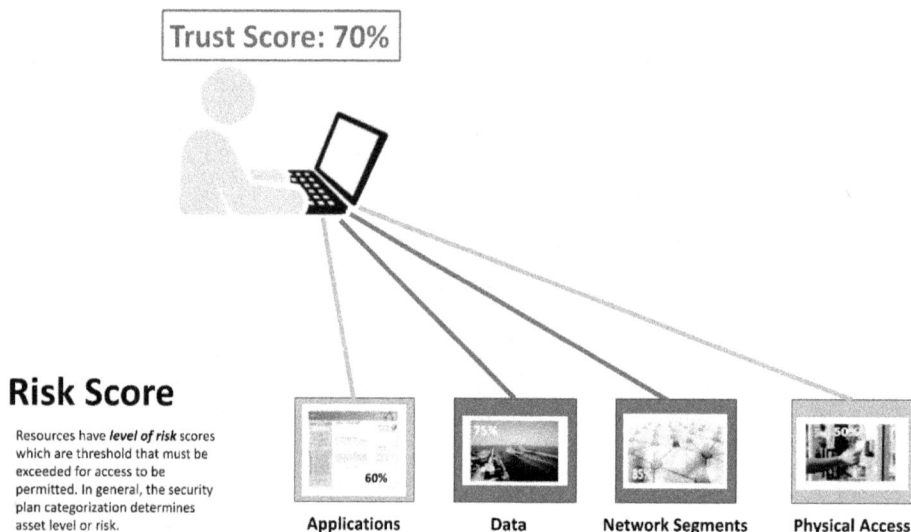

**Trust Score: 70%**

**Risk Score**

Resources have *level of risk* scores which are threshold that must be exceeded for access to be permitted. In general, the security plan categorization determines asset level or risk.

| Applications | Data | Network Segments | Physical Access |

**Figure 4-5**  *Zero Trust Architecture Concept of Trust Score vs. Risk Score*

The following factors and concepts that are shown in Figure 4-4 and Figure 4-5 all equate to the authorization for access for that specific user and device:

- **Static factors:** These are items that we know and can preemptively base access and authorization on. The most common of these factors are credentials but could also include the level of confidence, device trust, network, physical location, biometrics, and device orientation.

- **Dynamic factors:** These are sources of data that can be analyzed at the time of access to change what level of access and authorization (i.e., the trust score of the transaction in question) is being provided. The most common is threat intelligence, but can also be geo velocity (the difference between your current location and where you last logged in), GPS coordinates, and real-time data analytics around the transaction.

- **Entitled to access:** Users have various roles and, based on those roles, are entitled to specific access to complete their job. A financial user would need a different level of access than a human resource user. These two users should not have the same access or authorization. They may have overlapping access to resources that they both need to complete their job functions. Further examples of roles include contractors, affiliates, and foreign nationals. This is not limited to people but also is applied to devices. For example, printers would have a different entitled to access level than Internet of Things devices would.

- **Trust score:** This is a combination of factors, both static and dynamic, and is used to continually provide identity assurance. A trust score determines the level of access as required by the level of risk value of the asset being accessed.

- **Risk score:** Resources such as assets, applications, and networks have levels of risk scores, which are thresholds that must be exceeded for access to be permitted. In general, the security plan categorization determines asset level or risk.

- **Authorization for access:** For a resource, in this case a user or device, to be authorized for access to another resource, in this case an asset, application, or system, the trust score, and the entitlement level are combined to determine the authorization for access. Just because a trust score is high enough to access a resource, if that user or device doesn't have the correct entitlement, they will not have the appropriate authorization for access.

## Zero Trust Capabilities

The following are several capabilities within Zero Trust that you should know and understand from a network design perspective. Understand that while each of these capabilities is listed individually, vendor-specific solutions can provide multiple capabilities.

- **Inventory:** Single point of truth for all resources (all users, devices, workloads, and applications). This is an end-to-end inventory throughout the entire architecture/enterprise. This is also known as a policy information point (PIP).

- **Policy engine:** Location policy is implemented, rules are matched, and associated access (authorization) is pushed to the policy enforcement points. This is also known as a policy administration point (PAP) and a policy decision point (PDP).

- **Trust engine:** Dynamically evaluates overall trust by continuously analyzing the state of devices, users, workloads, and applications (resources). Utilizes a trust score that is built from static and dynamic factors, as previously shown in Figure 4-5. This is also known as a policy information point (PIP).

- **Endpoint device:** Any device an end user can leverage to access the enterprise network. Endpoint devices include business-owned assets and personally owned devices that are approved to access the enterprise network, potentially in a limited way like the common bring your own device (BYOD) deployment. In some vendor-specific implementations of Zero Trust, an endpoint device is also called a policy enforcement point (PEP).

■ **Infrastructure devices:** Networking devices within the enterprise, such as switches, routers, and firewalls. These devices are also called policy enforcement points (PEPs).

■ **Policy enforcement points (PEPs):** Locations where trust and policy are enforced.

■ **Feedback loop:** Continuous information sharing to allow for dynamic changes to policy based on constant analysis of new information via AI/ML, Big Data, and data lakes.

Figure 4-6 shows the Zero Trust Architecture components and Figure 4-7 highlights an example of a theoretical Zero Trust Architecture

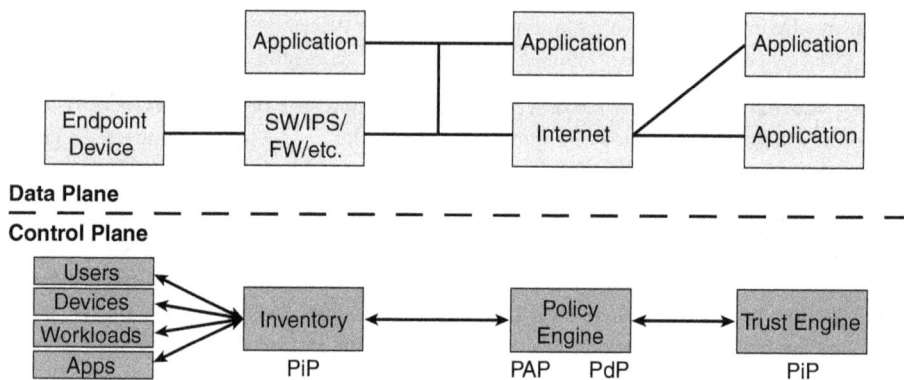

**Figure 4-6**  *Zero Trust Architecture Components*

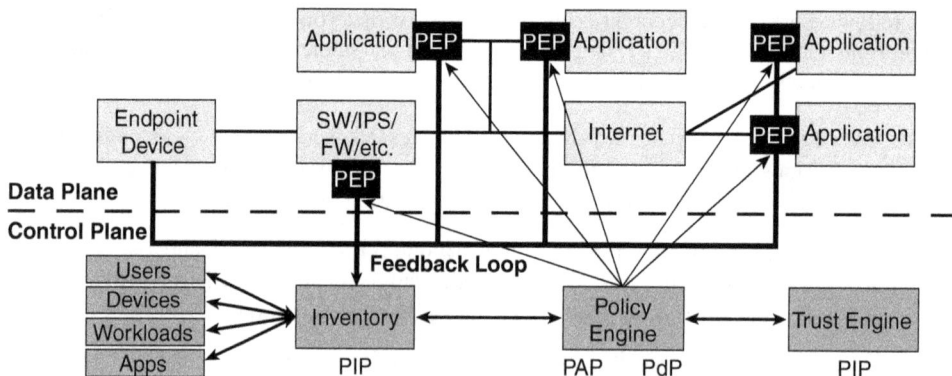

**Figure 4-7**  *Theoretical Zero Trust Architecture*

## Zero Trust Migration Steps

This section provides an example of how an organization can migrate from its current security model to a Zero Trust Architecture model. There are three high-level steps to take: establishing trust, providing trust-based access, and continuous verification of trust. The following list provides a breakdown of what might be included within each of these steps:

1. Establishing trust
   - Multiple factors of user identity
   - Device context and identity
   - Device posture and health
   - Location
   - Relevant attributes and context
   - Development, deployment

2. Providing trust-based access
   - Networks
   - Applications
   - Resources
   - Users and things

3. Continuous verification of trust
   - Original tenants used to establish trust are still true
   - Traffic is not threat traffic
   - Behavior for any risky, anomalous, or malicious actions
   - If compromised, then the trust is broken

This is not a fully inclusive section covering every aspect of Zero Trust, Zero Trust Architecture, and Zero Trust Networking. Giving the topic proper justice would require an entire book. The goal here is for you as a network designer to identify the different high-level capabilities that are included in a Zero Trust Architecture and how they impact the different design decisions and the overall state of the network.

## Security CIA Triad

Confidentiality, integrity, and availability, also known as the CIA triad, are considered the fundamentals of information security. In other words, these elements help to construct secure systems that protect information and resources (devices, communication, and data) from unauthorized access, modification, inspection, or destruction. Therefore, any breach of any of these three elements can lead to serious consequences. Network designers must be aware of these elements and how each impacts the network design, and at the same time how the network design can accommodate these principles as an integral part of it. Table 4-2 summarizes the characteristics of each of these elements and the common mechanisms used to achieve them.

**Key Topic**

**Table 4-2**  Confidentiality, Integrity, and Availability

| CIA Triad Security Element | Characteristic | Mechanisms to Achieve |
|---|---|---|
| **Confidentiality** | Protect against unauthorized access to information to maintain the desired level of secrecy of the transmitted information across the internal network or public Internet (in other words, identifying who should have access to this information). | Cryptography, encryption, user ID and password (two-factor authentication is more common today), or security tokens. Cryptography can provide data confidentiality by modifying the data into a format that can be understood only by the authorized entity. |
| **Integrity** | Maintain accurate information end to end by ensuring that no alteration is performed by any unauthorized entity. | Cryptography, data checksum. Cryptography enables the receiving entity to verify whether there was any alteration to the original, such as hashing. |
| **Availability** | Ensure that access to services and systems is always available and information is accessible by authorized users when required. Also, in modern networks, availability is measured not only by network or service up or down time, but also by its quality. (For example, malicious traffic may increase latency in the network, which leads to a degraded VoIP quality [users might not be able to make a voice call even though the voice system and the network are up].) | Systems designed with high availability at different levels (network, storage, compute, application) and that protect against attacks such as a DoS that attempts to stop access to resources. |

# Regulatory Compliance

A network designer should be able to take any compliance standard given to them and design a solution that fits into each of the specific constraints the standard governs. The goal of this section is to provide a quick overview of the most common compliance standards today, which will allow you to select design decisions that meet these constraints. This section does not provide an all-inclusive list of compliance regulations, nor does it cover every single aspect of each of the compliance regulations mentioned. Instead, it highlights two U.S. compliance standards, HIPAA and PCI DSS, and one EU compliance standard, GDPR, because they impact many organizations and are representative of the types of compliance standards network designers should be aware of. The section wraps up with a brief discussion of data sovereignty.

## HIPAA

The U.S. **Health Information Portability and Accountability Act (HIPAA)** was enacted in 1996 and is enforced by the Office of Civil Rights. HIPAA has a few rules that are important to know if you are designing a network for an organization that handles health records or information in any way:

- **The Privacy Rule:** Protects the privacy of individually identifiable health information, or protected health information (PHI).

- **The Security Rule:** Sets national standards for the security of electronic health information

- **The Patient Safety Rule:** Protects identifiable information being used to analyze patient safety events and improve safety

The gist of these HIPAA rules is that as we have to protect health information, patient information, and we as network designers have to ensure the design follows the security controls set forth by this law.

HIPAA is a compliance regulation that you will have to take into consideration when you are dealing with hospitals, doctors' offices, clinics, insurance agencies, and service providers that work closely with covered entities.

## PCI DSS

**Payment Card Industry Data Security Standard (PCI DSS)** compliance is mandated by credit card companies to help ensure the security of credit card transactions. This standard refers to the technical and operational standards that businesses must follow to secure and protect credit card data provided by cardholders and transmitted through card processing transactions. PCI DSS is developed and managed by the PCI Security Standards Council. Within PCI DSS there are 12 requirements that network designs should know:

1. Install and maintain a firewall configuration to protect cardholder data.
2. Do not use vendor-supplied defaults for system passwords and other security parameters.
3. Protect stored cardholder data.
4. Encrypt transmission of cardholder data across open, public networks.
5. Protect all systems against malware and regularly update anti-virus software programs.
6. Develop and maintain secure systems and applications.
7. Restrict access to cardholder data by business need to know.
8. Identify and authenticate access to system components.
9. Restrict physical access to cardholder data.
10. Track and monitor all access to network resources and cardholder data.
11. Regularly test security systems and processes.
12. Maintain a policy that addresses information security for all personnel.

As you can see, the rules of PCI DSS are more technical and security-focused than the rules of HIPAA, which are more generically stated.

PCI DSS is a compliance regulation that you will have to take into consideration when credit cards and the associated transactions are part of a business's functions. Always consider PCI DSS when you are dealing with a business that has a point of sale system or an online ordering system; in essence, PCI DSS applies to all entities that store, process, and transmit cardholder data—fast food restaurants, retail stores, online storefronts...the list is endless.

## GDPR

The **General Data Protection Regulation (GDPR)** is a European Union (EU) regulation for data protection that sets guidelines for the collection and processing of personal information from individuals. It applies to the processing of personal data of people in the EU by businesses that operate in the EU. It's important to note that GDPR applies not only to firms based in the EU, but any organization providing a product or service to residents of the EU. The regulation pertains to the full data life cycle, including the gathering, storage, usage, and retention of data. The following is a list of some of the types of privacy data GDPR protects:

- Basic identity information such as name, address, and ID numbers
- Web data such as location, IP address, cookie data, and RFID tags
- Health and genetic data
- Biometric data
- Racial or ethnic data
- Political opinions
- Sexual orientation

Any company that stores or processes personal information about EU citizens within EU states must comply with the GDPR, even if they do not have a business presence within the EU. Specific criteria for companies required to comply are

- A presence in an EU country
- No presence in the EU but processes personal data of EU residents

## Data Sovereignty

*Data sovereignty* is the requirement that information is subject to the location's regulations from where it was collected and processed. Sovereignty is a state-specific regulation requiring that information collected and processed in a country must remain within the boundaries of that country and must be safeguarded according to the laws of that country. This is most often seen today when businesses are sending data into other countries where their server or data centers physically reside. This is also common when migrating to cloud providers. In these circumstances, the data in question cannot leave the country and must be stored properly following the rules and regulations that country has set forth.

# Summary

Security is truly pervasive today. It is in each place in the network, in each product architecture, and horizontally interlocked across an end-to-end enterprise. From a network design perspective, security is one of those areas that has a number of constraints we have to comply with when making our design decisions. This can be as simple as determining whether we have to use encryption (IPsec or MACsec) over transport circuits or as complicated as developing and implementing a full Zero Trust Architecture in the enterprise network.

Zero Trust Architecture is a critical shift in how network security is both thought of and implemented in today's networks. In this chapter, we covered the Zero Trust Architecture pillars, capabilities, and concepts in a vendor-agnostic fashion, to arm you as a network designer in this critical security transformation that is happening in the industry today. Trust score, entitlement level, and authorization of access are some of the key concepts that dictate what a user or device can access within a Zero Trust Architecture. No more are those days where every user and device in an enterprise network receives full, unrestricted east–west access. Zero Trust builds on the CIA triad of confidentiality, integrity, and availability, which are considered the fundamentals of information security.

The final topic covered in this chapter was regulatory compliance standards. This section provided an overview of HIPAA, PCI DSS, and GDPR. Again, these were quick overviews to give you a general idea of how a network designer should understand a given standard and apply that understanding as they make design decisions. The goal for all network designers is to meet the standards in question, to allow the business and its networks to stay online. These were just three examples of a growing landscape of security standards that impact the overall design and architecture of modern enterprise networks.

## Reference

■ Al-shawi, Marwan, *CCDE Study Guide* (Cisco Press, 2015)

## Exam Preparation Tasks

As mentioned in the section "How to Use This Book" in the Introduction, you have a couple of choices for exam preparation: the exercises here, Chapter 18, "Final Preparation," and the exam simulation questions in the Pearson Test Prep Software Online.

## Review All Key Topics

Review the most important topics in this chapter, noted with the Key Topic icon in the outer margin of the page. Table 4-3 lists a reference of these key topics and the page numbers on which each is found.

**Table 4-3**  Key Topics for Chapter 4

| Key Topic Element | Description | Page Number |
|---|---|---|
| List | The pillars of Zero Trust Architecture | 91 |
| List | Zero Trust factors and concepts | 92 |
| List | Zero Trust capabilities | 93 |
| List | Zero Trust migration steps | 95 |
| Table 4-2 | Confidentiality, Integrity, and Availability | 96 |

## Complete Tables and Lists from Memory

Print a copy of Appendix D, "Memory Tables" (found on the companion website), or at least the section for this chapter, and complete the tables and lists from memory. Appendix E, "Memory Tables Answer Key," also on the companion website, includes completed tables and lists to check your work.

## Define Key Terms

Define the following key terms from this chapter and check your answers in the glossary:

static factors, dynamic factors, entitled to access, trust score, risk score, authorization for access, inventory, policy engine, trust engine, endpoint device, infrastructure devices, policy enforcement point (PEP), feedback loop, confidentiality, integrity, availability, Health Information Portability and Accountability Act (HIPAA), Payment Card Industry Data Security Standard (PCI DSS), General Data Protection Regulation (GDPR).

# CHAPTER 5

# Reference Architecture Models and Frameworks

## This chapter covers the following topics:

■ **Reference Architecture Models and Frameworks:** This section covers business architecture, enterprise architecture, and the current common architecture frameworks.

How do you know what you are choosing from a network design perspective is going to work? Are you making the right network design decisions? How do you define the phrase "architecture"? Who defines your architecture? If you asked another member of your organization, would they define architecture the same way you do?

This is where reference architectures, models, and frameworks come in to save the day. These items help guide your decisions by leveraging the different aspects that an organization has to help determine the proper way forward for that business.

This chapter highlights these reference architectures, models, and frameworks so you as a network designer understand how to properly map the decisions and the overarching architecture you are building to the specific business elements. These topics will not be covered end to end, as each individual framework and reference architecture warrants its own book in itself. This chapter will cover examples of these frameworks to show how they can dictate a design or architecture for a business.

This chapter covers the following "CCDE v3.0 Core Technology List" section:

■ 4.8 Reference models and paradigms

## "Do I Know This Already?" Quiz

The "Do I Know This Already?" quiz allows you to assess whether you should read this entire chapter thoroughly or jump to the "Exam Preparation Tasks" section. If you are in doubt about your answers to these questions or your own assessment of your knowledge of the topics, read the entire chapter. Table 5-1 lists the major headings in this chapter and their corresponding "Do I Know This Already?" quiz questions. You can find the answers in Appendix A, "Answers to the 'Do I Know This Already?' Quizzes."

**Table 5-1** "Do I Know This Already?" Section-to-Question Mapping

| Foundation Topics Section | Questions |
|---|---|
| Reference Architecture Models and Frameworks | 1–10 |

1. Which of the following options enables everyone from strategic planning teams to implementation teams to be synchronized, enabling them to address challenges and meet business objectives?

   **a.** Business architecture

   **b.** Enterprise architecture

   **c.** Business solution

   **d.** Business outcome

2. Which of the following options is a process of organizing logic for business processes and IT infrastructure to reflect the integration and standardization requirements of the company's operating model?

   **a.** Business architecture

   **b.** Enterprise architecture

   **c.** Business solution

   **d.** Business outcome

3. Which of the following options is a set of interacting business capabilities that delivers specific, or multiple, business outcomes?

   **a.** Business architecture

   **b.** Enterprise architecture

   **c.** Business solution

   **d.** Business outcome

4. Which of the following options is a specific measurable result of an activity, process, or event within the business?

   **a.** Business architecture

   **b.** Enterprise architecture

   **c.** Business solution

   **d.** Business outcome

5. Within Business Architecture, which of the following levels of alignment is a domain-specific architecture?

   **a.** Technology specific

   **b.** Technology architecture

   **c.** Business solutions

   **d.** Business transformation

**6.** Within business architecture, which of the following levels of alignment is a multi-domain architecture?

   **a.** Technology specific

   **b.** Technology architecture

   **c.** Business solutions

   **d.** Business transformation

**7.** Within business architecture, which of the following levels of alignment is a partial business architecture?

   **a.** Technology specific

   **b.** Technology architecture

   **c.** Business solutions

   **d.** Business transformation

**8.** Within business architecture, which of the following levels of alignment is a business-led architecture?

   **a.** Technology specific

   **b.** Technology architecture

   **c.** Business solutions

   **d.** Business transformation

**9.** Which of the following options is an enterprise architecture methodology that incorporates a high-level framework for enterprises that focuses on designing, planning, implementing, and governing enterprise information technology architectures?

   **a.** ITIL

   **b.** RMF

   **c.** TOGAF

   **d.** FEAF

   **e.** DODAF

**10.** Which of the following options is a set of best practice processes for delivering IT services to your organization's customers?

   **a.** ITIL

   **b.** RMF

   **c.** TOGAF

   **d.** FEAF

   **e.** DODAF

## Foundation Topics

## Reference Architecture Models and Frameworks

This covers the most common reference architectures, models, and frameworks (at the time of writing) that guide network design decisions to make a business successful.

## Business architecture

**Business architecture (BA)** enables everyone, from strategic planning teams to implementation teams, to get "on the same page," enabling them to address challenges and meet business objectives. The people, processes, and technology (tools) that align with business priorities enable business outcomes. Furthermore, a **business solution** is actually a set of interacting business capabilities that delivers specific, or multiple, business outcomes. A **business outcome** is a specific measurable result of an activity, process, or event within the business, traditionally following the SMART principle: specific, measurable, attainable, realistic, and time-bound.

Here are a few examples of real business outcomes (real statistics) that business leaders have had in the "real production" world today:

- Increase same day of order shipments by 20 percent by Q1

- Improve customer lifetime value by 10 in 24 months

- Accelerate new product time to market by 25 percent by Q3

- Drive sales growth of *solution* by 10 percent in the current fiscal year from the previous fiscal year

- Enable an increase of delivery efficiency by 25 percent in the current fiscal year

Figure 5-1 shows how it all fits together, specifically how a technology solution, in our case a network design decision we made, fits in with a business capability of automating business processes, which transforms the business to provide the business solution of customer care, which then delivers the business outcome of an improved customer lifetime value by 10 percent in 24 months.

## Fitting It All Together

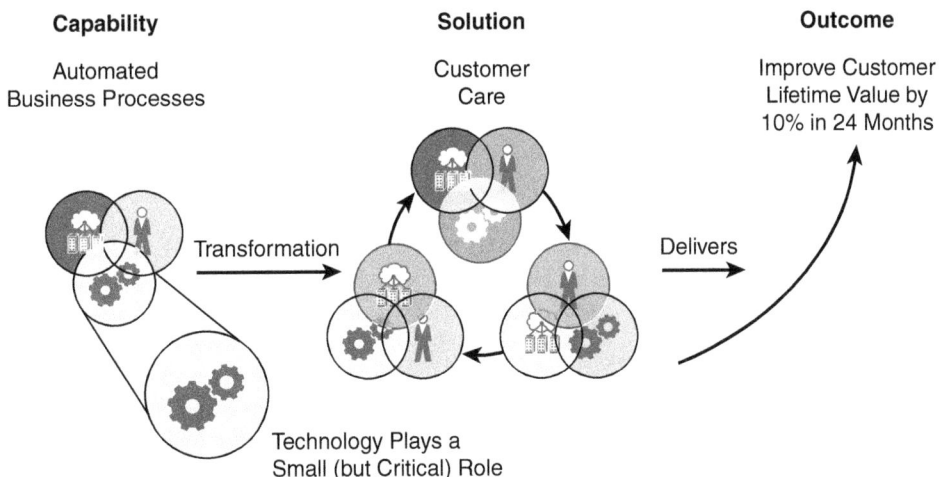

**Figure 5-1**  *Fitting It All Together*

The business transformation shown in Figure 5-1 implies a shift in business capabilities. In some cases, this means adding a new business capability, removing redundant business capabilities, or changing existing business capabilities.

Where you are located within the organization, or at what level you are partnering with another organization, will determine what type of **business architecture scope** you will have. This scope is a combination of you as the network designer working with the network architects and business stakeholders to achieve the ultimate outcomes the business wants to achieve. There are four levels of alignment:

- **Technology specific:** A domain-specific architecture. Within this scope, the business is requiring help with finding and purchasing the right product or group of products in an architecture focus area. This might be data center, security, or enterprise networking focused, but it doesn't cross between the different architecture focus areas.

- **Technology architecture:** A multi-domain architecture (MDA), also referred to as cross-architecture. In this scope the business needs help understanding the benefits of multi-domain technology architecture and how to show the value it provides to the business. With this scope, two or more architecture focus areas are incorporated.

- **Business solutions:** A partial business architecture scope for a business that requires expertise to help solve its business problems and determine how to measure the business impact of its technology investments (CAPEX, OPEX, ROI, TCO, etc.).

- **Business transformation:** A business-led architecture scope for a business that requires help with transforming its business capabilities to facilitate innovation to accelerate the company's digitization.

## Business Architecture Guiding Principles

Figure 5-2 highlights how the business architecture Guiding Principles, which is a combination of **The Open Group Architecture Framework (TOGAF)** and **Information Technology Infrastructure Library (ITIL)**, work together within business architecture.

The business architecture guiding principles are explained briefly in the following list:

- **Strategy and vision:** Aligns architecture capabilities to business objectives and outcomes

- **Business architecture:** Consistent with the business model being leveraged and the corresponding market/vertical (healthcare, financial, manufacturing, retail, government, etc.) perspective overlaid on top of it

- **Application architecture:** Co-existent with the hybrid cloud model (SaaS, PaaS, IPaaS, ERP, CRM) and supports composite three-tier, N-tier, architecture standards and capabilities

- **Technology architecture:** Supports automation, provisioning, orchestration, analytics, and telemetry capabilities

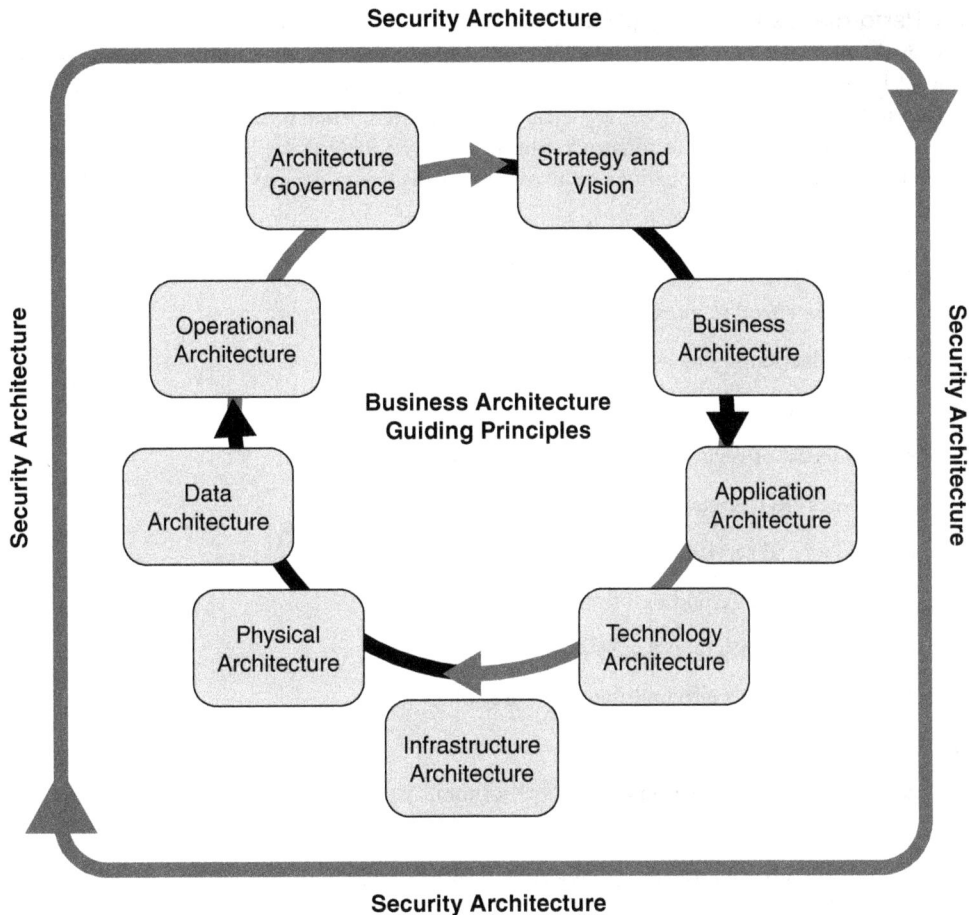

**Figure 5-2**  *Business Architecture Guiding Principles Workflow*

- **Infrastructure architecture:** Supports data center, cloud, collaboration, security, SASE, enterprise networking, and IoT capabilities

- **Physical architecture:** Supports the compute, storage, and virtualization aspects of the business

- **Data architecture:** Supports business intelligence, data integration, data security, business information, and data administration and operations capabilities

- **Operational architecture:** Supports ITIL service capabilities aligned to day 0, day 1, and day 2 managed services

- **Architecture Governance:** Supports architecture management office services aligned to the plan, design, implement, and operate life cycle

### Key Performance Indicators (KPIs)

How do we know when something is successful from a framework, reference architecture, or model perspective? We cannot just arbitrarily state that a network design is successful. We have to have a process with proper metric thresholds that, when reached, validated the success of leveraging a specific construct. This is where we identify key performance indicators (KPIs). The following list provides some example KPIs from different businesses in different markets:

**Key Topic**

- $x$% decrease in IT CAPEX and OPEX

- Increase speed to market and business agility by $x$%

- $x$% reduction in time to market of new IT services

- Reduction in secure Ops MTTD from $x$ days to $x$ hours

- Reduction of secure Ops MTTR from $x$ days to $x$ hours

- Business ROI contributes to IT solutions

- Product and service pull-through ROI

- ATR, ARR, AOV, and pull-through ROI

- $x$ services attach rate % upside

- Growth extends into multi-architecture solutions

- Growth extends into multi-architecture services

- Architecture automation margin %pProfitability influence on as-sold vs. as-delivered margin

- Architecture simplification and standardization margin % influence on as-sold vs. as-delivered margin

## Enterprise architecture

**Enterprise architecture (EA)** is a process of organizing logic for business processes and IT infrastructure reflecting the integration and standardization requirements of the company's operating model. With this perspective, an EA outcome might be reducing complexity, simplifying deployment, or being perceived as a service from an industry perspective. The direct business benefit an enterprise architecture provides is the ability for that business to respond to new market opportunities, justify its investments, reduce costs (CAPEX and OPEX), and enable quality. Figure 5-3 highlights the different aspects of enterprise architecture.

**Enterprise Architecture**

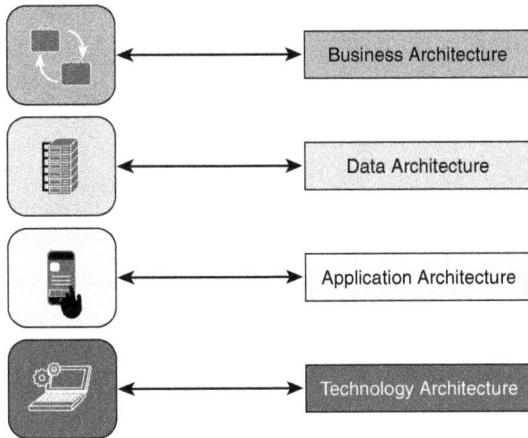

**Figure 5-3**   *Enterprise Architecture*

## Enterprise Architecture Methodology

Figure 5-4 walks through the Enterprise Architecture methodology steps.

**Enterprise Architecture Methodology**

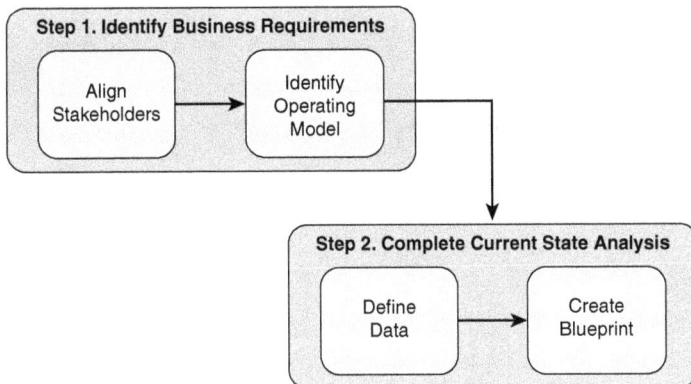

**Figure 5-4**   *Enterprise Architecture Methodology Steps*

The first step in this process is to identify the business requirements, which is achieved by aligning stakeholders and determining the operating model.

### Aligning Stakeholders

Who are the stakeholders we should align with? There are four groups of stakeholders we should look to align in this process: the IT steering committee, architects, finance and purchasing, and enterprise-wide process owners. The following list provides some examples of what roles, titles, and groups might fit into each of these four categories:

- **IT steering committee:** Could include members of an architecture management office (AMO), vendor-agnostic groups, and senior management within the business

- **Architects:** Could include business architects, technology-specific architects, and enterprise architects

- **Finance and purchasing:** Could include the chief finance officer (CFO), the procurement team (or a team member as a liaison), and supplier relations in some form

- **Enterprise-wide process owners:** Could include any sort of chief of *fill-in-the-blank*, user groups, business unit leaders, and business lines of effort owners

Keep in mind that these are example roles and titles, and not a pre-scripted or all-inclusive list. As a network designer, it is ultimately up to you to identify what stakeholders you need to include in this process.

### Identifying the Operating Model

Identifying, aligning, or defining the business operating model is the next step in identifying the business requirements. Keep in mind that the business as a whole may very well have to shift or change its current operating model once it's been identified. The following list provides a number of focus areas to help you identify the operating model being leveraged and what needs to change in the future state:

- **Standardization and integration:** To what extent does one business's success rely on the availability, accuracy, and timelines of others' data? To what extent does the company benefit from having business units run operations similarly?

- **Coordination:** High integration with little standardization of processes. Data is shared across the portfolio. Business units releasing innovation.

- **Unification:** Centralized management and process design, centrally mandated databases, and centralized IT organization.

- **Diversification:** Independence with shared services, few common customers or business practices, and shared IT services for economies of scale.

- **Replication:** Autonomy to business units, running operations in a standardized fashion, and growth through acquisitions.

### Identifying the Current State

The second step in this process is to complete a current state analysis, which basically means defining the data. What goes into data is called attributes and where the data is stored is called a configuration management database (CMDB). Starting with the latter, the CMDB is the single repository that stores information about all hardware and software assets in the environment. In some vendor-specific deployments of a CMDB, services can also be stored in the repository. The attributes that go into making the data itself are derived from incident management systems, escalation engines, reports, and service-level agreements. In addition to these items, there are other resources that can be leveraged to pull in more information and data. These include, but are not limited to, the following:

- Hardware inventory

- Software inventory

- Contract management

- Network discovery

- Administrative data

- User data

- Event management data (IT Operations Management [ITOM])

Having identified the business requirements in step one and completing a current state analysis in step two, we have a blueprint of the current state on top of which we can align the processes (create a common operating language), people (create the artifacts), and technology (the standard tooling that should be used) as an overlay.

## Architecture Frameworks for Consideration

This section briefly highlights some of the most common reference architectures, frameworks, and models at the time of writing.

- **The Open Group Architecture Framework (TOGAF):** TOGAF is an enterprise architecture methodology that incorporates a high-level framework for enterprises that focuses on designing, planning, implementing, and governing enterprise information technology architectures. TOGAF helps businesses organize their processes through an approach that reduces errors, decreases timelines, maintains budget requirements, and aligns technology with the business to produce business-impacting results. Figure 5-5 provides a high-level view of the steps in this framework.

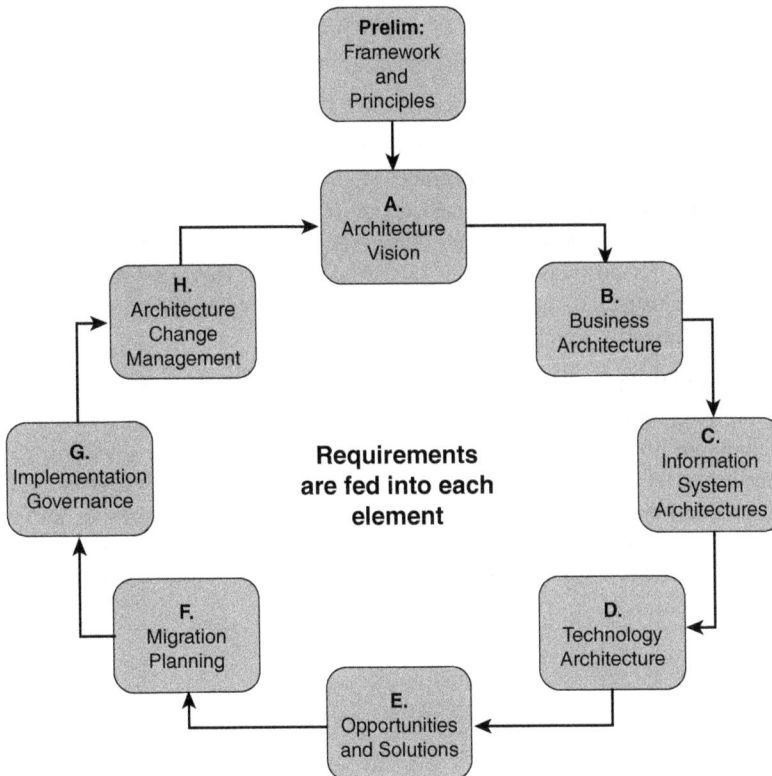

**Figure 5-5**   *The Open Group Architecture Framework High-Level View*

**Key Topic**

- **Zachman Framework:** This framework provides a means to classify a business's architecture in a structured manner. This is a proactive business tool that is used to model a business's functions, elements, and processes to help the business manage change throughout the organization. Figure 5-6 highlights the elements that this framework covers.

**Key Topic**

|  | Why | How | What | Who | Where | When |
|---|---|---|---|---|---|---|
| **Contextual** | Goal List | Process List | Material List | Organizational Unit and Role List | Geographical Locations List | Event List |
| **Conceptual** | Goal Relationship | Process Model | Entity Relationship Model | Organizational Unit and Role Relationship Model | Locations Model | Event Model |
| **Logical** | Rules Diagram | Process Diagram | Data Model Diagram | Role Relationship Diagram | Locations Diagram | Event Diagram |
| **Physical** | Rules Specification | Process Function Specification | Data Entity Specification | Role Specification | Location Specification | Event Specification |
| **Detailed** | Rules Details | Process Details | Data Details | Role Details | Location Details | Event Details |

**Figure 5-6** *Zachman Framework Elements*

- **Federal Enterprise Architecture Framework (FEAF):** FEAF is the industry standard framework for government enterprise architectures. Within this framework, the focus is on guiding the integration of strategic, business, and technology management architecture processes. One of the primary benefits of this framework is that it focuses on a common approach to technology acquisition within all U.S. federal agencies. Figure 5-7 shows how the different drivers (architecture, business, and design) are incorporated into the standards and transitional processes to create the strategic direction.

- **Risk Management Framework (RMF):** The RMF, created by NIST, focuses on the integration of security, privacy, and cyber supply chain processes into the system development life cycle. This risk-based approach to control the selection process considers effectiveness, efficiency, and legal constraints and directives. Managing organizational risk is critical to the security, safety, and privacy of information systems. The RMF can be applied to new and legacy systems, leveraging any technology, and within any specific market or sector. Figure 5-8 shows the RMF step-by-step process.

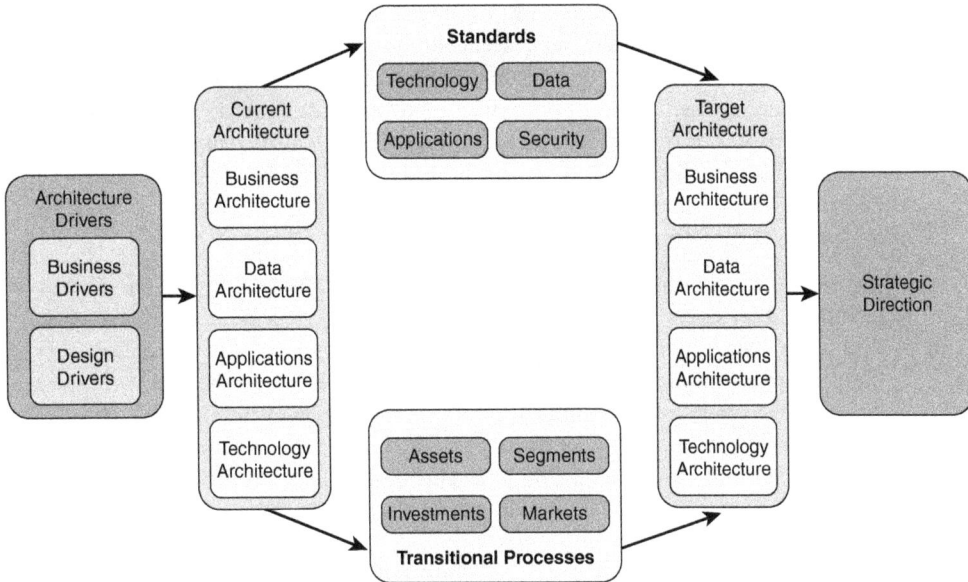

**Figure 5-7**  *The Federal Enterprise Architecture Framework*

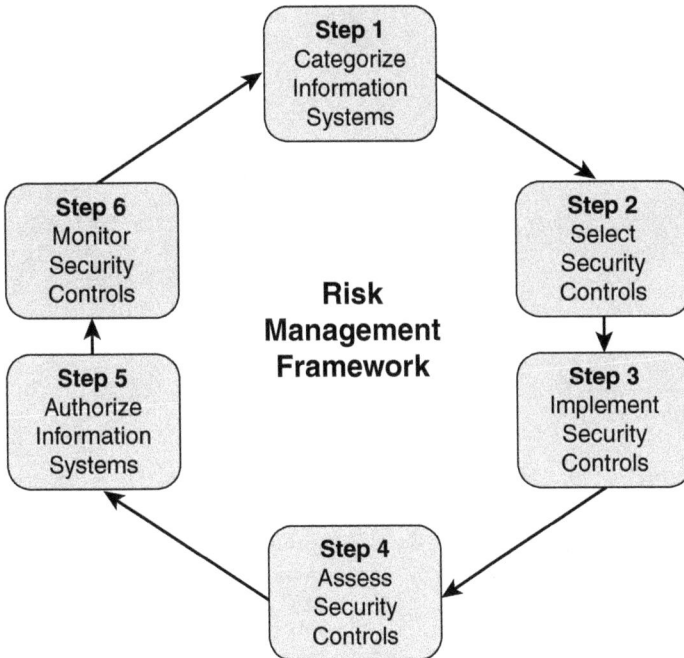

**Figure 5-8**  *Risk Management Framework*

- **Department of Defense Architecture Framework (DODAF):** DODAF defines a common approach for presenting, describing, and comparing DoD enterprise architectures across organizational, joint, or multinational boundaries. DODAF leverages common terminology, assumptions, and principles to allow for better integration between DoD elements. This framework is suited to large systems with complex integration and interoperability challenges. One element of this framework that is unique is its use of views. Each view offers an overview of a specific area or function and provides details for specific stakeholders within the different domains.

- **Business, Operations, Systems, and Technology (BOST):** The BOST framework provides the structure for enterprise models, their elements, and relationships. Each of the four elements of BOST has its own views. In this framework, the requirements flow downward through the four views. The capabilities flow upward in response to these requirements, creating a mapping between the requirements and capabilities. The success of this framework is based on the ability of a business to align its capabilities with the constantly changing requirements in all four views. Figure 5-9 highlights the four views and how the requirements move downward while the capabilities flow upward.

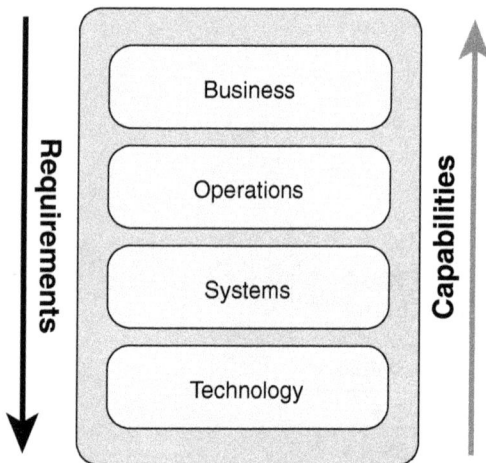

**Figure 5-9** *The BOST Framework*

- **ITIL:** Information Technology Infrastructure Library is a set of best practice processes for delivering IT services to your organization's customers. ITIL focuses on ITSM and ITAM, and includes processes, procedures, tasks, and checklists that can be applied by any organization. The three focus areas of ITIL are change management, incident management, and problem management.

Table 5-2 compares the most common advantages of the previously mentioned architecture frameworks.

**Key Topic**

**Table 5-2**  Architecture Framework Advantages Comparison

| Framework | Advantages |
|---|---|
| TOGAF | Reduce cost, time, and risk in the development of enterprise infrastructure architectures. |
| | Realize quick benefits from implementation due to improved flexibility and freedom. |
| | Allows for consistent business growth and simple restructuring. |
| Zachman | Improves communications around systems and processes. |
| | Provides an informed decision-making process that includes a holistic view of the organization. |
| | Increased efficiency from standardization. |
| | Reduced costs by increases in system performance. |
| RMF | Minimize risk. |
| | Protection of assets. |
| | Reputation management. |
| | Optimization of data management. |
| FEAF | Collaborative planning process. |
| | Common approach for complex environments. |
| | Leveraged within and between all U.S. federal agencies. |
| DODAF | Perspectives for every aspect. |
| | Common terminology, semantics, and viewpoints. |
| ITIL | Improves service quality. |
| | Cost optimization. |
| | Higher customer satisfaction. |
| | Better risk management. |
| | Alignment of business goals. |
| BOST | Four architecture views. |
| | Interlinked planning models. |
| | IT investments are tracked to ensure they map to a business requirement and capability. |
| | Business capabilities are aligned to the current state. |

5

## Summary

Typically, no single reference architecture, framework, or model will apply to the situation you are in as a network designer. In most cases, you may have to merge or adopt aspects of multiple frameworks to help guide you and the business to success. As you continue this journey, here are three guidelines to leverage when dealing with frameworks:

- Encourage architecture reviews regularly

- Avoid big-bang implementations (doing everything at once)

- Use your operating model as a North Star, following a path toward a strategic goal

This chapter started with a number of questions to really set the stage for the discussion of architectures, frameworks, and models. How do you know what you are choosing is going to work? Are you making the right decisions? How do you define the phrase "architecture"? Who defines your architecture?

To answer these questions, this chapter first covered business architecture, defining what business priorities, drivers, outcomes, and capabilities are and, more importantly, how they interact with one another. In this same context, we described the different levels of alignment, or scope, that business architecture includes and how you can identify which scope you are in. How do we know if the network design we are proposing will be successful was our next point of interest and is where the value of key performance indicators was brought forth.

Then we covered enterprise architecture, where we determine business requirements by identifying stakeholders and the operating model being leveraged. Finally, we identified the current state by defining the data and creating a blueprint.

The end of this chapter provided a list of the most common reference architectures, frameworks, and models at the time of writing. This list is a summary and is not meant to be inclusive of all the different components within each of these frameworks listed, nor is the list a fully inclusive list of all frameworks leveraged today.

## Reference

- Al-shawi, Marwan, *CCDE Study Guide* (Cisco Press, 2015)

## Exam Preparation Tasks

As mentioned in the section "How to Use This Book" in the Introduction, you have a couple of choices for exam preparation: the exercises here, Chapter 18, "Final Preparation," and the exam simulation questions in the Pearson Test Prep Software Online.

## Review All Key Topics

Review the most important topics in this chapter, noted with the Key Topic icon in the outer margin of the page. Table 5-3 lists a reference of these key topics and the page numbers on which each is found.

**Table 5-3**  Key Topics for Chapter 5

| Description | Key Topic Element | Page Number |
|---|---|---|
| List | Example business outcomes as SMART goals | 105 |
| List | Business architecture Guiding Principles | 106 |
| List | Example key performance indicators | 108 |
| Paragraph | The Open Group Architecture Framework (TOGAF) | 111 |
| Figure 5-5 | The Open Group Architecture Framework High-Level View | 111 |
| Paragraph | Zachman Framework | 112 |
| Figure 5-6 | Zachman Framework Elements | 112 |

| Description | Key Topic Element | Page Number |
|---|---|---|
| Paragraph | Information Technology Infrastructure Library (ITIL) | 114 |
| Table 5-2 | Architecture Framework Advantages Comparison | 115 |

## Complete Tables and Lists from Memory

Print a copy of Appendix D, "Memory Tables" (found on the companion website), or at least the section for this chapter, and complete the tables and lists from memory. Appendix E, "Memory Tables Answer Key," also on the companion website, includes completed tables and lists to check your work.

## Define Key Terms

Define the following key terms from this chapter and check your answers in the glossary:

business architecture (BA), business solution, business outcome, business architecture scope, The Open Group Architecture Framework (TOGAF), Information Technology Infrastructure Library (ITIL), enterprise architecture (EA), Zachman Framework, Federal Enterprise Architecture Framework (FEAF), Risk Management Framework (RMF), Department of Defense Architecture Framework (DODAF), Business, Operations, Systems, and Technology (BOST)

5

# CHAPTER 6

# Transport Technologies

### This chapter covers the following topics:

- **Layer 2 Media Access:** This section covers Layer 2 transport and connectivity model.

- **Virtual Private Wire Service Design Considerations:** This section covers VPLS models, concepts, and the corresponding network design elements.

- **EVPN Design Model:** This section covers EVPN concepts, protocols, and the corresponding network design elements.

This chapter covers multiple transport technologies that can be leveraged as network design options from an enterprise perspective. Not all transport technology options are covered in this chapter; some are covered in later chapters in this book, and some are not included because they are beyond the scope of the CCDE exam. The transport technologies included here at the time of writing are the most common options seen today in production networks.

This chapter covers the following "CCDE v3.0 Core Technology List" section:

- 1.0 Transport Technologies

## "Do I Know This Already?" Quiz

The "Do I Know This Already?" quiz allows you to assess whether you should read this entire chapter thoroughly or jump to the "Exam Preparation Tasks" section. If you are in doubt about your answers to these questions or your own assessment of your knowledge of the topics, read the entire chapter. Table 6-1 lists the major headings in this chapter and their corresponding "Do I Know This Already?" quiz questions. You can find the answers in Appendix A, "Answers to the 'Do I Know This Already?' Quizzes."

**Table 6-1** "Do I Know This Already?" Section-to-Question Mapping

| Foundation Topics Section | Questions |
|---|---|
| Layer 2 Media Access | 1–4 |
| Virtual Private Wire Service Design Considerations | 5–7 |
| EVPN Design Model | 8, 9 |

**CAUTION** The goal of self-assessment is to gauge your mastery of the topics in this chapter. If you do not know the answer to a question or are only partially sure of the answer, you should mark that question as wrong for purposes of the self-assessment. Giving yourself credit for an answer you correctly guess skews your self-assessment results and might provide you with a false sense of security.

1. Which of the following options is the correct port-based Metro Ethernet transport mode for an E-Line service type?

   **a.** Ethernet private line

   **b.** Ethernet virtual private line

   **c.** Ethernet private LAN

   **d.** Ethernet virtual private LAN

2. Which of the following options is the correct VLAN-based Metro Ethernet transport mode for an E-LAN service type?

   **a.** Ethernet private line

   **b.** Ethernet virtual private line

   **c.** Ethernet private LAN

   **d.** Ethernet virtual private LAN

3. Which of the following options is the correct VLAN-based Metro Ethernet transport mode for an E-Line service type?

   **a.** Ethernet private line

   **b.** Ethernet virtual private line

   **c.** Ethernet private LAN

   **d.** Ethernet virtual private LAN

4. Which of the following options is the correct port-based Metro Ethernet transport mode for an E-LAN service type?

   **a.** Ethernet private line

   **b.** Ethernet virtual private line

   **c.** Ethernet private LAN

   **d.** Ethernet virtual private LAN

5. Which of the following options is the correct transport mode over pseudowire for a Frame Relay access connection?

   **a.** AAL5 protocol data units over pseudowire cell relay over pseudowire

   **b.** Protocol-based per VLAN

   **c.** Port-based per DLCI

6. Which of the following options is the correct transport mode over pseudowire for an ATM access connection?

   **a.** AAL5 protocol data units over pseudowire cell relay over pseudowire

   **b.** Protocol-based per VLAN

   **c.** Port-based per DLCI

7. Which of the following options is the correct transport mode over pseudowire for an Ethernet access connection?

   **a.** AAL5 protocol data units over pseudowire cell relay over pseudowire

   **b.** Protocol-based per VLAN

   **c.** Port-based per DLCI

**8.** Which Layer 2 transport options are the best options for a very large enterprise Data Center Interconnect (DCI) solution where there are many data centers with many interconnection points with only a few thousand MAC addresses in total? (Choose two.)

   **a.** VPLS

   **b.** EVPN

   **c.** H-VPLS

   **d.** PBB-EVPN

   **e.** PBB-VPLS

**9.** Which Layer 2 transport options are best options for an optimized MAC mobility requirement? (Choose two.)

   **a.** VPLS

   **b.** EVPN

   **c.** H-VPLS

   **d.** PBB-EVPN

   **e.** PBB-VPLS

# Foundation Topics

## Layer 2 Media Access

Operators understand the importance of and the real need for an infrastructure that offers flexible and responsive network architecture, which supports current and future requirements without requiring any major change or upgrade to the network in the future. Therefore, in today's market, almost all service providers have either partially or fully migrated their core infrastructure to be MPLS enabled. This enables them to take advantage of the MPLS infrastructure and offer the same services to their customers (Layer 2 over legacy and modern WAN protocols) over one unified MPLS core infrastructure. This is a big incentive for those service providers that already offer MPLS L3VPN, because it will significantly reduce capital expenditures and operational expenses where the same core infrastructure and control plane can be used to transport both L2 and L3 services. Commonly, modern service providers that offer Layer 2 Ethernet services (for example, Metro Ethernet) are referred to as Carrier Ethernet providers. Based on that, the following are the primary business drivers for service providers to adopt and invest in modern Layer 2 services (for example, L2VPN and Metro Ethernet):

- **Flexible offerings:** For example, customized solutions with a flexible mix of services, data rates, and revenue (Ethernet private line [EPL], Ethernet virtual private line [EVPL], Ethernet LAN [E-LAN]).

- **Flexible bandwidth:** For example, service providers can provision 1G or 10G Ethernet physical links and offer fractional line access rates based on the agreed SLA. This provides operators and customers seamless roadmap upgrades in the future without changing any link or device (as compared to legacy services, such as a time-division multiplexing [TDM]-based WAN). Also, with today's flexible Ethernet nodes, an access port upgrade from 1G to 10G can be as simple as a Small Form-factor Pluggable (SFP) replacement.

- **Flexible media access:** A variety of access networks that leverage Ethernet, such as WiMAX, IP digital subscriber line access multiplexers (DSLAMs), Ethernet over Fiber, Ethernet over Copper, and so on. Also supports interworking between legacy access media, such as ATM and Frame Relay, which might need to be retained for some remote areas where Metro Ethernet (ME) access coverage is not available.

- **Cost savings:** Moving to an Ethernet infrastructure offers lower CAPEX compared to legacy hardware such as ATM switches. In addition, bandwidth flexibility, discussed earlier in this list, offers an optimized CAPEX and investment protection for both service providers and customers.

- **Scalable services that can increase revenue:** Support a large number of customers over one common infrastructure.

**NOTE**   This list covers the technologies and designs from a service provider point of view to provide the benefits to the enterprise network design if one of these technologies is chosen from a transport perspective. In addition, an enterprise could also deploy any of these services across its own networks, such as a self-deployed Metro Ethernet implementation to interconnect multiple campus locations over a Layer 3 WAN core network.

Figure 6-1 illustrates the different Layer 2 media access methods and topologies that can be transported over an MPLS-enabled backbone.

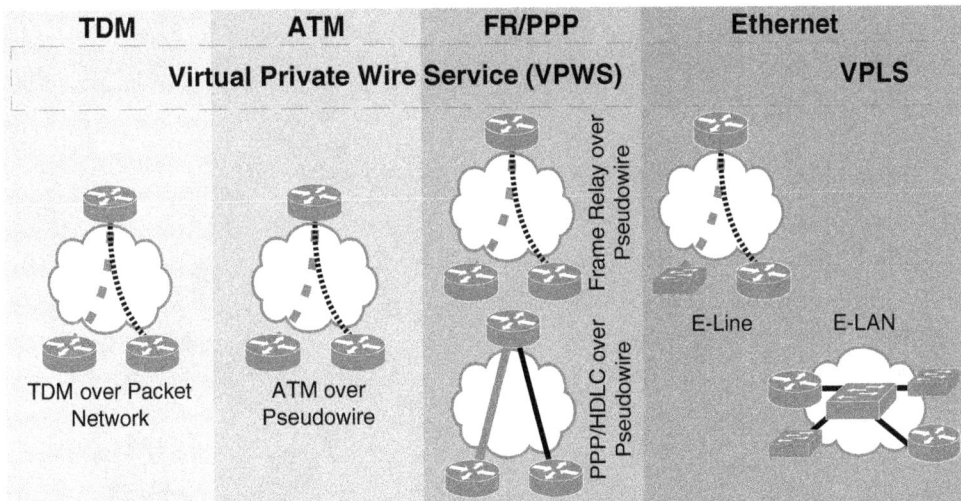

**Figure 6-1**   *Layer 2 Media Access Methods*

This section focuses on the L2VPN Ethernet-based models of modern service provider networks, as depicted in Figure 6-2.

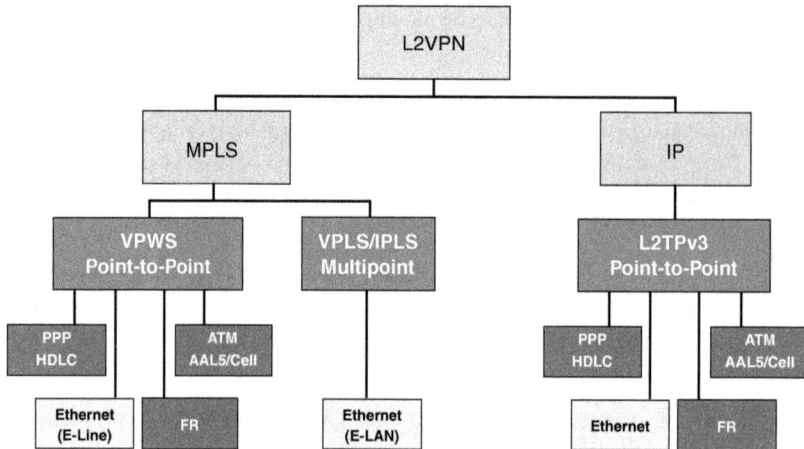

**Figure 6-2**  *L2VPN Ethernet-Based Models*

Moving forward we are going to talk about the different Carrier Ethernet options that are available to network designers. *Carrier Ethernet* refers to the modern service provider architectures that offer a cost-effective converged network infrastructure capable of delivering new and future services by providing scalable and flexible designs and technologies that can meet the different bandwidth and network requirements. As a result, Carrier Ethernet offers a wide range of modern services, including the following:

■ Business services such as Metro Ethernet enterprise connectivity services (E-Line, E-LAN)

■ Consumer and residential services (VoD, IPTV)

■ Business services over DOCSIS

■ Broadband mobility services

■ Wholesale services

Figure 6-3 shows a high-level view of the Carrier Ethernet architecture.

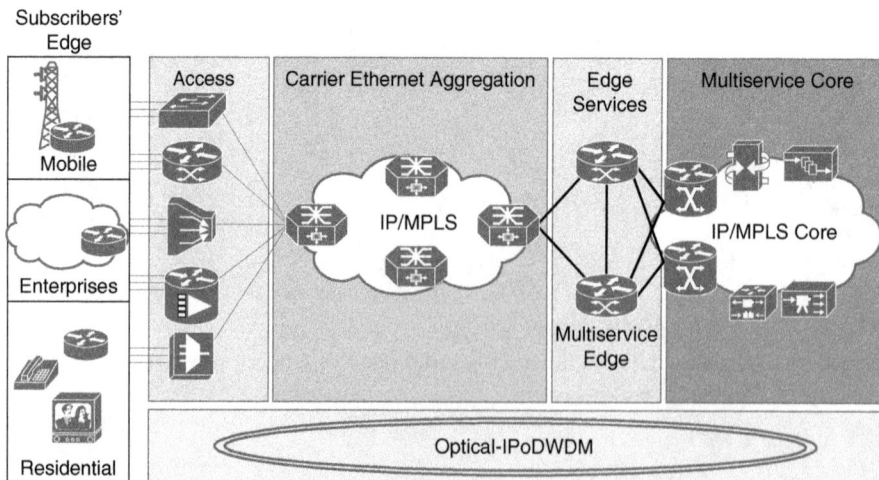

**Figure 6-3**  *Carrier Ethernet Architecture*

**Key Topic**

The following components are primary elements that comprise the architecture shown in Figure 6-3:

- **Access:** Provides access to residential and business customers over digital subscriber line (DSL), fiber, cable, or wireless.

- **Carrier Ethernet aggregation:** Aggregates the access network across a Carrier Ethernet network and provides interconnectivity to the IP/MPLS edge and IP/MPLS core.

- **IPoDWDM optical network:** Enables optical aggregation services with intelligent Ethernet multiplexing using MPLS/IP over dense wavelength-division multiplexing (IPoDWDM).

- **Multiservice edge:** Interface services with the IP/MPLS core; this is the provider edge for both residential and business subscriber services.

- **IP/MPLS core:** Provides scalable IP/MPLS routing in the core network.

Metro Ethernet services can be classified into the following two general categories:

- **Point to point (P2P):** A P2P Ethernet circuit is provisioned between two user network interfaces (UNIs).

- **Multipoint to multipoint (MP2MP):** A multipoint-to-multipoint Ethernet circuit is provisioned between two or more UNIs. If there are only two UNIs in the circuit, more UNIs can be provisioned to the same Ethernet virtual connection if required, which distinguishes this from the point-to-point type.

Metro Ethernet Forum (MEF), however, defines the two categories just discussed as two main types of Layer 2 Ethernet services:

- **Ethernet line service (E-Line):** Point to point, such as Virtual Private Wire Service (VPWS)

- **Ethernet LAN service (E-LAN):** Multipoint to multipoint such as Virtual Private LAN Services (VPLS)

There is also a P2P Metro Ethernet service known as E-Tree, as an abstraction of E-LAN, in which the spoke "leaves" can communicate with the hub or "root" location but not with each other, as shown in Figure 6-4. The typical application for E-Tree is in franchise operations. A great comparison for E-Tree is that of private VLANs, discussed in Chapter 10, "Security," as they have the same conceptual model.

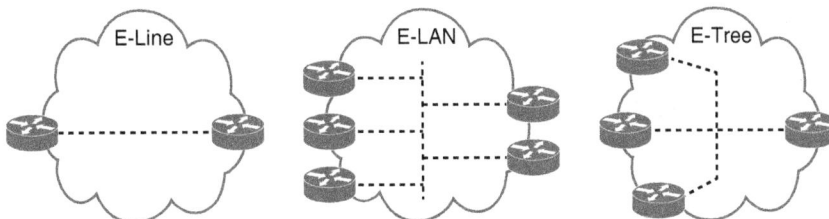

**Figure 6-4**  *Metro Ethernet Models*

Metro Ethernet services can be created by assigning values to a set of attributes grouped according to the following:

- **User network interface (UNI):** Represents a physical demarcation point between the connection and the subscriber. Also known as port-based ME.

- **Ethernet virtual connection (EVC):** Represents the association of two or more UNIs, which limits the exchange of service frames to UNIs within the EVC in Ethernet. When multiple EVCs could exist per single UNI and each EVC is distinguished by 802.1Q VLAN tag identification, this is known as VLAN based (EVPL).

Table 6-2 summarizes the relationship between the transport model and the Metro Ethernet service definitions.

**Key Topic**

**Table 6-2**   Metro Ethernet Transport Models

| Service Type | Port Based | VLAN Based |
|---|---|---|
| E-Line | Ethernet private line (EPL) | Ethernet virtual private line (EVPL) |
| E-LAN | Ethernet private LAN (EPLAN) | Ethernet virtual private LAN (EVPLAN) |

In addition to E-Line and E-LAN services, services are available for Layer 2 that are mainly to facilitate carrying legacy WAN transport over MPLS networks, such as the following:

- Frame Relay over MPLS (FRoMLS)

- ATM over MPLS (ATMoMPLS)

**NOTE**   Cisco's implementation of VPWS is known as Any Transport over MPLS (AToM) and delivers what is known as Ethernet over MPLS (EoMPLS). L2TPv3, however, can be used as an analogous service to AToM over any IP transport. Keep in mind here that Cisco's AToM includes more services (PPP, HDLC, FR, and ATM) as its "any transport" than EoMPLS. EoMPLS and Metro Ethernet services cannot provide this same level of flexibility from a connectivity and mixed circuits perspective.

## Virtual Private Wire Service Design Considerations

VPWS is based on the concept of **pseudowire** (PW) described in (RFC 3916), which can be defined as a connection between two edge nodes connecting two attachment circuits (ACs) to emulate a direct AC connection of a packet-switched network (PSN), as shown in Figure 6-5. The typical implementations of PWs are carried either over a native IP, such as L2TP, or over an MPLS network, such as EoMPLS. Within an MPLS-enabled network, the core visibility is limited to the transport MPLS labels (LSP) that each PW uses to communicate with the remote end, while the remote Provider Edge (**PE**) relies on the virtual circuit (VC) label to identify the intended AC per **pseudowire** (PW-to-AC mapping), as shown in Figure 6-5.

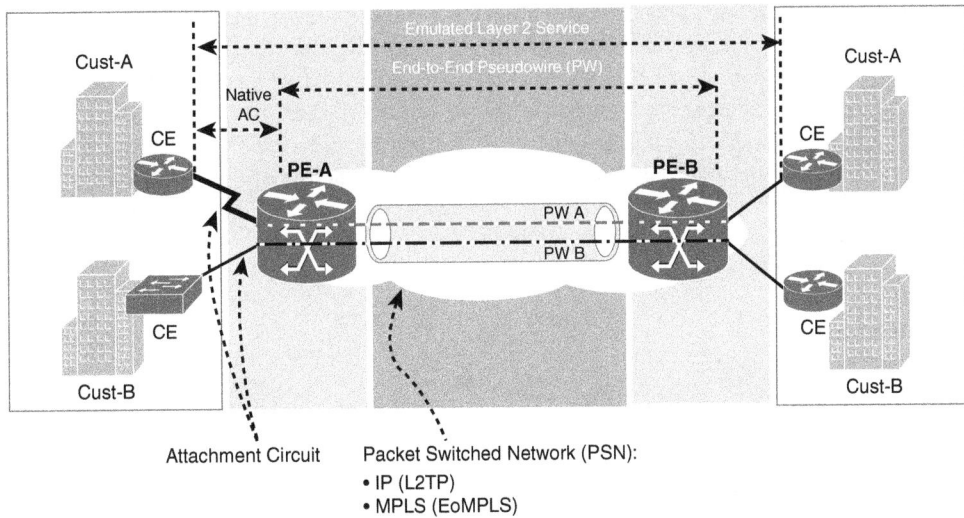

**Figure 6-5** *Virtual Private Wire Service*

As covered earlier in this chapter, the primary goal of service providers is to offer services that can satisfy their customer requirements; therefore, their core **public services network (PSN)** is the main business enabler. Pseudowire over PSN gives the service provider flexibility to handle customer connectivity using primary transport connectivity models, each of which can serve a different set of requirements:

■ **VLAN-based model:** As its name implies, it is a VLAN-based transport model. VLANs defined on the physical port can be switched to different remote PEs (for example, the scenario in Figure 6-6 where enterprise customers are seeking a service from the service provider to be analogous to their current WAN service [traditional Frame Relay or ATM], in which no routing and IP addressing changes are required).

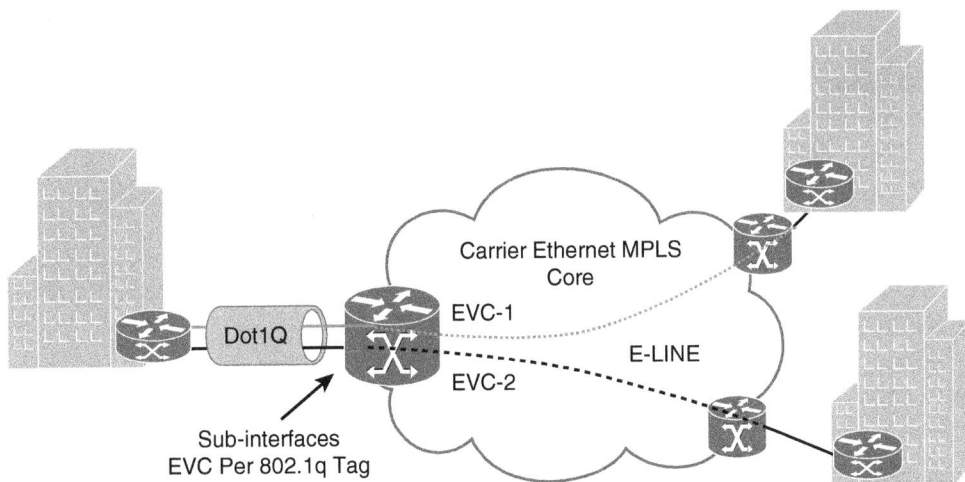

**Figure 6-6** *PW VLAN-Based Model*

■ **Port-based model:** This transport model tunnels all traffic entering the AC physical port and transports it to the targeted remote PE without any change. This model can be applicable in scenarios where the Layer 3 customer network is traditional WAN serial links and needs to be replaced with higher-speed P2P links. Another possible usage of this model is the case of a customer that needs to interconnect two LAN islands over the service provider's Layer 2 P2P PW (whether it is native IP core or MPLS enabled). Furthermore, this model is commonly used to provide Layer 2 data center interconnection to support the extending of multiple Layer 2 VLANs over the same link, as shown in Figure 6-7.

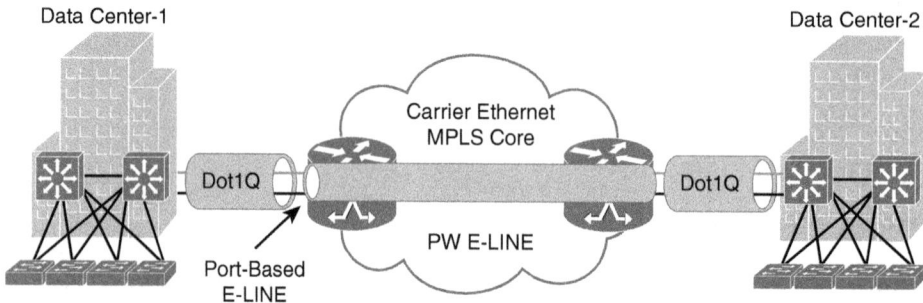

**Figure 6-7**  *PW Port-Based Model*

Furthermore, in some situations, the user access side may not always be provisioned as Ethernet. Instead, customers with legacy WAN media access such as ATM and Frame Relay might be provisioned with L2VPN services. With the VPWS (L2VPN MPLS based), carriers still can maintain customer access as it is and offer the same service capability. In addition, service providers can even maintain the connectivity between sites that have different media access methods, such as Frame Relay, ATM, and standard Ethernet. Table 6-3 compares and summarizes the primary L2 WAN media access technologies along with their transport modes over PW.

**Table 6-3**  PWs L2 Media Access

| L2 Technology | Transport Mode over PW |
|---|---|
| Frame Relay | Port based |
| | Per Data Link Connection Identifier (DLCI) |
| ATM | AAL5 protocol data units (PDUs) over PW cell relay over PW |
| Ethernet | Protocol based |
| | VLAN based |

## Virtual Private LAN Service Design Considerations

In today's market there is a significantly increasing demand for interconnected multiple sites over one common Layer 2 network. Along with that, there is an increasing demand to extend multiple modern data centers over Layer 2 transport. This supports the requirements of the next generation of data centers and the mobility of applications, such as virtual machine mobility and distributed workload, which supports the overall business continuity plan of the customer's business. This architecture is known as Virtual Private LAN Service (VPLS), or in Metro Ethernet Forum (MEF) terms, E-LAN.

With the VPLS architecture, end users see their network nodes as interconnected directly over a shared Ethernet LAN segment; this shared segment is an emulated LAN created by the VPLS domain (emulated switched network), as shown in Figure 6-8. In addition, it does support legacy media access methods that support Ethernet encapsulation, such as Frame Relay.

As with L3VPN, in a typical VPLS architecture service providers are normally required to update the PE or PEs, to which the customer needs to be connected to the same VPLS domain of a certain customer (equivalent to the concept of adding a new L3VPN customer to an existing VPN customer). In contrast, with the P2P Layer 2 VPN solutions, the service provider must reconfigure each of the peering PEs every time a change is required (typical scalability limitation of a P2P architecture). This optimized Layer 2 VPN architecture, along with its simplified operation compared to legacy Layer 2 WAN services, is a key business driver that attracts many Carrier Ethernet providers to adopt it and offer it to their customers quickly. As a result, VPLS in today's market is one of the most common reliable and mature Layer 2 WAN services.

**NOTE**  Although this section always refers to the network as a service provider network, in fact all the concepts are applicable, to a certain extent, to large-scale enterprise networks that have self-deployed Layer 2 VPN services, such as VPLS.

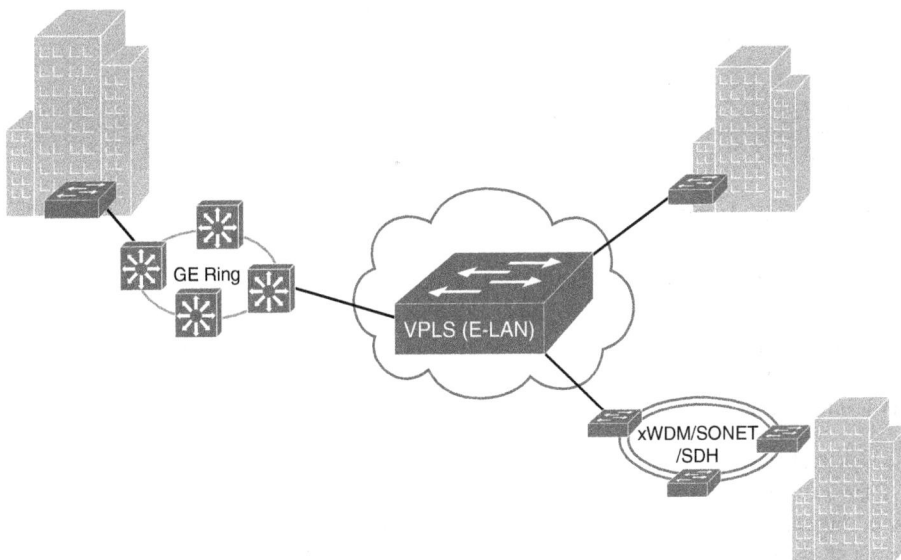

**Figure 6-8**  *VPLS Conceptual View*

## VPLS Architecture Building Blocks

In general, VPLS brings the standard IEEE 802.1 MAC address learning, flooding, and forwarding concept over the MPLS-based packet-switched network extended by the pseudo-wires (PWs), which commonly interconnect different customer LANs. In other words, VPLS is like the glue that interconnects the multiple LANs across one common packet-switched network over MPLS. Like any system architecture, multiple components construct the overall VPLS architecture. Each component performs specific functions to achieve the desired end-to-end result. In addition, as VPLS becomes a more mature architecture, there are multiple design models of the VPLS-induced architecture, with the components and their functions differing slightly based on the architecture used. This section discusses these different design models.

## VPLS Functional Components

Figure 6-9 illustrates the common functional components that construct a typical VPLS architecture. However, these components may vary slightly based on the VPLS model used, as covered in the following section:

- Network facing Provide Edge (**N-PE**): Provides VPLS termination/L3 services

- User facing Provider Edge **U-PE**: Provides customer UNI

- Customer Edge (**CE**): The customer device

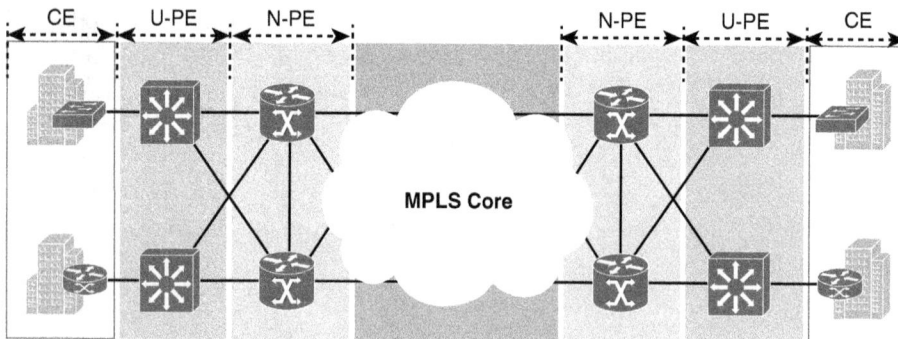

**Figure 6-9** *VPLS Architecture Functional Component*

## Virtual Switching Instance

The virtual switching instance (VSI) performs standard LAN (that is, Ethernet) bridging functions, including the participation in the learning and forwarding process based on MAC addresses and VLAN tags. Each VSI defined at each provider edge node usually handles the forwarding decisions of a single VPLS instance.

## VPLS Control Plane

To a large extent, the concept of the VPLS control plane is the same as VPWS (covered earlier in this chapter). However, the primary difference with VPLS is the need of full-mesh PWs among all the participating PEs in any given VPLS instance. With VPWS, in contrast, the typical scenario is P2P PWs between two sites or a few sites in a full-mesh or hub-and-spoke connectivity model. Consequently, with VPLS, the automation of the PE discovery and the PWs set up among the relevant PEs are important factors to consider. Otherwise, the design will encounter increased operational complexity when the network grows with many customers (VPLS instances) and remote sites (PEs per VPLS instance).

Nonetheless, as covered earlier, VPLS can be deployed by enterprises across their backbone to form an overlaid L2VPN connectivity (self-deployed VPLS). In such an environment, the number of PEs and VPLS instances can be limited compared to large-scale networks. Therefore, using a manual mechanism to set up the PWs, such as Label Distribution Protocol (LDP), can be a viable solution in this type of scenario. Furthermore, LDP is accustomed to more people and is a widely deployed control plane for L2VPN, even though it may not be the optimal scalable choice. However, upgrading this control plane may not be an option for those operators because of certain constraints and concerns, such as the operator cannot afford any service interruption to its existing VPLS customers or the existing PE nodes do not support BGP as the L2VPN control plane protocol. In these scenarios, network designers are forced to accommodate these design constraints in their design decisions.

In general, designing a typical VPLS with automated PE discovery and signaling can offer a more scalable and simple operational solution. This can be achieved by BGP as an L2VPN control plane for signaling and auto discovery (RFC 4761), or by a suboptimal solution that offers an optimized version of the only LDP control plane that retains the signaling to be based on LDP while automating the discovery of the participating PEs in each VPLS instance using BGP (as specified in RFC 6074).

## VPLS Design Models

One of the main drivers in the selection of one VPLS design model over others is the level of scalability possible. The primary factors that influence VPLS solution scalability are the supported number of VLANs (customers or end-user networks) and the number of PWs established between the PE nodes. Therefore, network designers should consider the answers to the following questions, which can help to build the foundation of the design decision during the planning phase:

- What is the goal of the solution (enterprise grade to interconnect multiple data centers of an enterprise; offer E-LAN services by a carrier Ethernet service provider grade)?

- Number of customers?

- Number of participating PEs?

- Scale of the interconnected sites in terms of the number of MAC addresses? (For instance, hosting cloud providers usually requires support for an extremely huge number of MACs.)

- PW termination between N-PEs versus between U-PEs along with the VSI?

- Any platform limitation, such as supporting VPLS VSI configuration or hardware resource capacity?

- Is traffic load-balancing/sharing for the multihomed sites required (active/active versus active/standby)?

- What is the targeted resiliency level? And how can it be achieved (for example, MPLS-TE FRR, redundant PWs)?

## Flat VPLS Design Model

The flat VPLS design model is the classic VPLS design model. It is the simplest design model among other VPLS models, with the least functional components, as shown in Figure 6-10.

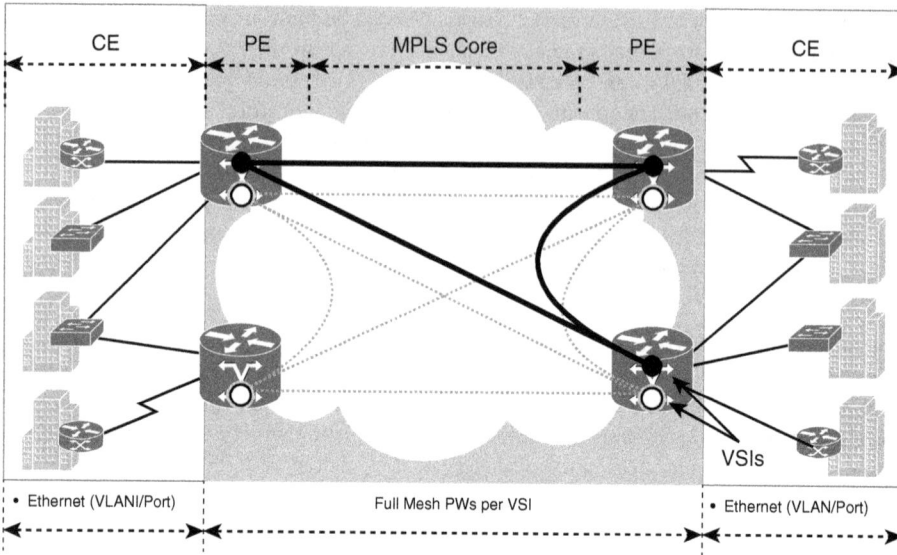

**Figure 6-10** *Flat VPLS*

With this design model, each PE usually hosts the VSIs of VPLS domains, along with a full mesh of PWs among the participating PEs per VPLS domain. Based on this, each PE will maintain a P2MP perspective of all other PEs per VPLS domain, and each PE will be seen as the root bridge of the PW's mesh to other PEs, as shown in Figure 6-11.

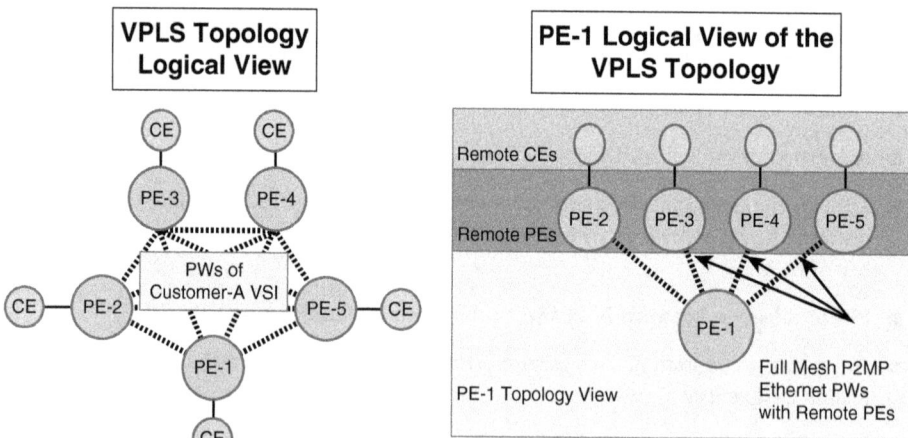

**Figure 6-11** *PE PW Topology View*

When PEs maintain a full mesh of PW connectivity, it eliminates the need to rely on or enable Spanning Tree Protocol (STP) across the PSN MPLS network. However, from the customer point of view, their STP (bridge protocol data units [BPDUs]) can be transparently carried over the emulated VPLS LAN between the different sites without impacting the service provider network. The split-horizon concept, however, is used with the model to block any potential loop of traffic received over one core/PE PW into another core/PE PW. Based on that, the following summarizes the characteristics of this VPLS design model:

- Limited scalability. The more PEs and VSIs there are, the more network hardware resources are consumed per PE associated with operational complexity:

  - Many PWs because of the nature of this model, where a full mesh of directed LDP sessions is required ($N \times (N - 1) / 2$ PWs required)

  - Potential signaling and packet replication overhead, when the number of PWs increases across multiple PEs to cover multiple remote sites (CEs) per customer VSI using LDP as the control plane protocol

  - Large amount of multicast replication, which may result in inefficient network bandwidth utilization (unless some mechanism is used to mitigate it such as Interior Gateway Management Protocol [IGMP] snooping with VPLS)

  - CPU overhead for replication

  - MAC distribution across the network limitations

  - Support limited number of customers/VLANs (maximum 4096)

- VLAN and port-level support (no **QinQ**)

- Support multihomed CEs in active-standby manner

- Suitable for simple and small deployments, such as a self-deployed enterprise VPLS solution

## Hierarchical VPLS Design Model

The hierarchical VPLS (H-VPLS) design model aims to optimize the high complexity of PW meshes and the limited scalability of the flat VPLS design model by introducing a hierarchical structure to the VPLS design. This hierarchical structure consists of hub-and-spoke and full-mesh networks, as shown in Figure 6-12.

The design model shown in Figure 6-12 provides a more structured multistage PW connectivity model, where the hub or core of fully meshed PWs in each PE node form a multipoint-to-multipoint forwarding relationship with all other PE nodes at the same level within the VPLS domain. At the hub-and-spoke level, each PE node needs a single PW to the hub PE node and can operate in a non-split-horizon mode. This allows inter-VC connectivity between PEs at the same level connected to the same hub PE node. In other words, the hub PE nodes perform PW aggregation of the edge node PWs.

Consequently, this hierarchical VPLS model offers significant reductions to the number of mesh PWs, as shown in Figure 6-13. The model also overcomes the scalability and performance efficiency limitations of the flat VPLS model to a great extent.

As shown in Figure 6-13, H-VPLS has some additional terms for its functional components:

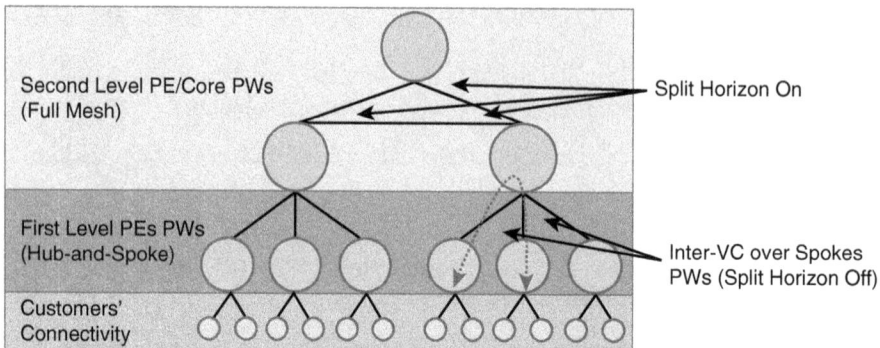

**Figure 6-12** *H-VPLS Structure*

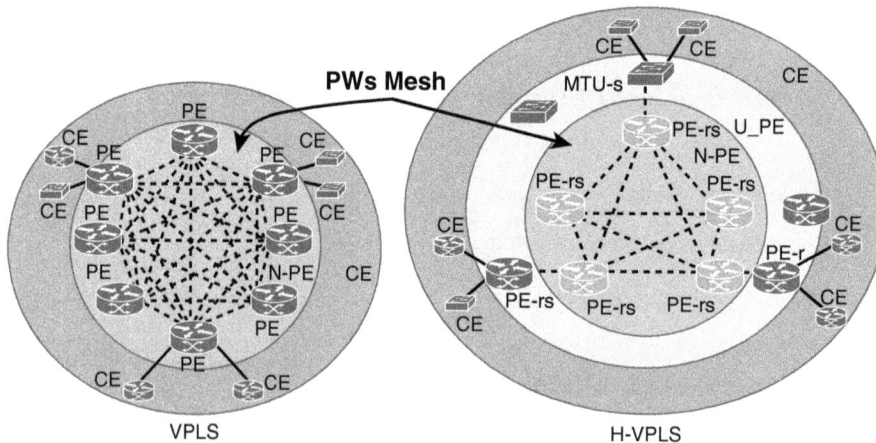

**Figure 6-13** *PE H-VPLS PW Aggregation*

- **MTU-s:** Multitenant unit switch capable of bridging (U-PE)

- **PE-r:** Nonbridging PE router

- **PE-rs:** Bridging- and routing-capable PE

## Provider Backbone Bridging with H-VPLS

The H-VPLS model offers an optimized version of the flat VPLS in terms of scalability and performance efficiency. However, there are still scalability and performance limitations with this design model for Carrier Ethernet, which offers services to a large number of customers with large-scale networks; in particular, large-scale modern data centers (virtualized and cloud-based data centers) with layer interconnect between them, where the service provider may need to carry millions of MAC addresses across their backbone. With H-VPLS, the following limitations are going to become constraints for the service provider from achieving this goal:

- In both flat and hierarchical VPLS models, each PE that performs MAC switching has to learn customer MAC addresses, which often leads to performance and scalability deficiencies.

- With the H-VPLS Ethernet access model, each access network is limited to about 4094 customer instances. With the MPLS access model, the larger the number of customers, the larger the number of PWs and VSIs to be handled by the N-PE (limited performance and scalability).

Therefore, the IEEE standard was developed to provide an optimized approach that can scale to a very large number of customer instances and MAC addresses (millions) without compromising network performance efficiency. This standard is called provider backbone edge (PBB). This section covers the high-level architecture of the PBB and how it can be integrated with VPLS/H-VPLS (when combined like this it is known as Provider Backbone Bridging with Virtual Private LAN Server [PBB-VPLS]) to enable service providers to meet the requirements of today's large-scale enterprise customers and cloud-based data center interconnect requirements.

### Provider Backbone Bridging Overview

PBB is an IEEE (802.1ah-2008) technology that defines an architecture based on a MAC tunneling mechanism, which enables service providers to build a large-scale Ethernet bridged network. The primary concept behind PBB is the hierarchical MAC address learning and forwarding in a similar manner to the MPLS label stacking concept, where the typical top label used for the transport within the SP core network and the bottom labels are relevant to the customer VPN or VCID in L2VPN. Instead, PBB uses multitier MAC address learning and forwarding in which the core PEs, also known as backbone core bridges (BCBs), communicate based on backbone MAC addresses of the core components. Customer MAC frames (C-MAC), however, are encapsulated (tunneled) at the edge of the bridged network into the backbone MAC, which enables Carrier Ethernet to scale a large number of C-MACs without impacting core components, because the communication (including broadcast, unicast, and multicast flooding BUM [broadcast, unknown destination address, multicast]) will be based only on the backbone MACs.

With PBB architecture, besides MAC tunneling, the network provides VLAN tunneling as well, at two stages. First, at the edge of the bridged network, where customer VLANs (C-VLAN) are tunneling using 802.1ad IEEE standard (QinQ), in this architecture referred to as service provider tag or S-VLAN. The second stage is where the S-VLANs are mapped into the PBB backbone VLAN (B-tag or B-VLAN).

Across the PBB bridged network, the end-to-end mapping and identification of each customer instance is based on the service instance ID (I-SID), which must be unique across the entire bridged network and is usually performed at the edge of the network and not within the backbone.

Furthermore, the flexibility of the PBB architecture offers multiple options to network designers to map between customer VLANs and the service interface, which is more flexible compared to the typical L2VPN VLAN and port-based forwarding modes discussed earlier in this chapter.

PBB with this hierarchal architecture can offer the following:

- Up to 224 service instances per bridge domain by defining a 24-bit service identification field (I-SID). In other words, 224 service instances per B-VLAN, also known as MAC-in-MAC.

- PBB provides the MAC address hiding (tunneling) technique from the SP core, which can significantly increase the solution scalability by confining customer MAC address learning to the edge and mapping customer MAC addresses to the backbone MAC addresses on the backbone edge bridges (BEB). (In this architecture, all the intelligence is on the BEBs, because they are responsible for translating frames to/from the new PBB format.)

# EVPN Design Model

Ethernet VPN (EVPN [Next Generation MPLS L2VPN]) technology is a next-generation Ethernet L2VPN (defined in RFC 7432), which uses the same principle of MPLS L3VPN, in which the BGP control plane for Ethernet segment and MAC address signaling and learning over the service provider MPLS core, in addition to being an access topology and VPN end-point discovery, will eliminate the need to use PWs (thereby avoiding its operational complexities and scalability limitations). In fact, the introduction of EVPN marks a significant milestone for the industry, because it aligns the well-understood technical and operational principles of IP VPNs to Ethernet services. Operators can now leverage their experience and the scalability characteristics inherent in IP VPNs to their Ethernet offerings. In addition, EVPN supports various L2VPN Ethernet over MPLS topologies, including Ethernet multipoint (E-LAN), Ethernet P2P (E-Line), and Ethernet rooted-multipoint (E-Tree).

## Business Drivers

One of the main drivers toward EVPN is the increased demand on distributed virtualized and cloud-based data centers, which commonly require a scalable and reliable stretched layer concavity among them. Recently, Data Center Interconnects (DCI) has become a leading application for Ethernet multipoint L2VPNs. Virtual machine (VM) mobility, storage clustering, and other data center services require nodes and servers in the same Layer 2 network to be extended across data centers over the WAN. Consequently, these trends and customer needs add new requirements for L2VPN operators to meet, such as the following:

- Flow-based load balancing (PE based and multipathing) across the PSN

- Fast convergence (avoid C-MAC flushing)

- Support of large-scale, virtualized data centers

In addition, with the increased demand and expectations from enterprise customers and cloud hosting providers, operators also need to consider a solution that offers the following:

- Flexible forwarding policies and topologies

- A scalable solution that supports a very large number of customers and MAC addresses

- Operational simplicity

## EVPN Business Strengths

Based on the market trends and new requirements and expectations of modern enterprise customers, EVPN can offer the following advantages for Carrier Ethernet:

**Key Topic**

- Ability to meet very strict customer SLA requirements for business continuity by offering the ability of all-active (per-flow) access load balancing and fast convergence (at different levels: link, node, and MAC moves), in addition to optimizing the customer ROI of their multiple links (multihomed).

- Scalable design that avoids any PW limitations and supports extremely large numbers of customers and MAC addresses.

- Simplified operations, such as EVPN. Uses the same L3VPN address learning and forwarding principle where no additional staff is required to manage or maintain the network because the same control plane (MP-BGP) is used for both. Furthermore, with the autodiscovery of PEs adding new nodes, it becomes very simplified and eliminates all the complexities of PWs that network operators need to deal with.

- Topology flexibility because EVPN supports the primary Ethernet services topologies (E-LAN, E-Line, E-Tree, and VLAN-aware bundling).

- Seamless interworking between TRILL, 802.1Qaq, and 802.1Qbp (draft-ietf-l2vpn- trill-evpn-01).

- EVPN offers investment protection to service providers because it is open standard technology and supported by multiple vendors (RFC 7432).

From the enterprise (customer) point of view, the greatest advantage of EVPN is that the customer can have a flexible and reliable L2 WAN/MAN service that meets their new requirements, such as large-scale next-generation data center interconnects (virtualized and cloud based).

With a solution like that of EVPN and VXLAN, for example, a network design can allow for Layer 2 mobility across disparate locations, like that of multiple data centers, while limiting the inherent drawbacks of the legacy Layer 2 protocols such as STP and First Hop Redundancy Protocol (FHRP). With such a solution, the network design can leverage the Layer 3 architecture to provide a scalable Layer 2 overlay that mitigates fate sharing for the traditional stretching of Layer 2 (with STP and FHRP) in a dual data center design.

## Provider Backbone Bridging with Ethernet VPN

By combining the principle of MAC tunneling and hiding of Provider Backbone Bridging (PBB) (802.1ah) with the principle of MP-BGP-based MAC learning and PE discovery of the Ethernet VPN (EVPN), network designers can achieve the optimal L2VPN architecture, which supports an extremely large number of customers and MACs. At the same time, it can reduce control plane overhead and scale to a larger extent than EVPN alone. In addition, it will help operators to offer L2VPN with faster convergence time. As a result, combining PBB with EVPN (PBB-EVPN) can achieve a superior solution for modern DCIs (virtualized and cloud based) and next-generation E-LAN offerings. The drawbacks of MAC in MAC tunneling like this are the increase in encapsulation requirements, the substantial increase in the amount of configuration needed, and the increased design complexity.

In other words, both LDP and BGP are valid and proven choices. The selection of one over the other should be based on the target environment (VPWS versus VPLS) and the scale of the network (Service Provider (SP) versus enterprise), considering design constraints.

For instance, if the SP is offering both MPLS L2 and L3VPN services and there are many clients who have full mesh or several sites to be connected using either VPLS or VPWS, the use of the BGP for L2VPN signaling with autodiscovery can be considered a scalable solution (especially if the same PEs are involved in the Layer 3 VPN service and Layer 2VPN service). Consequently, the same BGP sessions can be used for both the L3VPN network layer reachability information (NLRI) and the L2VPN NLRI, and the same BGP architecture can be used (for example, BGP RRs). This solution also simplifies the design and operations of the network because there is no need to maintain multiple separate protocols (BGP and LDP) for the control plane of L2VPN and L3VPN services.

## EVPN Architecture Components

The architecture of EVPN has three primary foundational building blocks, as described in more detail in the sections that follow:

- EVPN instance (EVI)
- Ethernet segment (ES)
- EVPN BGP routes and extended communities

### EVPN Instance

An EVI represents an L2VPN instance on a PE node. Similar to the VRF in MPLS L3VPN, import and export **Route Targets** (RTs) are allocated to each EVI. In addition, a bridge domain (BD) is associated with each EVI. Mapping traffic to the bridge domain, however, is dependent on the multiplexing behavior of the user to network interface (UNI). Typically, any given EVI can include one or more BDs based on the PE's service interface deployment type, as summarized in Figure 6-14.

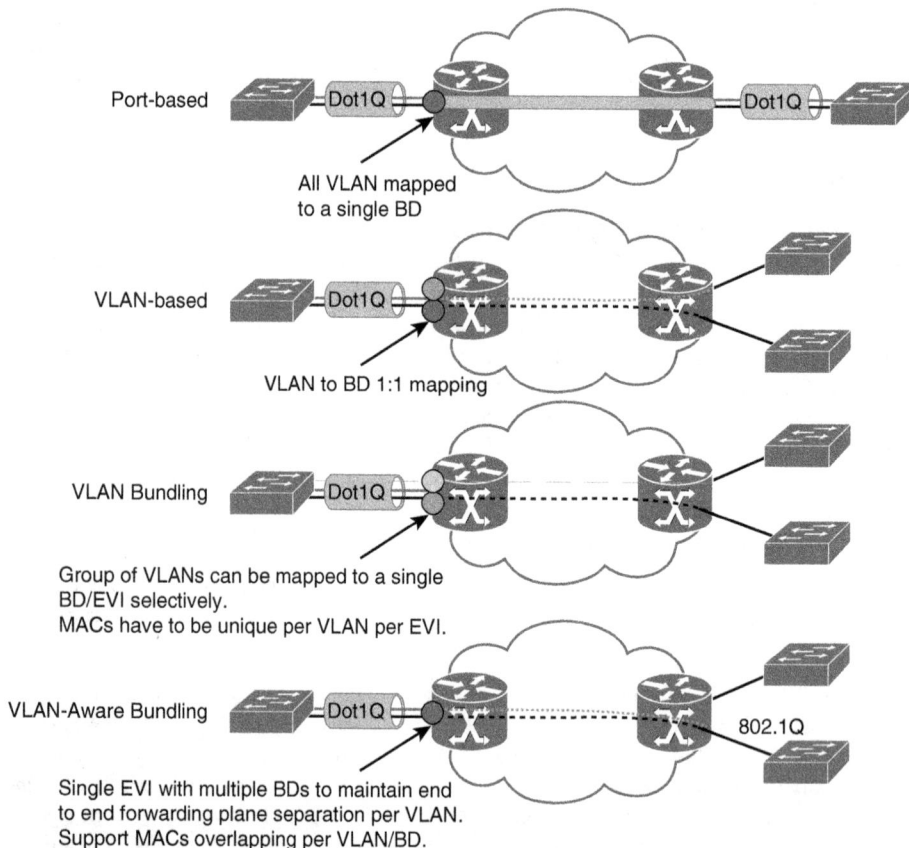

**Figure 6-14** *EVPN EVI Models*

For instance, you can use the VLAN bundling model in environments that require multiple VLANs to be carried transparently across the EVPN cloud between two or more sites. In contrast, the VLAN-aware bundling is more feasible for multitenant data center environments where multiple VLANs have to be carried over the DCI over a single EVI with multiple DBs (VLAN to BD 1:1 mapping), because the overlapping of tenant MAC addresses across different VLANs is supported.

### Ethernet Segment

*Ethernet segment (ES)* refers to a site that is connected to one or more PEs. (An ES can be either a single CE or an entire network.) Typically, each network segment is assigned to a single unique identifier, commonly referred to as an *Ethernet segment identifier (ESI)*, which can eliminate any STP type of protocol used for loop prevention and normally will add limitations to the design, especially for the multihomed CE scenarios. Based on this identifier, EVPN can provide access redundancy that offers georedundancy and multihoming, where a site (CE or entire network with multiple CEs) can be attached to one or multiple PEs that connect to the same provider core using multiple combinations of the CE-PE connectivity. Figure 6-15 illustrates the various EVPN-supported access connectivity models:

- Single-homed device (SHD)

- Multihomed device (MHD) with Multi-Chassis Link Aggregation (mLAG)

- Single-homed network (SHN)

- Multihomed network (MHN)

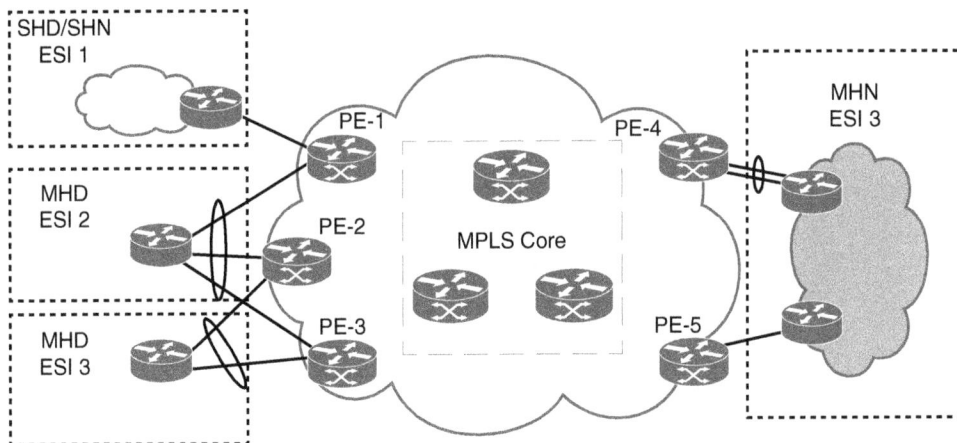

**Figure 6-15**  *EVPN Access Connectivity Models*

**NOTE**  In EVPN, the PE advertises in BGP a split-horizon label (ESI MPLS label) associated with each multihomed ES to prevent flooded traffic from echoing back to a multihomed Ethernet segment.

### EVPN BGP Routes and Extended Communities

Because both EVPN and PBB-EVPN PEs signal and learn MAC addresses over the core via BGP, a new MP-BGP address family and BGP extended communities were created to allow PE routers to advertise and learn prefixes that identify MAC addresses and Ethernet segments over the network. Therefore, this control plane learning significantly enhanced E-LAN capability over EVPN architecture to address the VPLS (data plane learning) shortcomings discussed earlier, such as supporting multihoming with per-flow load balancing. In addition, the sequence number BGP extended community attribute offers optimized MAC mobility control plane learning and forwarding to the EVPN solution, because it facilitates and enhances the triggering of the advertising PE to withdraw its MAC advertisement in case of a MAC move (mobility) to a new network with a different ESI. This can offer a reliable solution for customers of a DCI with a stretched Ethernet segment.

Furthermore, by using BGP as a common VPN control plane, providers can now leverage their operational experience and the scalability characteristics inherent to IP VPNs for their Ethernet offerings.

EVPN and PBB-EVPN support two primary load-balancing models for multihomed devices (CEs):

- **All-active load balancing (per-flow LB):** This connectivity model, shown in Figure 6-16, supports multihomed devices with per-flow load balancing. Access devices are connected over a single mLAG to multiple PEs, and traffic of the same VLAN can be sent and received from all the PEs by the same Ethernet segment.

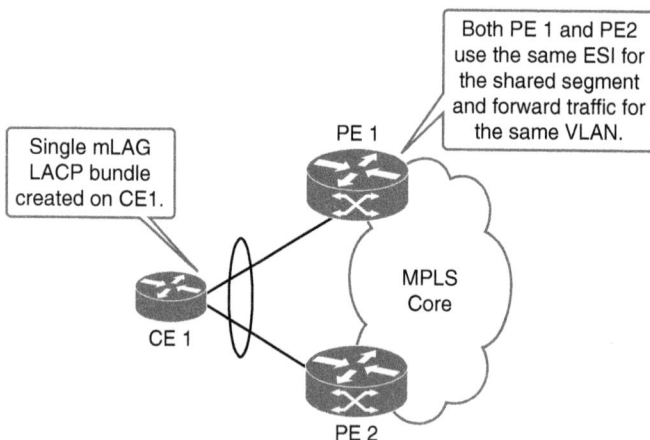

**Figure 6-16** *All-Active Load-Balancing EVPN*

- **Single-active load balancing (per-service LB):** This connectivity model, shown in Figure 6-17, supports multihomed devices with a per-VLAN load-balancing model. The access devices in this model connect over separate Ethernet bundles to multiple PEs. PE routers, in turn, automatically perform service carving to divide VLAN forwarding responsibilities across the PEs in the Ethernet segment. The access device learns, via the data plane, which Ethernet bundle to use for a given VLAN. This is unlike the traditional VPLS model, where manual administration is required to compensate for the lack of access, autodiscovery, and automatic service carving mechanisms.

**NOTE**  RFC 6391 describes a mechanism that introduces a flow label that allows P routers to distribute flows within a PW.

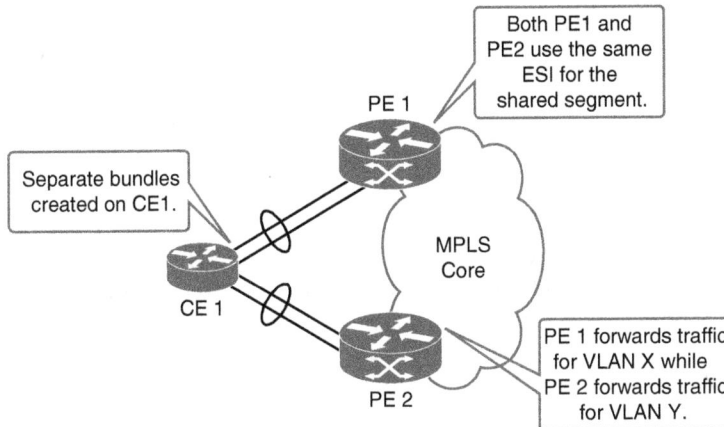

**Figure 6-17**  *Per VLAN Load-Balancing EVPN*

## VXLAN Design Model

The IETF standard VXLAN was created and commonly used as a data center fabric overlay protocol to facilitate and optimize modern data center performance and agility in response to virtualized and cloud-based data center requirements. In VXLAN, VMs can be moved anywhere within the DC, because of abstraction between the location of the VM/application within the DC and the physical access/leaf switch and port VM mobility. VXLAN adds simplicity, transparency to applications, and flexibility to modern Clos fabric architectures; in addition, it enhances its ability to scale an extremely large number of VMs/MACs, especially when an MP-BGP EVPN control plane is used for VXLAN.

BGP as a control plane, together with VXLAN headend replication, can optimize VXLAN behavior with regard to multi-destination traffic for MAC learning (broadcast and unknown unicast), especially as the behavior has become like Layer 3 routing (MAC learning is based on MAC address advertisement). Flooding will be eliminated unless there is a host that has not advertised or learned MAC, and only in this case will the typical Layer 2 flooding be used. At the time of this writing, VXLAN began to be considered a DCI solution, especially the unicast mode of VXLAN, because it can be more controllable at the DCI edge (can be policed more deterministically than multicast mode). This is applicable to any of the VXLAN design models (host based or network based), as shown in Figure 6-18.

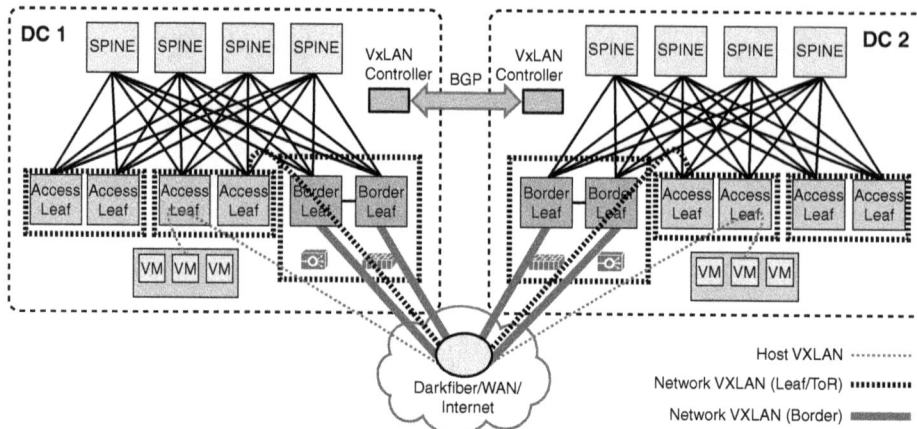

**Figure 6-18**  *VXLAN-Based DCI*

> **NOTE**  The placement of the functionality of a VXLAN controller with BGP (EVPN) as a control plane (where all the hosts' MAC addresses are hosted and updated) can vary from vendor to vendor and from solution to solution. For example, it can be placed at the spine nodes of the data center architecture, as well as at the virtualized (software-based) controller. It can also be used as a BGP route reflector to exchange Virtual Tunnel End Point (VTEP) list information between multiple VSMs.

> **NOTE**  Technically, you can use all the VXLAN models discussed earlier (host and network based) at the same time. However, from a design point of view, this can add design complexity and increase operational complexity. Therefore, it is always recommended to keep it simple and start with one model that can achieve the desired goal.

Currently, one of the challenges of using VXLAN as a DCI solution is the behavior of the distributed controller across the two DCs following a DCI failure event (split-brain) and after DCI link recovery. In addition, considerations about the VXLAN gateway location (localized versus at a remote location) can impact the design. However, at the time of this writing, advancements in VXLAN are increasing rapidly. Therefore, it is quite possible that new features and capabilities will be available in the near future to make VXLAN a more robust DCI solution.

The scope of each DC VXLAN can be contained within each DC locally, while the terminated VXLAN VTEPs at the DC/SP border nodes can be extended across the SP network that has EVPN enabled. Each VXLAN VNI is mapped to a unique EVPN instance (VNI-to-EVI 1:1 mapping) to maintain end-to-end path separation per VXLAN. This design offers a scalable, integrated DCI solution that takes advantage of both solutions in scenarios where EVPN is already in place as a DCI and the DC operator is deploying VXLAN within each DC, as shown in Figure 6-19.

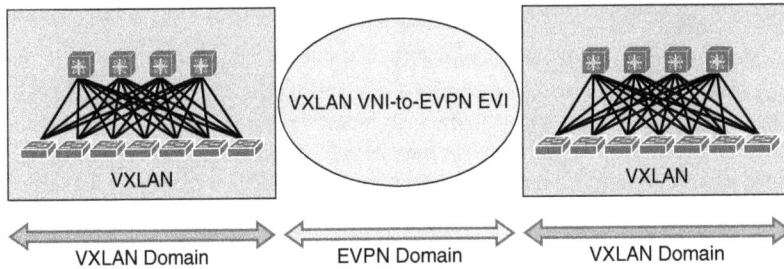

**Figure 6-19**  *VXLAN over EVPN*

Table 6-4 provides a summarized comparison of the different VPLS design models discussed in this chapter.

**Table 6-4**  Layer 2 Access Solution Comparison

|  | VPLS | H-VPLS | PBB-VPLS | EVPN | PBB-EVPN |
|---|---|---|---|---|---|
| Scalability | Very limited | Limited | Highly scalable | Scalable | Highly scalable |
| Control plane protocol | BGP/LDP | BGP/LDP | BGP/LDP | BGP | BGP |
| Network core control plane complexity | Very high | High | Moderate | High | Low |
| Flow-based load balancing (CE-PE) | No | No | No | Yes | Yes |
| Flow-based multipathing in the core | Yes | Yes | Yes | Yes | Yes |
| Operational complexity | High* | High* | High | Low | Moderate |
| Service interface VLAN-aware bundling | No | No | No | Yes | Yes |
| Loop prevention with multihomed CEs | STP | STP | STP | Split-horizon label (ESI MPLS label) | Split-horizon label (ESI MPLS label) |
| Fast convergence with local repairer | No | No | No | Yes | Yes |
| MAC mobility | Yes | Yes | Yes | Yes | Yes |
| Optimized control plane learning of MAC mobility | No | No | No | Yes, by the sequence number attribute | Yes, by the sequence number attribute |
| Targeted DCI solution** | Small (for example, enterprise controlled) | Medium | Very large | Large | Very large |

*If BGP is used as the control plane for VPLS, operational complexity will be reduced.

**What determines small, medium, and large DCI solutions is the number of interconnected sites per customer, scale of the VMs/MACs, and the number of customers; therefore, the suggestion here can be considered as generic and not absolute.

## Summary

This chapter covered the various transport technologies design models, protocols, and approaches, along with the characteristics of each. All these design options and protocols are technically valid and proven solutions and still in use by many operators today. However, as a network designer, you must evaluate the scenario that you are dealing with, ideally using the top-down design approach, where business goals and requirements are at the top, followed by the application requirements that should collectively drive the functional and technical requirements.

For instance, if an enterprise needs a basic self-deployed Layer 2 DCI solution between three distributed data centers with a future plan to add a fourth data center within two years, a flat VPLS solution can be cost-effective and simple to deploy and manage, in addition to scalable enough for this particular scenario. By contrast, if a service provider is already running flat VPLS and experiencing high operational complexity and scalability challenges and the SP is interested in a solution that supports its future expansion plans with regard to the number of L2VPN customers while minimizing the current operational complexity, then H-VPLS with BGP signaling, H-VPLS with PBB, EVPN, or EVPN PBB are possible solutions here. More detail gathering would be required to narrow down the selection. For example, is this operator offering MPLS L3VPN? Does it plan to offer multihoming to the L2VPN customers with active-active forwarding? If the answer to any of these questions is yes, EVPN (with or without PBB) can be an optimal solution. Because if the SP is offering L3VPN, that means the same control plane (BGP) can be used for both MPLS VPN services (simplifies operational complexity and offers a more scalable solution).

Adding PBB to this solution will optimize its scalability to a large extent, especially if this operator provides L2VPN connectivity to cloud-based data centers where a large number of virtual machine MAC addresses is expected to be carried over the L2VPN cloud. If multihoming with active-active forwarding is required, EVPN here will be a business enabler, along with optimized scalability and simplified operation as compared to the existing flat VPLS. However, there might be some design constraints here; for example, if the current network nodes do not support EVPN and the business is not allocating any budget to perform any hardware or software upgrade; or if this provider has existing an interprovider L2VPN link with a global Carrier Ethernet, to extend its L2VPN connectivity for some large enterprise customers with international sites, and this global Carrier Ethernet does not support EVPN.

In both situations, the network designer is forced to look into other suitable design alternatives such as H-VPLS. Again, the design requirements and constraints must drive the decision for which solution is the suitable or optimal one by looking at the bigger picture and not focusing only on the technical characteristics of the design option or protocol.

## Reference

- Al-shawi, Marwan, *CCDE Study Guide* (Cisco Press, 2015)

# Exam Preparation Tasks

As mentioned in the section "How to Use This Book" in the Introduction, you have a couple of choices for exam preparation: the exercises here, Chapter 18, "Final Preparation," and the exam simulation questions in the Pearson Test Prep Software Online.

## Review All Key Topics

Review the most important topics in this chapter, noted with the Key Topic icon in the outer margin of the page. Table 6-5 lists a reference of these key topics and the page numbers on which each is found.

**Key Topic**

**Table 6-5**  Key Topics for Chapter 6

| Key Topic Element | Description | Page Number |
|---|---|---|
| List | Carrier Ethernet architecture primary elements | 123 |
| Table 6-2 | Metro Ethernet Transport Models | 124 |
| Table 6-3 | PWs L2 Media Access | 126 |
| List | EVPN business strengths | 134 |

## Complete Tables and Lists from Memory

Print a copy of Appendix D, "Memory Tables" (found on the companion website), or at least the section for this chapter, and complete the tables and lists from memory. Appendix E, "Memory Tables Answer Key," also on the companion website, includes completed tables and lists to check your work.

## Define Key Terms

There are no key terms for this chapter.

**8.** Which of the following is the primary design concern for a Looped Triangle Topology?

    **a.** Introduces a single point of failure to the design if one distribution switch or uplink fails.

    **b.** A significant amount of access layer traffic might cross the interswitch link to reach the active FHRP.

    **c.** Inability to extend the same VLANs over more than a pair of access switches.

    **d.** STP limits the ability to utilize all the available uplinks within a VLAN or STP instance.

**9.** Which of the following is the primary design concern for a Loop-Free Inverted U Topology?

    **a.** Introduces a single point of failure to the design if one distribution switch or uplink fails.

    **b.** A significant amount of access layer traffic might cross the interswitch link to reach the active FHRP.

    **c.** Inability to extend the same VLANs over more than a pair of access switches.

    **d.** STP limits the ability to utilize all the available uplinks within a VLAN or STP instance.

**10.** Which of the following is the primary design concern for a Looped Square Topology?

    **a.** Introduces a single point of failure to the design if one distribution switch or uplink fails.

    **b.** A significant amount of access layer traffic might cross the interswitch link to reach the active FHRP.

    **c.** Inability to extend the same VLANs over more than a pair of access switches.

    **d.** STP limits the ability to utilize all the available uplinks within a VLAN or STP instance.

**11.** Which of the following is the primary design concern for a Loop-Free U Topology?

    **a.** Introduces a single point of failure to the design if one distribution switch or uplink fails.

    **b.** A significant amount of access layer traffic might cross the interswitch link to reach the active FHRP.

    **c.** Inability to extend the same VLANs over more than a pair of access switches.

    **d.** STP limits the ability to utilize all the available uplinks within a VLAN or STP instance.

## Foundation Topics

# Layer 2 Core Technologies

For a network designer to properly design a network, they need to know, understand, and be able to properly make design decisions with the following list of core Layer 2 technologies:

- Spanning Tree Protocol (STP)

- Virtual local-area networks (VLANs)

- Trunking

- Link aggregation

- Multichassis link aggregation (mLAG)

- First Hop Redundancy Protocol (FHRP)

There are more Layer 2 technologies than this core list. We will cover a number of them in subsequent chapters.

## Spanning Tree Protocol

As a Layer 2 network control protocol, the **Spanning Tree Protocol (STP)** is considered the most proven and commonly used control protocol in classical Layer 2 switched network environments, which include multiple redundant Layer 2 links that can generate loops. The basic function of STP is to prevent Layer 2 bridge loops by blocking the redundant L2 interface to a level that can provide a loop-free topology. There are multiple flavors or versions of STP. The following are the most commonly deployed versions:

- **802.1D:** The traditional STP implementation.

- **802.1W:** Rapid STP (RSTP) supports large-scale implementations with enhanced convergence time.

- **802.1S: Multiple Spanning Tree (MST)** permits very large-scale STP implementations. MST is a spanning tree protocol that reduces the total number of spanning-tree instances that match the physical topology of the network, by aggregating multiple VLANs into the same STP instance, which directly reduces the CPU load.

In addition, there are some features and enhancements to STP that can optimize the operation and design of STP behavior in a classical Layer 2 environment. The following are the primary STP features:

- **Loop Guard:** Prevents the alternate or root port from being elected unless bridge protocol data units (BPDUs) are present

- **Root Guard:** Prevents external or downstream switches from becoming the root

- **BPDU Guard:** Disables a PortFast-enabled port if a BPDU is received

- **BPDU Filter:** Prevents sending or receiving BPDUs on PortFast-enabled ports

Figure 7-1 briefly highlights the most appropriate place where these features should be applied in a Layer 2 STP-based environment.

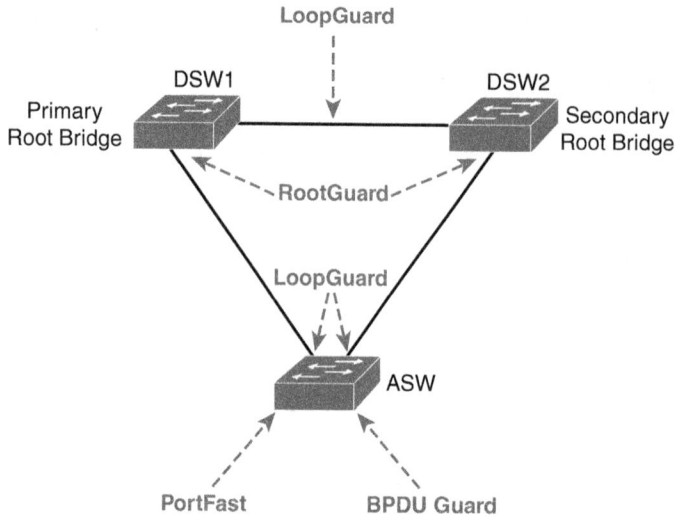

**Figure 7-1** *STP Features Locations*

> **NOTE** Cisco has developed enhanced versions of STP. It has incorporated a number of the preceding features into them using different versions of STP that provide faster convergence and increased scalability, such as Per-VLAN Spanning Tree Plus (PVST+) and Rapid PVST+.

## VLANs and Trunking

A Layer 2 **virtual local-area network (VLAN)** is considered as a type of network virtualization technique that provides logical separation with broadcast domains and policy control implementation. In addition, VLANs offer a degree of fault isolation at Layer 2 that can contribute to the optimization of network performance, stability, and manageability. Trunking, however, refers to the protocols that enable the network to extend VLANs across Layer 2 uplinks between different nodes by providing the ability to carry multiple VLANs over a single physical link.

From a design best practices perspective, VLANs should not span multiple access switches; however, this is only a general recommendation. For example, some designs dictate that VLANs must span multiple access switches to meet certain application requirements. Consequently, understanding the different Layer 2 topologies and the impact of spanning VLANs across multiple switches is a key aspect for Layer 2 design. Keep in mind that although spanning VLANs might not always be the best option from a network design perspective, that does not mean it's the incorrect decision to make, as there are numerous designs that require Layer 2 spanning to be in place. An example of this is the requirement in an architecture with multiple data centers where the network must allow for virtual machine mobility between the different data centers while maintaining the same IP address.

## Link Aggregation

The concept of *link aggregation* refers to the industry standard IEEE 802.3ad, in which multiple physical links can be grouped together to form a single logical link. This concept offers a cost-effective solution by increasing cumulative bandwidth without requiring any hardware upgrades. The IEEE 802.3ad **Link Aggregation Control Protocol (LACP)** offers several other benefits, including the following:

■ An industry standard protocol that enables interoperability of multivendor network devices

■ The optimization of network performance in a cost-effective manner by increasing link capacity without changing any physical connections or requiring hardware upgrades

■ Eliminates single points of failure and enhances link-level reliability and resiliency

Although link aggregation is a simple and reasonable mechanism to increase bandwidth capacity between network nodes, each individual flow will be limited to the speed of the utilized member link by that flow, based on the load-balancing hashing algorithm used.

**NOTE**   In addition to LACP (the industry standard link aggregation control protocol), Cisco has developed a proprietary link aggregation protocol called Port Aggregation Protocol (PAgP). Both protocols have different operational modes, which the network designer must be aware of.

There are two primary types of link aggregation connectivity models:

■ **Single-chassis link aggregation:** This is the typical link aggregation type of connectivity that connects two network nodes in a point-to-point manner.

■ **Multichassis link aggregation (mLAG):** This type of link aggregation connectivity is most commonly used when the upstream switches (typically two) are deployed in "switch clustering" mode. This connectivity model offers a higher level of link and path resiliency than the single-chassis link aggregation.

Figure 7-2 illustrates these two link aggregation connectivity models.

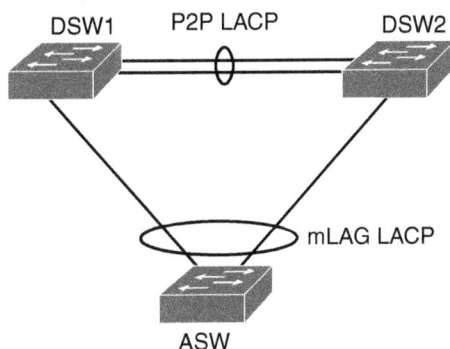

**Figure 7-2**  *P2P and mLAG LACP Link Aggregation*

**NOTE**  Cisco has created a "switch-clustering" solution called Virtual Switching System (VSS), which solves the Spanning Tree Protocol looping problem by converting the distribution switching pair into a logical single switch. From a design perspective, VSS removes the need for both STP and FHRP in the Layer 2 design.

## First Hop Redundancy Protocol and Spanning Tree

First-hop Layer 3 routing redundancy is designed to offer transparent failover capabilities at the first-hop Layer 3 IP gateways, where two or more Layer 3 devices work together in a group to represent one virtual Layer 3 gateway. The **First Hop Redundancy Protocol (FHRP)** options include **Hot Standby Router Protocol (HSRP), Virtual Router Redundancy Protocol (VRRP)**, and **Gateway Load Balancing Protocol (GLBP)**. These are the primary and most commonly used protocols to provide a resilient default gateway service for endpoints and hosts.

Understanding what drives the need for a FHRP is critical to know when to leverage one. Below is a list of characteristics for all FHRP options:

- Where is the current default gateway? Is the location suitable for FHRP? If not, can the default gateway be moved?

- Provide routing redundancy for access layer

- Independent of routing protocols

- Capable of providing subsecond failover

Table 7-2 summarizes and compares the main capabilities and functions of these different FHRP protocols.

**Table 7-2**  Business Priorities, Drivers, Outcomes, and Capabilities Relationship Mapping

|  | HSRP | VRRP | GLBP |
|---|---|---|---|
| Standard | Cisco proprietary | IEEE | Cisco proprietary |
| IPv6 support | Yes (v2) | Yes (v3) | Yes |
| Authentication | Yes | Yes | Yes |
| Load sharing techniques | Multiple HSRP groups per interface | Multiple VRRP groups per interface | Natively supported |
| Default hello timer | 3 | 1 | 3 |
| Virtual IP (VIP) | Different from interface IP | Can be different from or the same as the interface IP of the master router | Different from the interface IP |
| Bidirectional Forwarding Detection (BFD) support for sub second convergence | Yes | Yes | Yes |

**Key Topic**

One of the typical scenarios in classical hierarchical networks is when FHRP works in conjunction with STP to provide redundant Layer 3 gateway services. This is most often the first time a network designer will run into designing multiple technologies that impact each other directly. The following design model (depicted in Figure 7-3) is considered one of the common design models that has a proven ability to provide the most resilient design when FHRP is applied to an STP-based Layer 2 network (such as VRRP or HSRP). This design model has the following characteristics:

- The interswitch link between the distribution switches is configured as a Layer 3 link.

- No VLAN spanning across switches.

- The STP root bridge is aligned with the active FHRP instance for each VLAN.

- Uplinks from the access layer to the distribution layer are both forwarding from STP point of view.

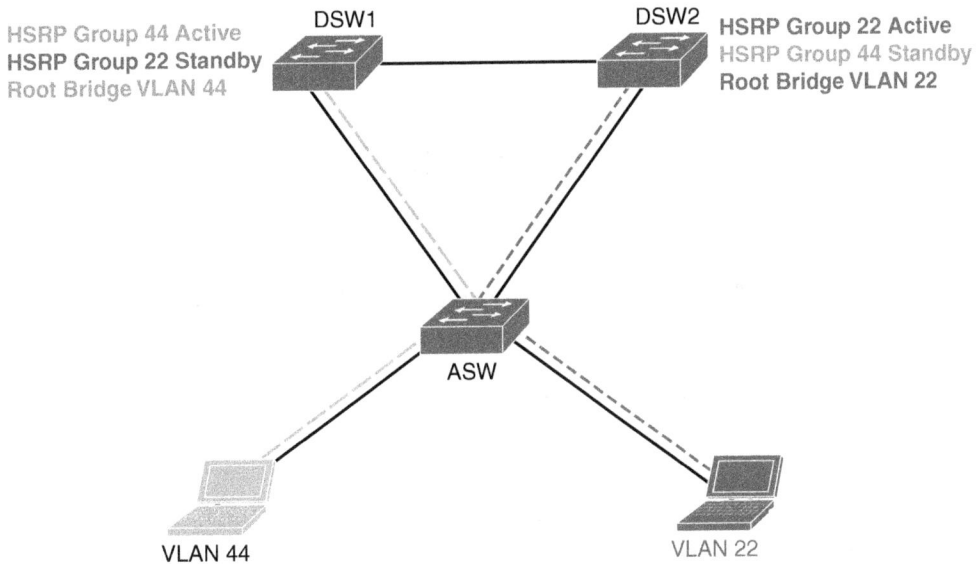

HSRP Group 44 Active
HSRP Group 22 Standby
Root Bridge VLAN 44

DSW1   DSW2

HSRP Group 22 Active
HSRP Group 44 Standby
Root Bridge VLAN 22

ASW

VLAN 44   VLAN 22

**Figure 7-3**  *FHRP and STP Common Design Model*

**NOTE**  In the design illustrated in Figure 7-3, when GLBP is used as the FHRP, it is going to be less deterministic compared to HSRP or VRRP because the distribution of Address Resolution Protocol (ARP) responses is going to be random.

If a network designer doesn't properly align the STP root bridge with the active FHRP instance for the corresponding VLAN, there will be a suboptimal impact to traffic traversing that VLAN. An example of this is shown in Figure 7-4.

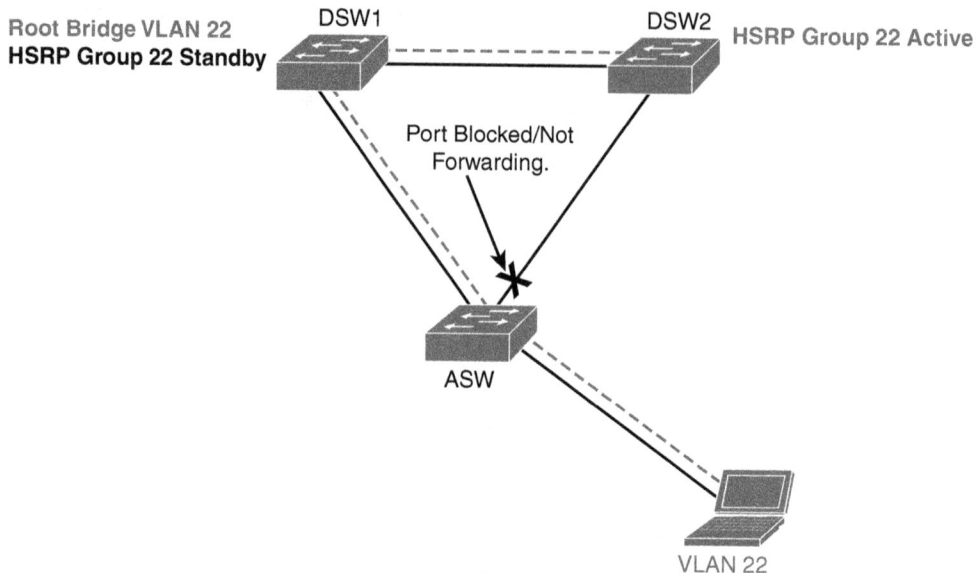

**Figure 7-4** *FHRP and STP Design Alignment*

While the design in Figure 7-4 functionally works, it is not optimal and in most cases would be called a poor design. This is because FHRP and STP have not been properly aligned from a design perspective, causing traffic sourced from the client device on VLAN 22 to traverse an extra hop through DSW1 to then get to DSW2.

# Layer 2 LAN Common Design Models

Network designers have many design options for Layer 2 LANs. This section will help network designers by highlighting the primary and most common Layer 2 LAN design models used in traditional LANs and today's LANs, along with the strengths and weaknesses of each design model.

## STP-Based Models

In classical Layer 2 STP-based LAN networks, the connectivity from the access layer switches to the distribution layer switches can be designed in various ways and combined with Layer 2 control protocols and features (discussed earlier) to achieve certain design functional requirements. In general, there is no single best design that someone can suggest that can fit every requirement, because each design is proposed to resolve a certain issue or requirement. However, by understanding the strengths and weaknesses of each topology and design model (illustrated in Figure 7-5 and compared in Table 7-2), network designers may then always select the most suitable design model that meets the requirements from different aspects, such as network convergence time, reliability, and flexibility. This section highlights the most common classical Layer 2 design models of LAN environments with STP, which can be applied to enterprise Layer 2 LAN designs. Figure 7-5 highlights the different STP-based LAN connectivity topologies.

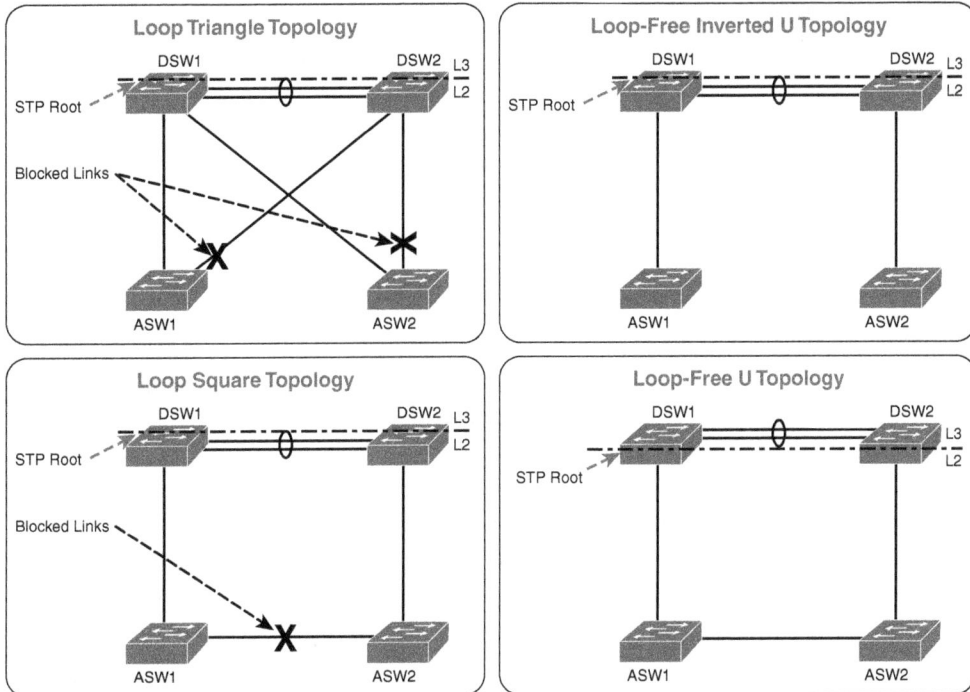

**Figure 7-5**  *Primary and Common Layer 2 (STP-Based) LAN Connectivity Topologies*

Table 7-3 contains a design comparison summary for the STP-based LAN connectivity topologies highlighted in Figure 7-5.

**Table 7-3**  Primary and Common Layer 2 (STP-Based) LAN Connectivity Models Comparison

| Topology Model | Design Concern | When to Use |
|---|---|---|
| Looped Triangle Topology | STP limits the ability to utilize all the available uplinks, which makes it inefficient and not a flexible option. | Single-homed endpoints without high-bandwidth capacity and fast convergence requirements. |
| Loop-Free Inverted U Topology | Introduces a single point of failure to the design if one distribution switch or the uplink between access and distribution switch fails. | Dual-homed endpoints with NIC teaming or endpoints that do not require low MTTR. |
| Looped Square Topology | A significant amount of access layer traffic might cross the interswitch link to reach the active FHRP. In case of a distribution switch failure, the oversubscription of the second access to distribution uplink will significantly increase, making it not a very reliable or scalable option. | Noncritical single- or dual-homed endpoints without high-bandwidth capacity requirements. |

| Topology Model | Design Concern | When to Use |
|----------------|----------------|-------------|
| Loop-Free U Topology | Inability to extend same VLANs over more than a pair of access switches; otherwise, a loop could be formed between multiple access pairs carrying same VLANs. | Single- or dual-homed endpoints with high forwarding capacity and relatively fast convergence requirements. Spanning VLANs across multiple access switches is not a requirement. |

**NOTE** All the Layer 2 design models in Figure 7-5 share common limitations: the reliance on STP to avoid loss of connectivity caused by Layer 2 loops and the dependency on Layer 3 FHRP timers, such as VRRP, to converge. These dependencies naturally lead to an increased convergence time when a node or link fails. Therefore, as a rule of thumb, tuning and aligning STP and FHRP timers is a recommended practice to overcome these limitations to some extent.

## Switch Clustering Based (Virtual Switch)

The concept of switch clustering significantly changed the Layer 2 design model between the access and distribution layer switches. With this design model, a pair of upstream distribution switches can appear as one logical (virtual) switch from the access layer switch point of view. Consequently, this approach transformed the way access layer switches connect to the distribution layer switches, because there is no longer a reliance on STP and FHRP, which means the elimination of convergence delays associated with STP and FHRP. In addition, from the uplinks and link aggregation perspective, one access switch can be connected (multihomed) to the two clustered distribution switches as one logical switch using one link aggregation bundle over multichassis link aggregation (mLAG), as illustrated in Figure 7-5.

As Figure 7-6 shows, all uplinks will be in a forwarding state across both distribution switches from a Layer 2 point of view. There will be one virtual IP gateway that should permit the forwarding across both switches from the forwarding plane perspective. It is obvious that this design model can enhance network resiliency and convergence time, and maximize bandwidth capacity, by utilizing all uplinks. In addition, this design model supports the extension of the Layer 2 VLAN across access switches safely, without any concern about forming any Layer 2 loop. This makes the design model simple, reliable, easy to manage, and more scalable as compared to the classical STP-based design model.

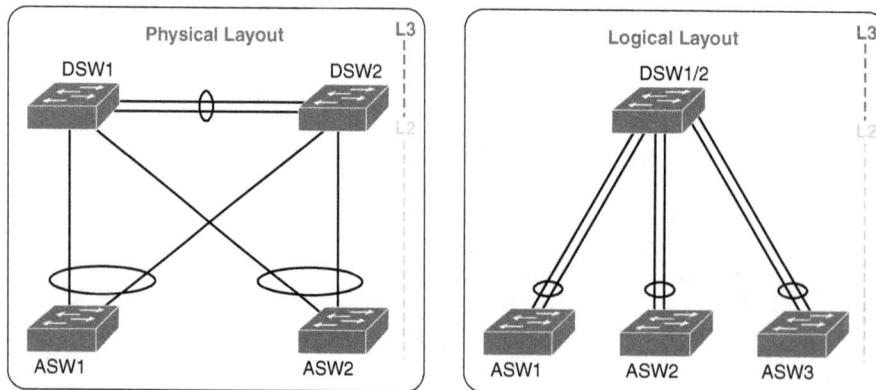

**Figure 7-6** *Switch Clustering Layouts*

Although a virtual switch adds many positive design attributes, it isn't without its own limitations that all network designers should know about. There are several different virtual switching protocols today, but no matter the protocol, they all face the same potential limitation of a *split brain*. A split-brain situation is where two or more devices in the cluster act as if they are the primary member of the cluster at the same time as the others. This usually occurs when there is an outage with the cluster link(s), which severs all cluster communication between the corresponding cluster members. Each device in the cluster is actually acting the way it should, the way the protocol is intended to work. To mitigate the occurrence of a split-brain scenario, network designers should incorporate what's known as Dual Active Detection mechanisms within the clustering protocol being selected. Figure 7-7 shows where the Dual Active Detection links would be in a Cisco VSS cluster design.

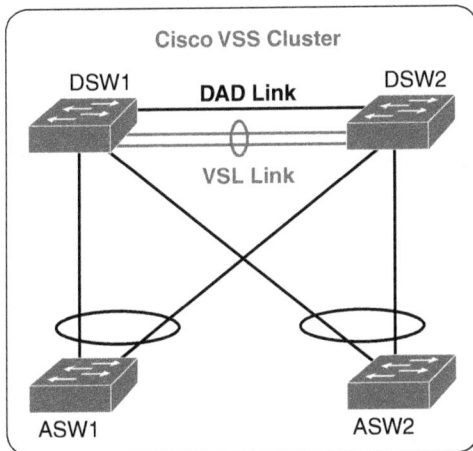

**Figure 7-7**   *Cisco VSS Cluster Design with Dual Active Detection*

Stacking switches is another form of clustering that is like the virtual switch architecture. Stacking has a number of the same attributes and design elements but is achieved by joining multiple physical switches into a single logical switch. From a Cisco product implementation perspective (Cisco-proprietary StackWise), switches are interconnected by StackWise interconnect cables, and a master switch is selected. The switch stack is managed as a single object and uses a single IP management address and a single configuration file. This reduces management overhead. Furthermore, the switch stack can create an EtherChannel connection, and uplinks can form Multichassis EtherChannel (MEC) with an upstream distribution architecture.

## Daisy-Chained Access Switches

Although this design model might be a viable option to overcome some limitations, network designers commonly use it as an interim solution. This design can introduce undesirable network behaviors. For instance, the design shown in Figure 7-8 can introduce the following issues during a link or node failure:

■ Dual active HSRP (split-brain situation)

■ Possibility of 50 percent loss of the returning traffic for devices that still use the distribution switch-1 as the active Layer 3 FHRP gateway

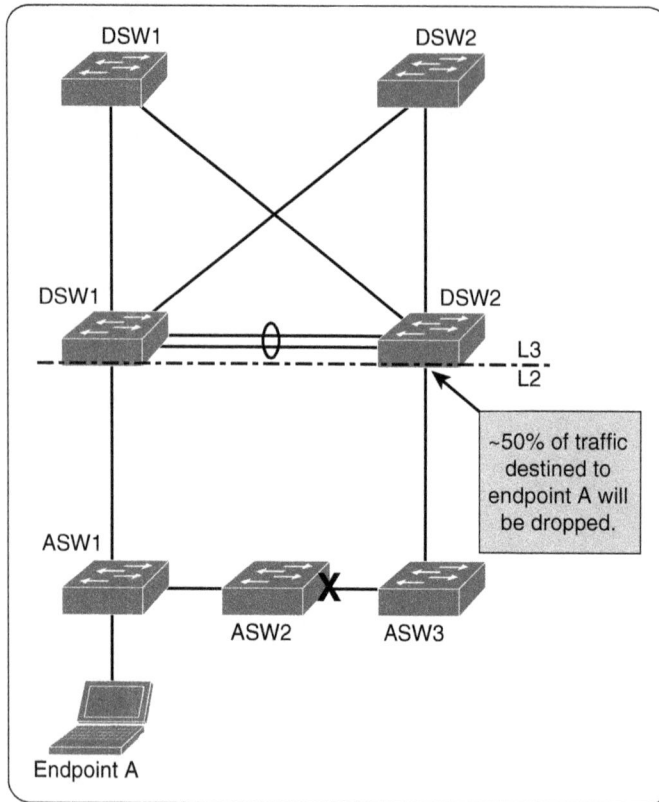

**Figure 7-8** *Daisy-Chained Access Switches*

When suggesting an alternative solution to overcome a given design issue or limitation, it is important to make sure that the suggested design option will not introduce new challenges or issues during certain failure scenarios. Otherwise, the newly introduced issues will outweigh the benefits of the suggested solution. In some situations, a network designer has no other choices available, in which case the designer may have to leverage a less than ideal solution as a short-term fix while the long-term solution is procured and deployed.

## Summary

In this chapter, we focused on the Layer 2 technologies that all network designers need to know. This included STP, VLANs, trunking, link aggregation, Multichassis link aggregation (mLAG), and FHRP. We then discussed and compared the most common LAN design models, when to use them, and why to use them. As a network designer, you need to have various technologies in place that properly protect the Layer 2 domain and facilitate having redundant paths to ensure continuous connectivity to the end user and their respective application, in the event of a failure scenario. Understanding the characteristics of these core Layer 2 technologies and their respective behaviors is critical to successfully designing a reliable and highly available Layer 2 network.

## Reference

Al-shawi, Marwan, *CCDE Study Guide* (Cisco Press, 2015)

## Exam Preparation Tasks

As mentioned in the section "How to Use This Book" in the Introduction, you have a couple of choices for exam preparation: the exercises here, Chapter 18, "Final Preparation," and the exam simulation questions in the Pearson Test Prep Software Online.

## Review All Key Topics

Review the most important topics in this chapter, noted with the Key Topic icon in the outer margin of the page. Table 7-4 lists a reference of these key topics and the page numbers on which each is found.

**Table 7-4**   Key Topics for Chapter 7

| Key Topic Element | Description | Page Number |
|---|---|---|
| Paragraph | FHRP and STP common design model | 151 |
| Figure 7-5 | Primary and Common Layer 2 (STP-Based) LAN Connectivity Topologies | 153 |
| Table 7-3 | Primary and Common Layer 2 (STP-Based) LAN Connectivity Models Comparison | 153 |

## Complete Tables and Lists from Memory

Print a copy of Appendix D, "Memory Tables" (found on the companion website), or at least the section for this chapter, and complete the tables and lists from memory. Appendix E, "Memory Tables Answer Key," also on the companion website, includes completed tables and lists to check your work.

## Define Key Terms

Define the following key terms from this chapter and check your answers in the glossary:

Spanning Tree Protocol (STP), Multiple Spanning Tree (MST), Root Guard, BPDU Guard, BPDU Filter, virtual local-area network (VLAN), Link Aggregation Control Protocol (LACP), First Hop Redundancy Protocol (FHRP), Hot Standby Routing Protocol (HSRP), Virtual Router Redundancy Protocol (VRRP), Gateway Load Balancing Protocol (GLBP)

# CHAPTER 8

# Layer 3 Technologies

## This chapter covers the following topics:

- **Enterprise Layer 3 Routing:** This section covers refresher Layer 3 routing concepts, link-state routing protocols, EIGRP, route summarization, traffic engineering and path selection options, and the corresponding network design elements within each of these items and between.

- **BGP Routing:** This section covers the basics of BGP routing, BGP as the core routing protocol, BGP scalability options, route redistribution, and the corresponding network design elements within each of these items.

- **Enterprise Routing Design Recommendations:** This section covers the enterprise routing design recommendations and the corresponding network design elements that go with them, such as how to select the proper routing protocol in the first place.

In network design, it is common that a certain design goal can be achieved "technically" using different approaches. While from a technical deployment point of view this can be seen as an advantage, from a network design perspective, the question is which design option should be selected and why? To answer this as a network designer, you must be aware of the different design options and protocols as well as the advantages and limitations of each. Therefore, this chapter will concentrate specifically on highlighting, analyzing, and comparing the various design options, principles, and considerations with regard to Layer 3 control plane protocols from different design aspects, focusing on enterprise-grade networks.

This chapter covers the following "CCDE v3.0 Core Technology List" sections:

- 3.0 Layer 3 Control Plane
- 3.1 Network hierarchy and topologies
- 3.2 Unicast routing protocol operation
- 3.3 Fast convergence techniques and mechanism
- 3.4 Factors affecting convergence
- 3.5 Route aggregation
- 3.6 Fault isolation and resiliency
- 3.7 Metric-based traffic flow and modification

# "Do I Know This Already?" Quiz

The "Do I Know This Already?" quiz allows you to assess whether you should read this entire chapter thoroughly or jump to the "Exam Preparation Tasks" section. If you are in doubt about your answers to these questions or your own assessment of your knowledge of the topics, read the entire chapter. Table 8-1 lists the major headings in this chapter and their corresponding "Do I Know This Already?" quiz questions. You can find the answers in Appendix A, "Answers to the 'Do I Know This Already?' Quizzes."

**Table 8-1** "Do I Know This Already?" Section-to-Question Mapping

| Foundation Topics Section | Questions |
|---|---|
| Enterprise Layer 3 Routing | 1–9 |
| BGP Routing | 10–11 |
| Enterprise Routing Design Recommendations | — |

**CAUTION** The goal of self-assessment is to gauge your mastery of the topics in this chapter. If you do not know the answer to a question or are only partially sure of the answer, you should mark that question as wrong for purposes of the self-assessment. Giving yourself credit for an answer you correctly guess skews your self-assessment results and might provide you with a false sense of security.

1. What is used by OSPF to share routing and topology information?
   a. Link-state PDU
   b. Distant vector update
   c. Link-state advertisement
   d. Link-state hellos

2. What type of link-state advertisement is generated by area border routers, which advertise networks from one area to the another?
   a. Summary LSA (type 3)
   b. Network LSA (type 2)
   c. Summary ASBR (type 4)
   d. Not-so-stubby area LSA (type 7)

3. Which OSPF network type will not elect a DR and BDR and have support for multi-access networks?
   a. Broadcast
   b. Point-to-multipoint
   c. Point-to-point
   d. Nonbroadcast

4.  Which EIGRP packet types are sent using multicast? (Choose all that apply.)

    a.  Hello

    b.  Reply

    c.  Acknowledgement

    d.  Update

    e.  Query

5.  True or false: An EIGRP stub router will respond to queries and is a transit router.

    a.  True

    b.  False

6.  Which EIGRP packets use the Reliable Transport Protocol?

    a.  Hello

    b.  Reply

    c.  Acknowledgement

    d.  Update

    e.  Query

7.  What does IS-IS use to make sure there is efficient flooding of LSPs in multiaccess networks?

    a.  Designated Intermediate System (DIS)

    b.  Designated router (DR)

    c.  LDP synchronization

    d.  Overload bit

8.  In IS-IS, a level 2 (L2) area router is considered the same as what router in OSPF?

    a.  Area border router

    b.  Backbone router

    c.  Autonomous system border router

    d.  Internal router

9.  What is the benefit of setting the overload bit in IS-IS?

    a.  Increase scaling

    b.  Force DIS election

    c.  Black hole avoidance

    d.  Enable pseudo-node LSP

10. In BGP, what nontransitive BGP attribute that is also standards-based is commonly leveraged on ingress to influence egress traffic flows?

    a.  AS Path Prepend

    b.  Weight

    c.  Route-Map

    d.  Local Preference

**11.** How would you influence traffic inbound to your AS? (Choose all that apply.)

    **a.** AS Path Prepend

    **b.** Weight

    **c.** Multi-Exit Discriminator?

    **d.** Local Preference

## Foundation Topics

## Enterprise Layer 3 Routing

This section covers the various routing design considerations and optimization concepts that pertain to enterprise-grade routed networks.

### IP Routing and Forwarding Concept Review

The main goal of routing protocols is to serve as a delivery mechanism to route packets to reach their intended destination. The end-to-end process of packets routing across the routed network is facilitated and driven by the concept of distributed databases. This concept is typically based on having a database of IP addresses (typically IPs of hosts and networks) on each Layer 3 node in the packet's path, along with the next-hop IP addresses of the Layer 3 nodes that can be used to reach each of these IPs. This database is known as the *routing information base (RIB)*. In contrast, the *forwarding information base (FIB)*, also known as the *forwarding table*, contains the destination addresses and the interfaces required to reach those destinations, as depicted in Figure 8-1. In general, routing protocols are classified as either link-state, path-vector, or distance-vector protocols. This classification is based on how the mechanism of the routing protocol constructs and updates its routing table, and how it computes and selects the desired path to reach the intended IP destination.

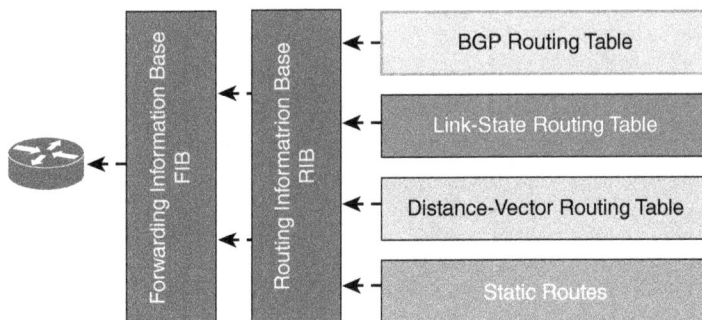

**Figure 8-1**   *RIB and FIB*

As illustrated in Figure 8-2, the typical basic forwarding decision in a router is based on three processes:

- Routing protocols
- Routing table
- Forwarding decision (switches packets)

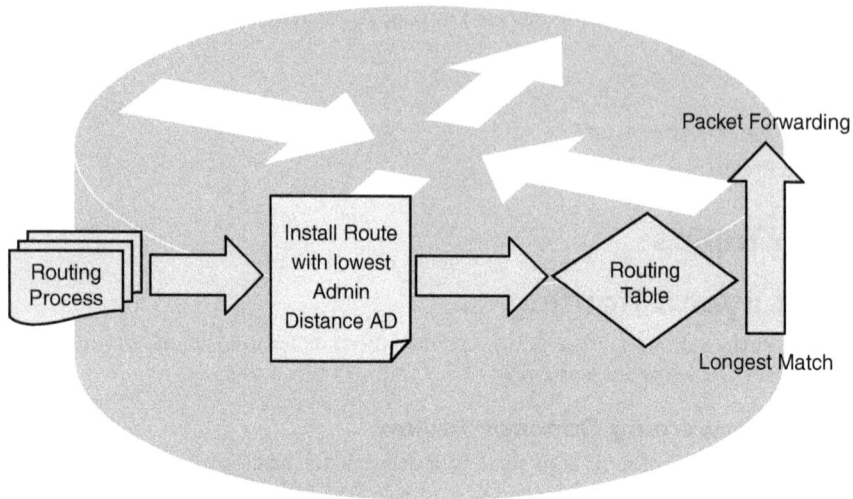

**Figure 8-2**  *Router's Forwarding Decision*

## Link-State Routing

**Link-state routing protocols** use Dijkstra's shortest path algorithm to calculate the best path. **Open Shortest Path First (OSPF)** and **Intermediate System-to-Intermediate System (IS-IS)** protocols are link-state routing protocols that have a common conceptual characteristic in the way they build, interact, and handle L3 routing to some extent. A **link-state advertisement (LSA)** is a message that is used to communicate network information such as router links, interfaces, link states, and costs within a link-state routing protocol. Figure 8-3 illustrates the process of building and updating a link-state database (LSDB).

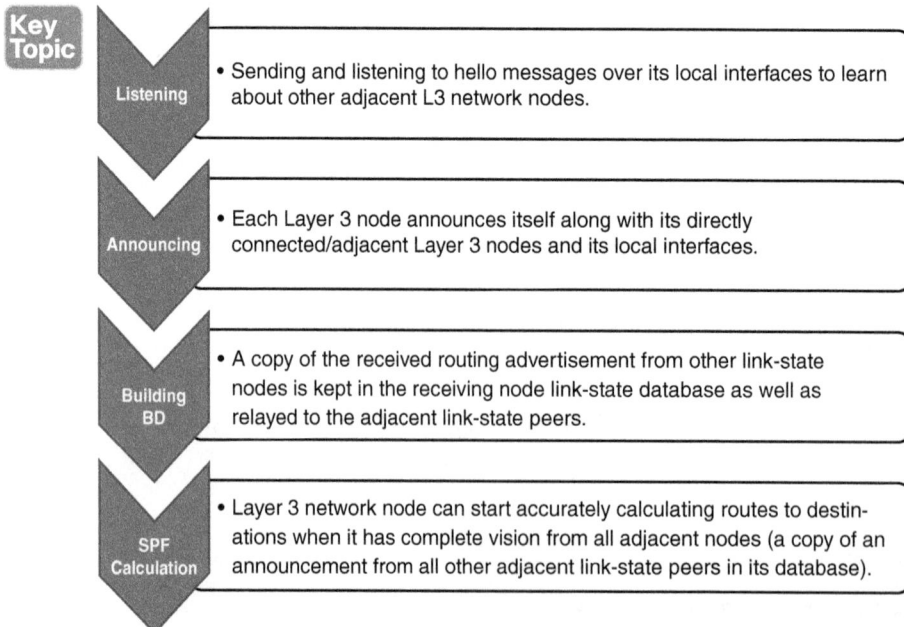

**Listening**
- Sending and listening to hello messages over its local interfaces to learn about other adjacent L3 network nodes.

**Announcing**
- Each Layer 3 node announces itself along with its directly connected/adjacent Layer 3 nodes and its local interfaces.

**Building BD**
- A copy of the received routing advertisement from other link-state nodes is kept in the receiving node link-state database as well as relayed to the adjacent link-state peers.

**SPF Calculation**
- Layer 3 network node can start accurately calculating routes to destinations when it has complete vision from all adjacent nodes (a copy of an announcement from all other adjacent link-state peers in its database).

**Figure 8-3**  *Process of Building an LSDB*

It is important to remember that although OSPF and IS-IS as link-state routing protocols are highly similar in the way they build the LSDB and operate, they are not identical! This section discusses the implications of applying link-state routing protocols (OSPF and IS-IS) on different network topologies, along with different design considerations and recommendations.

## Link-State over Hub-and-Spoke Topology

In general, some implications should be considered when link-state routing protocols are applied on a hub-and-spoke topology, including the following:

- There is a concern with regard to scaling to a large number of spokes because each spoke node typically will receive all other spoke nodes' link-state information, because there are no effective means to control the distribution of routing information among these spokes.

- Special consideration must be taken to avoid suboptimal routing, in which traffic can use remote sites (spokes) as a transit site to reach the hub or other spokes.

For instance, summarization of routing flooding domains in a multi-area/flooding domain design with multiple border routers requires specific routing information between the border routers (area border routers [ABRs] in OSPF or L1/L2 in IS-IS) over a non-summarized link, to avoid using spoke sites as a transit path, as illustrated in Figure 8-4. An OSPF ABR is a link-state router that is connected to more than one OSPF area.

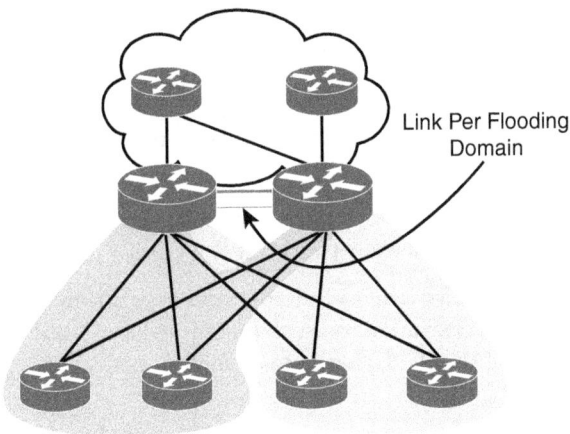

Link Per Flooding Domain

**Figure 8-4**  *Multi-Area Link State: Hub and Spoke*

So, for each hub-and-spoke flooding domain to be added to the hub routers, you need to consider an additional link between the hub routers in that domain. This is a typical use case scenario to avoid suboptimal routing with link-state routing protocols. However, when the number of flooding domains (for example, OSPF areas) increases, the number of VLANs, subinterfaces, or physical interfaces between the border routers will grow as well, which will result in scalability and complexity concerns. One of the possible solutions is to have a single link with adjacencies in multiple areas (RFC 5185). For instance, in the scenario illustrated in Figure 8-5, there is a hub-and-spoke topology that uses OSPF multi-area design.

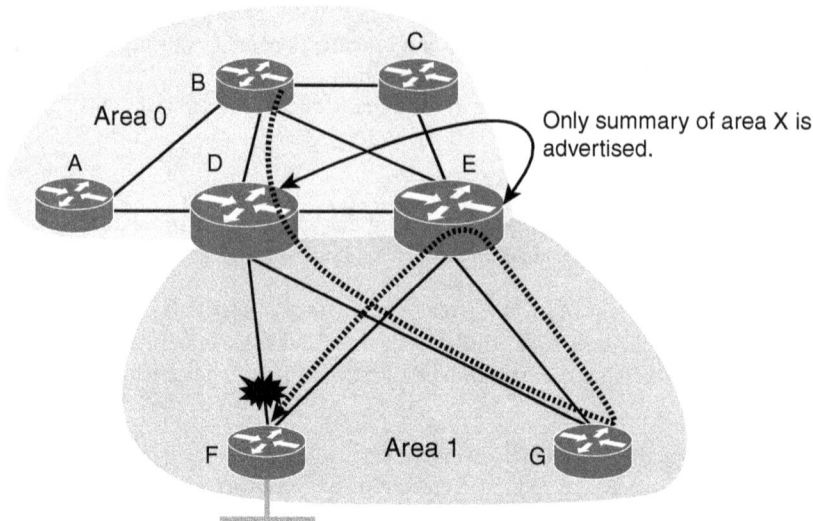

**Figure 8-5** *Multi-Area OSPF: Hub and Spoke*

If the link between router D and router F (part of OSPF area 1) fails, any traffic from router B destined to the LAN connected to router F going toward the summary advertised route by router D will traverse the more specific route over the path G, E, then F.

To optimize this design during this failure scenario, there are multiple possible solutions, and here network designers must decide which solution is the most suitable one with regard to other design requirements such as application requirements where the delay could affect critical business applications:

- Place the inter-ABR link (D to E) in area 1 (simple and provides "north to south" optimal routing in this topology).

- Place each spoke in its own area with LSA type 3 filtering. (May lead to complex operations and limited scalability; "depends on the network size.")

- Disable route summarization at the ABRs; for example, advertise more specific routes from ABR router E. (May not always be desirable because this means reduced scalability and the loss of some of the value of the OSPF multi-area design.)

**NOTE** The link between the two hub nodes (for example, ABRs) will introduce the potential of a single point of failure to the design. Therefore, link redundancy (availability) between the ABRs may need to be considered.

If IS-IS is applied to the topology in Figure 8-5 instead, using a similar setup where IS-IS L2 is to be used instead of the area 0 and IS-IS L1 is to be used by the spokes, the simplest way to optimize this architecture is to put the links between the border routers in IS-IS L1-L2 (overlapping levels capability), where we can extend L1 to overlap with L2 on the border router (ABR in OSPF), as illustrated in Figure 8-6. This will result in a topology that can support summarization with more optimal routing with regard to the failure scenario discussed previously.

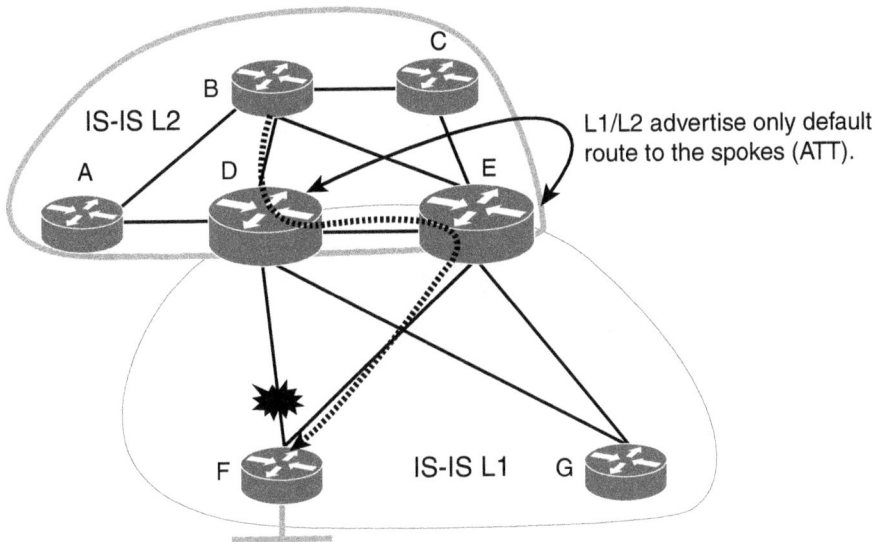

**Figure 8-6**  *Multilevel IS-IS: Hub and Spoke*

**NOTE**   OSPF is a more widely deployed and proven link-state routing protocol in enterprise networks compared to IS-IS, especially with regard to hub-and-spoke topologies. IS-IS has limitations when it works on nonbroadcast multiple access (NBMA) multipoint networks.

## OSPF Interface Network Type Considerations in a Hub-and-Spoke Topology

Figure 8-7 summarizes the different possible types of OSPF interfaces in a hub-and-spoke topology over NBMA transport (typically either Frame Relay or ATM), along with the associated design advantages and implications of each.

## Link-State over Full-Mesh Topology

Fully meshed networks can offer a high level of redundancy and the shortest paths. However, the substantial amount of routing information flooding across a fully meshed network is a significant concern. This concern stems from the fact that each router will receive at least one copy of every new piece of information from each neighbor on the full mesh. For example, in Figure 8-8, each router has four adjacencies. When a router's link connected to the LAN side fails, it must flood its LSA/LSP to each of the four neighbors. Each neighbor will then flood this LSA/LSP (link-state package) again to its neighbors. This process will culminate in a process like a broadcast being sent, due to this full-mesh connectivity and reflooding.

**NBMA/Broadcast**

- Single IP address
- Set OSPF Priority 0
- Spoke to Spoke

**Point-to-Multipoint**

- Single IP address P2MP interface
- P2P interface

**Point-to-Point**

- Per subinterface IP address P2P subinterface
- P2P interface

| Design Advantages | Design Limitations |
|---|---|
| Simplified IP addressing and smaller routing table size (smaller link-state database). | Manual configuration of each spoke with the right OSPF priority. No reachability between spokes or labor intensive at Layer 2 (frame-relay DCLI) configuration (high operation complexity). |
| Simplified IP addressing and operation that support small to medium networks (smaller link-state database than P2P). Simplified operations compared to other options. | Additional host routes inserted in the routing table, which may limit its scalability (depends on the number of prefixes and hardware resources). |
| Offer the capability to maintain end-to-end link state (signaling the down state). | Large IP addressing spaces. Larger routing table/larger link-state database. Overhead of subinterfaces operations. Limited scalability compared to other options. |

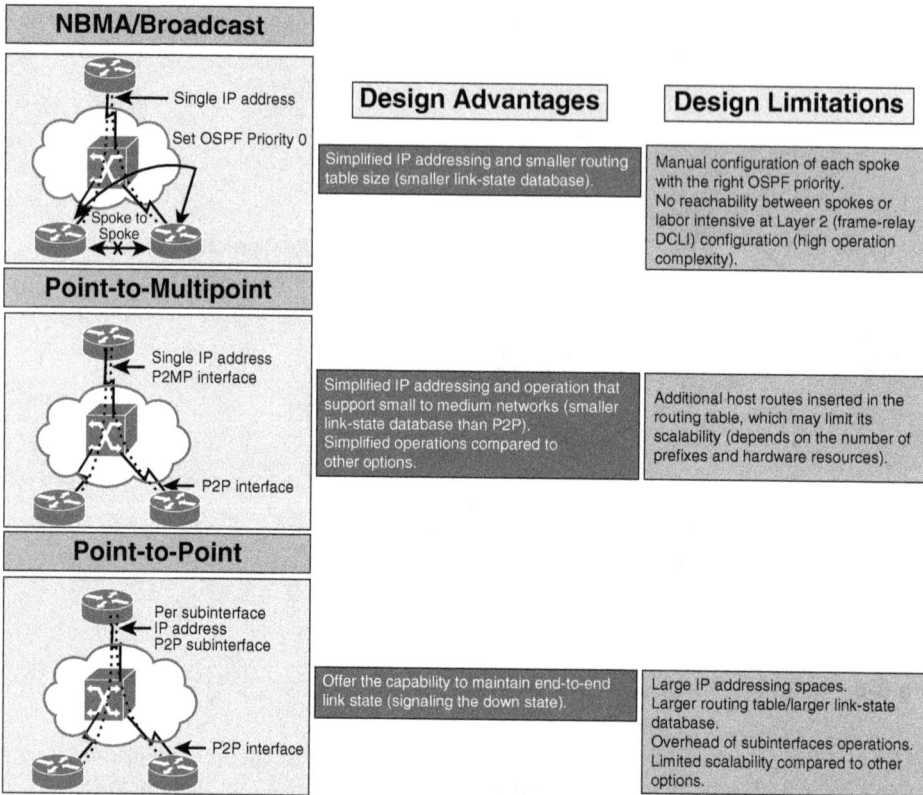

**Figure 8-7**  *OSPF Interface Types Comparison: Hub and Spoke*

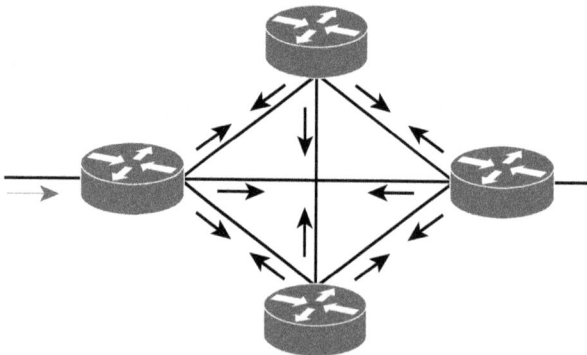

**Figure 8-8**  *Link State: Full-Mesh Topology*

With link-state routing protocols, you can use the mesh group technique to reduce link-state information flooding in a full-meshed topology. However, with link-state routing protocols in failure scenarios over a meshed topology, some routers may know about the failure before others within the mesh. This will typically lead to a temporarily inconsistent LSDB across the nodes within the network, which can result in transient forwarding loops. Even though the concept of a loop-free alternate (LFA) route can be considered to overcome situations like this, using LFA over a mesh topology will add complexity to the control plane.

**NOTE**   Later in this chapter, more details are provided about flooding domain and route summarization design considerations for link-state routing protocols, which can reduce the level of control plane complexity and optimize link-state information flooding and performance.

**NOTE**   Other mechanisms help to optimize and reduce link-state LSA/LSP flooding by reducing the transmission of subsequent LSAs/LSPs, such as OSPF floors reduction (described in RFC 4136). This is done by eliminating the periodic refresh of unchanged LSAs, which can be useful in fully meshed topologies.

### OSPF Area Types

Table 8-2 contains an overview of the different types of OSPF area types.

**Key Topic**

**Table 8-2**   Summary of OSPF Area Types

| Area Type | Advertised Route |
|---|---|
| Stubby | All routes except type 5, external routing information |
| Totally stubby | Internal area routes + default route (both type 3 and 5 LSAs are suppressed) |
| Not so stubby area (NSSA) | All routes with the ability to inject/originate external routing information (type 7 LSA) |
| Totally NSSA | Internal area routes + default route, with the ability to inject/originate external routing information (type 7 LSAs) |

Each of the OSPF areas allows certain types of LSAs to be flooded, which can be used to optimize and control route propagation across the OSPF routed domain. However, if OSPF areas are not properly designed and aligned with other requirements, such as application requirements, it can lead to serious issues because of the traffic black-holing and suboptimal routing that can appear as a result of this type of design. Subsequent sections in this book discuss these points in more detail.

**Key Topic**

Figure 8-9 shows a conceptual high-level view of the route propagation, along with the different OSPF LSAs, in an OSPF multi-area design with different area types.

The typical design question is, "Where can these areas be used and why?"

The basic standard answer is, "It depends on the requirements and topology."

For instance, if no requirement specifies which path a route must take to reach external networks such as an extranet or the Internet, you can use the "totally NSSA" area type to simplify the design. For example, the scenario in Figure 8-10 is one of the most common design models that use OSPF NSSA. In this design model, the border area that interconnects the campus or data center network with the WAN or Internet edge devices can be deployed as totally NSSA. This deployment assumes that no requirement dictates which path should be used. Furthermore, in the case of NSSA and multiple ABRs, OSPF selects one ABR to perform the translation from LSA type 7 to LSA type 5 and floods it into area 0 (normally the router with the highest router ID, as described in RFC 1587 [obsoleted by RFC 3101]). This behavior can affect the design if the optimal path is required.

8

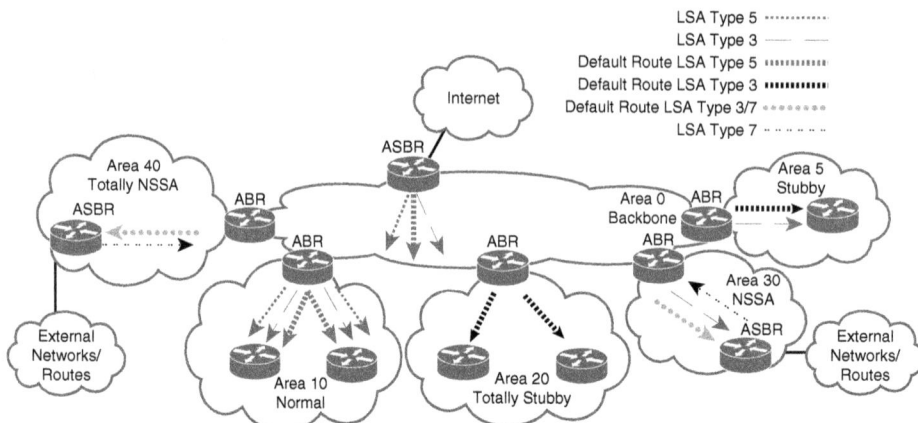

**Figure 8-9** *OSPF Route Propagation in Multi-Area Design*

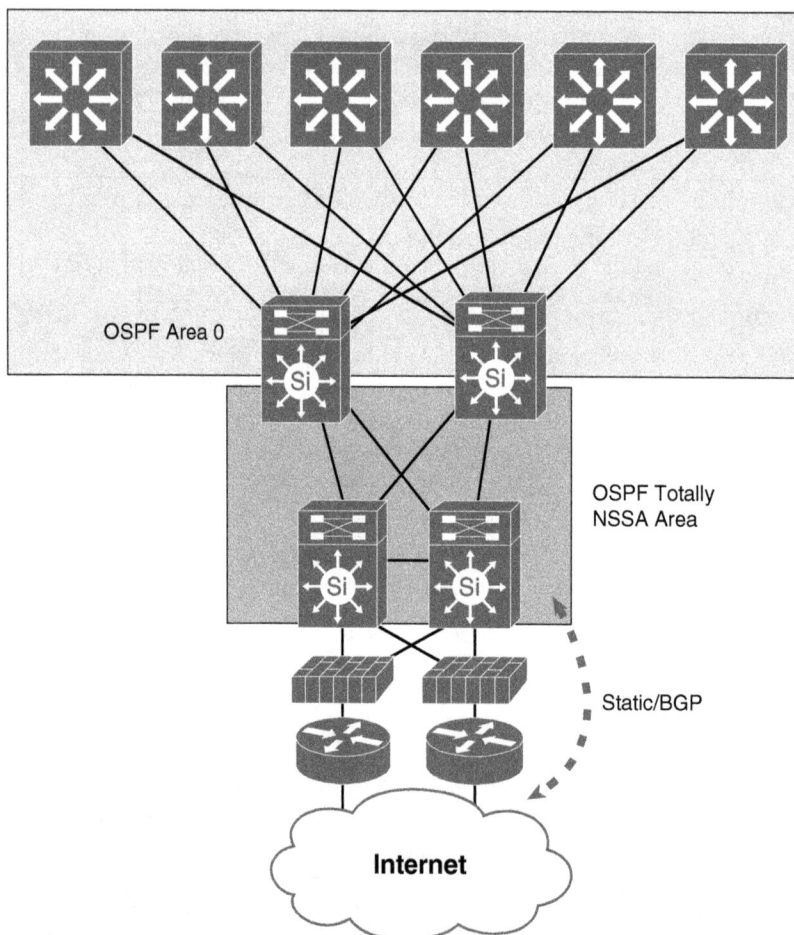

**Figure 8-10** *OSPF Totally NSSA*

**NOTE**   RFC 3101 introduced the ability to have multiple ABRs perform the translation from LSA type 7 to type 5. However, the extra unnecessary number of LSA type 7 to type 5 translators may significantly increase the size of the OSPF LSDB. This can affect the overall OSPF performance and convergence time in large-scale networks with a large number of prefixes.

Similarly, in the scenario depicted on the left in Figure 8-11, a data center in London hosts two networks (10.1.1.0/24 and 10.2.1.0/24). Both WAN/MAN links to this data center have the same bandwidth and cost. Based on this setup, the traffic coming from the Sydney branch toward network 10.2.1.0/24 can take any path. If this is not compromising any requirement (in other words, suboptimal routing is not an issue), the OSPF area 10 can be deployed as a "totally stubby area" to enhance the performance and stability of remote site routers.

In contrast, the scenario on the right side of Figure 8-11 has a slightly different setup. The data centers are located in different geographic locations with a data center interconnect (DCI) link. In a scenario like this, the optimal path to reach the destination network can be critical, and using a totally stubby area can break the optimal path requirement. To overcome this limitation, there are two simple alternatives to use: either "normal OSPF area" or the "stubby area" for area 10. This ensures that the most specific route (LSA type 3) is propagated to the Sydney branch router to select the direct optimal path rather than crossing the international DCI.

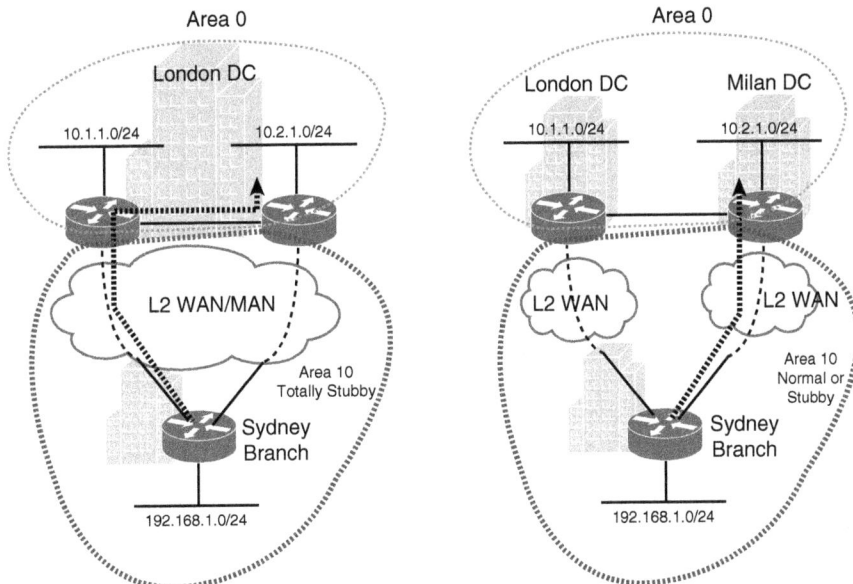

**Figure 8-11**   *OSPF Totally Stubby Area Versus Stubby Area Design*

In summary, the goal of these types of different OSPF areas is to add more optimization to the OSPF multi-area design by reducing the size of the routing table and lowering the overall control plane complexity by reducing the size of the fault domains (link-state flooding domains). This size reduction can help to reduce the overhead of the routers' resources, such as CPU and memory. Furthermore, the reduction of the flooding domains' size will help accelerate the overall network recovery time in the event of a link or node failure. However, in some scenarios where an optimal path is important, take care when choosing between these various area types.

**NOTE** In the scenarios illustrated in Figure 8-10 and Figure 8-11, asymmetrical routing is a possibility, which may be an issue if there are any stateful or stateless network devices in the path such as a firewall. However, this section focuses only on the concept of area design. Later in this book, you will learn how to manage asymmetrical routing at the network edge.

## OSPF Versus IS-IS

It is obvious that OSPF and IS-IS as link-state routing protocols are similar and can achieve (to a large extent) the same result for enterprises in terms of design, performance, and limitations. However, OSPF is more commonly used by enterprises as the interior gateway protocol (IGP), for the following reasons:

- OSPF can offer a more structured and organized routing design for modular enterprise networks.

- OSPF is more flexible over a hub-and-spoke topology with multipoint interfaces at the hub.

- OSPF naturally runs over IP, which makes it a suitable option to be used over IP tunneling protocols such as Generic Routing Encapsulation (GRE), Multipoint GRE (mGRE), Cisco Dynamic Multipoint Virtual Private Network (DMVPN), and Next Hop Resolution Protocol (NHRP), whereas with IS-IS, this is not a supported design.

- In terms of staff knowledge and experience, OSPF is more widely deployed on enterprise-grade networks. Therefore, compared to IS-IS, more people have OSPF knowledge and expertise.

However, if there is no technical barrier, both OSPF and IS-IS are valid options to consider.

## EIGRP Routing

**Enhanced Interior Gateway Routing Protocol (EIGRP)** is an enhanced **distance-vector routing protocol**, relying on the Diffusing Update Algorithm (DUAL) to calculate the shortest path to a network. A distance-vector protocol is a routing protocol that advertises the entire table to its neighbors. EIGRP, as a unique Cisco innovation, became highly valued for its ease of deployment, flexibility, and fast convergence. For these reasons, EIGRP is commonly considered by many large enterprises as the preferred IGP. EIGRP maintains all the advantages of distance-vector protocols while avoiding the concurrent disadvantages. For instance, EIGRP does not transmit the entire routing information that exists in the routing table following an update event; instead, only the "delta" of the routing information will be transmitted since the last topology update. EIGRP is deployed in many enterprises as the routing protocol for the following reasons:

- Easy to design, deploy, and support

- Easier to learn

- Flexible design options

- Lower operational complexities

- Fast convergence (subsecond)

- Can be simple for small networks while at the same time scalable for large networks

- Supports flexible and scalable multi-tier campus and hub-and-spoke WAN design models

Unlike link-state routing protocols, such as OSPF, EIGRP has no hard edges. This is a key design advantage because hierarchy in EIGRP is created through route summarization or route filtering rather than relying on a protocol-defined boundary, such as OSPF areas. As illustrated in Figure 8-12, the depth of hierarchy depends on where the summarization or filtering boundary is applied. This makes EIGRP flexible in networks structured as a multitier architecture.

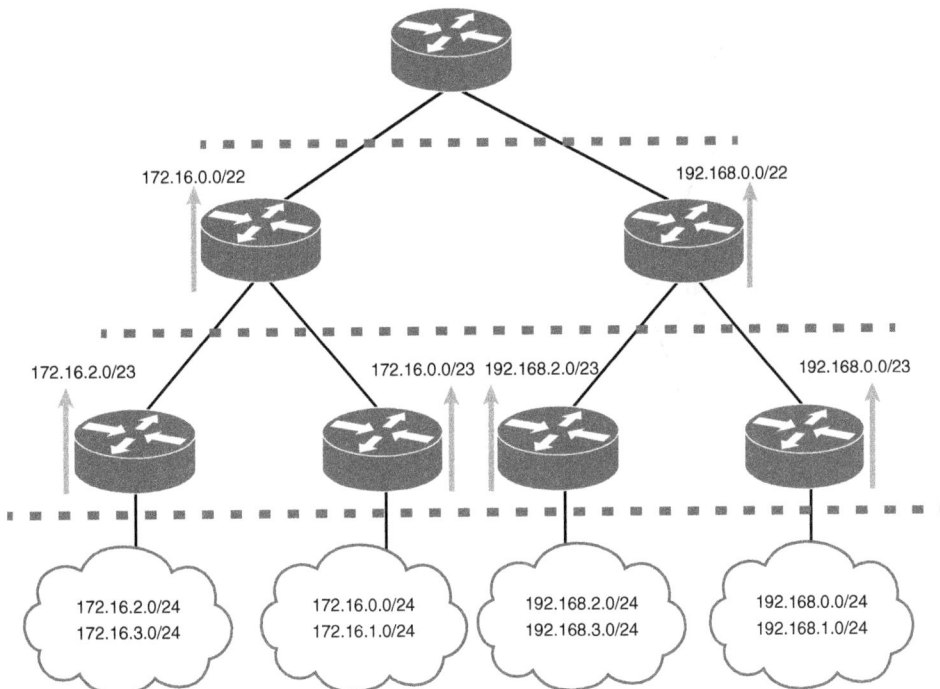

**Figure 8-12**  *EIGRP Domain Boundaries on a Multitier Network*

## EIGRP: Hub and Spoke

As discussed earlier, link-state routing protocols have some scaling limitations when applied to a hub-and-spoke topology. In contrast, EIGRP offers more flexible and scalable capabilities for the hub-and-spoke types of topologies. One of the main concerns in a hub-and-spoke topology is the possibility of a spoke or remote site being used as a transit path due to a configuration error or a link failure. With link-state routing protocols, several techniques to mitigate this type of issue were highlighted. However, there are still scalability limitations associated with it.

However, EIGRP offers the capability to mark the remote site (spoke) as a stub, which is unlike the OSPF stub (where all routers in the same stub area can exchange routes and propagate failure, and update information). With EIGRP, when the spokes are configured as a stub, it will signal to the hub router that the paths through the spokes should not be used as transit paths. As a result, there will be significant optimization to the design. This optimization results from the decrease in EIGRP query scope and the reduction of the unnecessary overhead associated with responding to queries by the spoke routers (for example, EIGRP stuck-in-active [SIA] queries).

In Figure 8-13, router B will see it has only one path to the LAN connected to router A, rather than four paths.

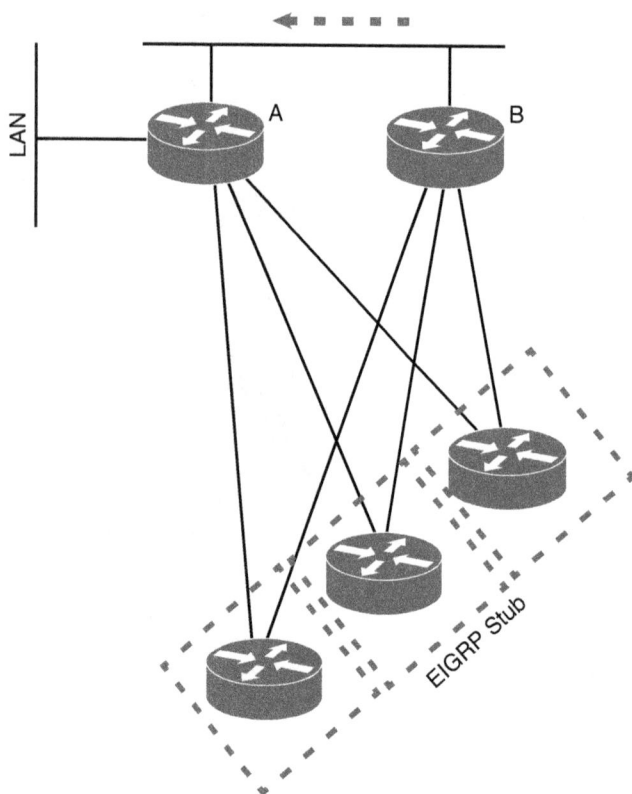

**Figure 8-13**  *EIGRP Stub*

Consequently, enabling EIGRP Stub over a "hub-and-spoke" topology helps to reduce the overall control plane complexity and increases the scalability of the design to support a large number of spokes without affecting its performance.

## EIGRP: Stub Route Leaking

You might encounter some scenarios like the one depicted in Figure 8-14, which is an extension to the EIGRP stub design with a backdoor link between two remote sites. In this scenario, the HQ site is connected to the two remote sites over an L2 WAN. These remote sites are also interconnected directly via a backdoor link. Remote sites are configured as EIGRP stubs to optimize the remote sites' EIGRP performance over the WAN.

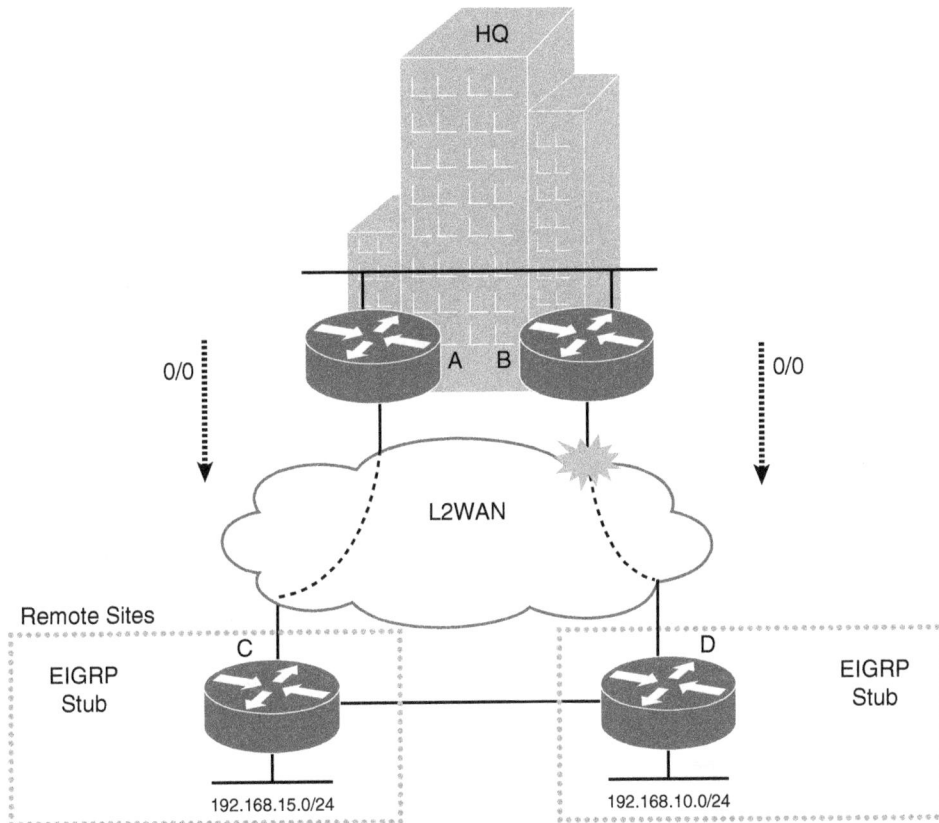

**Figure 8-14** *EIGRP Stub Leaking*

The issue with the design in this scenario is that if the link between router B and router D fails, the following will result as a consequence of this single failure:

- Router A cannot reach network 192.168.10.0/24 because router D is configured as a stub. Also, router C is a stub, which will not advertise this network to router A anyway.

- Router D will not be able to receive the default from router A because router C is a stub as well.

This means that the remote site connected to router D will be completely isolated, without taking any advantage of the backdoor link. To overcome this issue, EIGRP offers a useful feature called *stub leaking*, where both routers D and C in this scenario can advertise routes to each other selectively, even if they are configured as a stub. Route filtering might need to be incorporated in scenarios like this when an EIGRP leak map is introduced into the design to avoid any potential suboptimal routing that might happen as a consequence of route leaking.

## EIGRP: Ring Topology

Unlike link-state routing protocols, EIGRP has limitations with a ring topology. As depicted in Figure 8-15, the greater the number of nodes in the ring, the greater the number of queries to be sent during a link failure. As a general recommendation with EIGRP, always try to design in triangles where possible, rather than rings.

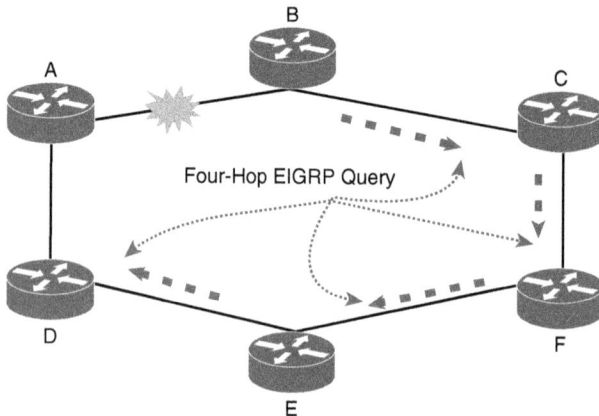

**Figure 8-15**  *EIGRP Queries on a Ring Topology*

## EIGRP: Full-Mesh Topology

EIGRP in a full-mesh topology (see Figure 8-16) is less desirable in comparison with link-state protocols. For example, with link-state protocols such as OSPF, network designers can designate one router to flood into the mesh and block flooding on the other routers, which can improve the topology. In contrast, with EIGRP, this capability is not available. The only way to mitigate the information flooding in an EIGRP mesh topology is by relying on route summarization and filtering techniques. To optimize EIGRP in a mesh topology, the summarization must be into and out of the meshed network.

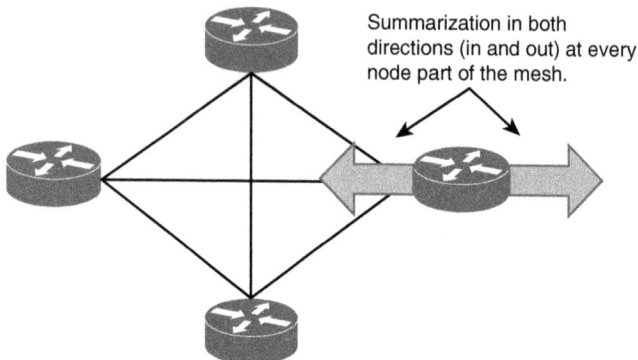

**Figure 8-16**  *EIGRP on a Mesh Topology*

**NOTE**  As discussed earlier, a link-state routing protocol can lead to transit forwarding loops in a ring and mesh topologies after a network component failure event. Therefore, both EIGRP and link-state routing protocols have limitations on these topologies, with different indications (fast and large number of EIGRP queries versus link-state transit loop).

**Key Topic**

## EIGRP Route Propagation Considerations

EIGRP offers a high level of flexibility to network designers because it can fit different types of designs and topologies. However, like any other protocol, some limitations apply to EIGRP (especially with regard to route propagation) and may influence the design choices.

Therefore, network designers must consider the following factors to avoid impacting the propagation of routing information, which can result in an unstable design:

- **EIGRP bandwidth:** By default, EIGRP is designed to use up to 50 percent of the main interface bandwidth for EIGRP packets; however, this value is configurable. The limitation of this concept occurs when there is a dialer or point-to-multipoint physical interface with several peers over one multipoint interface. In this scenario, EIGRP considers the bandwidth value on the main interface divided by the number of EIGRP peers on that interface to calculate the amount of bandwidth per peer. Consequently, when more peers are added over this multipoint interface, EIGRP will reach a point where it will not have enough bandwidth to operate over that dialer or multipoint interface appropriately. In addition, one of the common mistakes with regard to EIGRP and interface bandwidth is that sometimes network operators try to "influence" the best path selection decision in EIGRP DUAL by only tuning the bandwidth over an interface where the interface with the lowest bandwidth will be the least preferred. However, this approach can impact the EIGRP control plane peering functionality and scalability if it is tuned to a low value without proper planning.

  Therefore, the network designer must take this point into consideration and adopt alternatives, such as point-to-point subinterfaces under the multipoint interface. In addition, with overlay multipoint tunnel interfaces such as DMVPN, the bandwidth may be required to be defined manually at the tunnel interface when there is a large number of remote spokes.

- **Zero successor routes:** When EIGRP tries to install a route in the RIB table and it is rejected, this is called a zero successor route because this route simply will not be propagated to other EIGRP neighbors in the network. This behavior typically happens due to one of the following two primary reasons:

  - There is already the same route in the RIB table with a better **administrative distance (AD)**. (Administrative distance is a rating of the trustworthiness of a routing information source. A lower number is preferred.)

  - When there are multiple EIGRP autonomous systems (AS) defined on the same router, the router will typically install any given route learned via both EIGRP autonomous systems with the same AD from one EIGRP AS, while the other will be rejected. Consequently, the route of the other EIGRP AS will not be propagated within its domain.

## Hiding Topology and Reachability Information

Technically, both topology and reachability information hiding can help to improve routing convergence time during a link or node failure. Topology and reachability information hiding also reduces control plane complexity and enhances network stability to a large extent. For example, if there is a link flapping in a remote site, this might cause all other remote sites to receive and process the update information every time this link flaps, which leads to instability and increased CPU processing.

However, to produce a successful design, the design must first align with the business goals and requirements (and not just be based on the technical drivers). Therefore, before deciding how to structure IGP flooding domains, network architects or designers must first identify the business's goals, priorities, and drivers. Consider, for example, an organization that plans to merge with one of its business partners but with no budget allocated to upgrade any of the existing network nodes. When these two networks merge, the size of the network may increase significantly in a short period of time. As a result, the number of prefixes and network topology information will increase significantly, which will require more hardware resources such as memory or CPUs.

Given that this business has no budget allocated for any network upgrade, in this case introducing topology and reachability information hiding in this network can optimize the overall network performance, stability, and convergence time. This will ultimately enable the business to meet its goal without adding any additional cost. In other words, the restructuring of IGP flooding domain design in this particular scenario is a strategic business-enabler solution.

However, in some situations, hiding topology and reachability information may lead to undesirable behaviors, such as suboptimal routing. Therefore, network designers must identify and measure the benefits and consequences by following the top-down approach. The following are some of the common questions that need to be thought about during the planning phase of the IGP flooding domain design:

- What are the business goals, priorities, and directions?

- How many Layer 3 nodes are in the network?

- What is the number of prefixes?

- Are there any hardware limitations (memory, CPU)?

- Is optimal routing a requirement?

- Is low convergence time a requirement?

- What IGP is used, and what underlying topology is used?

Furthermore, it is important that network designers understand how each protocol interacts with topology information and how each calculates its path, so as to be able to identify design limitations and provide valid optimization recommendations.

Link-state routing protocols take the full topology of the link-state routed network into account when calculating a path. For instance, in the network illustrated in Figure 8-17, the router of remote site A can reach the HQ network (192.168.1.0/24) through the WAN hub router. Normally, if the link between the WAN hub router and router A in the WAN core fails, remote site A will be notified about this topology change "in a flat link-state design." In fact, in any case, the remote site A router will continue to route its traffic via the WAN hub router to reach the HQ LAN 192.168.1.0/24.

In other words, in this scenario, the link failure notifications between the WAN hub router and the remote site routers are considered unnecessary extra processing for the remote site routers. This extra processing could lead to other limitations in large networks with a large number of prefixes and nodes, such as network and CPU spikes. In addition, the increased size of the LSDB will impact routing calculation and router memory consumption. Therefore, by introducing the principle of "topology hiding boundary" at the WAN hub router (for example, by using OSPF multi-area design), the overall routing design will be optimized (different fault domains) in terms of performance and stability.

A path-vector routing protocol (Border Gateway Protocol [BGP]) can achieve topology hiding by simply using either route summarization or filtering, and distance-vector protocols, by nature, do not propagate topology information. Moreover, with route summarization, network designers can achieve "reachability information hiding" for all the different routing protocols.

**NOTE**   A link-state routing protocol can offer built-in information hiding capabilities (route suppression) by using different types of flooding domains, such as L1/L2 in IS-IS and stubby types of areas in OSPF.

The subsequent sections examine where and why to break a routed network into multiple logical domains. You will also learn summarization techniques and some of the associated implications that you need to consider.

**NOTE**   Although route filtering can be considered as an option for hiding reachability information, it is often somewhat complicated with link-state protocols.

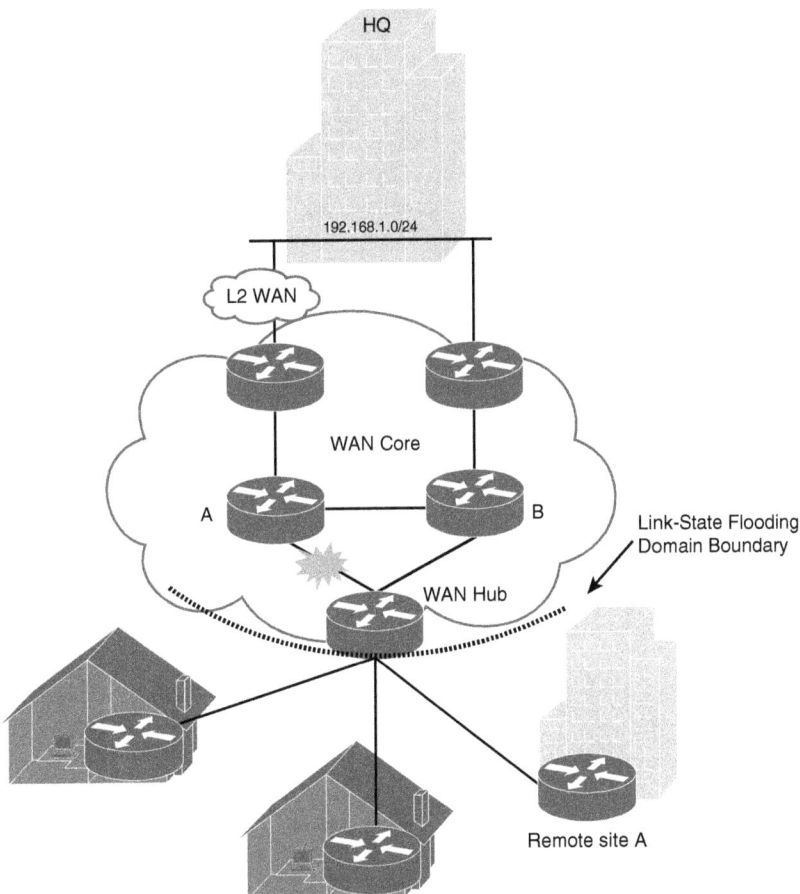

**Figure 8-17**   *Link-State Flooding Domain Boundaries*

## IGP Flooding Domains

As discussed earlier, modularity can add significant benefits to the overall network architecture. By applying this concept to the design of logical routing architectural domains, we can have a more manageable, scalable, and flexible design. To achieve this, we need to break a flat routing design into one that is more hierarchical and has modularity in its overall architecture. In this scenario, we may have to ask the following questions: How many layers should we consider in our design? How many modules or domains is good practice?

The simple answer to these questions depends on several factors, including the following:

- Design goal (simplicity versus scalability versus stability)

- Network topology

- Network size (nodes, routes)

- Routing protocol

- Network type (for example, enterprise versus service provider)

The following sections cover the various design considerations for IGP flooding domains, starting with a review of the structure of link-state and EIGRP domains.

### Link-State Flooding Domain Structure

Both OSPF and IS-IS as link-state routing protocols can divide the network into multiple flooding domains, as discussed earlier in this chapter. Dividing a network into multiple flooding domains, however, requires an understanding of the principles each protocol uses to build and maintain communication between the different flooding domains. In a multiple flooding domain design with OSPF, a backbone area is required to maintain end-to-end communication between all other areas (regardless of its type). In other words, area 0 in OSPF is like the glue that interconnects all other areas within an OSPF domain. In fact, non-backbone OSPF areas and area 0 (backbone area) interconnect and communicate in a hub-and-spoke fashion, as illustrated in Figure 8-18.

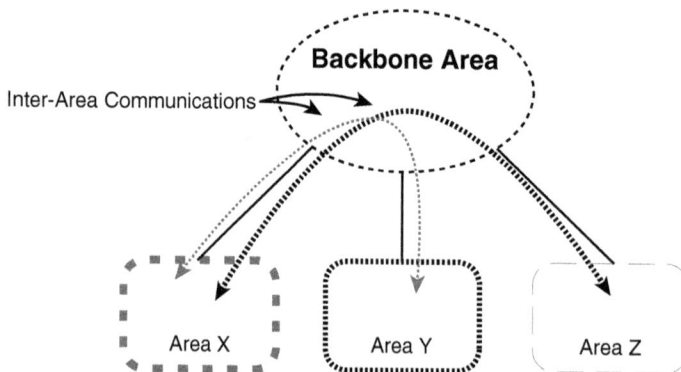

**Figure 8-18**  *OSPF Area Structure*

Similarly, with IS-IS, its levels chain (IS-IS flooding domains) must not be disjointed (L2 to L1/L2 to L1 and vice versa) for IS-IS to maintain end-to-end communications, where level 2 can be seen as analogous to area 0 in OSPF.

The natural communication behavior of link-state protocols across multiple flooding domains requires at least one router to be dually connected to the core flooding domain (backbone area) and the other area or areas, where an LSDB for each area is stored along with a separate shortest path first (SPF) calculation for each area. Moreover, the characteristic of the communication between link-state flooding domains (between border routers) is like a distance-vector protocol. In OSPF terminology, this router is called the **area border router (ABR)**. In IS-IS, the L1/L2 router is analogous to the OSPF ABR.

In general, OSPF and IS-IS are two-layer hierarchy protocols; however, this does not mean that they cannot operate well in networks with more hierarchies (as discussed later in this section).

In addition, although both OSPF and IS-IS are suitable for two-layer hierarchy network architecture, there are some differences in the way that their logical layout (flooding domains such as areas and levels) can be designed. For example, OSPF has a hard edge at the flooding domain borders. Typically, this is where routing policies are applied, such as route summarization and filtering, as shown in Figure 8-19.

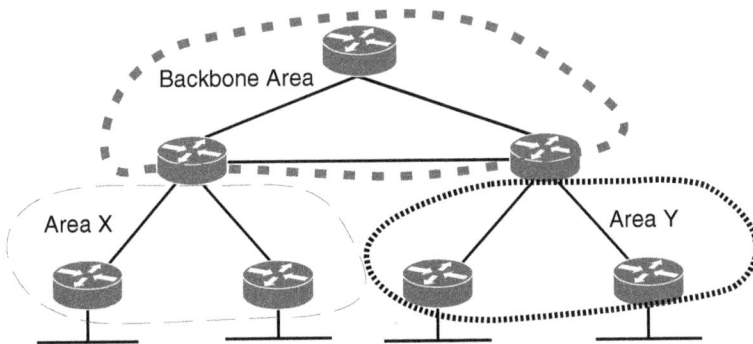

**Figure 8-19**  *OSPF Flooding Domain Borders*

By contrast, IS-IS routing information of the different levels (L1 and L2) is (technically) carried over different packets. This helps IS-IS have a softer edge at its flooding domain borders. This makes IS-IS more flexible than OSPF because the L2 routing domain can overlap with the L1 domains, as shown in Figure 8-20.

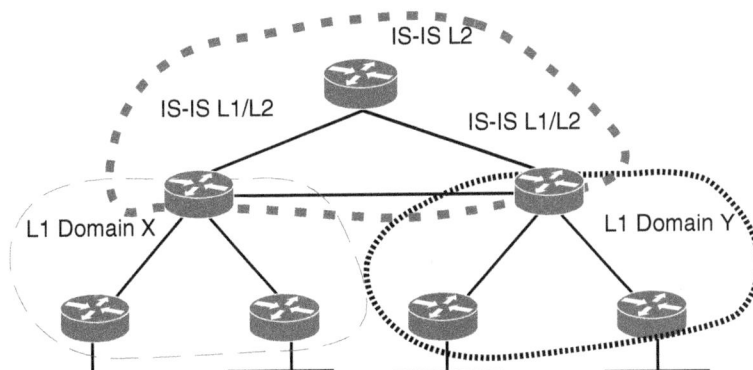

**Figure 8-20**  *IS-IS Flooding Domain Borders*

Consequently, IS-IS can perform better when optimal routing is required with multiple border routers, whereas OSPF requires special consideration with regard to the inter-ABR links (for example, which area to be part of, or in which direction is optimal routing more important).

Recommendation: With both OSPF and IS-IS, the design must always reflect that the backbone cannot be partitioned in case of a link or node failure. Although an OSPF virtual link can help to fix partitioned backbone area issues, it is not a recommended approach. Instead, redesign of the logical or physical architecture is highly desirable in this case. Nevertheless, an OSPF virtual link may be used as an interim solution (see the following example).

The scenario shown in Figure 8-21 illustrates poorly designed OSPF areas. It is considered a poor design because the OSPF backbone area has the potential to be partitioned if the direct interconnect link between the regional data centers (London and Sydney) fails. This will result in communication isolation between the London and Sydney data centers. However, let's assume that this organization needs to use its regional HQs (Melbourne, Amsterdam, and Singapore), which are interconnected in a hub-and-spoke fashion, as a backup transit path when the link between the London and Sydney sites is down.

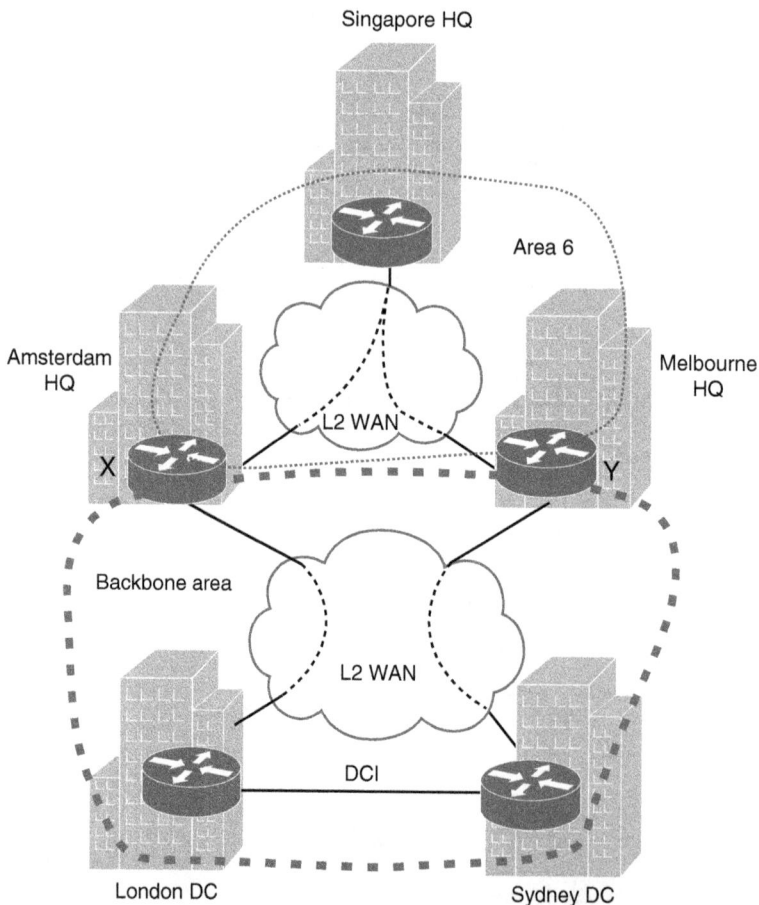

**Figure 8-21** *OSPF Poor Area Design*

Based on the current OSPF area design, a non-backbone area (area 6) cannot be used as a transit area. Figure 8-22 illustrates the logical view of OSPF areas before and after the failure event on the data center interconnect between London and Sydney data centers, which leads to a disjoint area 0 situation.

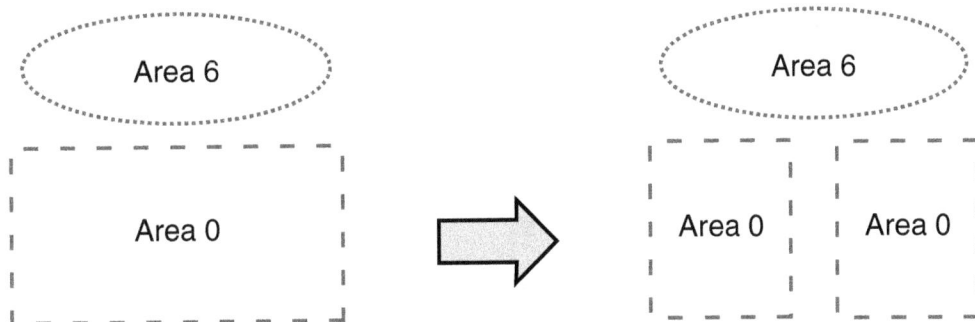

**Figure 8-22**  *Partitioned OSPF Backbone Area*

The ideal fix to this issue is to add redundant links from the London data center to WAN backbone router Y and/or from the Sydney data center to WAN backbone router X or to add a link between WAN backbone routers X and Y in area 0.

However, let's assume that the provisioning of the links takes a while and this organization requires a quick fix to this issue. As shown in Figure 8-23, if you deploy an OSPF virtual link between WAN backbone routers X and Y in Amsterdam and Melbourne, respectively (across the hub site in Singapore), OSPF will consider this link as a point-to-point link. Both WAN backbone routers (ABRs) X and Y will form a virtual adjacency across this virtual link. As a result, this path can be used as an alternate path to maintain the communication between London and Sydney data centers when the direct link between them is down.

**NOTE**   The solution presented in this scenario is based on the assumption that traffic flowing over multiple international links is acceptable from the perspective of business and application requirements.

You can use a GRE tunnel as an alternative method to the OSPF virtual link to fix issues like the one just described; however, there are some differences between using a GRE tunnel versus an OSPF virtual link, as summarized in Table 8-3.

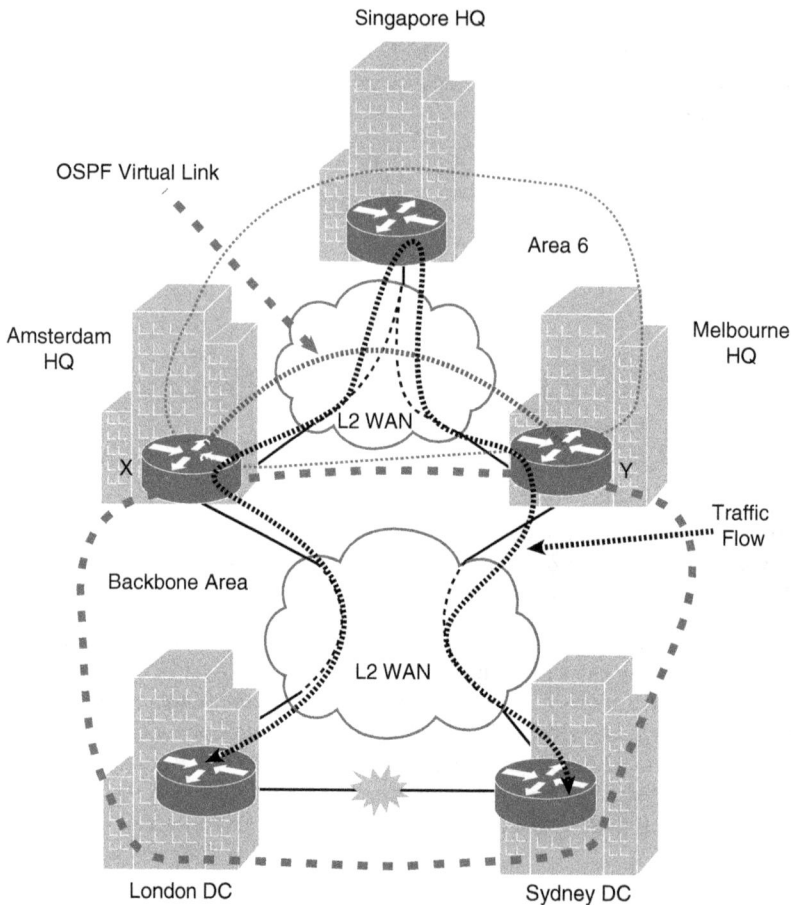

**Figure 8-23**  *OSPF Virtual Link*

**Table 8-3**  OSPF Virtual Link Versus GRE Tunnel

| GRE Tunnel | Virtual Link |
|---|---|
| May add tunnel overhead as all traffic is tunneled and encapsulated by the tunnel endpoints. | The routing updates are tunneled, but the data traffic is sent natively without tunnel overhead. |
| May add operational overhead; for example, IP addressing needs to be configured if not deployed as "unnumbered" and the tunnel interface/IP needs to be assigned manually to OSPF area 0. | Simplified operation; for example, no IP addressing needs to be configured manually, and it's under OSPF area 0 by default. |
| OSPF stub area can be used as a transit area for the tunnel. | The transit area cannot be an OSPF stub area. |

### Link-State Flooding Domains

One of the most common questions when designing OSPF or IS-IS is, "What is the maximum number of routers that can be placed within a single area?"

The common rule of thumb specifies between 50 and 100 routers per area or IS-IS level. However, in reality, it is hard to generalize the recommended maximum number of routers per area because the maximum number of routers can be influenced by several variables, such as the following:

- Hardware resources (such as memory, CPU)

- Number of prefixes (can be influenced by routes' summarization design)

- Number of adjacencies per shared segment

**NOTE**    The amount of available bandwidth with regard to the control plane traffic such as link-state LSAs/LSPs is sometimes a limiting factor. For instance, the most common quality of service (QoS) standard models followed by many organizations allocate one of the following percentages of the interface's available bandwidth for control (routing) traffic: 4-class model, 7 percent; 8-class model, 5 percent; and 12-class model, 2 percent. This is more of a concern when the interconnection is a low-speed link such as a legacy WAN link (time-division multiplexing [TDM] based, Frame Relay, or ATM) with limited bandwidth. Therefore, other alternatives are sometimes considered with these types of interfaces, such as a passive interface or static routing.

For instance, many service providers run thousands of routers within one IS-IS level. Although this may introduce other design limitations with regard to modern architectures, in practice it is proven as a doable design. In addition, today's router capabilities, in terms of hardware resources, are much stronger and faster than routers that were used five to seven years ago. This can have a major influence on the design, as well, because these routers can handle a high number of routes and volume of processing without any noticeable performance degradation.

In addition, the number of areas per border router is also one of the primary considerations in designing link-state routing protocols, in particular OSPF. Traditionally, the main constraint with the limited number of areas per ABR is the hardware resources. With the next generation of routers, which offer significant hardware improvements, ABRs can hold a greater number of areas. However, network designers must understand that additional areas to be added per ABR correlates to potential lower expected performance (because the router will store a separate LSDB per area).

In other words, the hardware capabilities of the ABR are the primary deterministic factor of the number of areas that can be allocated per ABR, considering the number of prefixes per area as well. Traditionally, the rule of thumb is to consider two to three areas (including the backbone area) per ABR. This is a foundation and can be expanded if the design requires more areas per ABR, with the assumption that the hardware resources of the ABR can handle this increase.

In addition to these facts and variables, network designers should consider the nature of the network and the concept of fault isolation and design modularity for large networks that can be designed with multiple functional fault domains (modules). For example, large-scale routed networks are commonly divided based on the geographic location of global networks or based on an administrative domain structure if they are managed by different entities.

8

### EIGRP Flooding Domain Structure

As discussed earlier, EIGRP has no protocol-specific flooding domains or structures. However, EIGRP with route summarization or filtering techniques can break the flooding domains into multiple hierarchies of routing domains, which can reduce the EIGRP query scope, as depicted in Figure 8-24. This concept is a vital contributor to the optimization of the overall EIGRP design in terms of scalability, simplicity, and convergence time.

In addition, EIGRP offers a higher degree of flexibility and scalability in networks with three or more levels in their hierarchies as compared to link-state routing protocols.

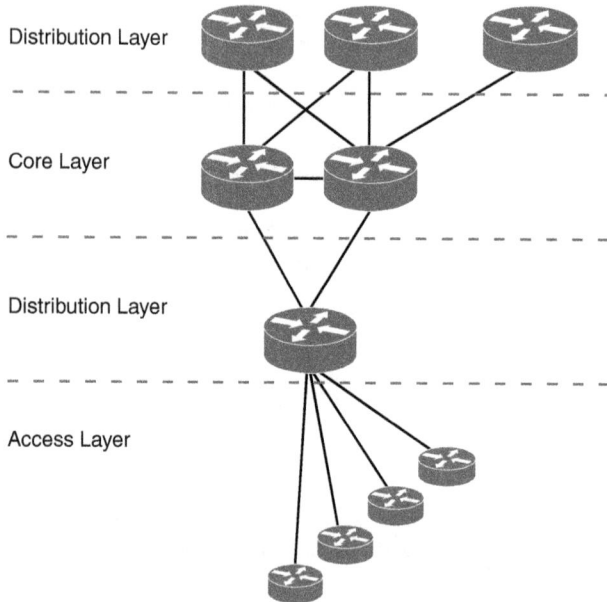

**Figure 8-24** *EIGRP Domain Boundaries*

### Routing Domain Logical Separation

The two main drivers for breaking a routed network into multiple logical domains (fault domains) are the following: to improve the performance of the networks and routers (fault isolation), and to modularize the design (to make it simpler, more stable, and scalable). These two drivers enhance network convergence and increase the overall routing architecture scalability. Furthermore, breaking the routed topology into multiple logical domains will facilitate topology aggregation and information hiding. It is critical to decide where a routing domain can be divided into two or multiple logical domains. In fact, several variables influence the location where the routing domains are broken or divided. The considerations discussed in the sections that follow are the primary influencers that help to determine the correct location of the logical routing boundaries. Network designers need to consider these when designing or restructuring a routed network.

#### Underlying Physical Topology

As discussed in Chapter 1, "Network Design," the physical network layout is like the foundation of a building. As such, it is the main influencer when designing the logical structure of a routing domain (for example, a hub-and-spoke versus ring topology). For instance, the level of the hierarchy held by a given network can impact the logical routing design if its structure includes two, three, or more tiers, as illustrated in Figure 8-25.

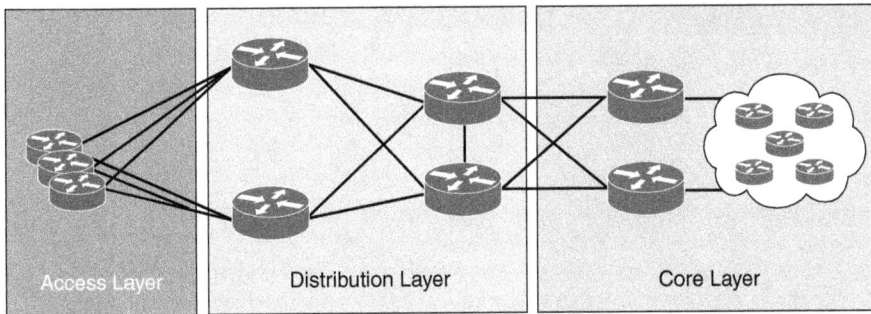

**Figure 8-25**  *Topology Depth*

Moreover, the points in the network where the interconnections or devices meet (also known as *chokepoints*) at any given tier within the network are good potential border locations of a fault domain boundary, such as ABR in OSPF. For instance, in Figure 8-26, the network is constructed of three-level hierarchies. Routers A and B and routers C and D are good potential points for breaking the routing domain (physical aggregation points). Also, these boundaries can be feasible places to perform route summarizations.

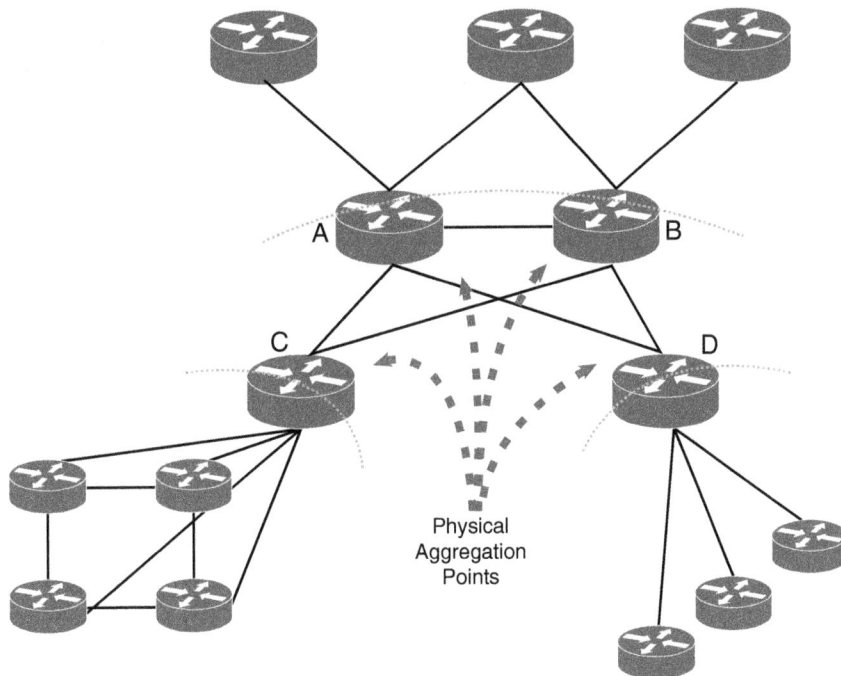

**Figure 8-26**  *Physical Aggregation Points*

The other important factor with regard to the physical network layout is to break areas that have a high density of interconnections into separate logical fault domains where possible. As a result, devices in each fault domain will have smaller reachability databases (for example, LSDB) and will only compute paths within their fault domain, as illustrated in Figure 8-27. This will ultimately lead to the reduction of the overall control plane design complexity. This concept will promote a design that can facilitate the support of other design principles, including simplicity, modularity, scalability, topology, and reachability of information hiding.

The network illustrated in Figure 8-27 has four different functional areas:

- The primary data center

- The regional data center

- The international WAN

- The hub-and-spoke network for some of the remote sites

From the perspective of logical separation, you should place each one of the large parts of the network into its own logical domain. The logical topology can be broken using OSPF areas, IS-IS levels, or EIGRP route summarization. The question you might be asking is, "Why has the domain boundary been placed at routers G and H rather than router D?" Technically, both are valid places to break the network into multiple logical domains. However, if we place the domain boundary at router D, both the primary data center network and regional data center will be under the same logical fault domain. This means the network may be less scalable and associated with lower control plane stability because routers E and F will have a full view of the topology of the regional data center network connected to routers G and H. In addition, routers G and H most probably will face the same limitations as routers E and F. As a result, if there is any link flap or routing change in the regional data center network connected to router G or H, it will be propagated across to routers E and F (unnecessary extra load and processing).

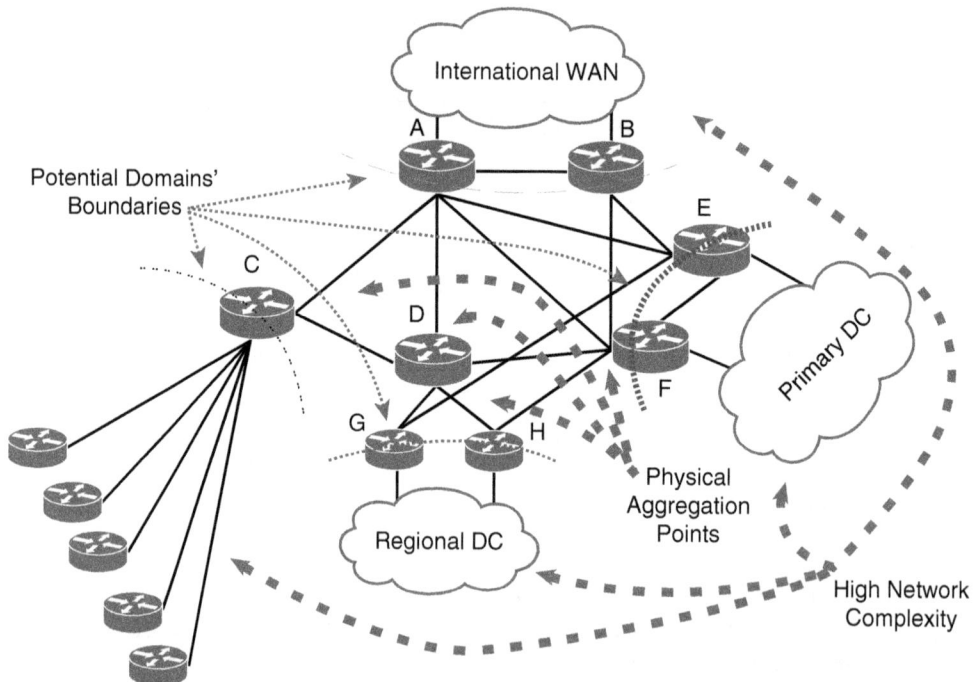

**Figure 8-27** *Potential Routing Domain Boundaries*

## Traffic Pattern and Volume

By understanding traffic patterns (for example, south–north versus east–west) and traffic volume trends, network designers can better understand the impact if a logical topology were to be divided into multiple domains at certain points (see Figure 8-28). For example, OSPF always prefers the path over the same area regardless of the link cost over other areas. (For more information about this, see the section "IGP Traffic Engineering and Path Selection: Summary" later in this chapter.) In some situations, this could lead to suboptimal routing, where a high volume of traffic will travel across low-capacity links or expensive links with strict billing that not every type of communication should go over them; this results from the poor design of OSPF areas that did not consider bandwidth or cost requirements.

Similarly, if the traffic pattern is mostly north–south, such as in a hub-and-spoke topology where no communication between the spokes is required, this can help network designers to avoid placing the logical routing domain boundary at points likely to use spoke sites as transit sites (suboptimal routing). For instance, the scenario depicted in Figure 8-29 demonstrates how the application of the logical area boundaries on a network can influence path selection. Traffic sourced from router B going to the regional data center behind router G should (optimally) go through router D, then across one of the core routers E or F, and finally to router C to reach the data center over one of the core high-speed links. However, the traffic is currently traversing the low-speed link via router A. This path (B-D-A-C-G) is within the same area (area 10), as shown in Figure 8-29.

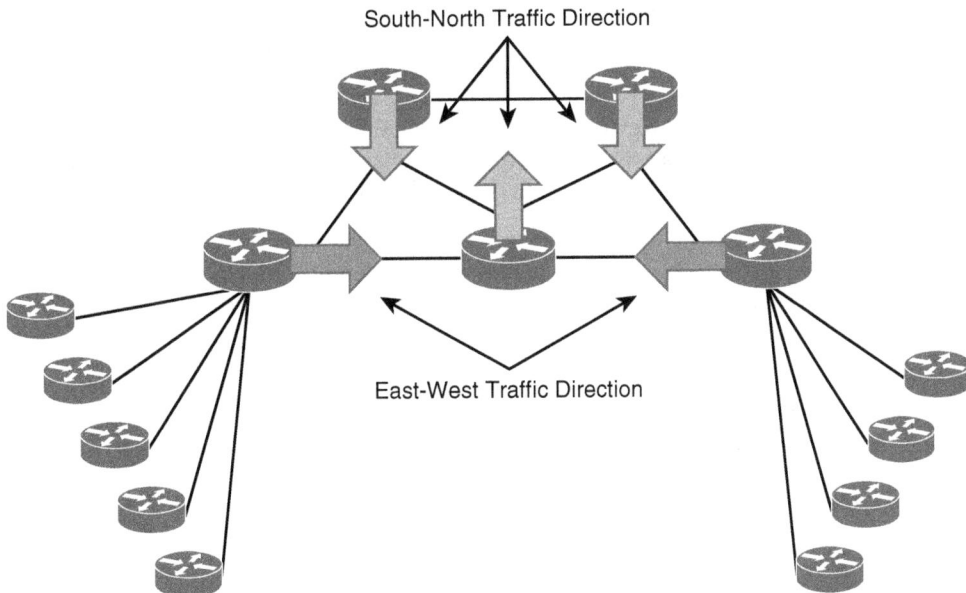

**Figure 8-28**  *Traffic Patterns*

Regional Data Center

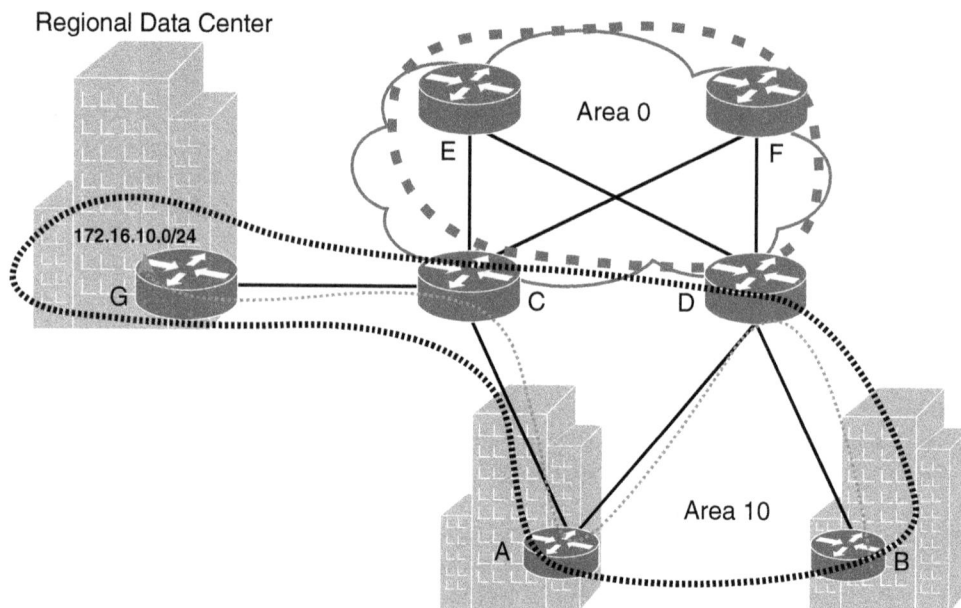

**Figure 8-29**  *OSPF Suboptimal Routing*

No route filtering or any type of summarization is applied to this network. This suboptimal routing results entirely from the poor design of OSPF areas. If you apply the concepts discussed in this section, you can optimize this design and fix the issue of suboptimal routing, as follows:

- First, the physical network is a three-tier hierarchy. Routers C and D are the points where the access, data center, and core links meet, which makes them a good potential location to be the area border (which is already in place).

- Second, if you divide this topology into functional domains, you can, for example, have three parts (core, remote sites, and data center), with each placed in its own area. This can simplify summarization and introduce modularity to the overall logical architecture.

- The third point here is the traffic pattern. It is obvious that there will be traffic from the remote sites to the regional data center, which needs to go over the high-speed links rather than going over the low-speed links by using other remote sites as a transit path.

Based on this analysis, the simple solution to this design is to either place the data center in its own area or to make the data center part of area 0, as illustrated in Figure 8-30, with area 0 extended to include the regional data center.

**NOTE**  Although both options are valid solutions, on the CCDE exam the correct choice will be based on the information and requirements provided. For instance, if one of the requirements is to achieve a more stable and modular design, a separate OSPF area for the regional data center will be the more feasible option in this case.

Regional Data Center

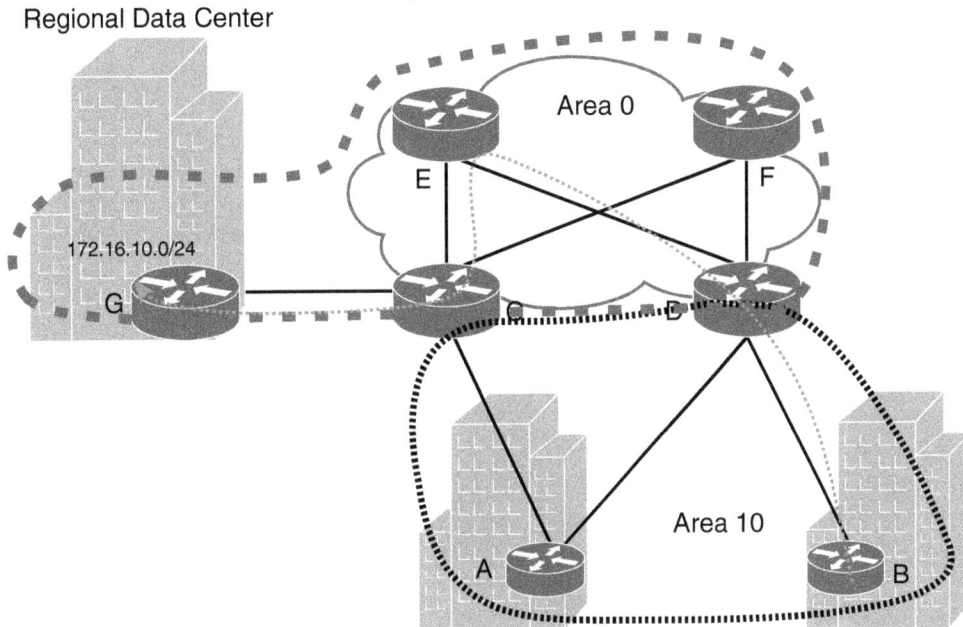

**Figure 8-30**   *OSPF Optimal Routing*

Similarly, if IS-IS is used in this scenario, as illustrated in Figure 8-31, router B will always use router A as a transit path to reach the regional data center prefix. Over this path (B-D-A-C-G), the regional data center prefix behind router G will be seen as IS-IS level 1, and based on IS-IS route selection rules, this path will be preferred compared to the one over the core, in which it will be announced as an IS-IS level 2 route. (For more information about this, see the section "IGP Traffic Engineering and Path Selection: Summary.") Figure 8-31 suggests a simple possible solution to optimize IS-IS flooding domain design (levels): including the regional data center as part of IS-IS level 2. This ensures that traffic from the spokes (router B in this example) destined to the regional data center will always traverse the core network rather than transiting any other spoke's network.

### Route Summarization

The other major factor when deciding where to divide the logical topology of a routed network is where summarization or reachability information hiding can take place. The important point here is that the physical layout of the topology must be considered. In other words, you cannot decide where to place the reachability information hiding boundary (summarization) without referring to what the physical architecture looks like and where the points are that can enhance the overall routing design if summarization is enabled. Subsequent sections in this chapter cover route summarization design considerations in more detail.

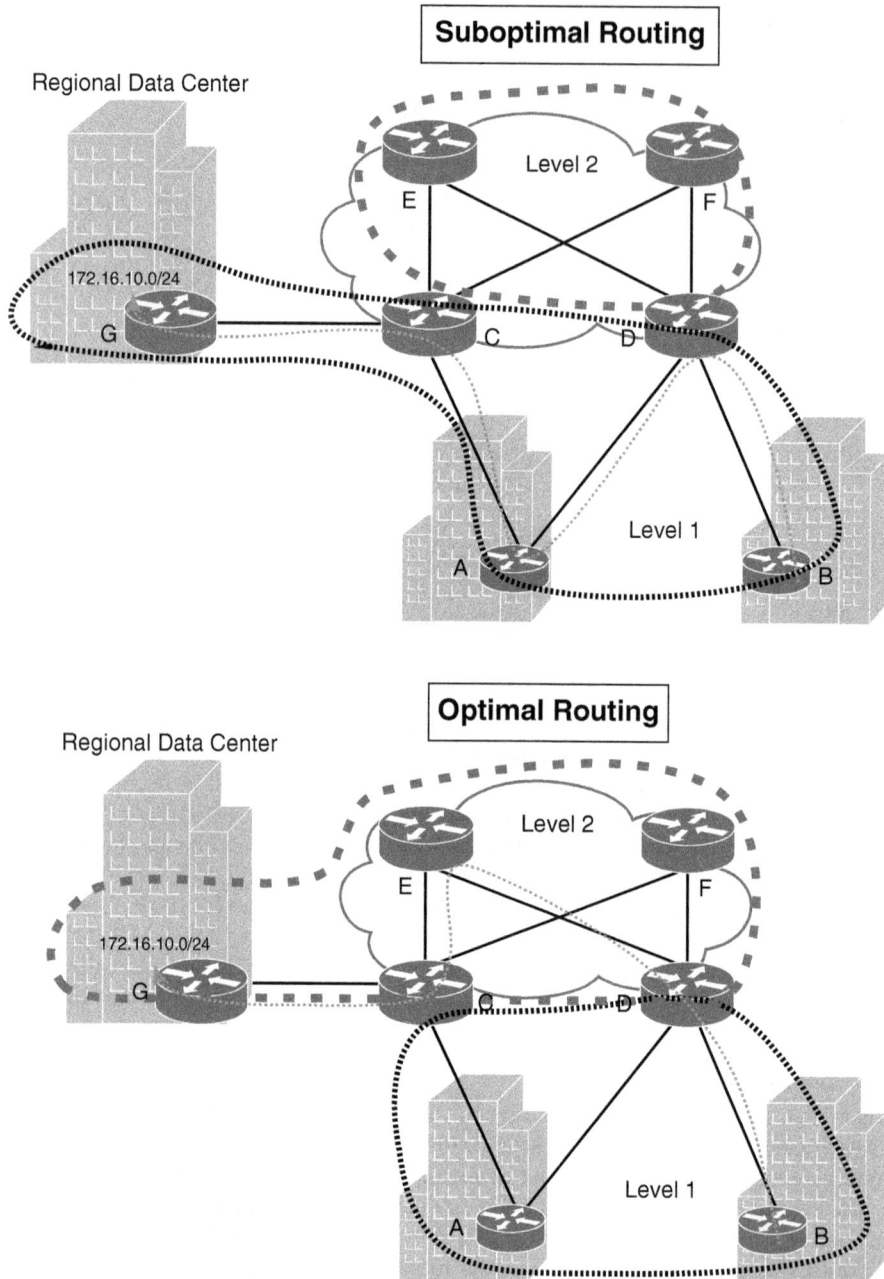

**Figure 8-31** *IS-IS Levels and Optimal Routing*

*Security Control and Policy Compliance*

This pertains more to what areas of a certain network have to be logically separated from other parts of the network. For example, an enterprise might have a research and development (R&D) lab where different types of unified communications applications are installed, including routers and switches. Furthermore, the enterprise security policy may dictate that this part of the network must be logically contained and only specific reachability information needs to be leaked between this R&D lab environment and the production network. Technically, this will lead to increased network stability and policy control.

## Route Summarization

By having a well-structured IP address align with the physical layout with reachability information hiding using route summarization, as shown in Figure 8-32, network designers can achieve an optimized level of network design simplicity, scalability, and stability.

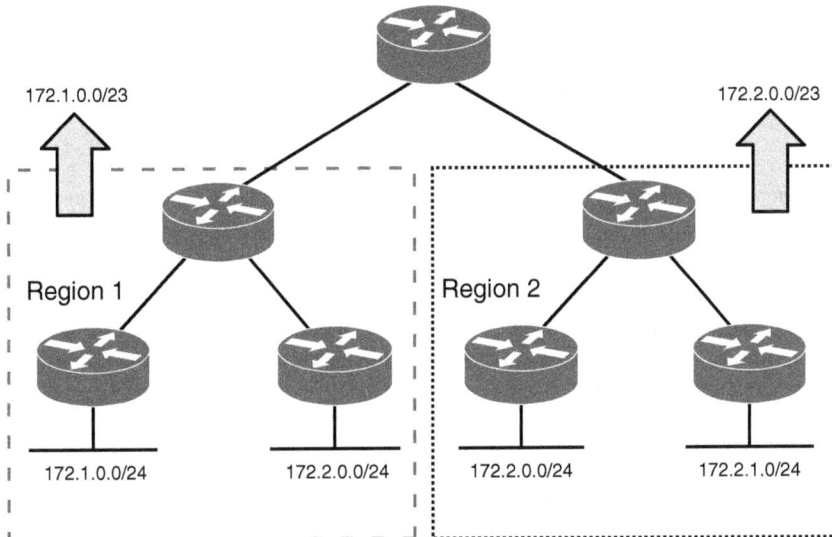

172.1.0.0/23

172.2.0.0/23

Region 1

Region 2

172.1.0.0/24      172.2.0.0/24      172.2.0.0/24      172.2.1.0/24

**Figure 8-32**   *Structured IP Addressing and Physical Connectivity*

For example, based on the routes' summarization structure illustrated in Figure 8-32, if there is any link flap in a remote site in region 2, it will not affect the remote site routers of region 1 in processing or updating their topology database (which in some situations might cause unnecessary path recalculation and processing, which in turn may lead to service interruption). Usually, route summarization facilitates the reduction of the RIB table size by reducing the number of route counts. This means less memory, lower CPU utilization, and faster convergence time during a network change or following any failure event. In other words, the boundary of the route summarization almost always overlaps with the boundary of the fault domain.

However, not every network has structured network IP addressing like the one shown in Figure 8-32. Therefore, network designers must consider alternatives to overcome this issue. In some situations, the solution is "not to summarize." For instance, Figure 8-33 illustrates a network with unstructured IP addressing, and the business may not be able to afford to change its IP scheme in the near future.

8

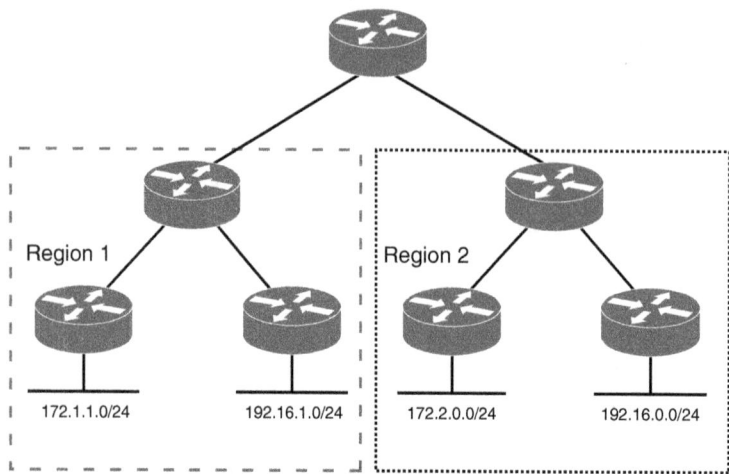

**Figure 8-33**  *Network with Unstructured IP Addressing*

Moreover, in some scenarios, the unstructured physical connectivity can introduce challenges with route summarization. For example, in Figure 8-34, summarization can lead to forcing all the traffic from the hub site to always prefer the high-cost and low-bandwidth link to reach the 172.2.0.0/24 network (more specific route over the high-cost non-summarized link), which may lead to undesirable outcome from the business point of view (for example, slow applications' response time over this link).

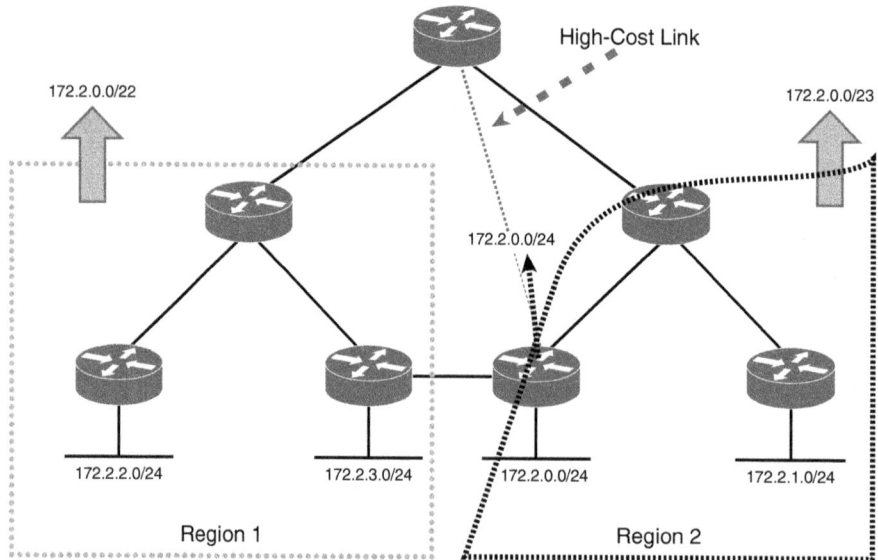

**Figure 8-34**  *Unstructured Physical Connectivity*

As a general rule of thumb (not always), summarization should be considered at the routing logical domain boundaries. The reason why summarization might not always be considered at the logical boundary domain is that in some designs it can lead to suboptimal routing or traffic black-holing (also known as summary black holes). The following subsections discuss summary suboptimal routing and summarization black-holing in more detail.

## Summary Black Holes

The principle of route summarization is based on hiding specific reachability information. This principle can optimize many network designs, as discussed earlier; however, it can lead to traffic black-holing in some scenarios because of the specific hidden routing information. In the scenario illustrated in Figure 8-35, routers A and B send the summary route only (172.1.0.0/21) with the same metric toward router C. Based on this design, in case of link failure between routers D and E, the routing table of router C will remain intact because it is receiving only the summary. Consequently, there is potential for traffic black-holing. For instance, traffic sourced from router C destined to network 172.1.1.0/24 landing at router B will be dropped because of this summarization black-holing. Moreover, the situation can become even worse if router C is performing per-packet load balancing across routers A and B. In this case, 50 percent of the traffic is expected to be dropped. Similarly, if router C is load balancing on a per-session basis, hypothetically some of the sessions will reach their destinations and others may fail. As a result, route summarization in this scenario can lead to serious connectivity issues in some failure situations.

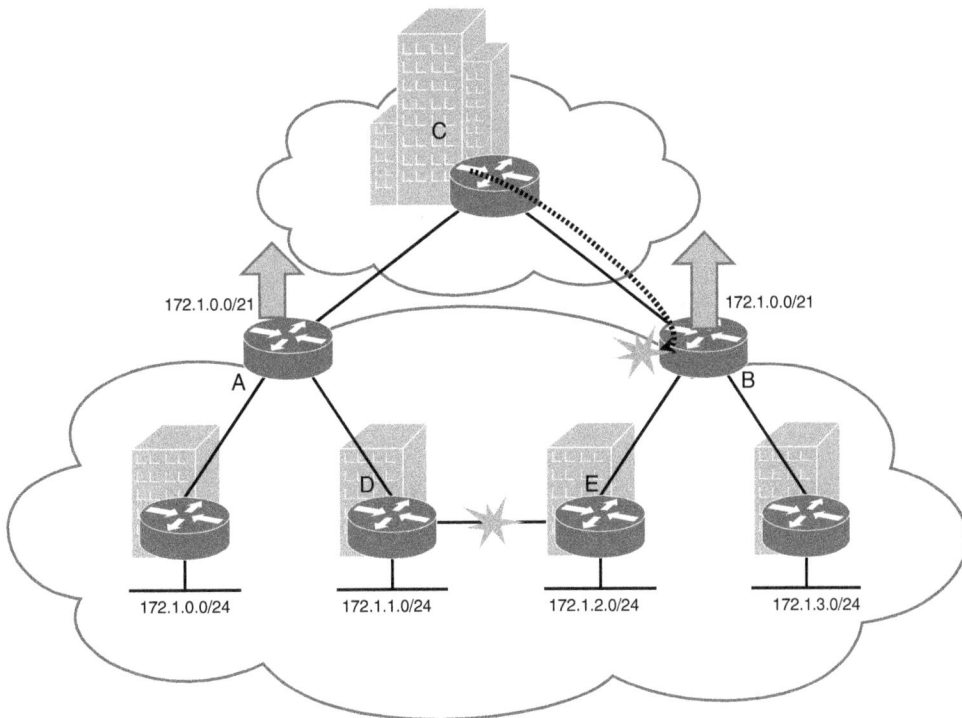

172.1.0.0/21

172.1.0.0/21

172.1.0.0/24          172.1.1.0/24          172.1.2.0/24          172.1.3.0/24

**Figure 8-35**   *Summary Black Hole*

To mitigate this issue and enhance the design in Figure 8-35, summarization either should be avoided (this option might not always be desirable because it can reduce the stability and scalability in large networks) or at least one non-summarized link must be added between the summarizing routers (in this scenario, between routers A and B, as illustrated in Figure 8-36). The non-summarized link can be used as an alternate path to overcome the route summarization black-holing issue described previously.

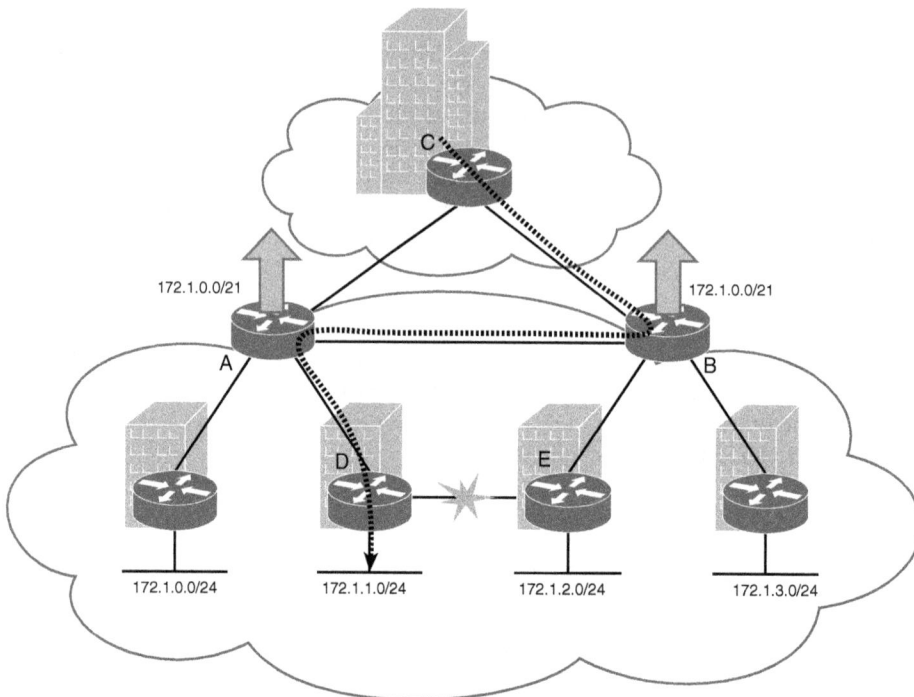

**Figure 8-36**  *Summary Black-Hole Optimization*

## Suboptimal Routing

Although hiding reachability information with route summarization can help to reduce control plane complexity, it can lead to suboptimal routing in some scenarios. This suboptimal routing, in turn, may lead traffic to use a lower-bandwidth link or an expensive link, over which the enterprise might not want to send every type of traffic. For example, if we use the same scenario discussed earlier in the OSPF areas, we then apply summarization to the data center edge routers of London and Milan and assume that the link between Sydney and Milan is a high-cost link that has a typically lower routing metric, as depicted in Figure 8-37.

**NOTE**  The example in Figure 8-37 is "routing protocol" neutral; it can apply to all routing protocols in general.

As illustrated in Figure 8-37, the link between the Sydney branch and the Milan data center is 10 Mbps, and the link to London is 5 Mbps. In addition, the data center interconnect between Milan and London data centers is only 2 Mbps. In this particular scenario, summarization of the Sydney branch from both data centers will typically hide the more specific route. Therefore, the Sydney branch will send traffic destined to any of the data centers over the high-bandwidth link (with a lower routing metric); in this case, the Sydney–Milan path will be preferred (almost always, higher bandwidth = lower path metric). This behavior will cause suboptimal routing for traffic destined to the London data center network. This suboptimal routing, in turn, can lead to an undesirable experience, because rather than having

5 Mbps between the Sydney branch and the London data center, their maximum bandwidth will be limited to the data center interconnect link capacity, which is 2 Mbps in this scenario. This is in addition to the extra cost and delay that will be from the traffic having to traverse multiple international links.

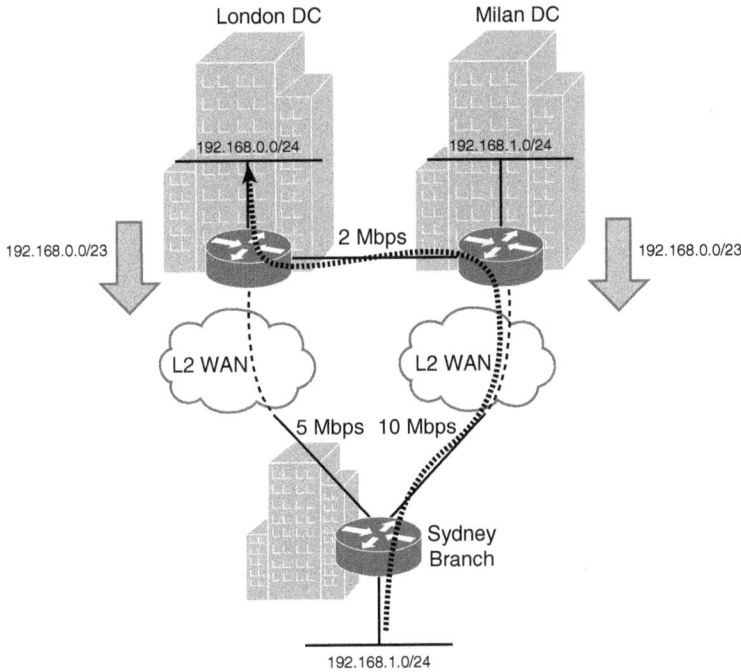

**Figure 8-37**  *Summary Route and Suboptimal Routing*

Even so, this design limitation can be resolved via different techniques based on the use of the routing protocol, as summarized in Table 8-4.

**Table 8-4**  Suboptimal Routing Optimization Techniques

| OSPF | ISIS | EIGRP | BGP |
|---|---|---|---|
| Using a normal OSPF area combined with LSA type 3 filtering at the ABRs to send 23 summary route from Milan side and the more specific route along with /23 summary route from London ABR where the optimal path needs to take place. | Using route leaking from L2 to L1 (RFC 5302); in this scenario, leaking the more specific route from London L1/L2 router. | Send two summary routes containing the more- and less-specific routes from the router that needs to be used for the more specific routes (London) Route leaking with the summary to send more specific routes from the desired router for optimal path (London router). | By sending the summary along with the more specific route (for example, using unsuppress-map with the BGP summary at the London router). |

Figure 8-38 illustrates link-state areas/levels application with regard to the discussed scenario and the suggested solutions because of the different areas/levels designs can have a large influence on the overall traffic engineering and path selection.

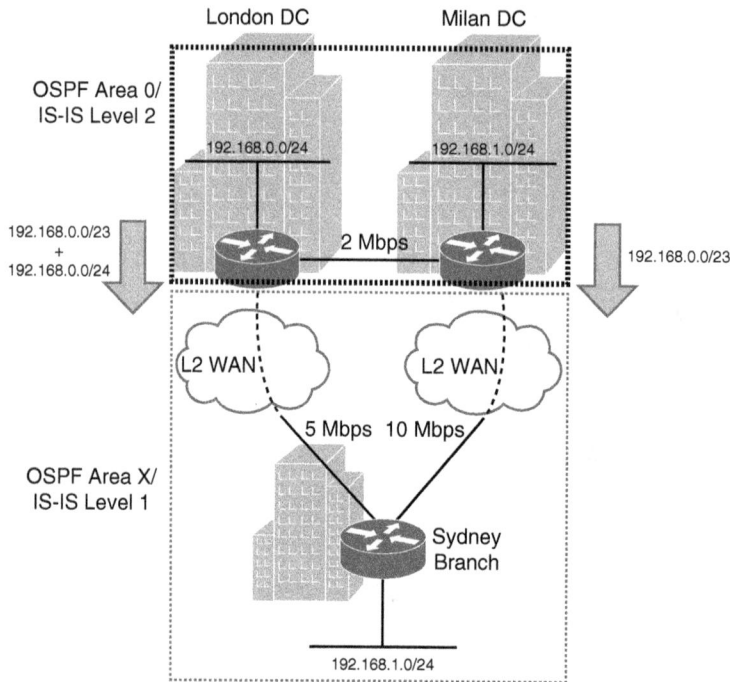

**Figure 8-38**  *Link-State Flooding Domain Applications and Optimal Routing*

**NOTE**   With IS-IS, L1-L2 (ABR) may send the default route toward the L1 domain, and the route leaking at the London ABR will leak/send the more specific local prefix for optimal routing.

Based on these design considerations and scenarios, we can conclude that although route summarization can optimize the network design for several reasons (discussed earlier in this chapter), in some scenarios, summarization from the core networks toward the edge or remote sites can lead to suboptimal routing. In addition, summarization from the remote sites or edge routers toward the core network may lead to traffic black holes in some failure scenarios. Therefore, to provide a robust and resilient design, network designers must pay attention to the different failure scenarios when considering route summarization.

## IGP Traffic Engineering and Path Selection: Summary

By understanding the variables that influence a routing protocol decision to select a certain path, network designers can gain more control to influence route preference over a given path based on a design goal. This process is also known as *traffic engineering*.

In general, routing protocols perform what is known as *destination traffic engineering*, where the path selection is always based on the targeted prefix and the attributes of the path to reach this prefix. However, each of the three IGPs discussed in this chapter has its own metrics, algorithm, and default preferences to select routes. From a routing point of view, they can be altered to control which path is preferred or selected over others, as summarized in the sections that follow.

## OSPF

If multiple routes cover the same network with different types of routes, such as inter-area (LSA type 3) or external (LSA type 5), OSPF considers the following list "in order" to select the preferred path (from highest preference to the lowest):

1. Intra-area routes

2. Inter-area routes

3. External type 1 routes

4. External type 2 routes

Let's take a scenario where there are multiple routes covering the same network with the same route type as well; for instance, both are inter-area routes (LSA type 3). In this case, the OSPF metric (cost) that is driven by the links' bandwidth is used as a tiebreaker to select the preferred path. Typically, the route with the lowest cost is chosen as the preferred path.

If multiple paths cover the same network with the same route type and cost, OSPF will typically select all the available paths to be installed in the routing table. Here, OSPF performs what is known as equal-cost multipath (ECMP) routing across multiple paths.

An OSPF router that injects external LSAs into the OSPF database is called an **autonomous system boundary router (ASBR)**. For external routes with multiple ASBRs, OSPF relies on LSA type 4 to describe the path's cost to each ASBR that advertises the external routes. For instance, in the case of multiple ASBRs advertising the same external OSPF E2 prefixes carrying the same redistributed metric value, the ASBR with the lowest reported forwarding metric (cost) will win as the preferred exit point.

## IS-IS

Typically, with IS-IS, if multiple routes cover the same network (same exact subnet) with different route types, IS-IS follows the sequence here "in order" to select the preferred path:

1. Level 1

2. Level 2

3. Level 2 external with internal metric type

4. Level 1 external with external metric type

5. Level 2 external with external metric type

Like OSPF, if there are multiple paths to a network with the same exact subnet, route type, and cost, IS-IS selects all the available paths to be installed in the routing table (ECMP).

## EIGRP

EIGRP has a set of variables that can solely or collectively influence which path a route can select. For more stability and simplicity, bandwidth and delay are commonly used for this purpose. Nonetheless, it is always simpler and safer to alter delay for EIGRP path selection, because of some implications associated with tuning bandwidth for EIGRP traffic engineering purposes discussed earlier in this chapter, which requires careful planning.

Like other IGPs, EIGRP supports the concept of ECMP; in addition, it does support "unequal cost load balancing," as well, with proportional load sharing.

### Summary of IGP Characteristics

As discussed in this chapter, each routing protocol behaves and handles routing differently on each topology. Table 8-5 summarizes the characteristics of the IGPs, taking into account the topology that is used.

**Table 8-5** IGP Characteristics Summary

|  | Link State | EIGRP |
|---|---|---|
| Hub-and-spoke scalability | Moderate scaling capability. | Excellent scaling capability. |
| Hub-and-spoke considerations | Care must be taken with summary black holes.<br><br>Consider stub areas with filtering to prevent transiting traffic via remote sites and large RIB tables. | Consider routes to address summary black holes.<br><br>Consider stub remote routers with filtering and summarization to prevent transiting traffic through remote sites. |
| Full-mesh scalability | Acceptable scaling capability, to a certain extent. | Acceptable scaling capability, to a certain extent. |
| Full-mesh considerations | Manually designate flooding points and increase scaling through a full mesh.<br><br>Potential of a temporary routing loop following a network failure event. | Summarize into and out of the full mesh.<br><br>Increased number of EIGRP queries following a network event failure. |
| Summarization | Only at border routers (ABRs). | At any place. |
| Filtering | Only at border routers (ABRs). | At any place. |
| Load balancing | Equal-cost load balancing. | Equal- and unequal-cost load balancing. |

**NOTE** In Table 8-5, link-state ABR refers to either OSPF ABR, ASBR, or IS-IS L1-L2 router.

**NOTE** As you'll notice, the full mesh in the preceding table has no excellent scalability among the IGPs. This is because the nature of full-mesh topology is not very scalable. (The larger the mesh becomes, the more complicated the control plane will be.)

## Route Redistribution Design Considerations

*Route redistribution* refers to the process of exchanging or injecting routing information (typically routing prefixes) between two different routing domains or protocols. However, route redistribution between routing domains does not always refer to the route redistribution between two different routing protocols. For example, redistribution between two

OSPF routing domains where the border router runs two different OSPF instances (process) represents the redistribution between two routing domains using the same routing protocol. Route redistribution is one of the most advanced routing design mechanisms commonly relied on by network designers to achieve certain design requirements, such as the following:

- In merger and acquisition scenarios, route redistribution can sometimes facilitate routing integration between different organizations.

- In large-scale networks, such as global organizations, where BGP might be used across the WAN core and different IGP islands connect to the BGP core, full or selective route redistribution can facilitate route injection between these protocols and routing domains in some scenarios.

- Route redistributions can also be used as an interim solution during the migration from one routing protocol to another.

**NOTE**    None of the preceding points can be considered as an absolute use case for route redistribution because the use of route redistribution has no fixed rule or standard design. Therefore, network designers need to rely on experience when evaluating whether route redistribution needs to be used to meet the desired goal or whether other routing design mechanisms can be used instead, such as static routes.

Route redistribution can sometimes be as simple as adding a one-line command. However, its impact sometimes leads to major network outages because of routing loops or the black-holing of traffic, which can be introduced to the network if the redistribution was not planned and designed properly. That is why network designers must have a good understanding of the characteristics of the participating routing protocols and the exact aim of route redistribution. In general, route redistribution can be classified into two primary models, based on the number of redistribution boundary points:

- Single redistribution boundary point

- Multiple redistribution boundary points

### Single Redistribution Boundary Point

This design model is the simplest and most basic route redistribution design model; it has minimal complexities, if any. Typically, the edge router between the routing domains can perform either one- or two-way route redistribution based on the desired goal without any concern, as depicted in Figure 8-39. This is based on the assumption that there is no other redistribution point between the same routing domains anywhere else across the entire network.

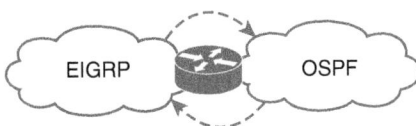

**Figure 8-39**  *Single Redistribution Boundary Point*

However, if the redistributing border router belongs to three routing domains, the route that is sourced from another routing protocol cannot be redistributed into a third routing protocol on the same router. For instance, in Figure 8-40, the route redistributed from EIGRP into OSPF cannot be redistributed again from OSPF into RIP.

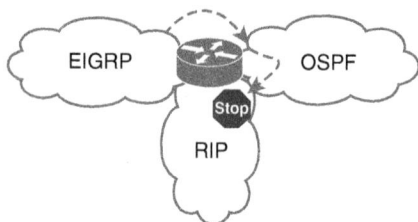

**Figure 8-40**  *Nontransitive Attribute of Route Redistribution*

## Multiple Redistribution Boundary Point

Networks with two or more redistribution boundary points between routing domains require careful planning and design prior to applying the redistribution into the production network, because it can lead to a complete or partial network outage. The primary issues that can be introduced by this design are as follows:

■ Routing loop

■ Suboptimal routing

■ Slower network convergence time

To optimize a network design that has two or more redistribution boundary points, network designers must consider the following aspects and how each may impact the network, along with the possible methods to address it based on the network architecture and the design requirements (for example optimal versus suboptimal routing):

■ Metric transformation

■ Administrative distance

### Metric Transformation

Typically, each routing protocol has its own characteristic and algorithm to calculate network paths to determine the best path to use based on certain variables known as *metrics*. Because of the different metrics (measures) used by each protocol, the exchange of routing information between different routing protocols will lead to metric conversion so that the receiving routing protocol can understand this route, as well as be able to propagate this route throughout its routed domain. Therefore, specifying the metric at the redistribution point is important, so that the injected route can be understood and considered.

For instance, a common simple example is a redistribution from RIP into OSPF. RIP relies on hop counts to determine the best path, whereas OSPF considers link cost that is driven by the link bandwidth. Therefore, redistributing RIP into OSPF with a metric of 5 (five RIP hops) has no meaning to OSPF. Hence, OSPF assigns a default metric value to the redistributed external route. Furthermore, metric transformation can lead to routing loops if not planned and designed correctly when there are multiple redistribution points. For example, Figure 8-41 illustrates a scenario of mutual redistribution between RIP and OSPF over two border routers. Router A receives the RIP route from the RIP domain with a metric of 5, which means five hops. Router B will redistribute this route into the OSPF domain with the default redistribution metrics or any manually assigned metric. The issue in this scenario is that when the same route is redistributed back into the RIP domain with a lower metric (for example, 2), router A will see the same route with a better metric from the second border router. As a result, a routing loop will be formed based on this design (because of metric transformation).

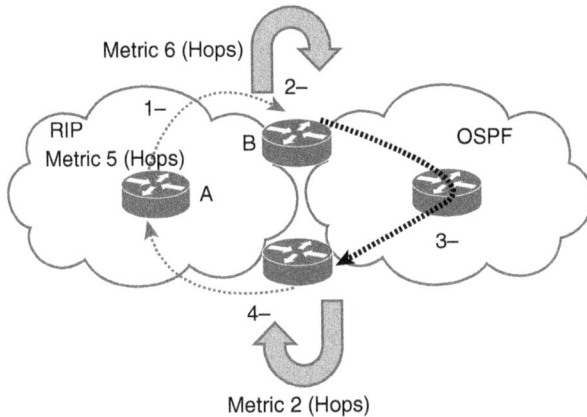

**Figure 8-41**  *Multipoint Routing Redistribution*

Hypothetically, this metric issue can be fixed by redistributing the same route back into the RIP domain with a higher metric value (for example, 7). However, this will not guarantee the prevention of routing loops because there is another influencing factor in this scenario, which is the administrative distance (discussed next in more detail). Therefore, by using route filtering or a combination of route filtering and tagging to prevent the route from being reinjected into the same domain, network designers can avoid route looping issues in this type of scenario.

### Administrative Distance

Some routing protocols assign a different administrative distance (AD) value to the redistributed route by default (typically higher than the locally learned route) to give it preference over the external (redistributed route). However, this value can be changed, which enables network designers and engineers to alter the default behavior with regard to route and path section. From the route redistribution design point of view, AD can be a concern that requires special design considerations, especially when there are multiple points of redistribution with mutual route redistribution.

To resolve this issue, either route filtering or route tagging jointly with route filtering can be used to avoid reinjecting the redistributed (external) route back into the same originating routing domain. You can tune AD values to control the preferred route. However, this solution does not always provide the optimal path when there are multiple redistribution border routers performing mutual redistribution. If for any reason AD tuning is used, the network designer must be careful when considering this option, to ensure that routing protocols prefer internally learned prefixes over external ones (to avoid unexpected loops or suboptimal routing behavior).

## Route Filtering Versus Route Tagging with Filtering

Route filtering and route tagging combined with route filtering are common and powerful routing policy mechanisms that you can use in many routing scenarios to control route propagation and advertisement and to prevent routing loops in situations where multiple redistribution boundary points are exits with mutual route redistribution between routing domains. However, these mechanisms have some differences that network designers must be aware of, as summarized in Table 8-6.

**Table 8-6**  Route Filtering Techniques Comparison

| Design Consideration | Route Filtering | Route Tagging with Filtering |
|---|---|---|
| Scalability | Low | High |
| Manageability | Complex | Simple |
| Multipoint redistribution loop prevention | Yes (complex) | Yes (simple) |
| Multipoint redistribution optimal routing | Yes (complex) | Yes (simple) |
| Flexibility | Limited | Flexible |

Based on the simple comparison in Table 8-6, it is obvious that route filtering is more suitable for small and simple filtering and loop-prevention tasks. In contrast, route filtering associated with route tagging can support large-scale and dynamic networks to achieve more scalable and flexible routing policies across routing domains.

For example, in the scenario illustrated in Figure 8-42, there are two boundary redistribution points with mutual redistribution between EIGRP and IS-IS in both directions deployed at R1 and 2. In addition, R10 is injecting an external EIGRP route for an organization to communicate with its business partner; this route will typically have by default an AD value of 170.

After this external route was injected into the EIGRP domain, internal users connected to the IS-IS domain started complaining that they could not reach any of the intended destinations located at their business partner network.

This design has the following technical concerns:

- Two redistribution boundary points

- Mutual redistribution at each boundary point from a high AD domain (external EIGRP in this case) to a lower AD domain

- Possibility of metric transformation (applicable to the external EIGRP route when redistributed back from IS-IS with better metrics)

As a result, a route looping will be formed with regard to the external EIGRP (between R1 and R2). With route filtering combined with tagging, as illustrated in Figure 8-43, both R1 and R2 can stop the reinjection of the redistributed external EIGRP route from IS-IS back into EIGRP again.

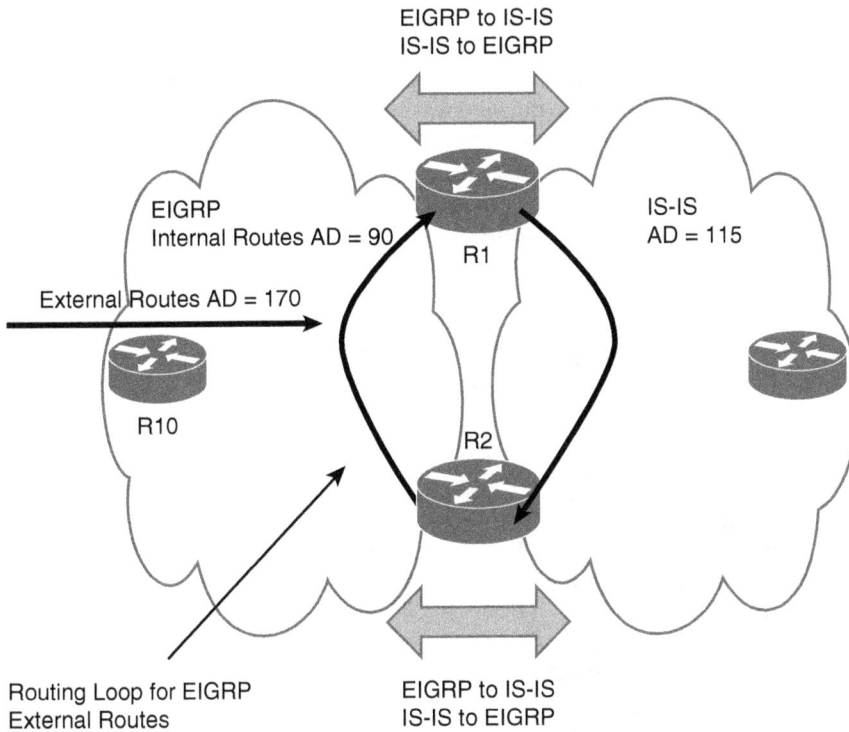

**Figure 8-42**  *Multipoint Route Redistribution: Routing Loop*

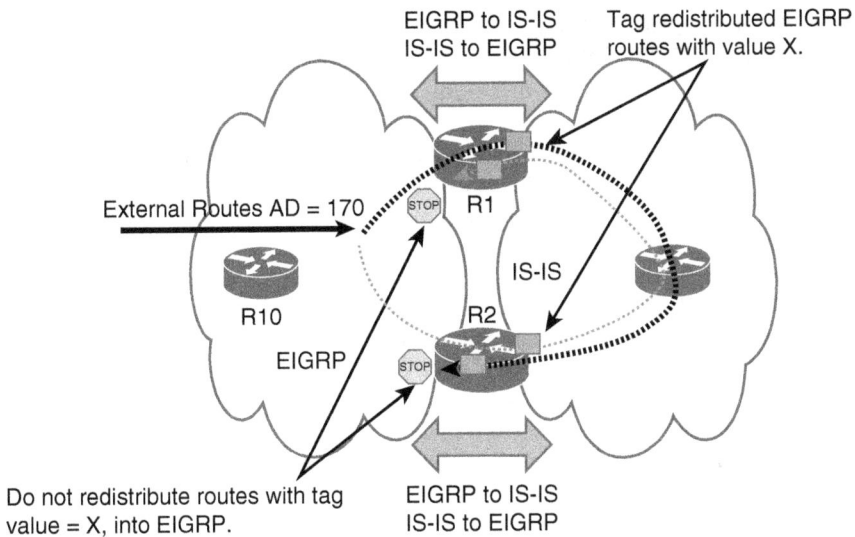

**Figure 8-43**  *Route Filtering with Route Tagging*

This is achieved by assigning a tag value to the EIGRP route when it is redistributed into IS-IS (at both R1 and R2). At the other redistribution boundary point (again R1 and R2), routes can be stopped from being redistributed into EIGRP again based on the assigned tag value. After you apply this filtering, the loop will be avoided, and path selection can be something like that depicted in Figure 8-44. With route tagging as in this example, network operators do not need to worry about managing and updating complicated access control lists (ACLs) to filter prefixes, because they can match the route tag at any node in the network and take action against it. Therefore, this offers simplified manageability and more flexible control.

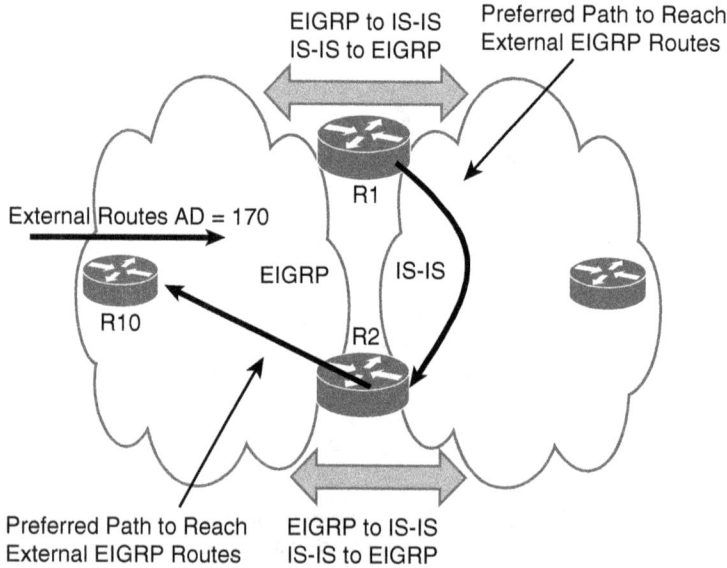

**Figure 8-44**  *Multipoint Route Redistribution: Routing After Filtering*

The optimal path, however, will not be guaranteed in this case unless another local filtering is applied to deny the EIGRP route from being installed in the local IS-IS routing table of the boundary routers. However, this must be performed only if the optimal path is a priority requirement, to avoid impacting any potential loss of path redundancy. For instance, if R1 in Figure 8-44 filters the redistributed EIGRP external routes by "R2" from being installed into the IS-IS local routing table (based on the assigned route tag by R2), the optimal path can be achieved. However, if there is a LAN or hosts connected directly to R1, R1 loses its connection to the EIGRP domain. In this case, any device or network that uses R1 as its gateway will not be able to reach the EIGRP external routes (unless there is a default route or a floating static route with higher AD that points to R2 within the IS-IS domain); in other words, to achieve the optimal path, a second filtering layer is required at the ASBRs (R1 and R2 in this example) to filter the "redistributed" external EIGRP routes by the other IS-IS ASBR from being reinjected into the IS-IS local routing table of the ASBR based on the route tag. Also, each ASBR should use a default route (ideally a static route, pointing to the other ASBR) to maintain redundancy to external prefixes in case of an ASBR link failure toward the EIGRP domain, as illustrated in Figure 8-45.

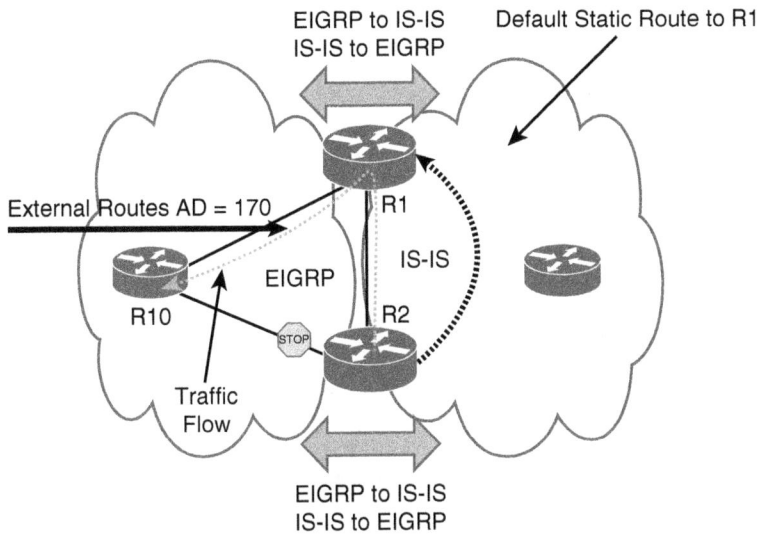

**Figure 8-45**  *Multipoint Route Redistribution with Optimal Path: Failure Scenario*

From a design point of view, achieving optimal network design does not mean an optimal path must always be considered. For example, as a network designer, you must look at the bigger picture using the "holistic approach" (highlighted previously in Chapter 1) to evaluate and decide what are the possible options to achieve the design requirements optimally, and what are the possible implications of each design option.

For instance, in the scenario discussed previously, if the IS-IS domain is receiving a default route from an internal node such as an Internet edge router, injecting a default route from the ASBRs (R1 and R2) most likely will break the Internet reachability for the IS-IS domain or any network directly connected to R1 and R2. Therefore, if both paths (over R1 and R2, with or without asymmetrical routing) technically satisfy the requirements for the communication between this organization and its partner network, in this case, from a network design perspective, "optimal path" is not a requirement to achieve "optimal design." Because optimal path can introduce design and operational complexity as well, it may break the Internet reachability in this particular scenario.

**NOTE**    Route tagging in some platforms requires the IS-IS "wide metric" feature to be enabled in order for the route tagging to work properly, where migrating IS-IS routed domain from "narrow metrics to wide metrics" must be considered in this case.

**NOTE**    If asymmetrical routing has a bad impact on the communication in the previous scenario, between EIGRP and IS-IS domains, it can be avoided by tuning EIGRP metrics such as delay, when the IS-IS route redistributed into EIGRP to control path selection from EIGRP domain point of view and align it with the selected path from IS-IS (to align both ingress and egress traffic flows).

# BGP Routing

**Border Gateway Protocol (BGP)** is an Internet Engineering Task Force (IETF) protocol and the most scalable of all routing protocols. As such, BGP is considered the routing protocol of the global Internet, as well as for service provider–grade networks. In addition, BGP is the desirable routing protocol of today's large-scale enterprise networks because of its flexible and powerful attributes and capabilities. Unlike IGPs, BGP is used mainly to exchange network layer reachability information (NLRI) between routing domains. (The routing domain in BGP terms is referred to as an autonomous system [AS]; typically, it is a logical entity with its own routing and policies and is usually under the same administrative control.) Therefore, BGP is almost always the preferred inter-AS routing protocol. A typical example is the global Internet, which is formed by numerous interconnected BGP autonomous systems.

There are two primary forms of BGP peering:

- **Interior BGP (iBGP):** The peering between BGP neighbors that is contained within one AS

- **Exterior BGP (eBGP):** The peering between BGP neighbors that occurs between the boundaries of different autonomous systems (interdomain)

## Interdomain Routing

Typically, eBGP is mainly used to determine paths and route traffic between different autonomous systems; this function is known as inter-domain routing. Unlike an IGP (where routing is usually performed based on protocol metrics to determine the desired path within an AS), eBGP relies more on policies to route or interconnect two or more autonomous systems. The powerful policies of eBGP allow it to ignore several attributes of routing information that typically an IGP takes into consideration. Therefore, an eBGP can offer simpler and more flexible solutions to interconnect various autonomous systems based on predefined routing policies.

Table 8-7 summarizes common AS terminology with regard to the interdomain routing concept and as illustrated in Figure 8-46.

**Key Topic**

**Table 8-7** Interdomain Routing Terminology

| Term | Description |
|------|-------------|
| Stub AS | An AS that has one connection to one upstream AS |
| Stub multihomed AS | An AS that has connections to more than one AS, and typically should not offer a transit path |
| Transit AS | An AS that connects two or more autonomous systems to provide a transit path for traffic sources from one AS and destined to another AS |

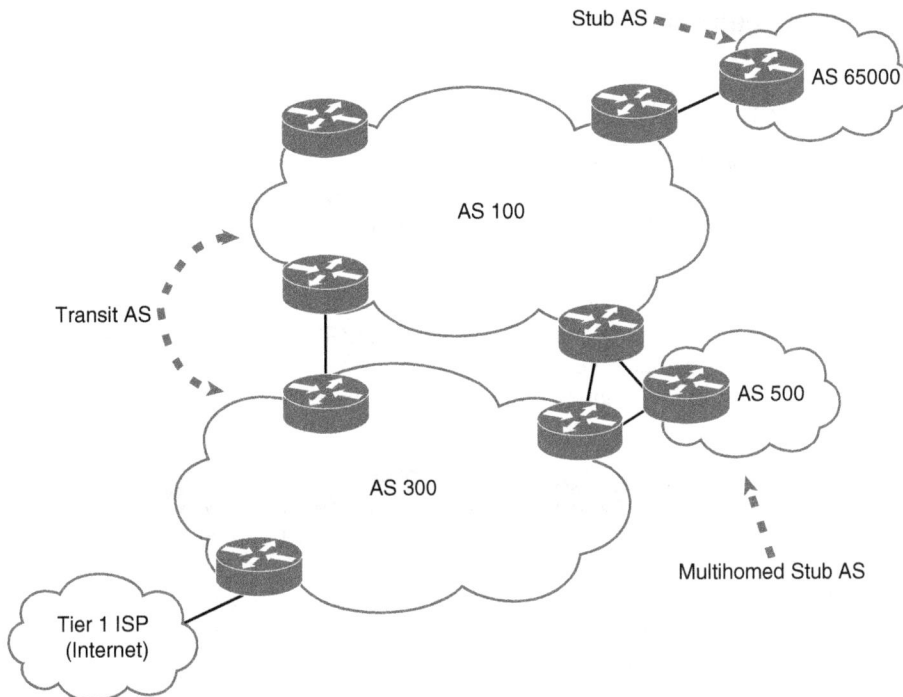

**Figure 8-46**  *Interdomain Routing*

Furthermore, normally, each AS has its own characteristic in terms of administrative boundaries, geographic restrictions, QoS scheme, cost, and legal constraints. Therefore, for the routing policy control to deliver its value to the business with regard to these variables, there must be a high degree of flexibility in how and where the policy control can be imposed. Typically, there are three standard levels where interdomain routing control can be considered (inbound, transit, and outbound):

- Inbound interdomain routing policy to influence which path egress traffic should use to reach other domains

- Outbound interdomain routing policy to influence which path ingress traffic sourced from other domains should use to reach the intended destination prefixes within the local domain

- Transportation interdomain routing policy to influence how traffic is routed across the transit domain as well as which prefixes and policy attributes from one domain are announced or passed to other neighboring domains, along with how these prefixes and policy attributes are announced (for example, summarized or non-summarized prefixes)

As a path-vector routing protocol, BGP has the most flexible and reliable attributes to match the various requirements of interdomain routing and control. Accordingly, BGP is considered the de facto routing protocol for the global Internet and large-scale networks, which require complex and interdomain routing control capabilities and policies.

## BGP Attributes and Path Selection

BGP attributes, also known as path attributes, are sets of information attached to BGP updates. The following excerpt from RFC 4271 describes the characteristics of a BGP prefix, either within an AS or between autonomous systems:

*BGP implementations MUST recognize all well-known attributes. Some of these attributes are mandatory and MUST be included in every UPDATE message that contains NLRI. Others are discretionary and MAY or MAY NOT be sent in a particular UPDATE message.*

Thus, BGP primarily relies on these attributes to influence the process of best-path selection. These attributes are critical and effective when designing BGP routing architectures. A good understanding of these attributes and their behavior is a prerequisite to producing a successful BGP design. There are four primary types of BGP attributes, as summarized in Table 8-8.

**Table 8-8**   BGP Attributes

| BGP Path Attribute | Characteristic |
| --- | --- |
| Well-known mandatory | Must appear in every update and must be supported by all speakers (for example, ORIGIN). |
| Well-known discretionary | May not be included in the update message but must be supported by all BGP speakers (for example, LOCAL_PREFERENCE). |
| Optional transitive | May be supported by BGP speakers, and they should be maintained and passed to other BGP AS peers whether or not they are supported (for example, COMMUNITY). |
| Optional nontransitive | May or may not be supported BGP speakers. If an update is received that includes an optional transitive attribute, it is not required that the router pass it on (for example, MULTI_EXIT_DISC). |

**Key Topic**

The following list highlights the typical BGP route selection (from the highest to the lowest preference):

1. Prefer highest weight (Cisco proprietary, local to router)
2. Prefer highest local preference (global within AS)
3. Prefer route originated by the local router
4. Prefer shortest AS path
5. Prefer lowest origin code (IGP < EGP < incomplete)
6. Prefer lowest MED (from other AS)
7. Prefer eBGP path over iBGP path
8. Prefer the path through the closest IGP neighbor
9. Prefer oldest route for eBGP paths
10. Prefer the path with the lowest neighbor BGP router ID

> **NOTE**   For more information about BGP path selection, refer to the document "BGP Best Path Selection Algorithm," at https://www.cisco.com/c/en/us/support/docs/ip/border-gateway-protocol-bgp/13753-25.html.

## BGP as the Enterprise Core Routing Protocol

Most enterprises prefer IGPs such as OSPF as the core routing protocol to provide end-to-end enterprise IP reachability. However, in some scenarios, network designers may prefer a protocol that can provide more flexible and robust routing policies and can cover single- and multi-routing domains with the ability to facilitate a diversified administrative control approach.

For example, an enterprise may have a large core network that connects different regions or large department networks, each with its own administrative control. To achieve that, we need a protocol that can provide interconnects between all the places in the network (PINs) and at the same time enable each group or region to maintain the ability to control its network without introducing any added complexity when connecting the PINs. Obviously, a typical IGP implementation in the core cannot achieve that, and even if it is possible, it will be very complex to scale and manage.

In other words, when the IGP of a large-scale global enterprise's network reaches the borderline of its scalability limits within the routed network, which usually contains a high number of routing prefixes, a high level of flexibility is required to support "splitting routed networks" into multiple failure domains with distributed network administration. In this scenario, BGP is the ideal candidate protocol as the enterprise core routing protocol.

BGP in the enterprise core can offer the following benefits to the overall routing architecture:

■ A high degree of responsiveness to new business requirements, such as business expansion, business decline, innovation (IPv6 over IPv4 core), and security policies like end-to-end path separation (for example, MP-BGP + MPLS in the core)

■ Design simplicity (separating complex functional areas, each into its own routed region within the enterprise)

■ Flexible domain control by supporting administrative control per routing domains (per region)

■ More flexible and manageable routing policies that support intra- and interdomain routing requirements

■ Improved scalability because it can significantly reduce the number of prefixes that regional routing domains need to hold and process

■ Optimized network stability by stressing fault isolation domain boundaries (for example, at IGP island edges), where any control plane instability in one IGP/BGP domain will not impact other routing domains (topology and reachability information hiding principle)

8

However, network designers need to consider some limitations or concerns that BGP might introduce to the enterprise routing architecture when used as the core routing protocol:

- **Convergence time:** In general, BGP convergence time during a change or following a failure event is slower than IGP. However, this can be mitigated to a good extent when advanced BGP fast convergence techniques are well-tuned, such as BGP Prefix Independent Convergence (BGP-PIC).

- **Staff knowledge and operational complexity:** BGP in the enterprise core can simplify the routing design. However, additional knowledge and experience for the operation staff are required because the network will be more complex to troubleshoot, especially if multiple control policies in different directions are applied for control and traffic engineering purposes.

- **Hardware and software constraints:** Some legacy or low-end network devices either do not support BGP or may require a software upgrade to support it. In both cases, there is a cost and the possibility of a maintenance outage for the upgrade. This might not always be an acceptable or supported practice by the business.

## Enterprise Core Routing Design Models with BGP

This section highlights and compares the primary and most common design models that network designers and architects can consider for large-scale enterprise networks with BGP as the core routing protocol (as illustrated in Figure 8-47 through Figure 8-50). These design models are based on the design principle of dividing the enterprise network into a two-tiered hierarchy. This hierarchy includes a transit core network to which a number of access or regional networks are attached. Typically, the transit core network runs BGP and glues the different geographic areas (network islands) of the enterprise regional networks. In addition, no direct link should interconnect the regional networks. Ideally, traffic from one regional network to another must traverse the BGP core.

However, each network has unique and different requirements. Therefore, all the design models discussed in this section support the existence or addition of backdoor links between the different regions; remember to always consider the added complexity to the design with this approach:

- **Design model 1:** This design model (see Figure 8-47) has the following characteristics:

    - iBGP is used across the core only.

    - Regional networks use IGP only.

    - Border routers between each regional network and the core run IGP and iBGP.

    - IGP in the core is mainly used to provide next-hop (NHP) reachability for iBGP speakers.

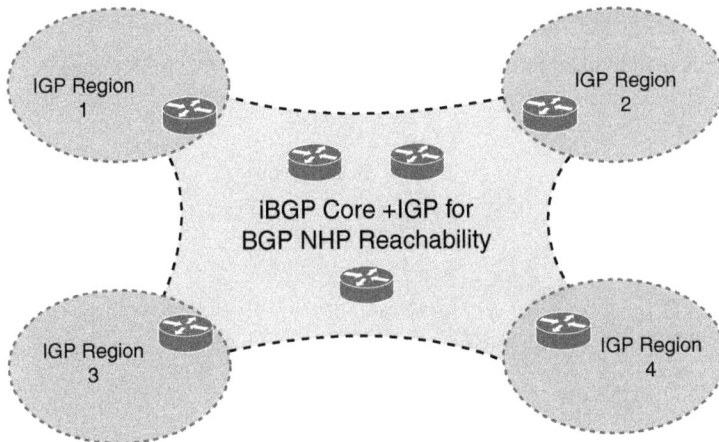

**Figure 8-47**   *BGP Core Design Module 1*

- **Design model 2:** This design model (see Figure 8-48) has the following characteristics:

  - BGP is used across the core and regional networks.

  - Each regional network has its own BGP AS number (ASN) (no direct BGP session between the regional networks).

  - Reachability information is exchanged between each regional network and the core over eBGP (no direct BGP session between regional networks).

  - IGP in the core as well as at the regional networks is mainly used to provide NHP reachability for iBGP speakers in each domain.

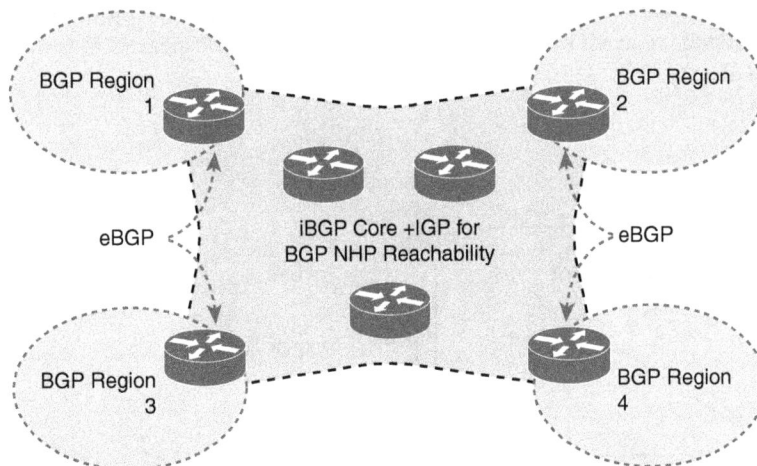

**Figure 8-48**   *BGP Core Design Module 2*

- **Design model 3:** This design model (see Figure 8-49) has the following characteristics:

  - MP-BGP is used across the core (MPLS L3VPN design model).

  - MPLS is enabled across the core.

- Regional networks can run either static IGP or BGP.

- IGP in the core is mainly used to provide NHP reachability for MP-BGP speakers.

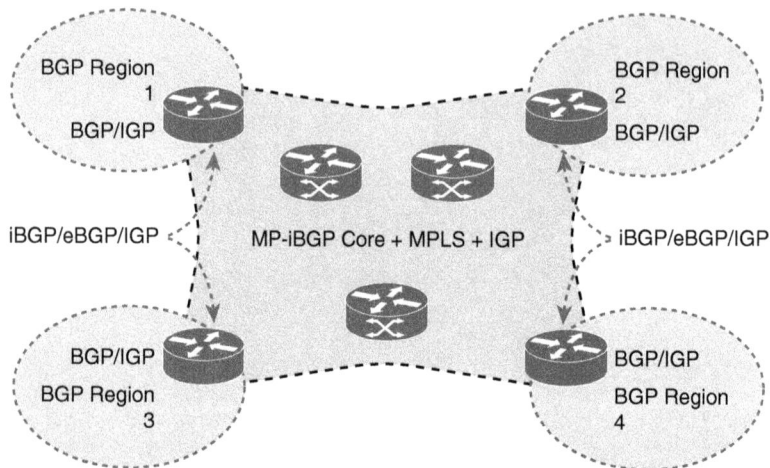

**Figure 8-49** *BGP Core Design Module 3*

- **Design model 4:** This design model (see Figure 8-50) has the following characteristics:

  - BGP is used across the regional networks.

  - In this design model, each regional network has its own BGP ASN.

  - Reachability information is exchanged between the regional networks directly over direct eBGP sessions.

  - IGP can be used at the regional networks to provide local reachability within each region and may be required to provide NHP reachability for BGP speakers in each domain (BGP AS).

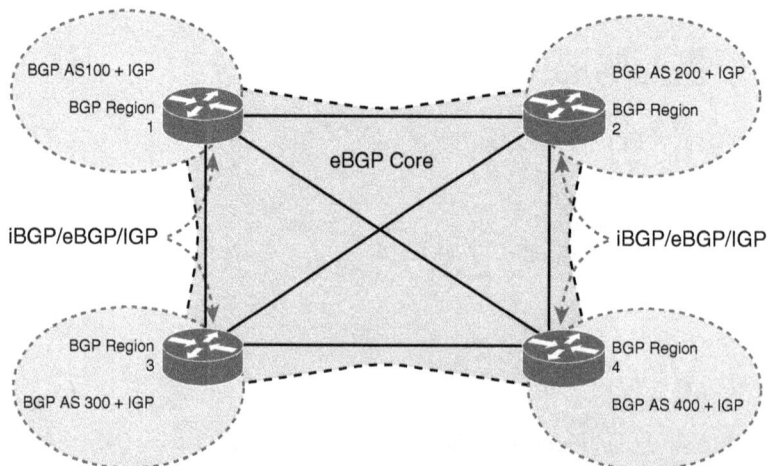

**Figure 8-50** *BGP Core Design Module 4*

These designs are all valid and proven design models; however, each has its own strengths and weaknesses in certain areas, as summarized in Table 8-9. During the planning phase of network design or design optimization, network designers or architects must select the most suitable design model as driven by other design requirements, such as business and application requirements (which ideally must align with the current business needs and provide support for business directions such as business expansion).

**Table 8-9** Comparing BGP Core Design Models

| Design Models | Core | Branches/Region | Design Model Attributes |
|---|---|---|---|
| Design Model 1 | iBGP | IGP only | This design model offers the least administrative domain control as compared to other models.<br><br>Can be suitable in large-scale environments to overcome IGP complexities in the core, and offers more control between regions compared to IGP-based only.<br><br>Moderate operational complexity. |
| Design Model 2 | iBGP | iBGP + IGP | This design model offers moderate administrative domain control between routing regions.<br><br>Can be suitable for environments under multiple admin domains and large-scale (global) enterprise WANs.<br><br>Low operational complexity. |
| Design Model 3 | MP-iBGP + MPLS | iBGP/eBGP/IGP or eiBGP + IGP | This design model offers the highest administrative domain control between routing regions, combined with the ability to control multiple routing islands in different places with end-to-end path isolation.<br><br>Can be suitable for environments under multiple admin domains and large-scale (global) enterprise WANs. Offers the highest flexibility and simplicity to introduce new capabilities across the entire enterprise or for specific regions only (for example, IPv6, multicast, and end-to-end traffic separation).<br><br>High operational complexity. |
| Design Model 4 | eBGP | iBGP + IGP | This design model offers high administrative domain control between routing regions.<br><br>Can be suitable for merging networks scenarios and environments under multiple admin domains (global organizations).<br><br>Moderate operational complexity. |

**NOTE** IGP or control plane complexity referred to in Table 8-9 is in comparison to the end-to-end IGP-based design model, specifically across the core.

## BGP Shortest Path over the Enterprise Core

BGP as a path-vector control plane protocol normally prefers the path with the smallest number of autonomous systems when traversing multiple autonomous systems when other attributes such as local preference are the same (classical interdomain routing scenarios). Typically, in interdomain routing scenarios, the different routed domains have their own policies, which do not always need to be exposed to other routing domains. However, in the enterprise core with BGP scenarios, when a router selects a specific path based on the BGP AS-PATH attribute, the "edge eBGP" nodes cannot determine which path within the selected core or transit BGP core AS is the shortest (hypothetically, the optimal path). For instance, the scenario in Figure 8-51 depicts design model 2 of BGP enterprise core. The question is, "How can router A decide which path is the shortest (optimal) within the enterprise core (AS 65000)?"

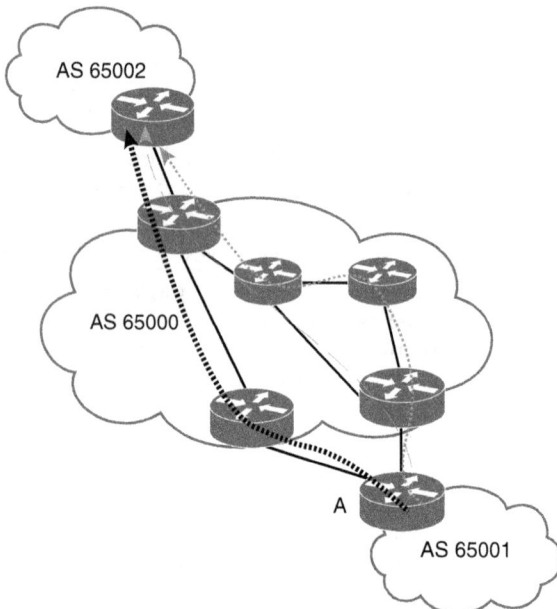

**Figure 8-51** *BGP AIGP*

Accumulated IGP Cost for BGP (AIGP) is an optional nontransitive BGP path attribute, designed to enhance shortest path selection in scenarios like the one in the example, where a large-scale network is part of a single enterprise with multiple administrative domains using multiple contiguous BGP networks (BGP core routing design model 2, discussed earlier in this section). Therefore, it is almost always more desirable than BGP considering the shortest path with the lowest metric across the transit BGP core. In fact, AIGP replicates the behavior of link-state routing protocols in computing the distance associated with a path that has routes within a single flooding domain. Although the BGP MED attribute can carry IGP metric values, MED comes after several BGP attributes in the path selection process.

In contrast, AIGP is considered before the AS-PATH attribute when enabled in the BGP path selection process, which makes it more influential in this type of scenario:

**Key Topic**

1. Prefer the highest weight (Cisco proprietary, local to router)
2. Prefer highest local preference (global within single AS)
3. Prefer route originated by the local router
4. Prefer lowest AIGP cost
5. Prefer shortest AS path
6. Prefer lowest origin code (IGP < EGP < incomplete)
7. Prefer lowest MED (from other AS)
8. Prefer eBGP path over iBGP path
9. Prefer the path through the closest IGP neighbor

It is obvious that AIGP can be a powerful feature to optimize the BGP path selection process across a transit AS. However, network designers must be careful when enabling this feature because when AIGP is enabled, any alteration to the IGP routing can lead to a direct impact on BGP routing (optimal path versus routing stability).

## BGP Scalability Design Options and Considerations

This section discusses the primary design options to scale BGP in general at an enterprise network grade. The natural behavior of BGP can be challenging when the size of the network grows to a large number of BGP peers, because it will introduce a high number of route advertisements, along with scalability and manageability complexities and limitations. According to the default behavior of BGP, any iBGP-learned route will not be advertised to any iBGP peer (the typical BGP loop-prevention mechanism, also known as the iBGP split-horizon rule). This means that a full mesh of iBGP peering sessions is required to maintain full reachability across the network. On this basis, if a network has 15 BGP routers within an AS, a full mesh of iBGP peering will require (15(15 − 1) / 2) = 105 iBGP sessions to manage within an AS. Consequently, it will be a network that has a large amount of configuration associated with a high probability of configuration errors, is complex to troubleshoot, and has very limited scalability. However, BGP has two main proven techniques that you can use to reduce or eliminate these limitations and complexities of the BGP control plane:

- Route reflection (described in RFC 4456)
- Confederation (described in RFC 3065)

**Key Topic**

### BGP Route Reflection

Route reflection is a BGP route advertisement mechanism based on relaying the iBGP-learned routes from other iBGP peers. This process involves a special BGP peer or set of peers called *route reflectors (RRs)*. These RRs can alter the classical iBGP split-horizon rule by re-advertising the BGP route that was received from iBGP peers to other iBGP peers, also known as *route reflector clients*, which can significantly reduce the total number of iBGP sessions, as illustrated in Figure 8-52. Moreover, RRs reflect routes to nonclient iBGP peers as well, in certain cases.

8

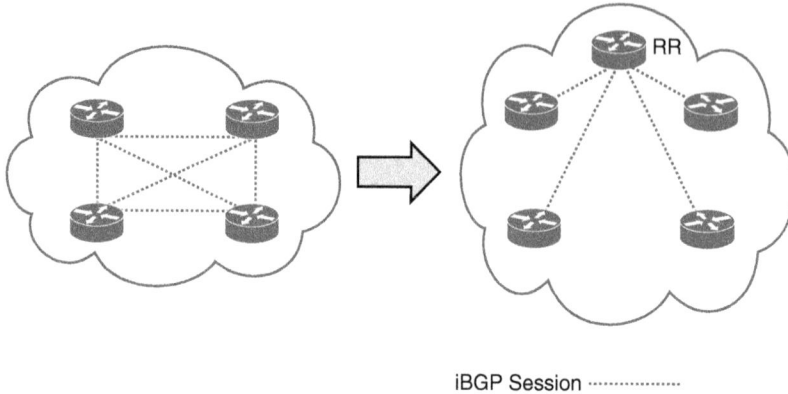

iBGP Session ·················

**Figure 8-52** *iBGP Session With and Without RR*

Figure 8-53 summarizes RR route advertisement rules based on three primary route sources and receivers in terms of the BGP session type (eBGP, iBGP RR client, and iBGP non-RR client).

## Route Source

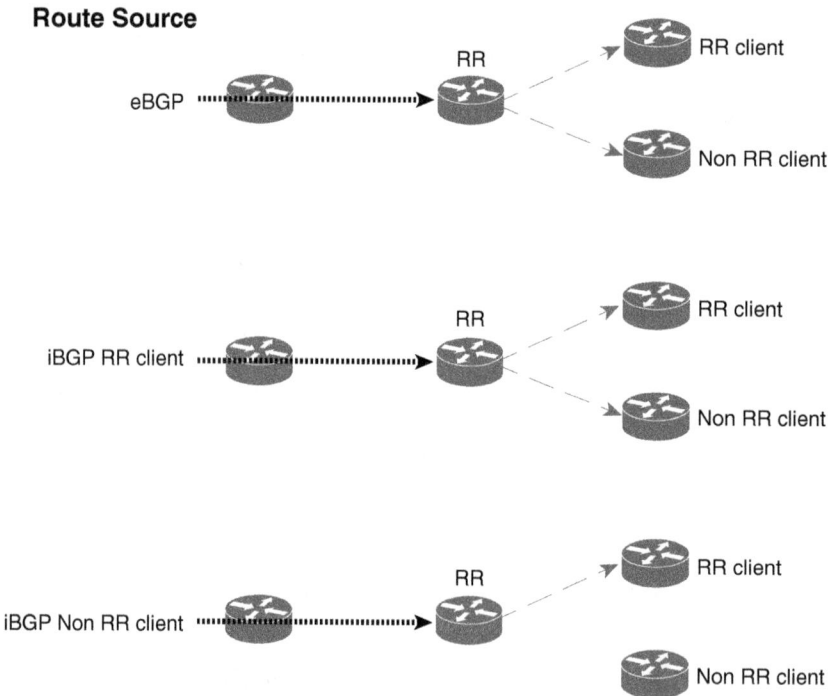

**Figure 8-53** *RR Route Advertisement Rules*

It is obvious from the figure that the route(s) sourced from an iBGP non-RR client peer(s) will not be re-advertised by the RR to another iBGP non-RR client peer(s).

As a result, the concept of RR can help network designers avoid the complexities and limitations associated with iBGP full-mesh sessions, where more scalable and manageable designs can be produced. However, BGP RR can introduce new challenges that network designers should take into account, such as redundancy, optimal path selection, and network convergence. These points are covered in the subsequent sections, as well as in other chapters throughout this book.

### Route Reflector Redundancy

In BGP environments, RRs can introduce a single point of failure to the design if no redundancy mechanism is considered. RR clustering is designed to provide redundancy, where typically two (or more) RRs can be grouped to serve one or more iBGP clients. With RR clustering, technically, BGP uses special 4-byte attributes called CLUSTER_ID. Each route exchanged between these RRs in the same cluster will be ignored and not installed in their BGP routing table if the corresponding route identified by the receiving RR has the same CLUSTER_ID attribute that is being used. However, in some situations, it is recommended that two redundant RRs be configured with different CLUSTER_IDs for an increased level of BGP routing redundancy.

For instance, the RR client in Figure 8-54 is multihomed to two RRs leveraging the link addresses (not lookback addresses) for the corresponding iBGP neighborships. If each RR is deployed with a different CLUSTER_ID, the RR client will continue to be able to reach prefix X, even after the link with RR 1 fails.

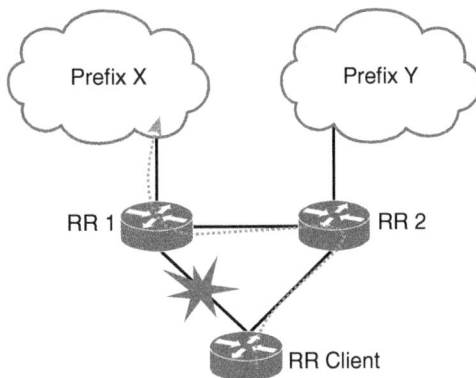

**Figure 8-54**   *RR Clustering*

In contrast, if RR 1 and RR 2 were deployed with the same CLUSTER_ID, after this failure event the RR client in Figure 8-54 would not be able to reach prefix X. This is because the CLUSTER_ID attribute mechanism will stop the propagation of a route from RR 1 to RR 2 with the same CLUSTER_ID.

Furthermore, two BGP attributes were created specifically to optimize redundant RR behavior, especially with regard to avoiding routing information loops (for example, duplicate routing information). If the redundant RRs are being deployed in different clusters, the two attributes are ORIGINATOR_ID and CLUSTER_LIST.

### RR Logical and Physical Topology Alignment

As discussed in Chapter 1, the physical topology forms the foundation of many design scenarios, including BGP RRs. In fact, with BGP RRs, the logical and physical topologies must be given special consideration. They should be as congruent as possible to avoid any undesirable behaviors, such as suboptimal routing and routing loops. For example, the scenario depicted in Figure 8-55 is based on an enterprise network that uses BGP as the core routing protocol (based on design model 1, discussed earlier in this chapter). In this scenario, the data center is located miles away from the campus core and is connected over two dark fiber links. The enterprise campus core routers C and D are configured as BGP RR (same RR cluster) to aggregate iBGP sessions of the campus buildings and data center routers. Data center aggregation router E is the iBGP client of core RR D, and data center aggregation router F is the iBGP client of core RR C.

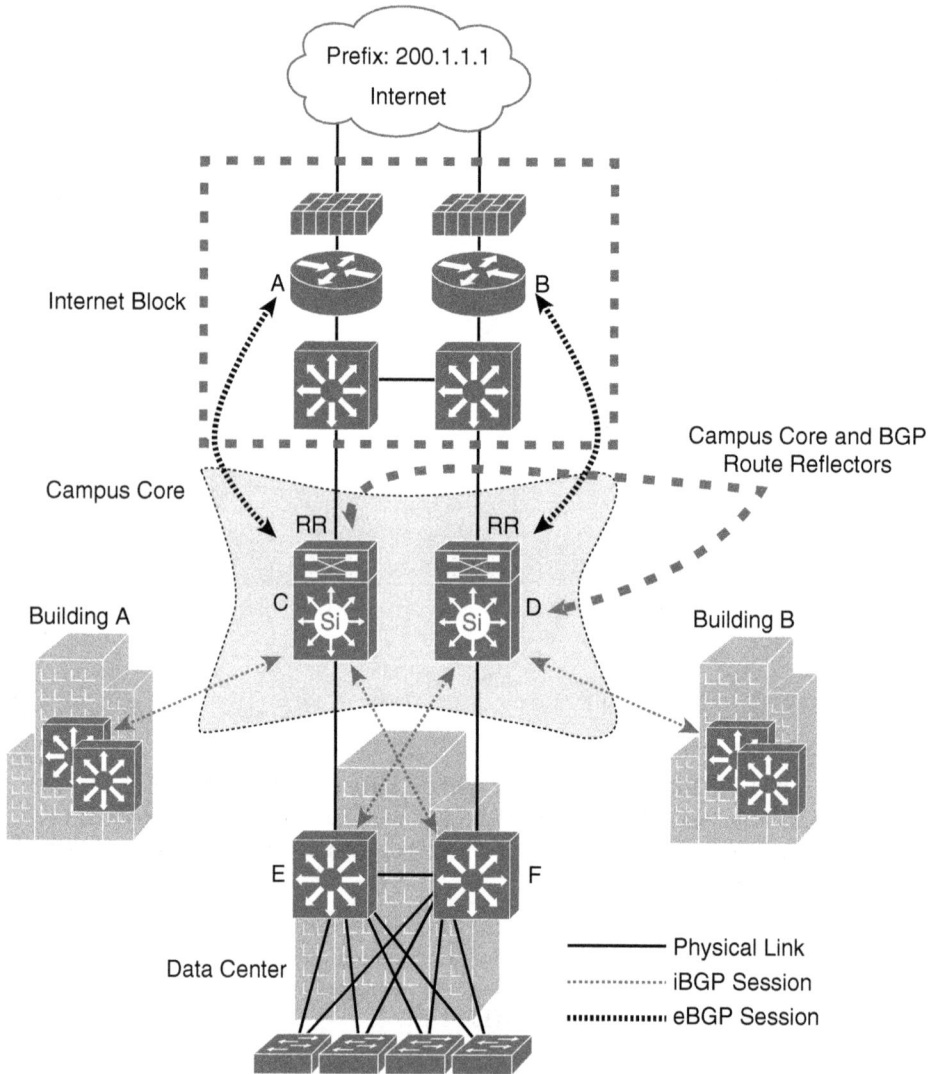

**Figure 8-55** *BGP RR Physical and Logical Topology Congruence*

If the prefix 200.1.1.1 is advertised by both Internet edge routers (A and B), typically router A will advertise it to core router C, and router B will advertise it to core router D over eBGP sessions. Then, each RR will advertise this prefix to its clients. (RR C will advertise it to data center aggregation router F, and RR D will advertise it to data center aggregation router E.) Up to this stage, there is no issue. However, when routers E and F try to reach prefix 200.1.1.1, a loop will be formed, as follows:

**NOTE**  For simplicity, this scenario assumes that both campus cores (RR) advertise the next-hop IPs of the Internet edge routers to all the campus blocks.

- Based on the design in Figure 8-55, data center aggregation router E will have the next hop to prefix 200.1.1.1 as Internet edge router B.

- Data center aggregation router F will have the next hop to prefix 200.1.1.1 as Internet edge router A.

- Data center aggregation router E will forward the packets destined to prefix 200.1.1.1 to data center aggregation router F. (Based on physical connectivity and IGP, the Internet edge router B is reachable via data center aggregation router F from the data center aggregation router E point of view.)

- Because data center aggregation router F has prefix 200.1.1.1, which is reachable through A, it will then send the packet back to data center aggregation router E, as illustrated in Figure 8-56.

This loop was obviously formed because there is no alignment (congruence) between iBGP-RR topology and the physical topology. The following are three simple possible ways to overcome this design issue and continue using RRs in this network:

- Add a physical link directly between E and D and between F and C, along with an iBGP session over each link to the respective core router. (It might take a long time to provision a fiber link, or it might be an expensive solution from the business point of view.)

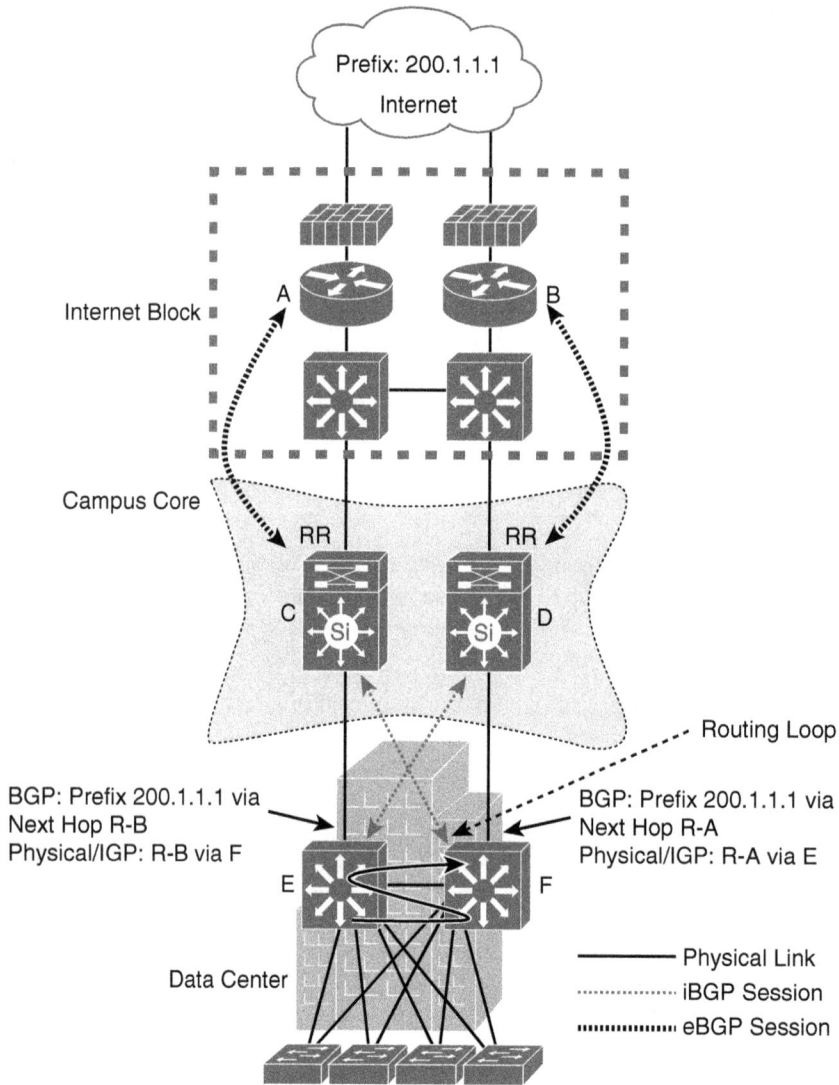

**Figure 8-56** *BGP RR and Physical Topology Congruence: Routing Loop*

- Align the iBGP-RR peering with physical topology by making E the iBGP client to RR C and F the iBGP client to RR D (the simplest solution), as illustrated in Figure 8-57.

- Add a direct link between core RRs and place each RR in a different RR cluster along with a direct iBGP session between them. (This might add control plane complexity in this particular scenario to align IGP and BGP paths without alignment between the physical topology and iBGP client to RR sessions.)

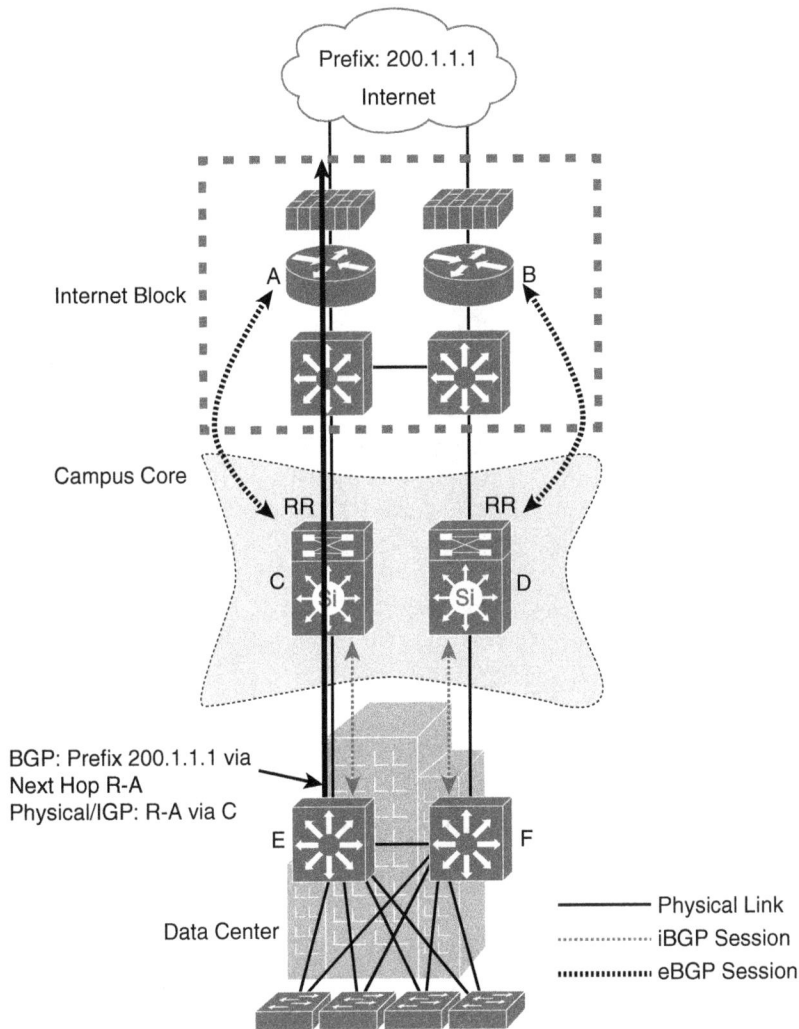

**Figure 8-57**  *BGP RR Alignment with the Physical Topology*

**NOTE**  One of the common limitations of the route reflection concept in large BGP environments is the possibility of suboptimal routing. This point is covered in more detail later in this book.

### Update Grouping

Update grouping helps to optimize BGP processing overhead by providing a mechanism that groups BGP peers that have the same outbound policy in one update group, and updates are then generated once per group. By integrating this function with BGP route reflection, each RR update message can be generated once per update group and then replicated for all the RR clients that are part of the relevant group, as depicted in Figure 8-58.

Without Update Grouping

With Update Grouping

Update Group 1

BGP RR Formats
the Update
Messages Per Peer

BGP RR Formats
the Update
Messages Once
Per Group

Update Group 2

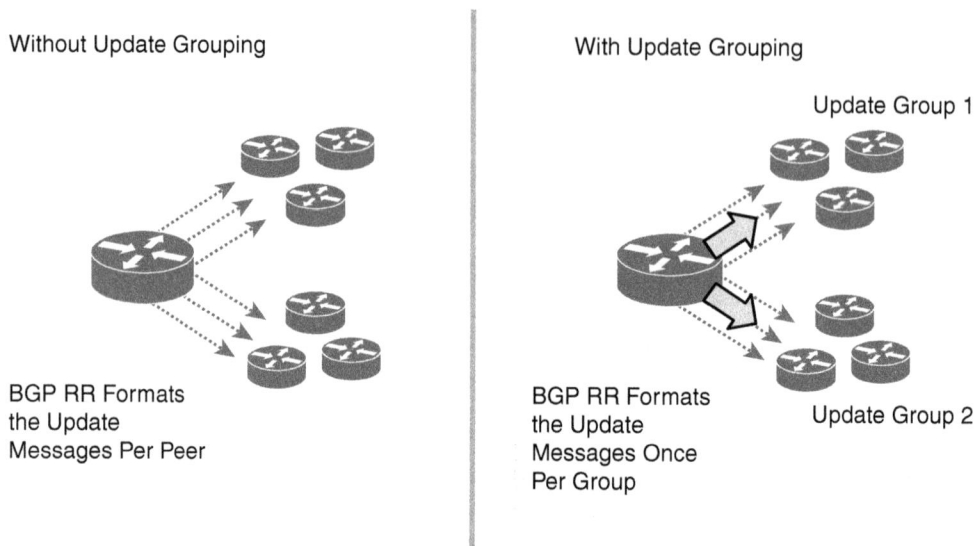

**Figure 8-58** *BGP Update Grouping*

Technically, update grouping can be achieved by using peer group or peer template features, which can enhance BGP RR functionality and simplify the overall network operations in large BGP networks by

■ Making the configuration easier, less error-prone, and more readable

■ Lowering CPU utilization

■ Speeding up iBGP client provisioning (because they can be configured and added quickly)

## BGP Confederation

**Key Topic**

The other option to solve iBGP scalability limitations in large-scale networks is through the use of confederations. The concept of a BGP confederation is based on splitting a large iBGP domain into multiple (smaller) BGP domains (also known as sub-autonomous systems). The BGP communication between these sub-autonomous systems is formed over eBGP sessions (a special type of eBGP session referred to as an intra-confederation eBGP session). Consequently, the BGP network can scale and support a larger number of BGP peers because there is no need to maintain a full mesh among the sub-autonomous systems; however, within each sub-AS iBGP, full mesh is required, as illustrated in Figure 8-59.

**NOTE** The intra-confederation eBGP session has a mixture of both iBGP and eBGP characteristics. For example, NEXT_HOP, MED, and LOCAL_PREFERENCE attributes are kept between sub-autonomous systems. However, the AS_PATH is changed with updates across the sub-autonomous systems.

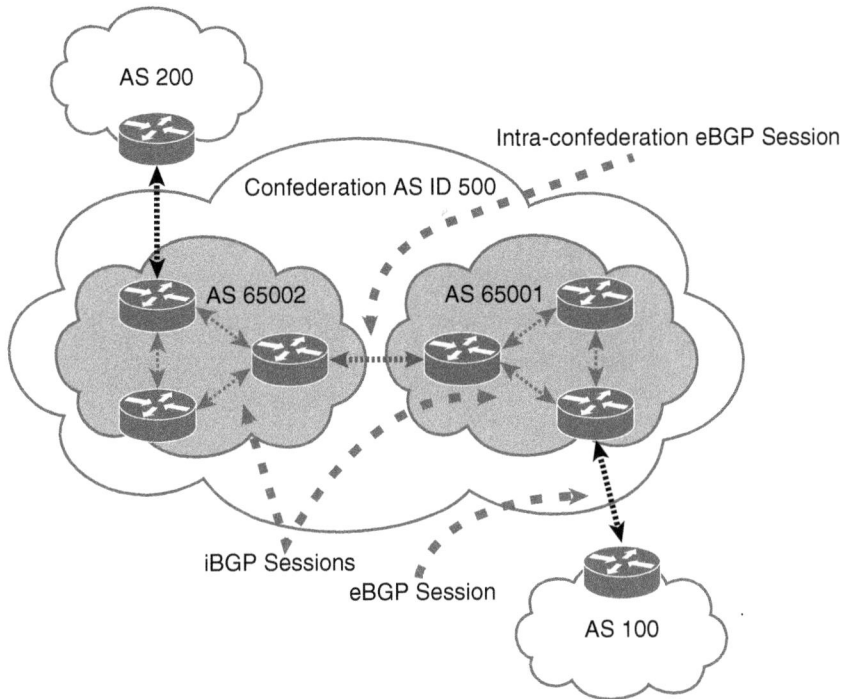

**Figure 8-59**  *BGP Confederation*

**NOTE**  The confederations appear as a single AS to external BGP autonomous systems. Because the sub-AS topology is invisible to external peering BGP autonomous systems, the sub-AS is also removed from the eBGP update sent to any external eBGP peer.

In large iBGP environments like a global enterprise (or Internet service provider [ISP] type of network), you can use both RR and confederation jointly to maximize the flexibility and scalability of the design. As illustrated in Figure 8-60, the confederation can help to split the BGP AS into sub-autonomous systems, where each sub-AS can be managed and controlled by a different team or business unit. At the same time, within each AS, the RR concept is used to reduce iBGP full-mesh session complexity. In addition, network designers must make sure that IGP metrics within any given sub-AS are lower than those between sub-autonomous systems to avoid any possibility of suboptimal routing issues within the confederation AS.

**NOTE**  To avoid BGP route oscillation, which is associated with RRs or confederations in some scenarios, network designers must consider deploying higher IGP metrics between sub-autonomous systems or RR clusters than those within the sub-AS or cluster.

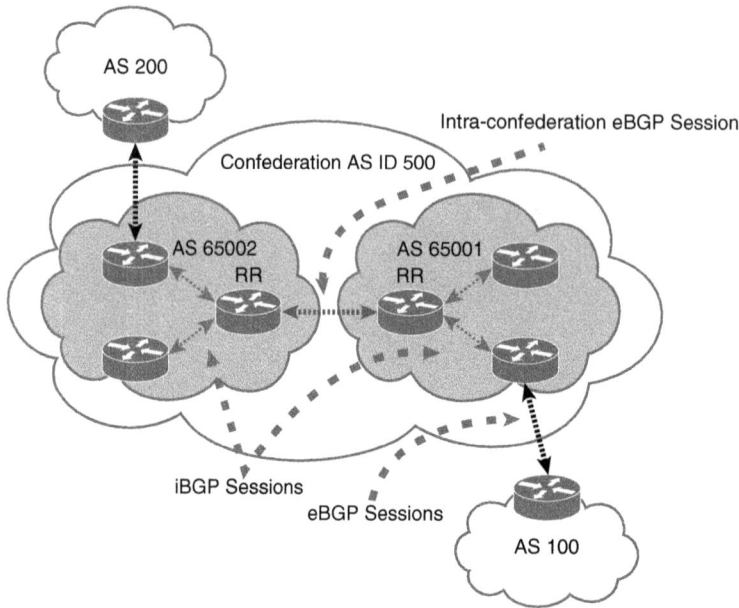

**Figure 8-60** *BGP Confederation and RR*

> **NOTE** Although BGP route reflection combined with confederation can maximize the overall BGP flexibility and scalability, it may add complexity to the design if the combination of both is not required. For instance, when merging two networks with a large number of iBGP peers in each domain, confederation with RR might be a feasible joint approach to optimize and migrate these two networks if it does not compromise any other requirements. However, with a large network with a large number of iBGP peers in one AS that cannot afford major outages and configuration changes within the network, it is more desirable to optimize using RR only rather than combined with confederation.

## Confederation Versus Route Reflection

The most common dilemma is whether to use route reflection or confederation to optimize iBGP scalability. The typical solution to this dilemma, from a design point of view, is "it depends." Like any other design decision, deciding what technology or feature to use to enhance BGP design and scalability depends on different factors. Table 8-10 highlights the different factors that can help you narrow down the design decision with regard to BGP confederation versus route reflection.

**Key Topic**

**Table 8-10** Confederation Versus RR

|  | Route Reflection (RR) | Confederation (Conf) | Conf + RR |
|---|---|---|---|
| IGP architecture | Ideally one IGP domain | Supports multiple IGP domains | Supports multiple IGP domains |
| Hierarchal topology | More flexible | Less flexible | Flexible within the sub-AS |
| Policy control | Less control | More control between domains | More control between domains |

| | Route Reflection (RR) | Confederation (Conf) | Conf + RR |
|---|---|---|---|
| Control Plane complexity | Moderate | The larger the sub-AS, the higher the control plane complexity | Low (optimized) |
| Optimal routing | May be effected | Maintained within and between sub-autonomous systems | May be effected |
| Integration with MPLS-TE | Simple | Simple within the same sub-AS, complex between sub-autonomous systems | Simple within the same sub-AS, complex between sub-autonomous systems |

Again, there is no 100 percent definite answer. As a designer, you can decide which way to go based on the information and architecture you have and the goals that need to be achieved, taking the factors highlighted in Table 8-10 into consideration.

## Enterprise Routing Design Recommendations

This chapter discussed several concepts and approaches pertaining to Layer 3 control plane routing design. Table 8-11 summarizes the main Layer 3 routing design considerations and recommendations in a simplified way that you can use as a foundation to optimize the overall routing design.

**Table 8-11**   Route Filtering Techniques Comparison

| Design Consideration | Design Recommendations |
|---|---|
| Scalability | Modular routing design (contain and optimize fault domains design). |
| | Reduce the number of prefixes (for example, suppress the advertisement of transport link IPs, routes summarization). |
| Resiliency and fast convergence | Reduce the number of prefixes (for example, suppress the advertisement of transport link IPs, routes summarization). Modular routing design. |
| | Fast detection, processing, and reaction to the failure. LFA can be used with link state in some scenarios. |
| | When possible, design in triangles rather than squares between the routed layers. |
| Control and security | Enable routing authentication. |
| | Suppress peering with end-host VLANs (passive interface). Route filtering and tagging between routing domains. |

In large-scale enterprise networks with different modules and many remote sites, selecting a routing protocol can be a real challenge. Therefore, network designers need to consider the answers to the following questions as a foundation for routing protocol selection:

- What is the underlying topology, and which protocol can scale to a larger number of prefixes and peers?

- Which routing protocol can be more flexible, considering the topology and future plans (for example, integrating with other routing domains)?

8

- Is fast convergence a requirement? If yes, which protocol can converge faster and at the same time offer stability enhancement mechanisms?

- Which protocol can utilize fewer hardware resources?

- Is the routing internal or external (different routing domains)?

- Which protocol can provide less operational complexity (for instance, easy to configure and troubleshoot)?

Although these questions are not the only ones, they cover the most important functional requirements that can be delivered by a routing protocol. Furthermore, there are some factors that you need to consider when selecting an IGP:

- Size of the network (for example, the number of L3 hops and expected future growth)

- Security requirements and the supported authentication type

- IT staff knowledge and experience

- Protocol's flexibility in the modular network such as support of flexible route summarization techniques.

Generally speaking, EIGRP tends to be simpler and more scalable in hub-and-spoke topology and over networks with three or more hierarchical layers, whereas link-state routing protocols can perform better over flat networks when flooding domains and other factors discussed earlier in this book are tuned properly. In contrast, BGP is the preferred protocol to communicate between different routing domains (external), as summarized in Figure 8-61.

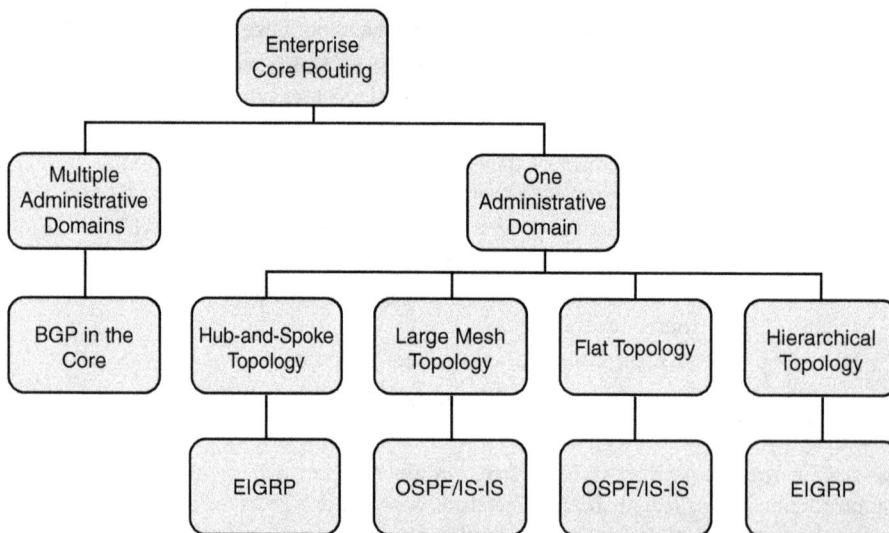

**Figure 8-61** *Routing Protocol Selection Decision Tree*

Moreover, the decision tree depicted in Figure 8-62 highlights the routing protocol selection decision to migrate from one routing protocol to another based on the topology used. This tree is based on the assumption that you have the choice to select the preferred protocol.

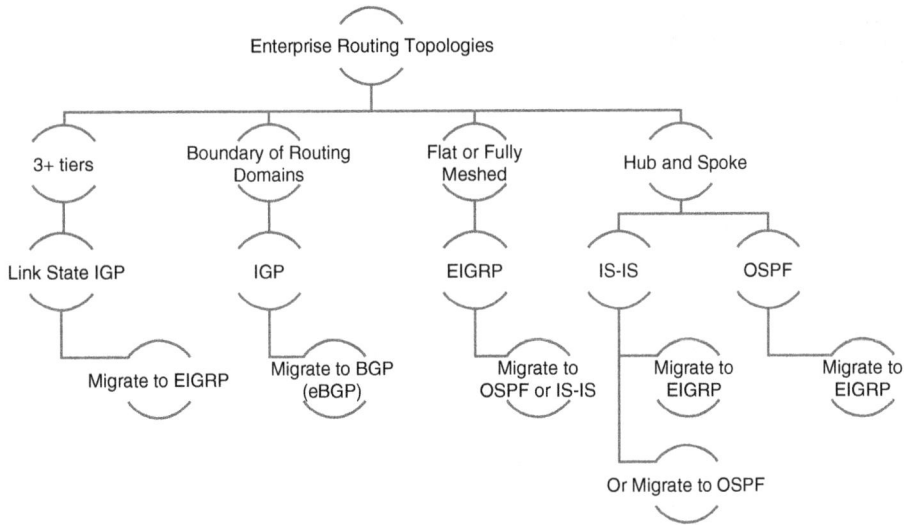

**Figure 8-62** *Routing Protocol Migration Decision Tree*

## Summary

For network designers and architects to provide a valid and feasible network design (including both Layer 2 and Layer 3), they must understand the characteristics of the nominated or used control protocols and how each behaves over the targeted physical network topology. This understanding will enable them to align the chosen protocol behavior with the business, functional, and application requirements, to achieve a successful business-driven network design. Also, considering any Layer 2 or Layer 3 design optimization technique, such as route summarization, may introduce new design concerns (during normal or failure scenarios), such as suboptimal routing. Therefore, the impact of any design optimization must be taken into consideration and analyzed, to ensure the selected optimization technique will not introduce new issues or complexities to the network that could impact its primary business functions. Ideally, the requirements of the business-critical applications and business priorities should drive design decisions.

## Reference

Al-shawi, Marwan, *CCDE Study Guide* (Cisco Press, 2015)

## Exam Preparation Tasks

As mentioned in the section "How to Use This Book" in the Introduction, you have a couple of choices for exam preparation: the exercises here, Chapter 18, "Final Preparation," and the exam simulation questions in the Pearson Test Prep Software Online.

## Review All Key Topics

Review the most important topics in this chapter, noted with the Key Topic icon in the outer margin of the page. Table 8-12 lists a reference of these key topics and the page numbers on which each is found.

**Key Topic**

**Table 8-12**  Key Topics for Chapter 8

| Key Topic Element | Description | Page Number |
|---|---|---|
| Figure 8-3 | Process of Building an LSDB | 162 |
| Table 8-2 | Summary of OSPF Area Types | 167 |
| Figure 8-9 | OSPF Route Propagation in Multi-Area Design | 167 |
| Section | OSPF Versus IS-IS | 170 |
| List | Reasons to deploy EIGRP in an enterprise network | 170 |
| Section | EIGRP Route Propagation Considerations | 174 |
| Table 8-3 | OSPF Virtual Link Versus GRE Tunnel | 182 |
| Table 8-4 | Suboptimal Routing Optimization Techniques | 195 |
| Section | Single Redistribution Boundary Point | 199 |
| Section | Multiple Redistribution Boundary Point | 200 |
| Section | Route Filtering Versus Route Tagging with Filtering | 202 |
| Table 8-7 | Interdomain Routing Terminology | 206 |
| List | Typical BGP route selection steps | 208 |
| List | AIGP BGP route selection steps | 215 |
| Section | BGP Route Reflection | 215 |
| Section | BGP Confederation | 222 |
| Table 8-10 | Confederation Versus RR | 224 |
| Figure 8-61 | Routing Protocol Selection Decision Tree | 226 |
| Figure 8-62 | Routing Protocol Migration Decision Tree | 227 |

## Complete Tables and Lists from Memory

Print a copy of Appendix D, "Memory Tables" (found on the companion website), or at least the section for this chapter, and complete the tables and lists from memory. Appendix E, "Memory Tables Answer Key," also on the companion website, includes completed tables and lists to check your work.

## Define Key Terms

Define the following key terms from this chapter and check your answers in the glossary:

link-state routing protocols, Open Shortest Path First (OSPF), Intermediate System-to-Intermediate System (IS-IS), link-state advertisement (LSA), Enhanced Interior Gateway Routing Protocol (EIGRP), distance-vector routing protocol, administrative distance (AD), area border router (ABR), autonomous system boundary router (ASBR), Border Gateway Protocol (BGP).

# CHAPTER 9

# Network Virtualization

## This chapter covers the following topics:

- **MPLS:** This section covers critical MPLS topics and network design elements for MPLS.

- **Software-Defined Networks:** This section covers SD-WAN and SD-LAN in a vendor-agnostic perspective to provide the corresponding network design elements associated with the inherent capabilities these solutions provide.

This chapter covers the different design options and considerations of network virtualization options, including the forwarding and control plane mechanisms of MPLS, MP-BGP, and software-defined networking.

This chapter covers the following "CCDE v3.0 Core Technology List" sections:

- 4.1 Multiprotocol Label Switching

- 4.2 Layer 2 and 3 VPN and tunneling technologies

- 4.3 SD-WAN

## "Do I Know This Already?" Quiz

The "Do I Know This Already?" quiz allows you to assess whether you should read this entire chapter thoroughly or jump to the "Exam Preparation Tasks" section. If you are in doubt about your answers to these questions or your own assessment of your knowledge of the topics, read the entire chapter. Table 9-1 lists the major headings in this chapter and their corresponding "Do I Know This Already?" quiz questions. You can find the answers in Appendix A, "Answers to the 'Do I Know This Already?' Quizzes."

**Table 9-1** "Do I Know This Already?" Section-to-Question Mapping

| Foundation Topics Section | Questions |
|---|---|
| MPLS | 1–8 |
| Software-Defined Networks | 9, 10 |

**CAUTION** The goal of self-assessment is to gauge your mastery of the topics in this chapter. If you do not know the answer to a question or are only partially sure of the answer, you should mark that question as wrong for purposes of the self-assessment. Giving yourself credit for an answer you correctly guess skews your self-assessment results and might provide you with a false sense of security.

1. Which of the following MPLS L3VPN RD allocation models is simple and manageable?

   a. Unique RD per VPN per interface of each PE

   b. Unique RD per VPN

   c. Unique RD per VPN per PE

2. Which of the following MPLS L3VPN RD allocation models provides active/active or active/standby connectivity to a multihomed remote site?

   a. Unique RD per VPN per interface of each PE

   b. Unique RD per VPN

   c. Unique RD per VPN per PE

3. What does EIGRP SoO specifically mitigate from a routing perspective?

   a. Route looping and racing issues when OSPF is used as a PE-CE routing protocol at sites that contain just an MPLS VPN link

   b. Route looping and racing issues when EIGRP is used as a PE-CE routing protocol at sites that contain both MPLS VPN and backdoor links

   c. Route looping and racing issues when OSPF is used as a PE-CE routing protocol at sites that contain both MPLS VPN and backdoor links

   d. Route looping and racing issues when EIGRP is used as a PE-CE routing protocol at sites that contain just an MPLS VPN link

4. What is used in OSPF for loop prevention in the super backbone when redistributing into OSPF as a PE-CE routing protocol?

   a. BGP SoO

   b. OSPF domain ID

   c. OSPF DN bit

   d. EIGRO SoO

5. What MPLS feature supports having overlapping address spaces of multiple customer VPNs/VRFs within the MPLS core?

   a. Route target

   b. Cost community

   c. VRF

   d. Route distinguisher

6. What MPLS feature is used to control which VPN route can be installed in a VRF?

   a. Route target

   b. Cost community

   c. VRF

   d. Route distinguisher

7. Which MPLS L3VPN topology should be used if an enterprise requires all remote site traffic to be inspected by a firewall in the data center?

    a. Full mesh

    b. Hub and spoke

    c. Multilevel hub and spoke

    d. Extranet and shared services

8. Which MPLS L3VPN topology should be used if an enterprise is required to share the corporate internet service with all MPLS customer VPNs?

    a. Full mesh

    b. Hub and spoke

    c. Multilevel hub and spoke

    d. Extranet and shared services

9. Which of the follow SDN mechanisms is defined by the physical switches and routers that are part of the local-area network?

    a. Underlay

    b. SDN controller

    c. Overlay

    d. VN

10. Which of the following is an SD-WAN overlay mechanism that provides segmentation much like a VRF?

    a. Underlay

    b. SDN controller

    c. Overlay

    d. VN

## Foundation Topics

## MPLS

Over the years Multiprotocol Label Switching (MPLS) has become more popular within the large enterprise space. The benefits and overall design goals are the same and are as follows.

**Key Topic**

- Support a large number of customer groups and a large number of sites per group (scalable).

- Provide customers with value-added services that can create new revenue-generation sources such as service differentiation–capable transport with end-to-end quality of service (QoS) (flexible and reliable).

- Support various services for all enterprise customers over one unified infrastructure. Having a single consolidated network reduces capital expenditure (CAPEX) and can offer a significant return on investment (ROI) to the large enterprise business on its hardware equipment (with the virtualized or overlaid architectures).

- Provide flexible service provisioning over various media access methods.

In addition, the MPLS peer model has proven its flexibility and reliability in fulfilling these goals for many large enterprise customers by offering

**Key Topic**

- Single infrastructure that can serve all VPN customers (as shown in Figure 9-1).

- The optimization of OPEX. For instance, adding a new customer or a new site for an existing customer will require simple changes to the relevant edge nodes (provider edge [PE] nodes) only as the core control plane intelligence is pushed to the provider cloud, as shown in Figure 9-1.

- The opening of new revenue-generation sources to the business by offering differentiated services for its customers, such as prioritization and expedited forwarding for voice.

- The optimization of time to market to introduce new services to the organization's L3VPN customers, such as IPv6 and multicast support.

- A high degree of flexibility by offering various media access methods for the organization's customers, such as legacy equipment, Ethernet over copper or fiber, and Long Term Evolution (LTE) or 5G.

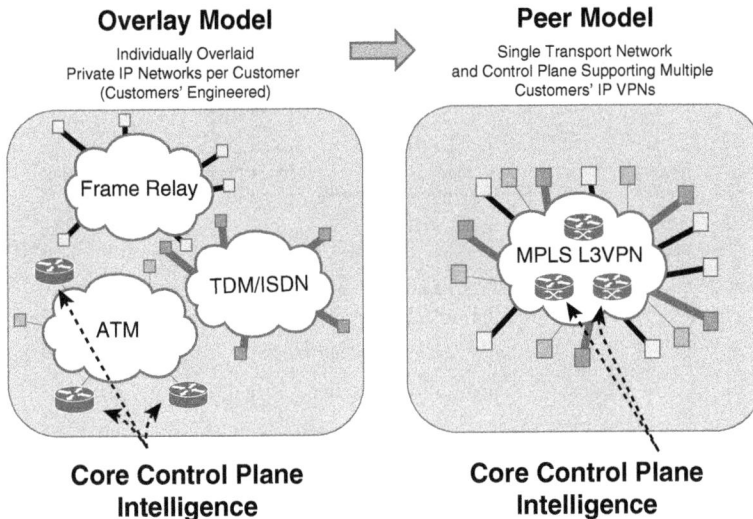

**Figure 9-1**  *MPLS Peer Model*

## MPLS Architecture Components

In a typical MPLS environment, the architecture is constructed of the following components:

- Customer edge (CE)

- Provider edge (PE)

- Provider nodes (P)

In the typical MPLS architecture, the provider edge nodes (PEs) carry customer routing information to inject customer routes from the directly connected customer edge nodes (CEs), each to the relevant Multiprotocol Border Gateway Protocol (MP-BGP) VPNv4/v6, along with the relevant VPN and transport MPLS labels (label edge router [LER]). This achieves the optimal routing of traffic that pertains to each customer within each VPN routing domain. However, provider routers (Ps) at the core of the network are mainly responsible for switching MPLS labeled packets. Therefore, they are also known as label switching routers (LSRs). Figure 9-2 illustrates the primary component of an MPLS architecture:

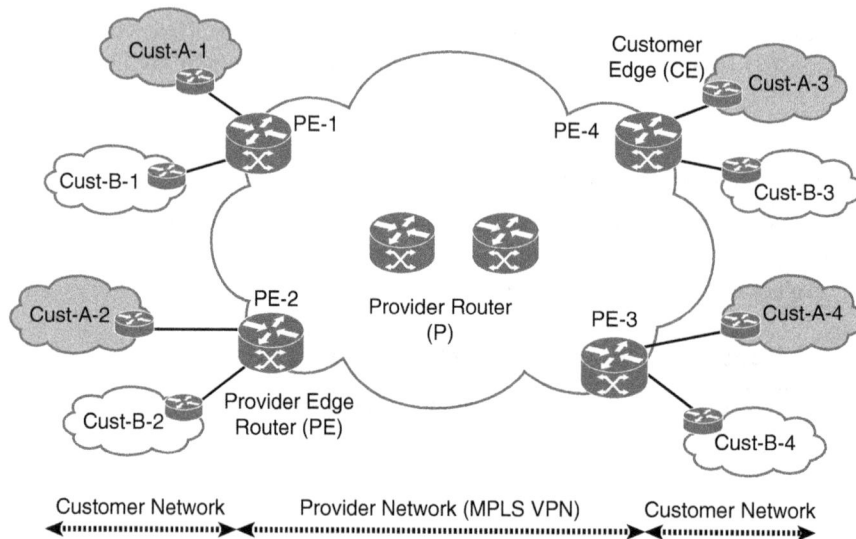

**Figure 9-2** *Primary Elements of an MPLS Architecture*

On top of the architectural components shown in Figure 9-2, the different control plane protocols are overlaid to construct the control and forwarding planes for each VPN network (per customer). Figure 9-3 shows the relationship between the different control and forwarding plane components in an MPLS architecture.

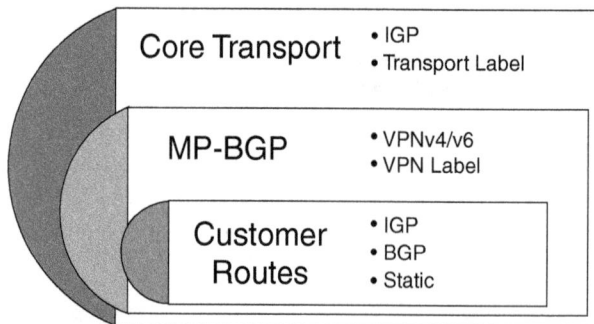

**Figure 9-3** *MPLS Control Plane Components*

The actual communication in a typical MPLS environment is driven by following three primary elements:

- Routing information isolation between different VPNs (for example, Virtual Routing and Forwarding [VF] instances)

- Controlled sharing of routing information to sites within a VPN or between different VPNs (for example, route distinguisher [RD] + route target [RT] + MP-BGP)

- MPLS traffic forwarding of packets across the MPLS core (for example, VPN and transport labels)

## MPLS Control Plane Components

This section covers the primary control plane elements of an MPLS environment.

### Virtual Routing and Forwarding

**Key Topic**

**Virtual Routing and Forwarding (VRF)** is one of the primary mechanisms used in today's modern networks to maintain routing isolation on a Layer 3 device level. In MPLS architecture, each PE holds a separate routing and forwarding instance per VRF per customer, as shown in Figure 9-4. Typically, each customer's VPN is associated with at least one VRF. Maintaining multiple VRFs on the same PE is similar to maintaining multiple dedicated routers for customers connecting to the provider network. In addition, maintaining multiple forwarding tables at the PE is essential to support overlapping address spaces. Normally, the routing information of each customer (VPN) is installed at the relevant VRF routing tables of a PE, either from directly connected CEs (using a VRF-aware interior gateway protocol [IGP], BGP, or static route) or routes of other CEs learned via remote PEs over MP-BGP VPNv4/v6.

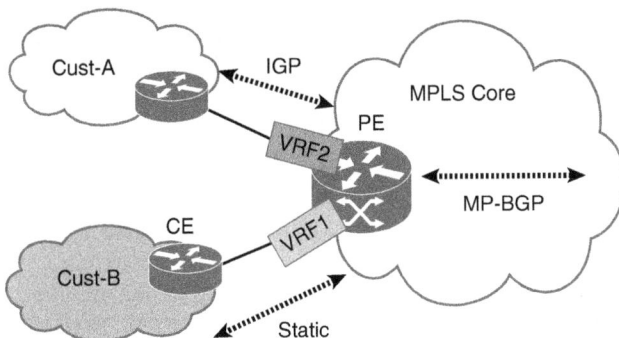

**Figure 9-4**   *MPLS L3VPN: Virtual Routing and Forwarding*

### Route Distinguisher

For an MPLS L3VPN to support having multiple customer VPNs with overlapping addresses and to maintain the control plane separation, the PE router must be capable of using processes that enable overlapping address spaces of multiple customers' VPNs. In addition, the PE router must also learn these routes from directly connected customer networks and propagate this information using the shared backbone. This is accomplished by using a **route distinguisher (RD)** per VPN or per VRF instance. As a result, the MPLS core can seamlessly transport customers' routes (overlapped and nonoverlapped) over one common infrastructure and control plane protocol to take advantage of the RD prepended per MP-BGP VPNv4/v6 prefix. Normally, the RD value can be allocated using different approaches. Each approach has its strengths and weaknesses, as covered in Table 9-2.

**Key Topic**

**Table 9-2** MPLS L3VPN RD Allocation Models

| RD Model | Strength | Weakness | Suitable Scenario |
|---|---|---|---|
| Unique RD per VPN | Simple to design and manage.<br><br>Lower hardware resource consumption compared to other models. | Lacks the ability to support load-balancing capability when VPN route reflection (RR) is used in the MPLS VPN network and there are customers that have multihomed CE routes. | Very-large-scale MPLS VPN without load-balancing or load-sharing requirements toward multihomed sites. |
| Unique RD per VPN per PE* | Offers the ability to load balance traffic toward multihomed sites that are part of the same VPN but connected to different PEs. | Requires more hardware resources, such as memory to store the additional VPN routes.<br><br>Higher design and operational complexity than the unique RD per VPN model. | Large-scale MPLS VPN with load-balancing or load-sharing requirements toward multihomed sites.** |
| Unique RD per VPN per interface of each PE | Simplifies identifying sites within a VPN. | Highest hardware resources utilization (a separate VRF per CE).<br><br>High design and operational complexity. | MPLS VPN of an enterprise with small number of VPNs with requirements to have a simplified mechanism for the operation team to identify the origin of any route (from which site per VPN) through its RD value.*** |

\* Load balancing or load sharing for multihomed sites using unique RD per VPN per PE is covered in more detail later in this chapter.

\*\* In large-scale networks with a large number of PEs and VPNs, unique RD per VPN RD allocation should be used. The unique RD per VPN per PE RD allocation model can be used only for multihomed sites if the customer needs to load balance/share traffic toward these sites.

\*\*\* BGP site of origin (SoO) can be used as an alternative to serve the same purpose without the need of a unique RD per interface/VRF.

Figure 9-5 illustrates these different RD allocation models discussed in Table 9-2.

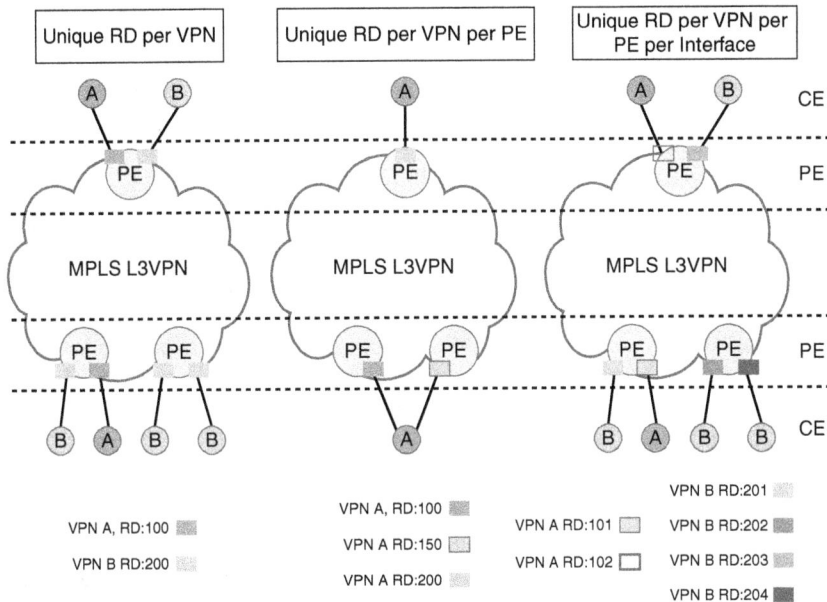

**Figure 9-5**  *MPLS L3VPN RD Allocation Models*

**NOTE**  Based on the RD allocation models covered in Table 9-2, a single VPN may include multiple RDs with different VRFs. However, the attributes of the VPN (per customer) will not change and is still considered an intra-VPN because technically the route propagation is controlled based on the import/export of the RT values.

## Route Targets

**Route targets (RTs)** are an additional identifier and considered part of the primary control plane elements of a typical MPLS L3VPN architecture because they facilitate the identification of which VRF instance can install which VPN routes. In fact, RTs represent the policies that govern the connectivity between customer sites. This is achieved via controlling the import and export RTs. Technically, in an MPLS VPN environment, the export RT is to identify a VPN membership with regard to the existing VRFs on other PEs, whereas the import RT is associated with each PE local VRF. The import RT recognizes and maps the VPN routes (received from remote PEs or leaked on the local PE from other VRF instances) to be imported into the relevant VRF instance of any given customer. In other words, RTs can offer network designers a powerful capability to control what MP-BGP VPN route is to be installed in any given VRF/customer routing instance. In addition, they provide flexibility to create various logical L3VPN (WAN) topologies for the enterprise customer, such as any to any, hub and spoke, and partially meshed, to meet different connectivity requirements.

## L3VPN Forwarding Plane

In addition to the control plane discussed earlier, the data or forwarding plane forms the other major component of typical MPLS L3VPN building blocks. In MPLS VPN environments, the data plane is based on forwarding packets based on labels (transport labels, VPN labels, MPLS Traffic Engineering [MPLS-TE] labels, and so on). This section focuses on the VPN label.

Typically, VPN traffic is assigned to a VPN label at the egress PE (LER) that can be used by the remote ingress PEs (LER), where the egress PE demultiplexes the traffic to the correct VPN customer egress interface based on the assigned VPN label. In other words, the **VPN label** is generated and assigned to every VPN route by the egress PE router, then advertised to the ingress PE routers over an MP-BGP update. Therefore, it is only understood by the egress PE node that performs demultiplexing to forward traffic to the respective VPN customer egress interface/CE based on its VPN label. This is true for all local labels regardless of which protocol allocates and binds them.

The VPN labels in MPLS VPN architectures can be allocated by the PE nodes using different models for MP-BGP L3VPN routes, based on the scenario and the design requirements. The VPN labels also offer network designers the flexibility to achieve a level of trade-offs between network performance and scalability when possible. The following are the common MPLS VPN label-allocation models:

**Key Topic**

- **Per prefix:** In this model, a VPN label is assigned for each VPN prefix. Although this model can generate a large number of labels, it is required in scenarios where the VPN packets sent between the PE and CE are label switched, such as in Carrier supporting Carrier (CsC) designs.

- **Per VRF:** In this model, a single label is allocated to all local VPN routes of any given PE in a given VRF. This model offers an efficient label space and BGP advertisements. In addition, some vendor platforms support the same per-VRF label for both IPv4 and IPv6 prefixes.

- **Per CE:** The PE router allocates one label for every immediate next hop; in most cases, this would be a CE router. This label is directly mapped to the next hop, so there is no VRF route lookup performed during data forwarding. However, the number of labels allocated is one for each CE rather than one for each VRF. Because BGP knows all the next hops, it assigns a label for each next hop (not for each PE-CE interface). When the outgoing interface is a multiaccess interface and the media access control (MAC) address of the neighbor is not known, Address Resolution Protocol (ARP) is triggered during packet forwarding.

Network designers must be careful if they plan to change the default label allocation behavior, because any inconsistency or simple error can lead to a broken forwarding plane that can easily bring down the entire network or a portion of the network. In a service provider network, a PE that goes down this way may result in several customer sites (usually single-homed ones) being out of service, which can impact the business significantly, especially if there is a strict service-level agreement (SLA) with its customers.

Figure 9-6 shows a summary of end-to-end forwarding and control planes of an MPLS L3VPN architecture.

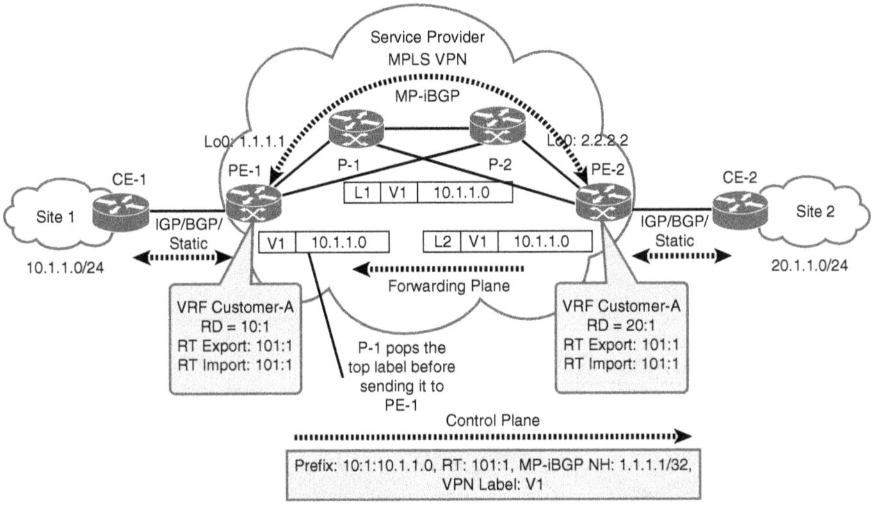

**Figure 9-6**   *Forwarding and Control Planes of MPLS L3VPN Architecture*

## L3VPN Design Considerations

This section discusses the primary design considerations common in an MPLS L3VPN environment, along with the possible design options of each. One of the common connectivity models of enterprise customers to the MPLS L3VPN-based WAN is multihoming. Enterprise customers with this connectivity model often need to load balance traffic across both WAN links (this model includes both, one CE with two links or one site with two CEs and links), as shown in Figure 9-7.

As shown in Figure 9-7, there is an MP-BGP route reflector (RR) part of the MP-BGP control plane architecture. Normally, the RR will advertise only the best route to its clients (other PEs) from the RR point of view to the topology, which will usually break the requirement of load balancing or sharing for those multihomed enterprise customers. One simple solution is to remove the RR and use a full mesh of MP-iBGP sessions.

However, this might not be an ideal solution for many carrier networks because it may introduce MP-BGP scalability limitations on the underlay network. The other common and simple solution to this requirement is to configure the multihomed VPNs/VRFs of the multihomed sites with different RDs, where each route will appear as a unique VPN route to the RR. Consequently, the RR will send these VPN routes to the other remote PEs (PE-1 in Figure 9-7, with PE-2 and PE-3 as the MP-BGP next hops).

9

**Unique RD per VPN**

RR

10.2.1.0

PE-2

iBGP

iBGP

PE-1

| 1:100:10.2.1.0, NHP PE-2 |
| 1:100:10.1.1.0, NHP PE-2 |

10.3.1.0

10.1.1.0

PE-3

Traffic Flow

MPLS L3VPN

VPN A RD 1:100

- - - - - - - - - - - - - - - - - - - - - - - - - - - - - - - - - - - - - - - - - - - - - - - -

**Unique RD per VPN per PE**

RR

10.2.1.0

PE-2

iBGP

iBGP

PE-1

| 1:100:10.2.1.0, NHP PE-2 |
| 1:100:10.1.1.0, NHP PE-2 |
| 1:200:10.2.1.0, NHP PE-3 |
| 1:200:10.1.1.0, NHP PE-3 |

10.3.1.0

10.1.1.0

PE-3

Traffic Flow

MPLS L3VPN

VPN A RD 1:100
VPN A RD 1:200

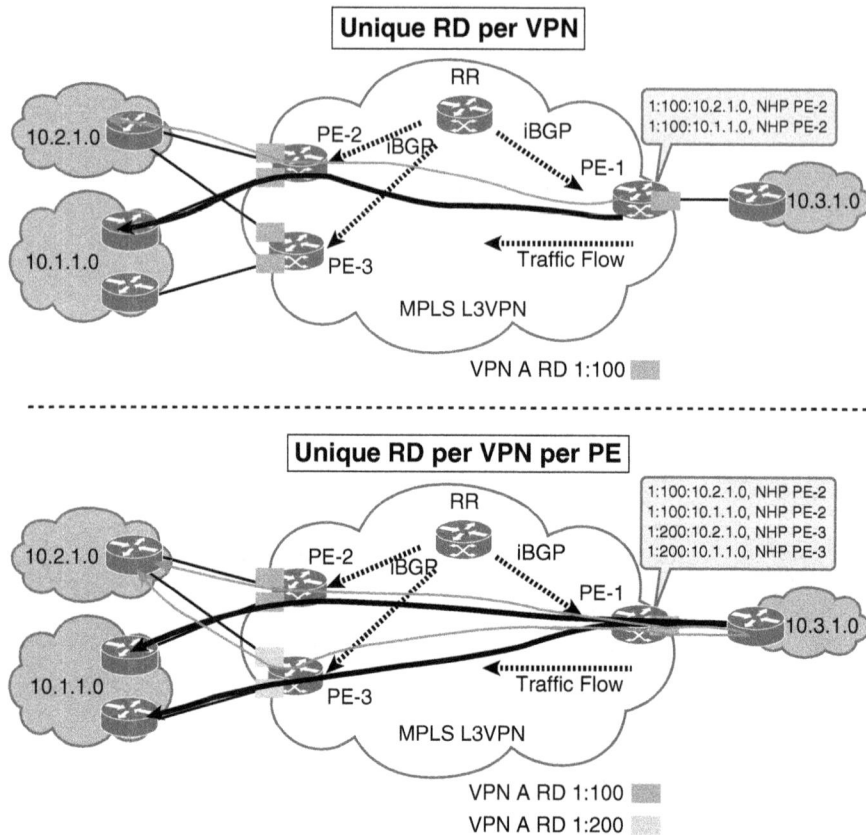

**Figure 9-7**  *Multihoming in MPLS L3VPN Environment*

**NOTE**  The BGP multipathing feature must be enabled within the relevant BGP VRF address family at the remote PE routers (for example, PE-1 in the preceding example). Similarly, enabling BGP multipathing is required in a single CE dual-attached use case to enable the load balancing/sharing from the CE end as well when BGP is used between the CE and PE.

## MPLS L3VPN Topologies

As covered earlier in this section, RT values enable you to control the import and export of VRF routes, which can control VPN membership per customer. This facilitates the creation of different L3VPN overlaid topologies based on customer requirements. The following are the most common L3VPN WAN topologies, which are controlled by RTs.

### Full Mesh

The full-mesh topology shown in Figure 9-8 is the simplest and most common topology that represents the typical MPLS L3VPN layout. Simply, the any-to-any communications model between different customer sites that normally belong to the same customer (under a single VPN or multiple VPNs) must carry the same RT values of the import and export among them (among the relevant PEs).

This design model logically can be shown as one large router with all other locations connected directly to it, as shown in Figure 9-9, where the big router in the middle is the MPLS L3VPN cloud and all other sites are directly attached to it in a start topology.

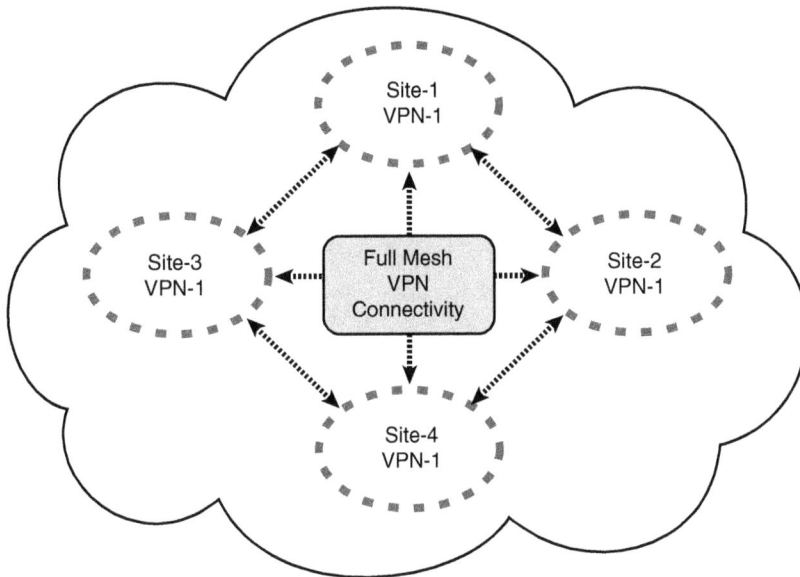

**Figure 9-8**   *MPLS L3VPN Full-Mesh Topology*

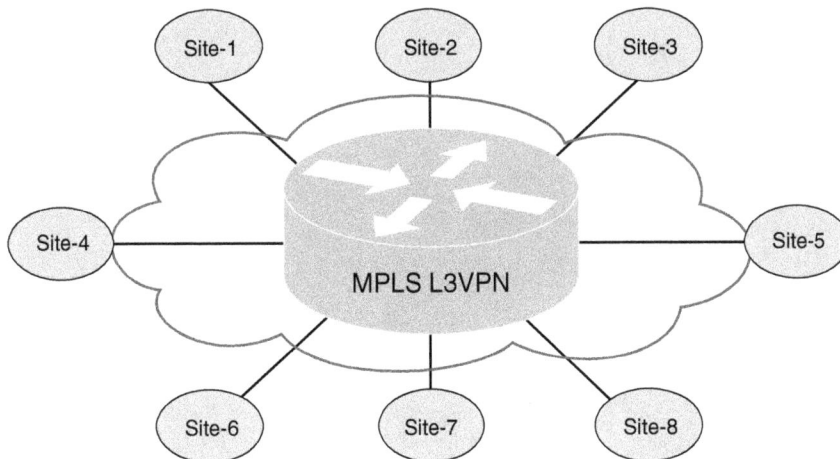

**Figure 9-9**   *MPLS L3VPN Conceptual View*

## Hub-and-Spoke L3VPN Service

In some situations, MPLS L3VPN customers require, for different reasons, that the communication between remote sites has to go through the main or hub site (for example, to align with the enterprise security policy requirements). From the L3VPN service provider point of view, a hub-and-spoke topology can be provisioned for this type of requirement by controlling MP-BGP VPN route propagation (using RT import and export), as shown in Figure 9-10. The most common and proven way to achieve a hub-and-spoke topology over an L3VPN network is to deploy two links with one VRF per link between the PE and the directly connected CE hub, as shown in Figure 9-11.

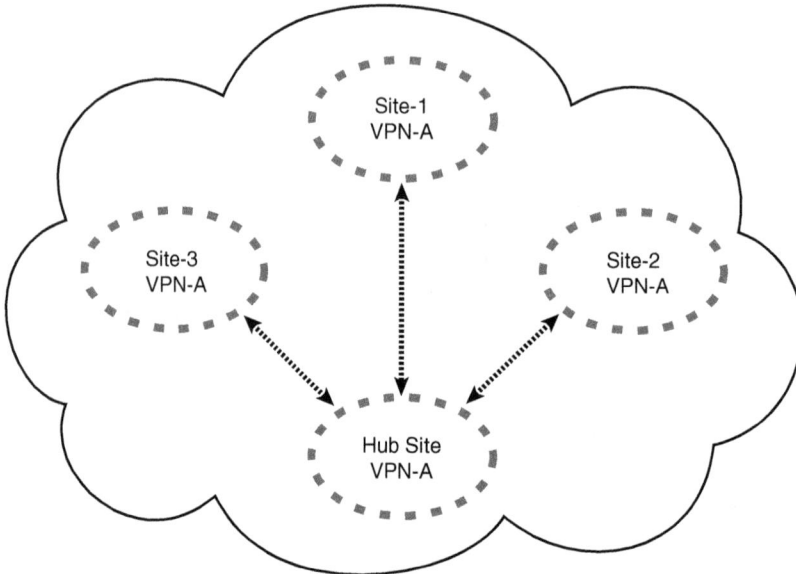

**Figure 9-10**  *MPLS L3VPN Hub-and-Spoke Topology*

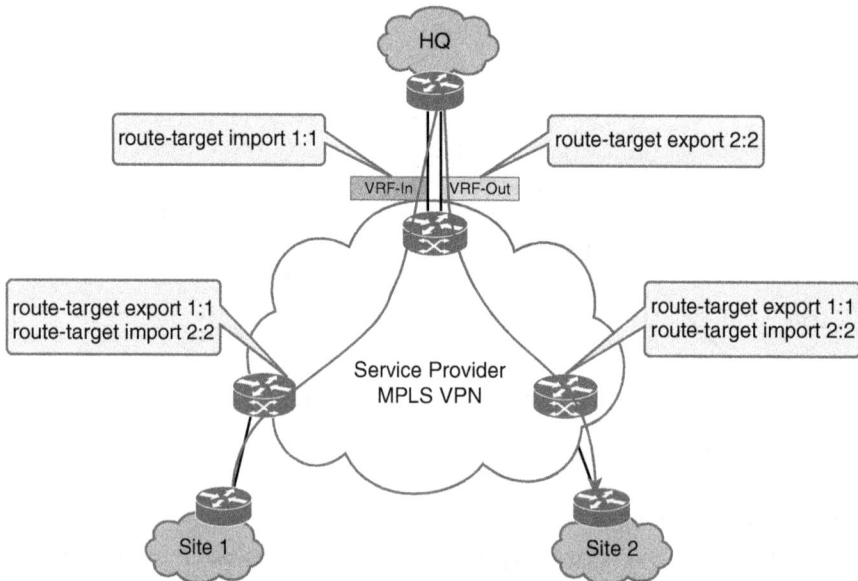

**Figure 9-11**  *MPLS L3VPN Design*

Achieving a hub-and-spoke topology in an MPLS L3VPN environment is as simple as con-trolling the import and export of RT values. However, network designers must be aware of the following design features to avoid breaking the communications across the overlaid hub-and-spoke topology:

- If the BGP used as the PE-CE routing protocol across the hub-and-spoke topology over L3VPN and each site uses the same BGP autonomous system number (ASN), BGP AS override should be used by the PE connected to the hub-and-spoke sites. This avoids blocking communication among the sites as a result of BGP loop-prevention behavior. Although BGP allows the AS-in feature to be used for the same purpose from the CE side, it must be planned carefully to avoid any unexpected BGP AS_PATH looping.

- If more than one spoke is connected to the same PE, VRF is required to avoid traffic bypassing the hub site.

- If the hub site has two edge CE routers connected to the MPLS L3VPN cloud, each CE must (ideally) be assigned the role of handling routing/traffic in one direction; one hub CE is connected to the receiving link, and the other hub CE is connected to the sending link.

### Multilevel Hub and Spoke

In this model, large enterprise customers (usually with distributed sites in multiple geographic areas) can take advantage of multitiered hub-and-spoke topology, as shown in the logical layout in Figure 9-11. For instance, remote sites distributed across different regions can be aggregated into first-level hub sites per region, while the first-level hub sites connect to each other in a hub-and-spoke topology and to a second-level hub site, such as a centralized data center.

In Figure 9-12, each group of hub and spoke is allocated its own MPLS VPN, which can represent the grouping of sites based on geographic location. With an architecture like this, global enterprises can achieve a more structured network design and traffic flow between different sites and geographic regions. At the same time, service providers can easily provi-sion this type of architecture by controlling routing information propagation among the dif-ferent VPNs by controlling the import and export of RT values between the VPNs.

### Extranet and Shared Services

In this particular design model, communication between one or more different VPN net-works and a centralized VPN network is required. This is achieved in the same manner as the previous models: by controlling the import and export RT values. In this particular model, the central VPN must have import RT values matching the different export RT values of the VPNs that require access, as shown in Figure 9-13.

Figure 9-14 illustrates a detailed view of how the import and export routes happen in a shared services VPN architecture.

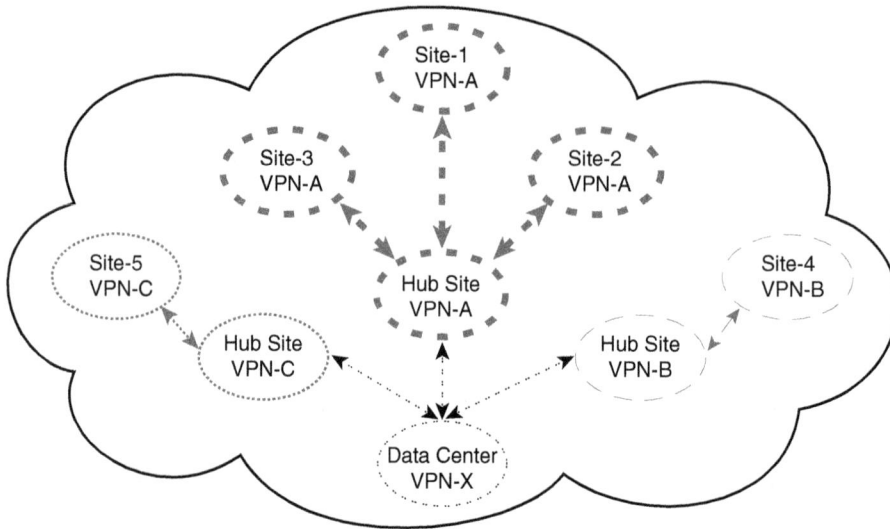

**Figure 9-12** *MPLS Multilevel Hub and Spoke*

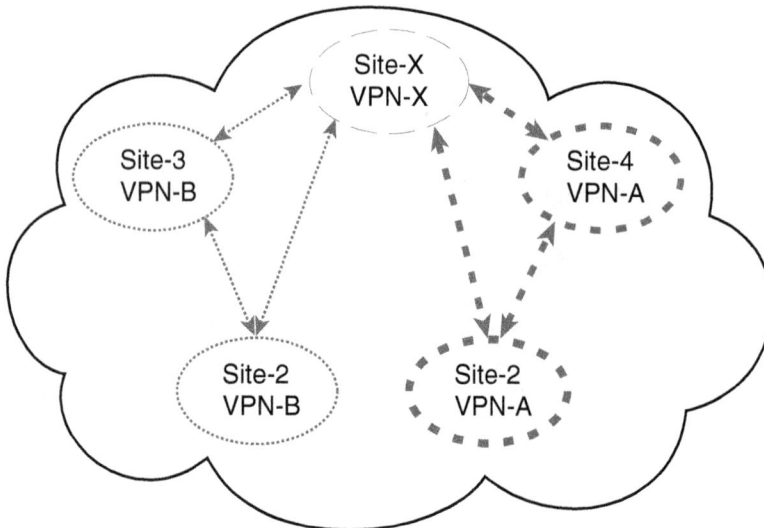

**Figure 9-13** *MPLS L3VPN Extranet Topology*

With this design, the shared services VPN will be accessed by the both VPN-A and VPN-B without compromising routing and reachability separation requirements between these VPNs (no communication between VPN-A and VPN-B). However, because the shared services VPN will have visibility on both VPNs' routes in this example, the IP prefixes of VPN-A and VPN-B must be unique; otherwise, VRF-aware Network Address Translation (NAT) should be deployed.

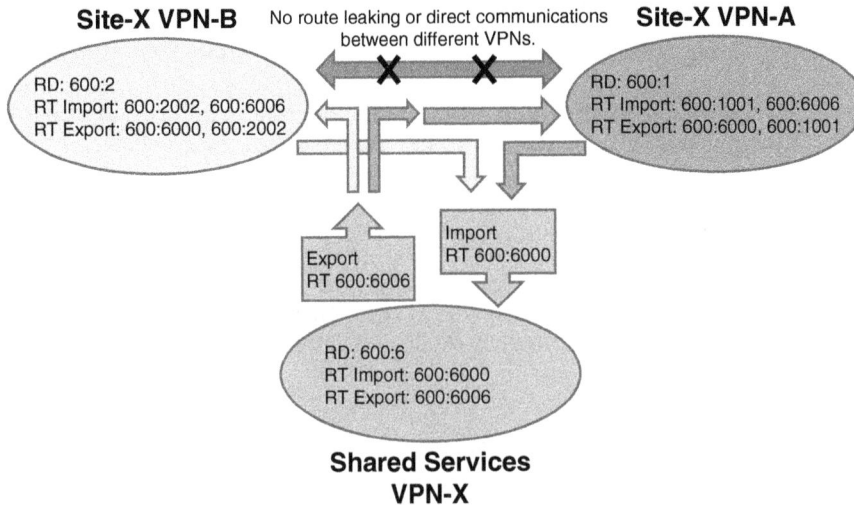

**Figure 9-14**  *MPLS L3VPN Shared Services Connectivity Model*

The following are the most common scenarios used with this design model:

- **Management:** Network operations center (NOC) management access. For example, the service provider can offer managed CE services to its clients. Accordingly, the NOC or management VPN of the service provider requires access to the relevant customer's VPN, to manage or monitor its CE routers.

- **Shared services:** In general, shared service access refers to several services that have to be accessible from different MPLS L3VPN networks, such as file services, Voice over IP (VoIP) gateways, and hosted applications (Software as a Service [SaaS]) or Internet connection.

- **Extranet or business-to-business (B2B) communication:** This scenario is common in modern enterprises where vendors and partners, for example, can share limited reachability between their networks to facilitate different types of communications, such as business-to-business telepresence.

- **Community communication:** In this model, a centralized entity provides central services access via a common MPLS VPN cloud. An example of this is an educational system, where commonly a centralized entity provides shared services access to the different schools across different locations while maintaining traffic separation between the schools.

**NOTE**  Depending on the network environment, the number of prefixes to be exported and imported can be limited in a controlled manner. For example, in the case of the managed services, only the loopback IP of each CE is exported from the customer VPN to the NOC/ management VPN for monitoring and remote-access purposes. At the same time, controlling the number of exported and imported prefixes prevents the leaking of extra prefixes, which can lead to other issues such as exposing customer internal routes and unnecessary extra overhead on the PE nodes.

## MPLS VPN Internet Access Design Options

The flexibility provided by MPLS-based infrastructures, specifically MPLS VPN, allows operators to offer more than basic private IP WAN connectivity to their customers. One of the primary services that today's MPLS VPN architectures offer is Internet access as a value-added service, taking advantage of the flexible MPLS VPN architecture that uses the same infrastructure to provide various connectivity models and services. In addition, it is an additional source of revenue, from the operator's business perspective.

However, from a network design perspective, Internet access for MPLS VPN customers can be provided using multiple design options. Typically, each option has its strengths and weaknesses. This section discusses and compares the most common design options used to provide Internet service from the MPLS VPN architecture perspective. The design models are categorized as follows:

**Key Topic**

- Non-MP-BGP VPN Internet routing

- MP-BGP VPN-based Internet routing

### Non-MP-BGP VPN Internet Routing

The primary concept of this design model is that Internet routes are carried across the carrier network using the global routing table. The following are the two common design options of the model:

- **Option 1: VRF-specific default route:** This design option is based on using the static default route to redirect traffic from a customer VRF to the Internet gateway (Autonomous System Boundary Router [ASBR]) using the PE's global routing table, as shown in Figure 9-15. For the traffic in the other direction (Internet to VPN customers), the same concept is used (static route to move Internet traffic from the global routing table to the corresponding VRF/VPN per customer).

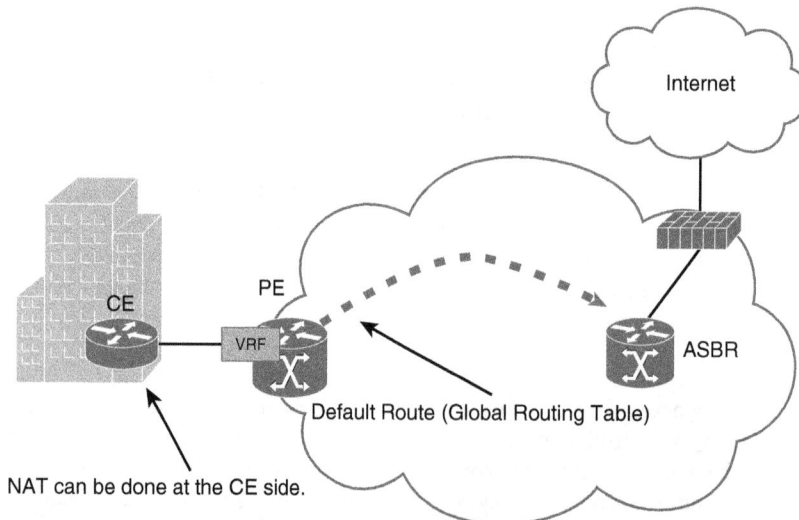

**Figure 9-15**  *MPLS L3VPN Internet Access Option 1*

- **Option 2: Separate PE-CE subinterface:** This design option uses the conceptual model of Option 1, but with the ability to separate Internet traffic and IP VPN traffic over two physical links or subinterfaces (such as 802.1Q based, or interface or data link connection identifier [DLCI] based in case of FR CE-PE physical connectivity). For the operators to propagate Internet routes (or default only) between a PE and the directly attached CE, either a static route or BGP is used over the Internet link (the non-VRF interface), as shown in Figure 9-16.

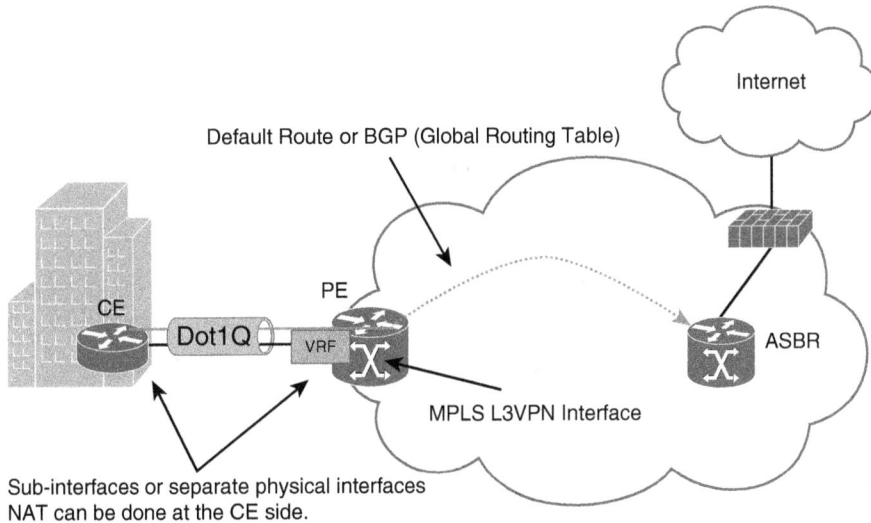

**Figure 9-16**   *MPLS L3VPN Internet Access Option 2*

### MP-BGP VPN Internet Routing

This design model relies primarily on MP-BGP VPN to carry Internet routes across the carrier network. The following are the two common design options of the model:

**NOTE**   The sequence of numbers of the following design options continues from the two other design options previously discussed, to facilitate the reference of each model in Table 9-3.

- **Option 3: Extranet with Internet-VRF:** This design option is based on the principle discussed earlier in this chapter, which is the extranet or shared VPN where the Internet gateway (ASBR) installs Internet routes in its own VPN. This VPN has bidirectional communication with the customer's VPN, a customer that requires Internet access over the same link/VPN (operators control the propagation of the Internet routes or default route into a customer's VRF routing table via the import and export RT values), as shown in Figure 9-17. In addition, if the full Internet routes are injected into the customer VPN, it is highly recommended to change MPLS VPN label allocation mode to either per CE or per VRF. This prevents the default behavior per-prefix allocation mode, which will usually allocate a label per prefix. Based on the size of the Internet routes, there can be noticeable performance inefficiencies in the case of the per-prefix label allocation model.

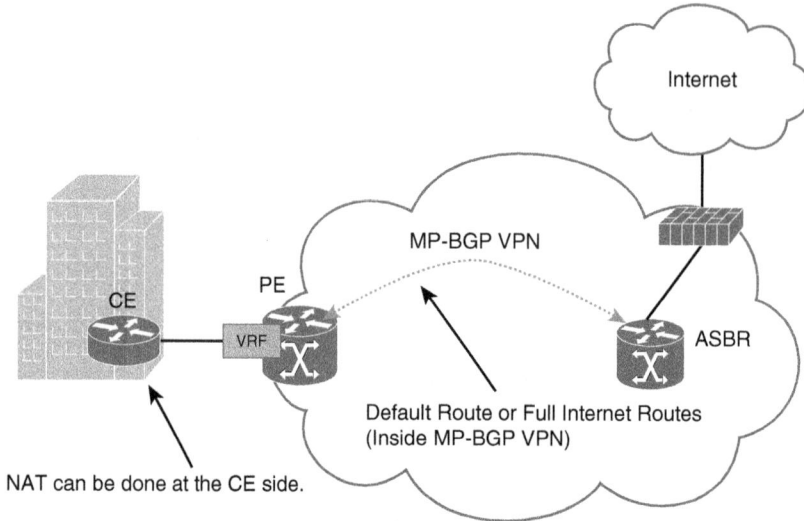

**Figure 9-17** *MPLS L3VPN Internet Access Option 3*

■ **Option 4: VRF-aware NAT:** This design option relies on MP-BGP VPN to customer Internet traffic, like Option 3. However, in this design option, the Internet gateway (ASBR) is not placed in a shared VPN. Instead, each VRF/VPN is created at the ASBR and each VRF is injected with a default route to be propagated to the corresponding VPN across the MPLS VPN network (toward the ASBR inside interface), while performing VRF-aware NATing on the outside interface (Internet facing) for both ingress and egress Internet traffic. As a result, with this design option, customers may retain their private IP addressing even if there is overlap with other customer IP addressing ranges, as shown in Figure 9-18.

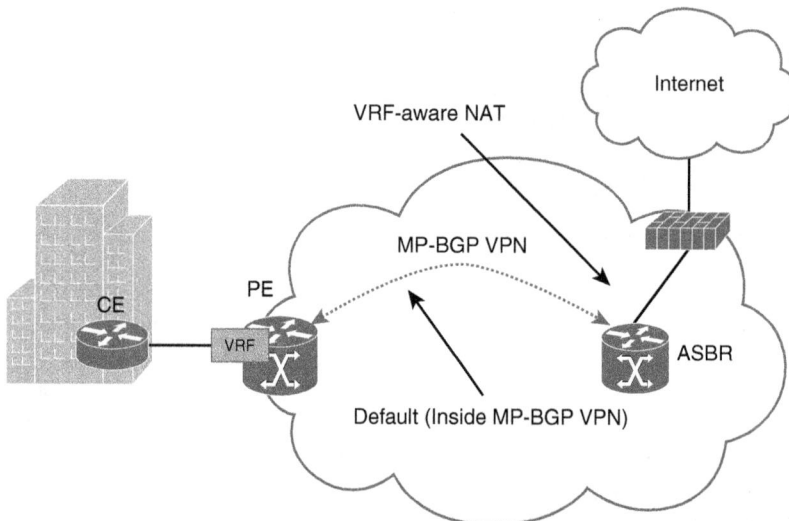

**Figure 9-18** *MPLS L3VPN Internet Access Option 4*

Table 9-3 compares these design options from different design aspects.

**Table 9-3**   MPLS L3VPN Internet Access Design Options Comparison

| Design Consideration | Option 1 | Option 2 | Option 3 | Option 4 |
|---|---|---|---|---|
| PE manageability | Moderate | Simple | Simple | Simple |
| ASBR manageability | Simple | Simple | Moderate | Complex |
| CE simplicity | Simple | Simple | Moderate | Moderate |
| Scalability | High | High | Moderate* | High** |
| Supports overlapping customers' IPs | No (should be direct public provider independent/ provider assigned [PI/PA]) | No (should be direct public PI/PA) | Yes | No |
| Full Internet routes at the PE | Yes | May | May | No |
| Design concerns | Static routing may lead to configuration errors and operational complexity. | May add routing/BGP configurations to the CE side. | Several VRFs on the same PE, each with full Internet routes, can add significate load on the PE. | VPN customers do not receive full Internet routes. |

\* If the full Internet routing table is installed per VRF, there can be scalability limitations at the PE level.

\*\* There might be limitations at the ASBR level (VRF routes or NATing entries/sessions).

The comparison in Table 9-3 might make it seem like one option is better than others for certain design requirements, but usually the network environment and the situation drive the design choices (for instance, if an Internet service provider [ISP] wants to start providing MPLS VPN to its customers but cannot afford any service interruption to its existing Internet customers). In this situation, keeping the Internet routing at the BGP global routing table can be a viable solution. Similarly, if an MPLS L3VPN provider wants to offer the Internet as a value-added service for its customers, adding the Internet as a new VPN (extranet VPN connectivity model) may be less interruptive. Therefore, always consider the other factors such as business priorities, design constraints, and the targeted environment, in addition to the technical aspects, before making any design decision.

## PE-CE L3VPN Routing Design

Designing routing between the PE and CE in an MPLS L3VPN environment can sometimes be challenging. This is applicable to any MPLS L3VPN environment, whether it is a service provider or an enterprise with an MPLS L3VPN core. Although the CE and PE sides are most commonly managed and deployed by different teams (usually the CE is controlled by the enterprise and the PE is controlled by the service provider side), the routing design of the CE to PE should (ideally) not be performed in isolation. It should be coordinated and aligned to achieve a successful design, because of dependencies in some scenarios that can impact the overall design and end-to-end communication. This section discusses the design considerations from both the CE side (usually the enterprise) and MPLS L3VPN side (either a service provider or self-deployed enterprise MPLS L3VPN).

## PE-CE Routing Design Considerations

For network designers to achieve a successful PE-CE routing design, they must follow the top-down approach in a structured manner to identify the goals and direction of the design during the planning phase. To achieve that, network designers must have a good understanding of the business, functional, and business-critical application requirements. The following questions can be considered the foundation, during the information-gathering and planning phases, for making a suitable design choice that can deliver value to the business:

**Key Topic**

- Business requirements of the design:

  - Is it to reduce the cost of existing expensive links?

  - Is it to reduce the time to expand the business (for example, to add new remote sites and integrate them into the network very quickly)?

  - Is it to increase the reliability of business applications and services with minimal cost (to be used as a backup path)?

  - Is it to optimize the ROI of the links by using multiple links in a load-balancing/ sharing manner?

- PE-CE functional and application requirements:

  - Provide primary WAN connectivity?

  - Provide backup WAN connectivity?

  - Provide WAN and Internet access?

  - Provide a primary path for applications that need high bandwidth or strict end-to-end QoS requirements?

  - Provide efficient bandwidth utilization over multiple paths (multihoming)?

  - Provide connectivity to remote areas with limited optical coverage (for example, remote sites connect over cellular 4G/5G)?

- Technical requirements (connectivity characteristics):

  - Is the CE single-homed or multihomed to the MPLS L3VPN network (PE)?

  - Is there any backdoor link from the CE side?

  - Is there any limitation to run any specific routing protocol (such as lack of staff knowledge, software limitations, or lack of support by the service provider side)?

After identifying the different requirements, network designers should have a good understanding of the design goal and direction, so as to achieve a successful design.

For example, it is always common for the MPLS L3VPN to be used as the primary path; however, in some scenarios, the backdoor link between any given sites might have a high-bandwidth capacity and offer lower latency and more control from the enterprise point of view. Therefore, in this case, the enterprise may prefer to use the MPLS L3VPN as a backup path only, which has to be reflected in the routing design.

### PE-CE Routing Protocol Selection

One of the critical parts of designing PE-CE is the routing protocol selection because each protocol has its strengths and weaknesses. In particular, in the PE-CE types of scenarios, the alignment of routing design between the service provider and the enterprise (CE side and PE side) is important. Otherwise, the lack of coordination and alignment between these two networks can lead to serious issues, such as suboptimal routing and routing loops. As shown in Figure 9-19, in the PE-CE type of scenario, what normally happens is that a route is received from one CE via a routing protocol. Next, this route is converted into an MP-BGP route, along with some protocol-related information that is normally transformed into BGP extended community values. These values can be used at the remote egress PE when the route is reconverted into the original routing protocol, such as Open Shortest Path First (OSPF) or Enhanced Interior Gateway Routing Protocol (EIGRP) (assuming both CEs are using the same routing protocol with the PEs; otherwise, this specific routing protocol information most likely will be invaluable). Nonetheless, each routing protocol behaves differently in these types of scenarios.

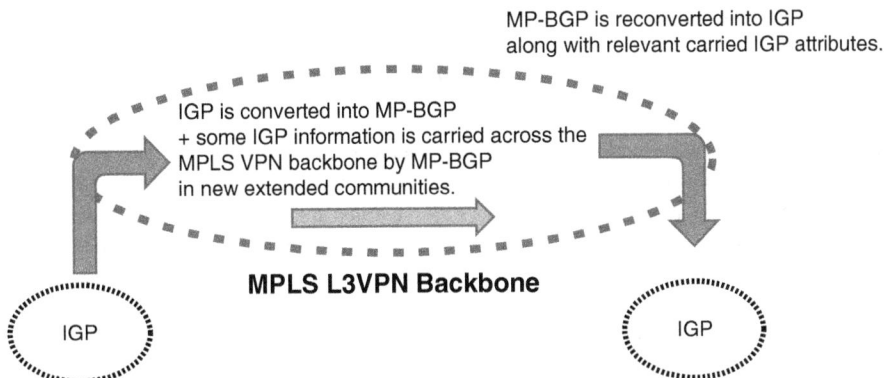

**Figure 9-19**   *PE-CE Routing Principle*

Therefore, understanding the business and functional requirements and the targeted topology (single-homed versus multihomed versus sites with backdoor links) can help you select the right routing protocol, to a large extent, taking into consideration the specific protocol attributes and behaviors with regard to the underlying network topology, WAN connectivity, and path selection requirements.

### PE-CE Design Options and Recommendations

As already mentioned, each routing protocol has its own strengths and weaknesses with regard to PE-CE routing design. This section discusses the technical design considerations and recommendations per routing protocol. Figure 9-20 shows the reference network architecture exemplifying various PE-CE connectivity models that will form the foundation of the design considerations and recommendations in this section.

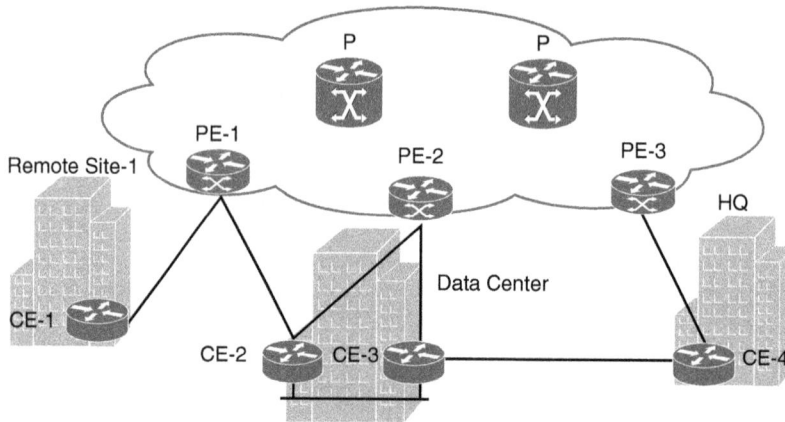

**Figure 9-20** *PE-CE Connectivity Model Reference Architecture*

**NOTE** This section covers the primary and most commonly used PE-CE routing protocols (static, OSPF, EIGRP, and BGP).

### Static Route PE-CE

Applying a static route to the reference architecture network shown in Figure 9-20 can lead to some design limitations and operational complexities. For example, CE-2, CE-3, and CE-4 have multiple links (direct PE-CE and backdoor links) using a static route, which can lead to limited design scalability and flexibility. In addition, managing these multiple edge devices with multiple links can be complicated and is associated with the high possibility of human (configuration) errors. However, the static route between CE-1 and PE-1 is a feasible design option because it is a single-homed site; even a single default route in CE-1 can be used, assuming the Internet is through the same link.

### Link State: OSPF as a PE-CE Routing Protocol

OSPF is not commonly used for PE-CE routing by service providers. As stated earlier in this section, however, this design concept is applicable not only to typical enterprise to service provider MPLS WAN connectivity but also to self-deployed MPLS L3VPN by enterprises (and OSPF is one of the most common routing protocols used by enterprises). The reason why OSPF is not commonly used as the PE-CE routing protocol is that the design can be more complex with OSPF, especially when there are sites with backdoor links or that are multihomed. As shown in Figure 9-21, by using the same reference architecture network in Figure 9-20 and applying different OSPF area designs, the implications of each differ significantly.

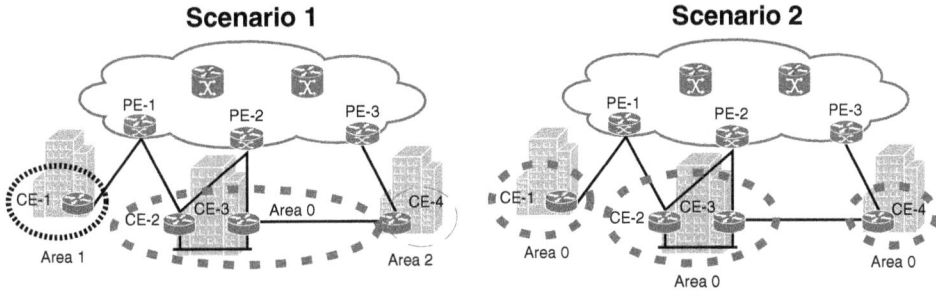

**Figure 9-21**  *PE-CE Connectivity Model OSPF*

**NOTE**  If you are unfamiliar with the OSPF terms used in this section, such as OSPF DN bit, OSPF domain identifier, and OSPF sham link, it is recommended that you refer to IETF RFC 4577 to build foundational knowledge about these terms before reading this section.

Scenario 1 in Figure 9-21 represents a multi-area OSPF design, where each site or branch is deployed with its own OSPF area. The service provider side is usually part of the super backbone area (area 0). Although this design scenario may look simple and straightforward, any lack of coordination between the service provider side and the enterprise (CE) side can break the functional requirements. For instance, if the CE side is configured with a different OSPF process ID than the PE side, traffic between the data center and the HQ will always prefer the backdoor link, as shown in Figure 9-22. If the functional requirements dictate that the traffic should always prefer the MPLS cloud as the primary path, this will lead to a design failure.

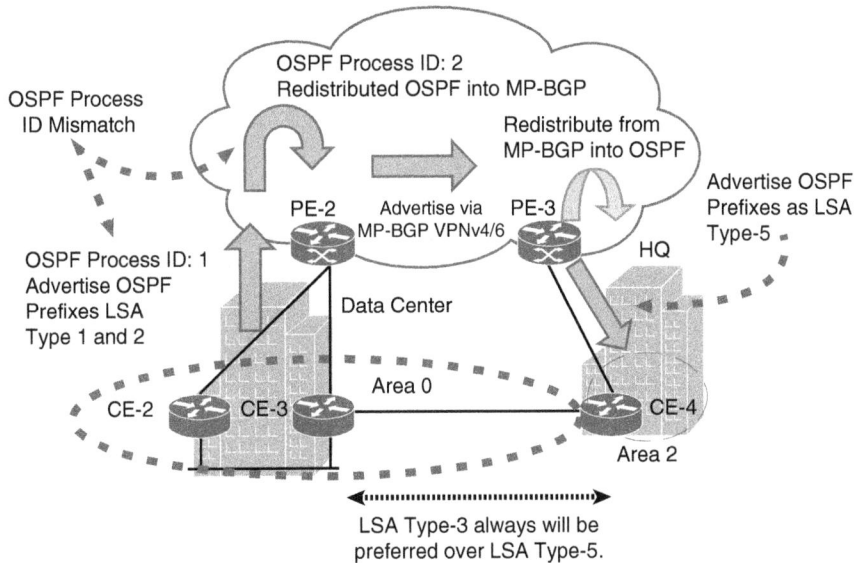

**Figure 9-22**  *PE-CE Connectivity Model OSPF with Backdoor Link*

This issue can be avoided by considering a single (matching) OSPF process ID between the CEs and the PEs. Nevertheless, practically speaking, this option can add operational complexity from the service provider's point of view. Alternatively, service providers (PE side) can overcome this issue by deploying the same OSPF identifier on both ingress and egress PEs, along with OSPF tuning and the OSPF cost metric (to ensure that the WAN link has a lower OSPF cost metric).

In fact, OSPF incorporates multiple attributes, such as the DN bit and route tag (for loop prevention) and the OSPF domain identifier. The technical reason behind these attributes is to help BGP carry these routes across the MP-BGP backbone and convert the route back to OSPF appropriately, in a transparent way to the OSPF peers at the edge of the network (CEs). For instance, in Figure 9-23, when PE-1 sets the DN bit to prefixes of remote site 1 (CE-1), when advertising it to CE-2 it will help PE-2 to stop re-advertising the same prefix back to the MPLS VPN super backbone when it is sourced from CE-2 or CE-3. In other words, when any of the PEs receives a type 3, 5, or 7 LSA from any CE with the DN bit set, the routing information from this LSA will not be considered in the OSPF route computation. As a result, this LSA will not be converted into a BGP route. This will ultimately help to avoid the potential loop in this scenario.

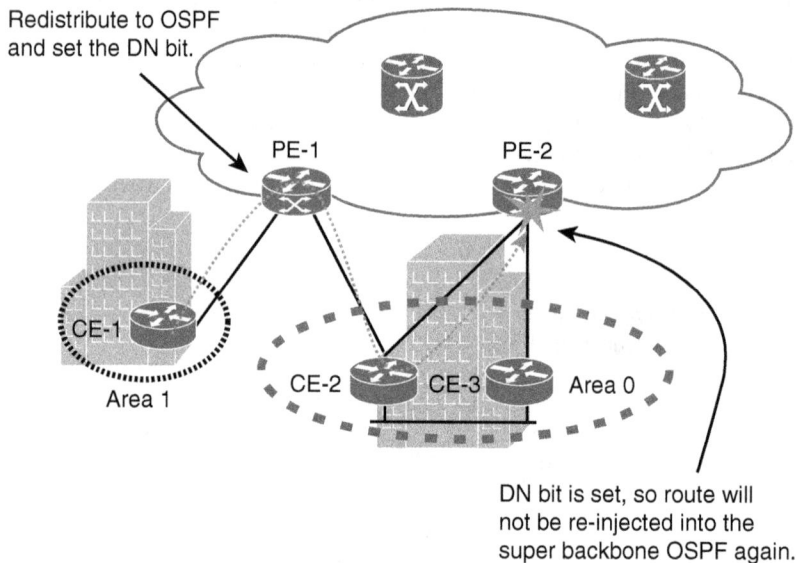

**Figure 9-23** *PE-CE Connectivity Model OSPF Loop Prevention*

**NOTE**   In some scenarios, such as a multi-VRF CE and a hub-and-spoke topology over MPLS L3VPN, covered earlier in this chapter (see Figure 9-12), when OSPF is used as the PE-CE routing protocol with the hub-and-spoke over MPLS L3VPN topology, there will usually be multiple PE nodes communicating with a central/hub PE router that connects to the hub CE router over dual interfaces/subinterfaces (each in a separate VRF). Technically in this scenario, when two remote sites (spokes) need to communicate, the OSPF link-state advertisements (LSAs) from each remote site will reach the central/hub PE, then the hub CE router, where the traffic flow loops and comes back into a different VRF (typical hub-and-spoke model). The issue here is that when these LSAs are type 3, 5, or 7, the LSAs will not be considered by the central/hub PE because they have the DN bit set. Therefore, the "DN bit ignore" feature is required in this scenario to "disable DN bit checking" at the hub PE node, in order to meet traffic flow and routing information distribution requirements (in which the route/LSA must be considered when it loops and comes back into a different VRF on the hub/central PE in order to reach other spokes). This OSPF feature is also known as capability VRF-lite. Ideally, before considering this feature, a careful analysis is required to avoid introducing any potential routing information loop.

However, these attributes (DN bit or route tag) will be stripped from the prefixes if the route is redistributed into another routing domain (such as EIGRP) and then redistributed back into OSPF and then to MPLS L3VPN from another PE. As discussed earlier in this book, route redistribution can cause metric transformation. This is a good example of how multiple redistributions may lead to a routing loop, as shown in Figure 9-24.

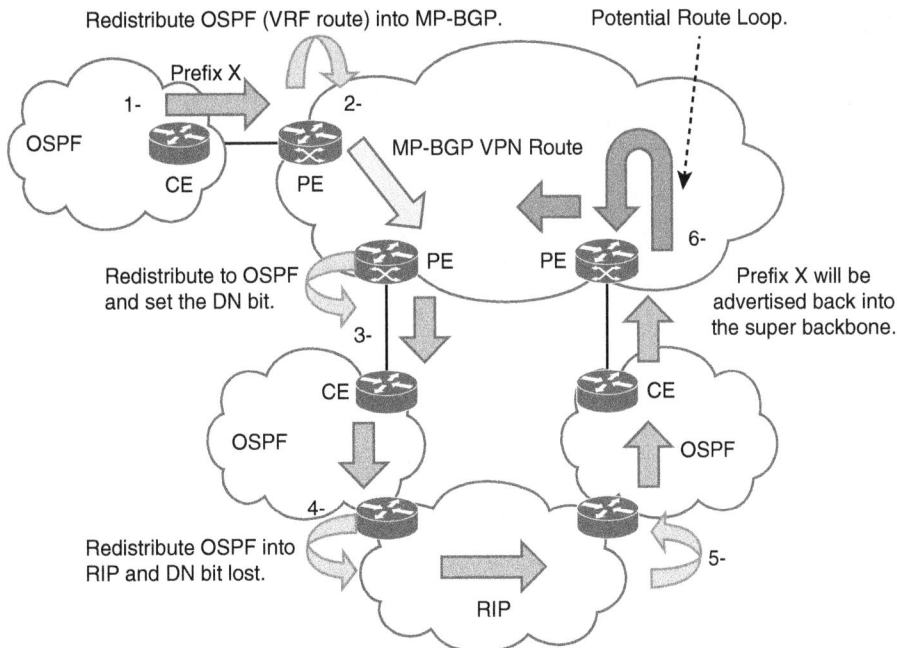

**Figure 9-24**   *PE-CE Connectivity Model OSPF with Multiple Redistributions*

Consequently, the network designer must be careful when there is a possibility of multiple redistribution points across multiple routing domains, because this scenario can break down the communication between OSPF islands across the MPLS L3VPN backbone.

In contrast, in Scenario 2 in Figure 9-25, all the sites are deployed in OSPF area 0. Although at a high level this design might look simpler, the most significant issue here is that all the routes between the data center and HQ site will be seen as OSPF intra-area routes. In other words, no matter what the WAN link cost metric is, the backdoor will always be the preferred path between the data center and the HQ (because the route from the MPLS L3VPN will be seen as either an inter-area route or external route). This might not always be a desirable design. To resolve this issue, based on the OSPF area design, network designers must make sure that the route coming from the MPLS L3VPN backbone is received as an intra-area route as well. To achieve this, the service provider must coordinate and set up the OSPF sham link between the relevant PEs (in this scenario, PE-2 and PE-3) to create a logical intra-area link between the ingress and egress PEs (area 0 in this example).

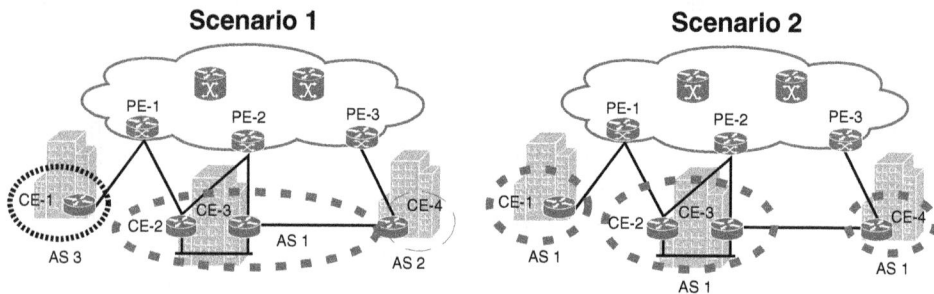

**Figure 9-25** *OSPF Sham Link*

After the OSPF sham link is set up and the OSPF adjacency is established over this logical link, the network can manipulate the OSPF cost metric on the relevant interface to make the MPLS L3VPN link the preferred path.

**NOTE** For the enterprise (CE side) to avoid the reliance on the SP side to set up a sham link, the OSPF areas design can be migrated to use a unique OSPF area per site (for the sites connected with a backdoor link), if this option is available.

Consequently, using OSPF as the PE-CE routing protocol can be challenging if there is any lack of coordination between the PE side (super backbone) and the CE side. Furthermore, even if the design is performed with good coordination, OSPF can still impose some design limitations in certain scenarios, including the following:

■ If any CE needs to send a summary route to other CEs or any given CE, this has to be deployed by the provider or PE side.

■ In scenarios like the one discussed in this chapter, the data center has multiple paths (links and routes) to the MPLS L3VPN. OSPF can offer network designers very limited control to achieve detailed load sharing over these paths. In addition, there is no such mechanism to influence the service provider's route selection from the CE, like BGP AS-PATH prepending.

Therefore, the protocol selection has to be aligned with all the different design requirements (business, functional, and application) to achieve a successful design.

**NOTE**   Intermediate System-to-Intermediate System (IS-IS) acts similarly to OSPF when there is a backdoor link, as the redistributed prefixes with the MPLS cloud (from MP-BGP into IS-IS) will be seen as an external route by the receiving CE while the same route over the backdoor link will be received as an internal route. This usually leads to always preferring the backdoor path. To overcome this issue, BGP supports carrying some of the critical IS-IS information as part of BGP extended communities, which can be converted back into an IS-IS link-state packet (LSP). For example, if the original route was received as level 1, it will be reconverted into an IS-IS level 1 route at the other end (PE). (For more details, refer to this IETF draft: draft-sheng-isis-bgp-mpls-vpn.)

### EIGRP as a PE-CE Routing Protocol

Using EIGRP as a PE-CE routing protocol may offer a simpler design compared to OSPF if it is designed properly. This is because there are certain scenarios, especially when there is a backdoor link between the CE sites, where any design mistake can lead to serious connectivity issues. Therefore, network designers must be aware of EIGRP behavior and the possible techniques and mechanisms to optimize the design.

As discussed earlier in this section, the typical routing behavior in PE-CE scenarios is that the CE routing protocols (IGP) are converted at the ingress PE router to MP-BGP routes, to be carried across the MPLS VPN backbone to the egress PE router, and to be reconverted back to the relevant PE-CE routing protocol on that VRF. As with link-state protocols, MP-BGP carries specific EIGRP information in new BGP extended communities set by the PEs (usually the ingress PE), which carries EIGRP ASNs and multiple EIGRP attributes across the MPLS L3VPN backbone, such as delay, bandwidth, hop count, and reliability. This can help the receiving PE EIGRP instance to have relevant and usable routing information.

However, as mentioned earlier, PE-CE EIGRP design requires special considerations from network designers in some scenarios to avoid undesirable behaviors. Figure 9-26 illustrates different EIGRP application scenarios using the same reference network architecture of this section.

In Scenario 1 in Figure 9-26, EIGRP is applied with different ASNs per site. Each site will receive the EIGRP route of the other site as an external route. This is because the ASN is carried in the new BGP extended community. When it is redistributed from MP-BGP into EIGRP at the egress PE with ASN mismatch, EIGRP will install the route as an external route. However, the concern is always with where the backdoor link exists between different sites, because EIGRP does not have a built-in mechanism like OSPF does to detect a loop. Therefore, there is potential that EIGRP information will circulate in this looped topology. It is also possible as well that EIGRP will introduce unpredictable behavior because of what is known as a *race condition*, where the route is accepted based on the timing of EIGRP and BGP updates with the local router decision.

9

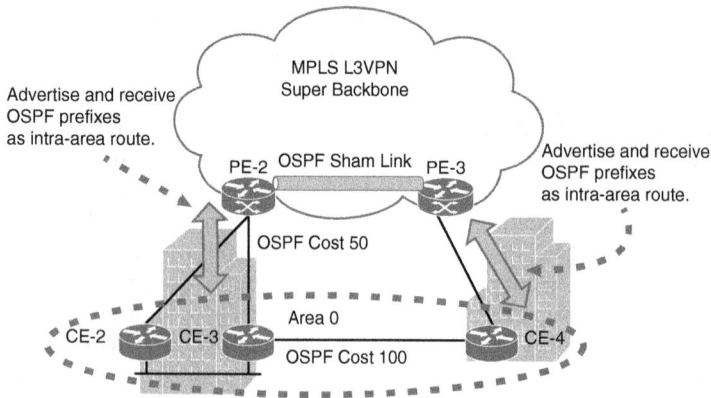

**Figure 9-26** *PE-CE Connectivity Model EIGRP*

Route racing in Scenario 1 can be mitigated to some extent by limiting the maximum number of EIGRP hop counts. However, with this approach, the network will continue to experience undesirable (looping) behavior because of the route racing condition until the EIGRP reaches its maximum configured hop counts. In addition, it is difficult to determine the right number of hops to be configured (and there is a high degree of human configuration errors and added operational complexities). Alternatively, you can use a route tagging–based solution to offer a more deterministic and simplified solution known as EIGRP site of origin (SoO), which is technically an extended BGP community associated with the route. When redistributed from EIGRP into MP-BGP, any route assigned an SoO value will not be advertised over links (interfaces) deployed with the same SoO value, which helps to avoid or mitigate the impact of routing loops in complex topologies, such as sites with EIGRP used as a PE-CE routing protocol that contain both MPLS VPN and backdoor links, as shown in Figure 9-27.

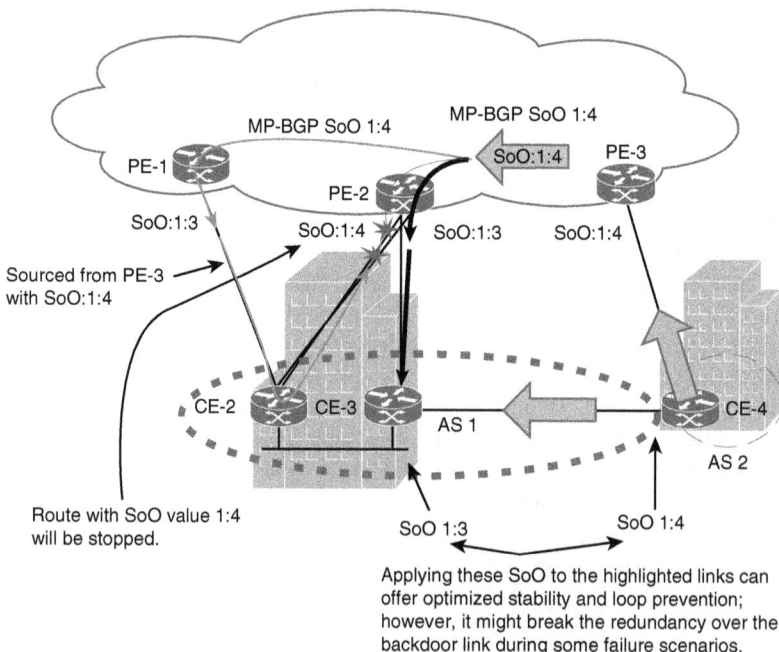

**Figure 9-27** *PE-CE Connectivity Model EIGRP SoO*

Table 9-4 summarizes the behavior of EIGRP SoO in different situations.

**Table 9-4**    EIGRP SoO Actions

| Received Route SoO Details | Action |
|---|---|
| SoO value matches the SoO value on the sending or receiving interface. | The route will be filtered out. |
| CE deployed with SoO value that does not match. | The route is added to the EIGRP topology table so that it can be redistributed into BGP and the SoO value preserved. |
| Does not contain an SoO value. | The route is accepted into the EIGRP topology table, and the SoO value from the interface that is used to reach the next-hop CE router is appended to the route before it is redistributed into BGP. |

In Figure 9-27, SoO helps to optimize the EIGRP looping (race condition) by preventing the route from being reinjected into the network based on the attached SoO value to the route and the deployed SoO value on the interface. For instance, traffic sourced from the HQ LAN (CE-4) passing through PE-3 will have an SoO value of 1:4 assigned to it. Then, any interface in the scenario shown in Figure 9-27 that has an SoO value of 1:4 will not pass this route information through.

Although in this type of scenario SoO can help to mitigate route looping and racing issues to a certain extent, it might sometimes be necessary to introduce other limitations, such as reduced redundancy. For example, if the SoO values are applied on the backdoor link (as covered in Figure 9-27) and the PE-3-CE-4 link goes down, any traffic with SoO value of 1:3 or 1:4 destined for the HQ (behind CE-4) will be isolated (because of the SoO filtering at the backdoor link), even though the backdoor link is available. Therefore, as a network designer, you must understand the design goals and priorities and what the impact is of applying SoO on the different interfaces/paths (for example, redundancy + suboptimal routing versus stability + optimal routing). In other words, if the time required for the EIGRP to stabilize following a failure event is acceptable, a simple SoO design should be sufficient, like the one shown in Figure 9-28, in which SoO stops the information feedback looping faster than relying on the hop count for EIGRP to stabilize after CE-2 failure.

However, Scenario 2 shown in Figure 9-26 is designed with the same EIGRP ASN on all sites, which is typical in that all the routes learned over MPLS L3VPN and the backdoor links will be an internal EIGRP route.

One of the common design concerns with this setup is when the backdoor link is intended to be used only as a backup path (because with this design there is a possibility that some remote sites will use the backdoor link to reach either the DC LAN or the HQ LAN). For instance, in Figure 9-29, the HQ LAN prefix is advertised in EIGRP to the MPLS VPN PE-3 and to CE-2 and CE-3 over the backdoor link. Likewise, CE-2 and CE-3 advertise this route to PE-1 and PE-2, respectively. Therefore, PE-1 in this case has two BGP paths available for the HQ LAN: the iBGP path via PE-2 and PE-3, and the locally redistributed BGP route from EIGRP advertisement via CE-2 EIGRP.

9

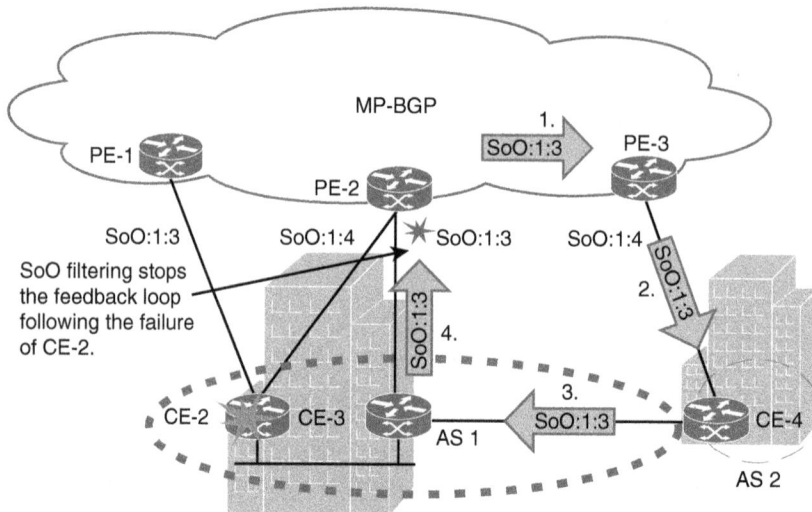

**Figure 9-28**  *PE-CE Connectivity Model EIGRP Loop Prevention with SoO: Failure Scenario*

Consequently, from the PE-1 point of view, the locally originated route will be preferred based on the BGP best-path selection. As a result, traffic from CE-1 destined for the HQ site (CE-4) will use the backdoor link as a primary path, as shown in Figure 9-29.

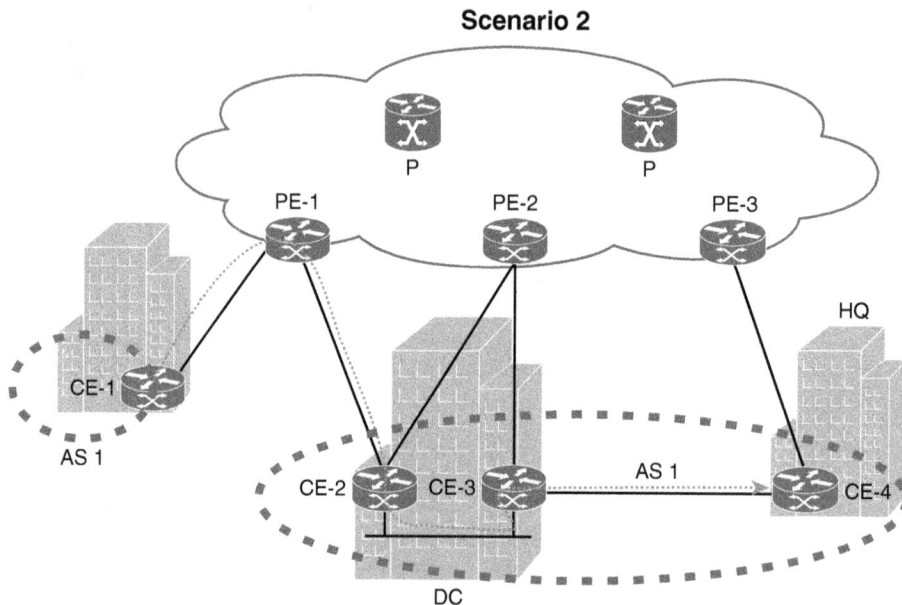

**Figure 9-29**  *PE-CE Connectivity Model EIGRP Suboptimal Routing*

There must be an attribute to influence PE-1 path selection in which the internal BGP (iBGP) learned routes are to be preferred over other routes. One possible and common solution that can be used in this scenario is the "BGP cost community" to influence BGP path selection. EIGRP routes injected into MP-BGP with a point of insertion (POI) value of 128, along with the cost set to the EIGRP composite metric (in this case, the community's value), must be considered before the typical BGP path selection algorithm (such as local preference and AS-PATH). The BGP cost community value can be any number in the range of zero through four billion (4,294,967,295) with all cost communities leveraging a default value of two billion (2,147,483,647). This is to say, if a BGP cost community is not manually set, it will still maintain a default value of 2,147,483,647. The assumption in this scenario is that the EIGRP composite metric is less than the default BGP cost community value, which would allow that corresponding path to be prioritized. Consequently, when the PEs redistribute the EIGRP routes into BGP, the BGP cost community attribute will be populated along with the accumulated EIGRP metrics. This means that each PE will have a level of visibility of the EIGRP path cost. As a result, PE-1 will prefer the HQ LAN route advertised by PE-3, because the associated EIGRP metric of this LAN route advertised by either CE-2 or CE-3 contains the accumulated path cost, including the backdoor link, while the BGP cost community of the iBGP (from PE-3) has a lower cost. Therefore, it will be the preferred path, as shown in Figure 9-30.

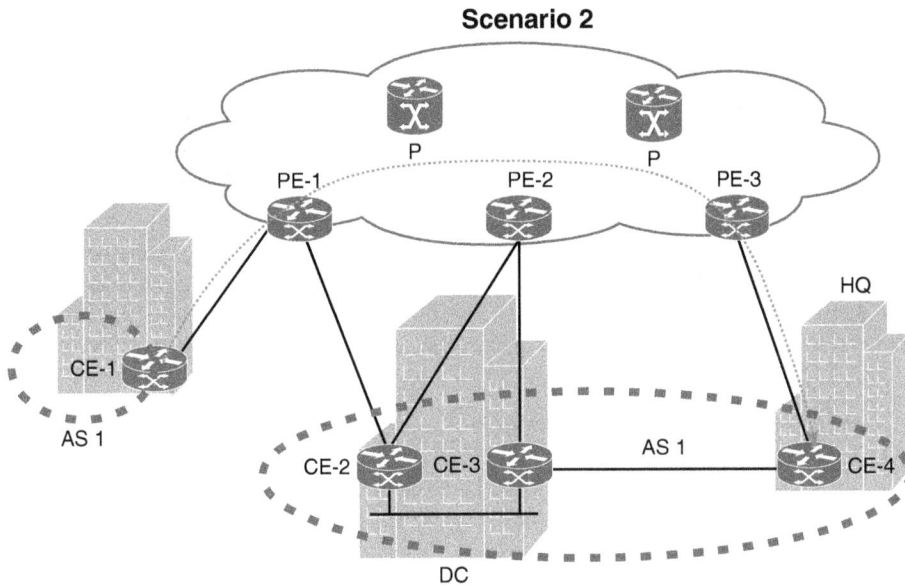

**Figure 9-30** *PE-CE Connectivity Model EIGRP Optimal Routing*

Furthermore, the BGP cost community helps to optimize optimal routing in scenarios like this (and scenarios like Scenario 1 shown in Figure 9-25). For example, if the route from site 1 (CE-1) is assigned a BGP cost community value, lower than the default value (2,147,483,647), at the ingress PE (PE-1), it will always be preferred by PE-2 and PE-3 over any other BGP advertisement that has no cost community attached to it, regardless of what BGP attributes it has. As a result, the site 1 route will always be preferred via PE-1 (directly connected PE to site 1), as shown in Figure 9-31.

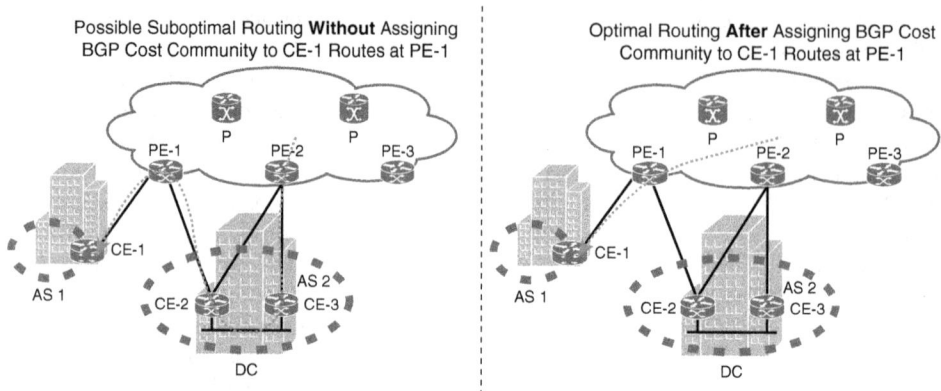

**Figure 9-31** *PE-CE Connectivity Model EIGRP Suboptimal Routing-2*

**NOTE** This cost community may transform BGP to act in a way it is not designed to (like IGP), which may lead to undesirable behaviors in some scenarios.

Consider, however, what happens when a new remote site is added (for example, in a new country where the current service provider does not have a presence and requires inter-AS communication to extend the MPLS-L3VPN reachability). In this scenario, the BGP cost community will not be a valid solution because it does not support propagation over external BGP (eBGP) sessions, as shown in Figure 9-32.

Consequently, using EIGRP as a PE-CE routing protocol may add simplicity to the enterprises that already use EIGRP as the enterprise routing protocol. However, when there are multihomed sites to the MPLS provider or sites with backdoor links, the design may prove too complicated, and overall flexibility and stability may be reduced. EIGRP Over the Top (OTP), however, can offer a more flexible PE-CE design and is independent of the service provider routing control.

### BGP as a PE-CE Routing Protocol

BGP, however, as a PE-CE routing protocol can achieve the optimal routing and traffic control over the most complex connectivity layouts because it has multiple powerful attributes that can influence inbound and outbound path selection. All the design considerations and options discussed earlier in this book are applicable to the design of BGP as a PE-CE routing protocol. In addition, the BGP SoO attribute uses the same concept of the EIGRP SoO to control route propagation when there is a backdoor link between the customer sites.

In other words, when possible, it is always desirable to consider BGP as the PE-CE routing protocol for multihomed sites to single- or multiple-provider networks. For network designers, this can facilitate the achievement of a very advanced level of traffic engineering with flexible BGP policies. This, in turn, offers the business adequate flexibility to use the available paths in a way that aligns with the business, functional, and application requirements in a more dynamic manner regardless of the extent of the physical connectivity's complexity. The flexibility of BGP can obviously provide an optimized ROI of the available paths and lower operational complexities when there is complicated connectivity with multiple links.

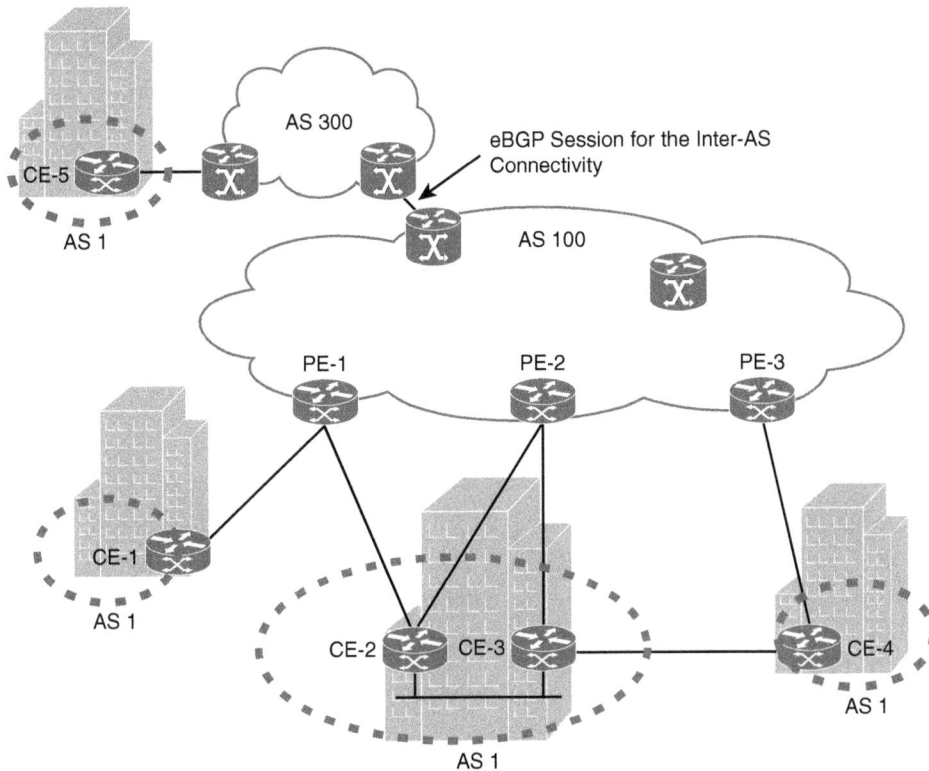

**Figure 9-32** *PE-CE Connectivity Model EIGRP: BGP Cost Community Limitation Scenario*

Normally, MPLS VPN providers consider the following two BGP ASN allocation models when BGP is used as a PE-CE routing protocol (see Figure 9-33):

- **Same ASN (ASN) per site:** With this model, the MPLS provider allocates the same ASN to all the customer sites. One of the main advantages of this model is reduced BGP ASN collisions.

- **Single ASN per site:** With this model, the MPLS provider allocates each of the customer sites a separate BGP ASN. This model offers network designers and operators the ability to identify the source of prefixes (from which site) in a simple way (based on BGP ASN in the AS-PATH attribute of each prefix). However, it may introduce scalability limitations with regard to the available ASNs.

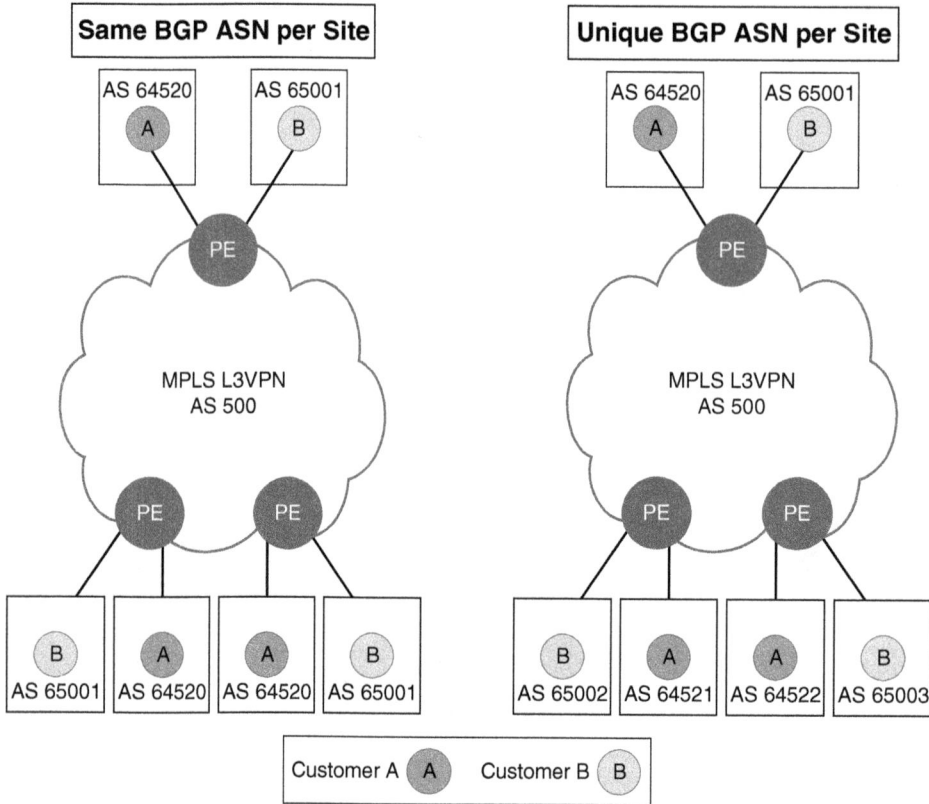

**Figure 9-33** *BGP ASN Allocation Models as an MPLS VPN PE-CE Routing Protocol*

Although allocating the same BGP ASN offers better scalability with regard to the number of ASNs/sites and reduced ASN collisions, this model introduces some design concerns for multihomed sites. For example, in Figure 9-34, the service provider must rewrite the customer ASN (AS override) to overcome the default BGP loop-prevention mechanism and allow all the sites with the same ASN to communicate. However, the primary issue with the rewriting of the BGP AS-PATH is that the CE routers will not be able to detect BGP looping. As shown in Figure 9-34, after Prefix X is received by the provider PE from CE-2, AS override applied to this prefix, and AS-PATH rewriting converted the original AS-PATH to become (300 300). Technically, when this prefix is received by CE-1, it will be accepted because the AS-PATH has been altered and the original AS was removed from the AS-PATH. As a result, a route loop will be formed in this case. To overcome this loop, the SoO concept discussed earlier with EIGRP can be used, and a BGP extended community attribute can be attached to BGP prefixes; network operators and PEs can use that to identify the actual prefix source, and based on the deployed SoO value, the route can be permitted or denied. As with EIGRP, if the SoO value of a prefix is equal to the deployed SoO for a BGP peer, the prefix will be stopped from being advertised. By applying this concept to the scenario shown in Figure 9-34, both PEs facing CE-1 and CE-2 configure the same SoO to its direct BGP peering (CE-1 and CE-2, respectively). The route looping will be stopped without impacting the communication with any other remote site belonging to the same customer (AS).

From a design perspective, rewriting the ASN along with SoO considerations can be seen as an added complexity to the design and operation of this model (same ASN per site).

Nevertheless, BGP cannot always practically be considered. Design constraints that may prevent network designers from considering BGP include BGP not being supported as PE-CE by the service provider or BGP not being supported by the software of the CE nodes. A lack of BGP knowledge among the enterprise's IT staff may also be a constraint. In these cases, the network designer has to find an alternative supported routing protocol that can achieve the intended goal.

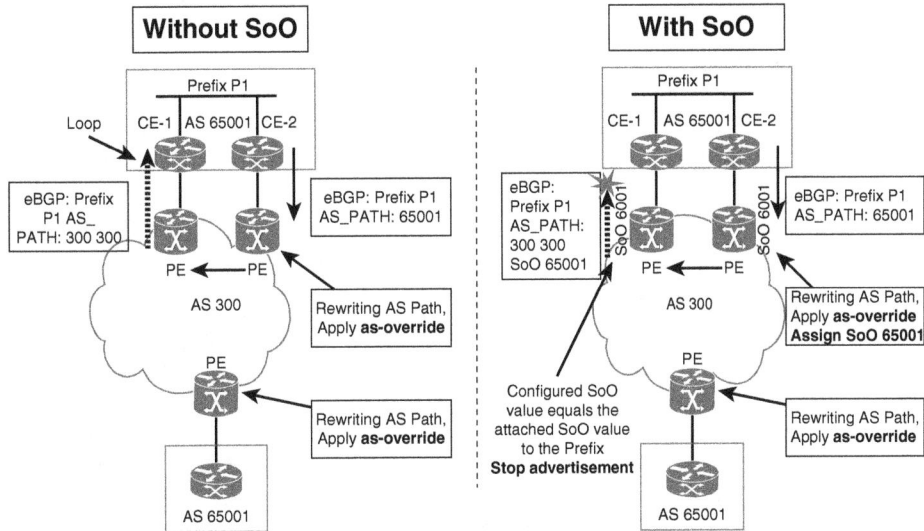

**Figure 9-34**   *BGP PE-CE Routing Protocol Loop Prevention*

Table 9-5 summarizes the characteristics (strengths and weaknesses) of each routing protocol with regard to the PE-CE routing design.

**Table 9-5**   Comparison of PE-CE Routing Protocols

| Routing Protocol | Strengths | Weaknesses |
|---|---|---|
| Static | Simple and reliable "when combined with IP SLA." Low operational complexity in small environments with a small number of prefixes. | Nonscalable. High operational complexity in large environments. Limited flexibility in multihomed scenarios with automatic failover limitations. |
| Link state | Is reliable to a certain extent (for example, supports built-in loop prevention). Supports multiple connectivity and flooding domain design scenarios. | High design and operational complexity. Limited flexibility in large environments with backdoor links and multihoming scenarios. |

| Routing Protocol | Strengths | Weaknesses |
|---|---|---|
| EIGRP | Reliable to a certain extent (topology dependent)<br><br>EIGRP OTP can simplify and optimize CE-PE designs to a large extent. | High design and operational complexity and limited flexibility in a large environment with multihoming scenarios.<br><br>EIGRP SoO may lead to inefficient use of available paths or lack of redundancy in multihoming scenarios with backdoor links. |
| BGP | Most powerful and flexible protocol that can support all types of connectivity. | For multihomed sites with complex policies, it requires advanced operational staff expertise.<br><br>May not be supported by some low-end routers (for very small remote sites). |

# Software-Defined Networks

With the advent of software-defined solutions, we now have more capabilities that can be leveraged from a network designer perspective. This section will compare and contrast why a network designer could leverage a software-defined solution in an overarching network design to meet multiple business requirements. Inherently, a software-defined solution is more complex, and there are more pieces to the solution that network designers, architects, and engineers need to understand. This complexity, though, is obfuscated by the additional capabilities a software-defined solution provides, assuming it is designed, deployed, and functioning properly. The following sections will highlight SD-WAN and SD-LAN in a vendor-agnostic perspective (leveraging vendor-specific examples as needed to provide context) and the corresponding design decisions and options around each solution.

## SD-WAN

From a vendor-agnostic perspective, software-defined wide-area networking (SD-WAN) is composed of separate orchestration, management, control, and data planes. The orchestration plane assists in the automatic onboarding of the edge (spoke) routers into the SD-WAN overlay. The management plane is responsible for central configuration and monitoring. The control plane builds and maintains the network topology and makes decisions regarding where traffic flows. The data plane is responsible for forwarding packets based on decisions from the control plane. Figure 9-35 shows the different SD-WAN planes and how they interact with one another.

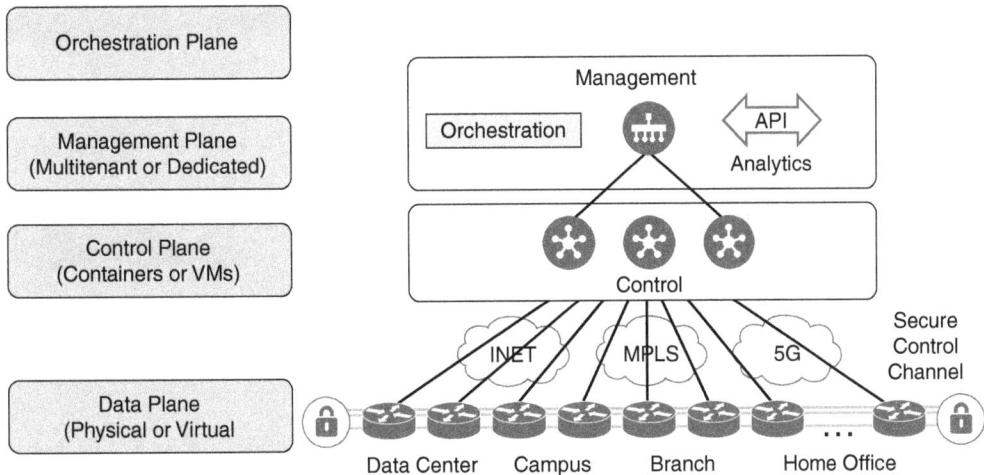

**Figure 9-35**  *SD-WAN Solution Planes*

## SD-WAN Components

The primary components of the SD-WAN consist of a network manager, the controller, the orchestrator, and the edge router. The following list provides an overview of each of these components and their functions:

- **Network manager:** This centralized network management system provides a GUI to easily monitor, configure, and maintain all SD-WAN devices and links in the underlay and overlay network.

- **Controller:** This software-based component is responsible for the centralized control plane of the SD-WAN fabric network. It establishes a secure connection to each edge router and distributes routes and policy information to it. It also orchestrates the secure data plane connectivity between the different edge routers by distributing crypto key information, allowing for a very scalable, IKE-less architecture.

- **Orchestrator:** This software-based component performs the initial authentication of edge devices and orchestrates controller and edge device connectivity. It also has an important role in enabling the communication of devices that sit behind NAT.

- **Edge routers:** These devices sit at a physical site or in the cloud and provide secure data plane connectivity among the sites over one or more WAN transports. They are responsible for traffic forwarding, security, encryption, QoS, routing protocols such as BGP and OSPF.

**NOTE**  Depending on the specific vendor implementation of SD-WAN, these components and capabilities can be integrated together into the same system or they can be integrated into dedicated individual systems. The key takeaway is that these capabilities are what make up an SD-WAN solution, and as a network designer you will have to know when to leverage a solution like SD-WAN to solve the underlying business requirements.

Figure 9-36 depicts the different SD-WAN components and capabilities.

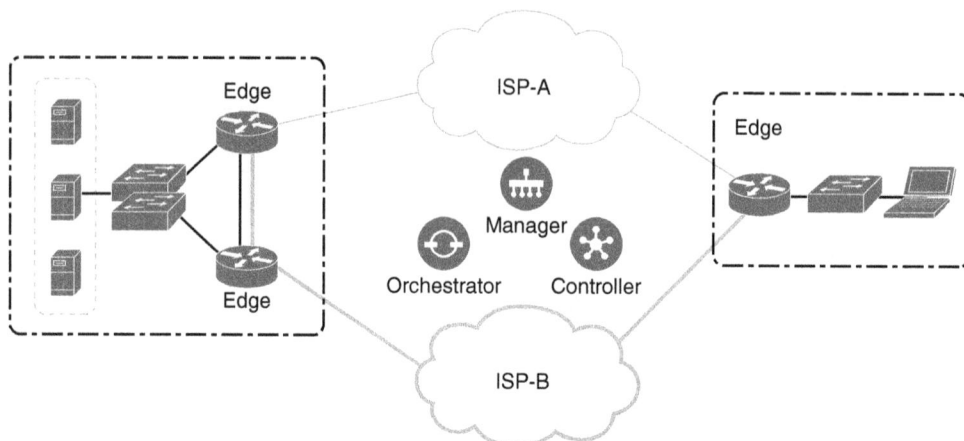

**Figure 9-36**  *SD-WAN Components and Capabilities*

## SD-WAN Management Protocol

The management protocol manages the SD-WAN overlay network. The protocol runs between the controllers and the edge routers where control plane information, such as route prefixes, next-hop routes, crypto keys, and policy information, is exchanged over a secure Datagram Transport Layer Security (DTLS) or Transport Layer Security (TLS) connection. The controller acts a lot like a BGP route reflector; it receives routes from the edge routers, processes and applies any policy to them, and then advertises the routes to other edge routers in the overlay network. If there is no policy defined, the default behavior is a full-mesh topology, where each edge can connect directly to another edge at another site and receive full routing information from each site.

The management protocol advertises three types of routes:

- **Management protocol routes:** Prefixes that are learned from the local site, or service side, of an edge router. The prefixes are originated as static or connected routes, or from within the OSPF or BGP protocol, and redistributed into the management protocol so they can be carried across the overlay. Management protocol routes advertise attributes such as transport location (TLOC) information, which is similar to a BGP next-hop IP address for the route, and other attributes such as origin, originator, preference, site ID, tag, and VPN. A management protocol route is only installed in the forwarding table if the TLOC to which it points is active.

- **TLOC routes:** The logical tunnel termination points on the edge routers that connect to a transport network. A TLOC route is uniquely identified and represented by a three-tuple, consisting of system IP address, link color, and encapsulation. TLOC routes also carry attributes such as TLOC private and public IP addresses, carrier, preference, site ID, tag, and weight. For a TLOC to be considered in an active state on a particular edge router, an active Bidirectional Forwarding Detection (BFD) session must be associated with that edge router TLOC.

■ **Service routes:** Represent services (firewall, IPS, application, optimization, etc.) that are connected to the edge device local-site network and are available for other sites for use with service insertion. In addition, these routes also include VPNs; the VPN labels are sent in this update type to tell the controllers what VPNs are serviced at a remote site.

### Virtual Networks

In the SD-WAN overlay, **virtual networks (VNs)** provide segmentation, much like Virtual Routing and Forwarding instances (VRFs). Each VN is isolated from other VNs and each has its own forwarding table. An interface or subinterface is explicitly configured under a single VN and cannot be part of more than one VN. Labels are used in the management protocol route attributes and in the packet encapsulation, which identifies the VN a packet belongs to. The VN number is a 4-byte integer with a value from 0 to 65530.

### TLOC Extension

A very common network setup in a site with two edge routers is for each edge router to be connected to just one transport. There are links between the edge routers, which allow each edge router to access the opposite transport through a TLOC extension interface on the neighboring edge router. **TLOC extensions** can be separate physical interfaces or subinterfaces.

### SD-WAN Policies

Policies are an important part of the SD-WAN architecture and are used to influence the flow of data traffic among the edge routers in the overlay network. Policies apply either to control plane or data plane traffic and are configured centrally on the controllers or locally on the edge device routers.

Centralized control policies operate on the routing and TLOC information and allow for customizing routing decisions and determining routing paths through the overlay network. These policies can be used in configuring traffic engineering, path affinity, service insertion, and different types of VPN topologies (full-mesh, hub-and-spoke, regional mesh, etc.). Another centralized control policy is application-aware routing, which selects the optimal path based on real-time path performance characteristics for different traffic types. Localized control policies enable routing policy at a local site, specifically through OSPF or BGP.

Data policies influence the flow of data traffic through the network based on fields in the IP packet headers and VPN membership. Centralized data policies can be used in configuring application firewalls, service chaining, traffic engineering, and QoS. Localized data policies allow data traffic to be handled at a specific site, such as ACLs, QoS, mirroring, and policing. Some centralized data policy may affect handling on the edge device itself, as in the case of application route policies or a QoS classification policy. In these cases, the configuration is still downloaded directly to the controllers, but any policy information that needs to be conveyed to the edge routers is communicated through the secure connection already established.

## SD-LAN

Software-defined local area network (SD-LAN) is an evolved evolution of existing campus LAN designs that introduces programmable overlays enabling easy-to-deploy network virtualization across the LAN, capable of supporting multiple enclaves. In addition to network virtualization, SD-LAN allows for software-defined segmentation and policy enforcement based on user identity, device, method of connectivity, and group membership. These are newer technologies and protocols that eliminate many of the issues and problems previously described. These capabilities also provide a significant reduction in operational expenses and an increased ability to drive business assurance and outcomes quickly with minimal risk, at the cost of increased complexity and staff expertise. Figure 9-37 highlights the different layers within SD-LAN today.

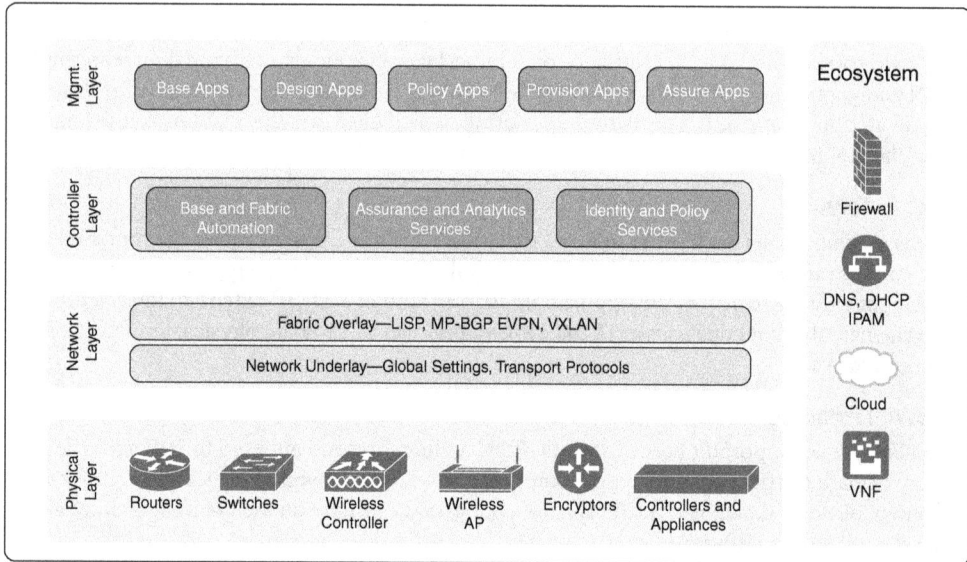

**Figure 9-37**  *SD-LAN Layers*

## SD-LAN Terminology

The following terms are used in SD-LAN and in other software-defined solutions today:

- Underlay network

- Overlay network

- SD-LAN data plane

- SD-LAN control plane

Similarly to SD-WAN, the SD-LAN architecture enables the use of virtual networks (overlay networks) running on a physical network (underlay network) in order to create alternative topologies to connect devices. Figure 9-38 depicts a conceptual view of the underlay and overlay network concepts. This section provides information about the architectural elements that define SD-LAN.

**Overlay Network**

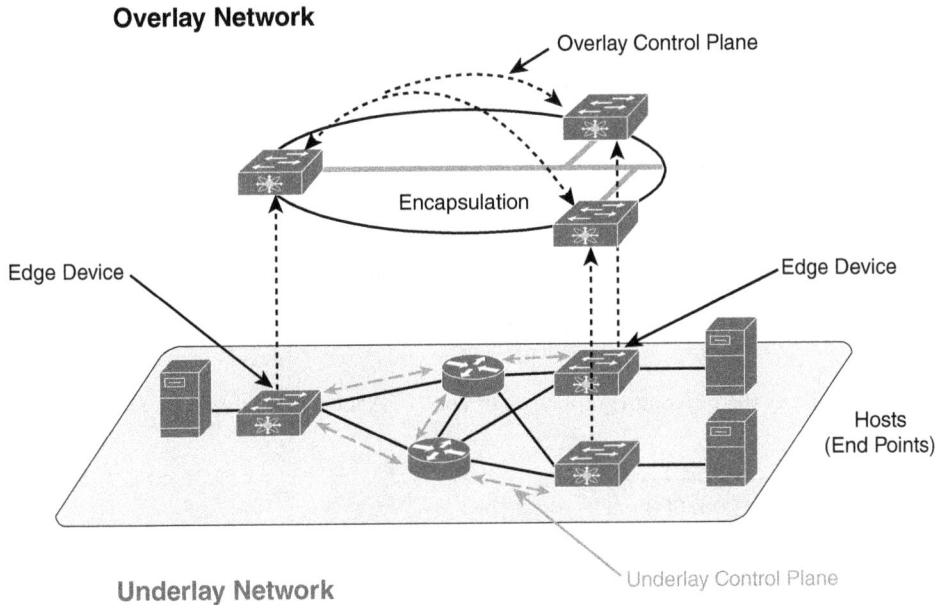

**Figure 9-38**   *Conceptual Underlay and Overlay Networks*

### Underlay Network

The **underlay network** is defined by the physical switches and routers that are part of the LAN. All network elements of the underlay must establish Internet Protocol (IP) connectivity via the use of a routing protocol. Theoretically, any topology and routing protocol can be used, but the implementation of a well-designed Layer 3 foundation to the LAN edge is highly recommended to ensure performance, scalability, and high availability of the network. In the SD-LAN, end-user subnets are not part of the underlay network but instead are part of the overlay network. The underlay is typically a Layer 3 fabric without any Layer 2. All Layer 2 requirements can be achieved in the overlay network.

### Overlay Network

An **overlay network** runs over the underlay in order to create a virtual network. Virtual networks isolate both data plane traffic and control plane behavior among the physical networks of the underlay. Virtualization is achieved inside SD-LAN by encapsulating user traffic over IP tunnels that are sourced and terminated at the boundaries of SD-LAN. Network virtualization extending outside of the SD-LAN is preserved using traditional virtualization technologies such as virtual routing and forwarding (VRF)-Lite, MPLS VPN, or SD-WAN. Overlay networks can run across all or a subset of the underlay network devices. Multiple overlay networks can run across the same underlay network to support multitenancy through virtualization.

## SD-LAN Components

SD-LAN is composed of several node types. This section describes the functionality of each node and how the nodes map to the physical campus topology.

**NOTE** To properly show the different capabilities and components of an SD-LAN solution, we will cover specific aspects of the Cisco SD-A solution. The goal of this comparison is not to teach you all about Cisco SD-A but rather about the capabilities that all SD-LAN solutions should have and how you as a network designer can properly leverage an SD-LAN solution to meet the needs and requirements of a business.

**NOTE** Cisco SD-A is a unique set of technologies, automation, and central control that doesn't necessarily fit perfectly into the SD-LAN category. In other words, SD-LAN represents a subset of Cisco SD-A, but it is a good representation for most of the capabilities.

Figure 9-39 shows the different components of an SD-LAN (specifically, Cisco SD-A) solution.

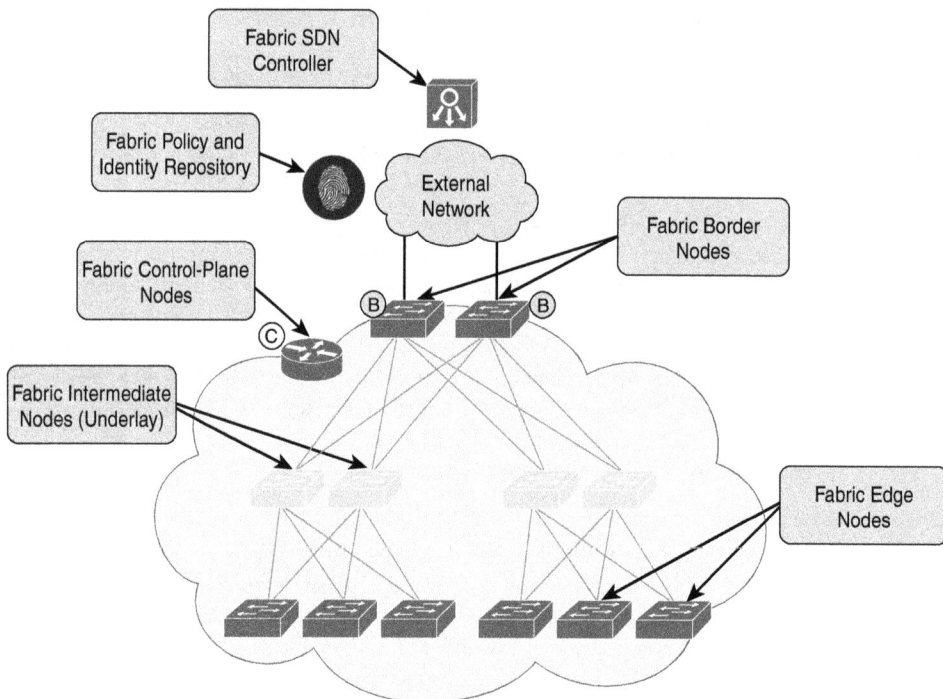

**Figure 9-39** *SD-LAN Components: Cisco SD-A*

### Fabric Control-Plane Node

When leveraging Locator/Identifier Separation Protocol (LISP) in the fabric control plane node, it is based on the LISP Map-Server (MS) and Map-Resolver (MR) functionality combined on the same node. The LISP fabric control plane node functionality can be instantiated on the fabric border node or a dedicated node. The LISP fabric control plane node enables the following functions:

- **Host tracking database (HTDB):** The host tracking database is a control repository of endpoint identifiers to fabric edge node bindings.

- **Map-Server (MS):** The LISP MS is used to populate the HTDB registration messages from fabric edge devices.

- **Map-Resolver (MR):** The LISP MR is used to respond to map queries from fabric edge devices looking to determine the routing locator (RLOC) mapping information for a destination endpoint identifier.

When leveraging Multi-Protocol Border Gateway Protocol with the Ethernet Virtual Private Network (MP-BGP EVPN) as the fabric control plane, MP-BGP peering relationships are established between fabric edge nodes and BGP route reflectors. Similar to the LISP MS and MR functionality, the route reflector can be instantiated on the fabric border node or a dedicated node. The MP-BGP EVPN fabric control plane node enables the following functions:

- Learning of endpoint identifier's Layer 2 and Layer 3 reachability information between fabric edge nodes. The exchange and advertisement of this information keeps the bindings current across the VXLAN overlays.

- Minimizes network flooding through protocol-based host MAC/IP route distribution and ARP suppression on the local VXLAN tunnel endpoint (VTEP).

### Fabric Edge Node

The fabric edge nodes are the equivalent of an access layer switch in a traditional campus design. The fabric edge nodes implement a Layer 3 access design with the addition of the following fabric functions:

- **Endpoint registration:** After an endpoint is detected by the fabric edge, it is added to a local host tracking database. The edge device also issues a LISP map-register message to inform the control plane node of the endpoint detected so that it can populate the host tracking database.

- **Mapping of the user to virtual network:** Endpoints get placed into virtual networks by assigning the endpoint to a VLAN, which is mapped to a LISP instance. The mapping of endpoints into VLANs can be done statically or dynamically using technologies like 802.1X. Additional policies can be assigned to provide segmentation and policy enforcement at the fabric edge.

- **Anycast Layer 3 gateway:** A common gateway (IP and MAC addresses) can be used at every node that shares a common subnet in order to provide for optimal forwarding and mobility across different edge nodes.

- **LISP forwarding (using LISP as control plane):** Instead of a typical routing table–based decision, the fabric edge nodes query the map server to determine the RLOC or edge node associated with the destination IP and use that information to encapsulate the traffic in VXLAN. In case of a failure to resolve the destination RLOC, the traffic is sent to the fabric border in which the global routing table is used for forwarding.

- **Routing-based forwarding (using MP-BGP EVPN as control plane):** The fabric edge nodes are part of the VXLAN control plane and distribute the MP-BGP EVPN routes among their MP-BGP EVPN peers. Fabric edge nodes will be MP-iBGP EVPN peers to route reflectors. The MP-BGP EVPN peers between fabric edge nodes and route reflectors will advertise end hosts behind all VTEPs.

9

### Fabric Intermediate Node

The fabric intermediate nodes are part of the Layer 3 network that interconnects the edge nodes to border nodes. In the case of a three-tier campus design using core, distribution, and access tiers, the fabric intermediate nodes are the equivalent of a distribution switch. Fabric intermediate nodes only route IP traffic inside the fabric. No VXLAN encapsulation/ de-encapsulation or LISP control-plane messages are required from the fabric intermediate node.

### Fabric Border Node

The fabric border nodes serve as the gateway between the fabric domain and the network outside of the fabric. The fabric border node is responsible for network virtualization inter-working from the campus fabric to the rest of the network. The fabric border nodes implement the following functions:

- **Advertisement of IP subnets:** The fabric border runs either an Interior Gateway Protocol (IGP) or BGP as a routing protocol to advertise the IP prefixes outside of the fabric, and traffic destined to IP subnets from outside the campus fabric goes through the border nodes. These IP prefixes appear only on the routing tables at the border. Throughout the rest of the fabric, the IP information is accessed using the fabric control plane node.

- **Fabric domain exit point:** The fabric border is the gateway of last resort for the fabric edge nodes.

- **Mapping of LISP instance to VRF:** The fabric border can extend network virtualization from inside the campus fabric to outside the campus fabric by using external VRF instances in order to preserve the virtualization.

### Fabric Policy and Identity Repository

The fabric policy and identity services are leveraged for dynamic users or endpoints to group mappings and policy definition. The services can be used to enforce different policy types within the campus fabric. These capabilities can enhance the security, onboarding of users, traffic treatment, and agility of the network. These different policy types include the following:

- **Access policy:** Determines how the user or endpoint is authenticated and authorized onto the campus fabric. In most cases, Active Directory will be used as the identity repository, with 802.1X and network access control solutions as the mechanism for authentication and authorization of users and endpoints. Based on the authentication and authorization policies, endpoints can be placed into virtual networks and/or security groups (SGs). The "group" context can be propagated throughout the network using security group tags (SGTs), which are defined in RFC 3514 and draft-Smith-Kandula standards.

- **Network segmentation policy:** Determines to which network overlay a user or endpoint should be assigned. The policy service can dynamically or statically assign network segmentation policies based on virtual networks and security group information.

Relying on both virtual networks and security groups allows for two levels of segmentation for role-based access control (RBAC). The security group segmentation layer can be referred to as *micro-segmentation.*

- **Access control policy:** Rules and policies governing who can access what. Role-based access control policies are enforced with the use of security group access control lists (SGACLs) for segmentation within VNs and dynamic VLAN assignment for mapping endpoints into VNs at the fabric edge node. Leveraging group-based policies will simplify access control rules and the end-to-end security policy enforcement by decoupling the user's identity from the network design or architecture.

- **Application policy:** Policies can be enforced based on traffic treatment such as QoS for applications, path optimization, and so forth.

### Fabric SDN Controller

The fabric SDN controller oversees the configurations and operations of its network elements, including the configuration of fabric elements and policies associated with users/devices/endpoints as they connect to the network. The controller offers a network abstraction layer to arbitrate the specifics of various network elements in this automation and toward the orchestration and analytics engines. The SDN controller exposes northbound Representational State Transfer (REST)-based application programming interfaces (APIs) that abstract out the network functionality and services available at a network level.

## Summary

This chapter covered the different design options and considerations of forwarding and control plane mechanisms of MPLS, MP-BGP, and software-defined networking. These services have become primary business enablers for enterprises by meeting customer connectivity requirements, whether it is a Layer 3 or Layer 2 type of connectivity. In addition, automation within the software-defined solutions discussed in this chapter brings new capabilities to businesses that they haven't seen before. By leveraging these automation and orchestration capabilities, businesses can redirect their staff members to focus on business initiatives rather than the day to day operations and maintenance of the infrastructure. These design models can directly or indirectly impact business effectiveness and overall efficiency.

In addition to the design models and considerations, this chapter covered the different design approaches that offer a scalable design to support enterprise businesses in this modern software-defined world with a very large number of nodes and prefixes. Last but not least, the design decision of selecting a certain design approach or protocol has to be based on the holistic approach, to avoid designing in isolation of other parts of the network, regardless of whether it is for a physical, virtual, underlay, or overlay entity.

## References

Al-shawi, Marwan, *CCDE Study Guide* (Cisco Press, 2015)

## Exam Preparation Tasks

As mentioned in the section "How to Use This Book" in the Introduction, you have a couple of choices for exam preparation: the exercises here, Chapter 18, "Final Preparation," and the exam simulation questions in the Pearson Test Prep Software Online.

## Review All Key Topics

Review the most important topics in this chapter, noted with the Key Topic icon in the outer margin of the page. Table 9-6 lists a reference of these key topics and the page numbers on which each is found.

**Key Topic**

**Table 9-6**   Key Topics for Chapter 9

| Key Topic Element | Description | Page Number |
|---|---|---|
| List | MPLS design benefits and goals | 232 |
| List | MPLS peer model benefits | 233 |
| Subsection | Virtual Routing and Forwarding | 235 |
| Table 9-2 | MPLS L3VPN RD Allocation Models | 236 |
| List | VPN label allocation models | 238 |
| List | Shared services use cases | 245 |
| List | MPLS VPN Internet access design options | 246 |
| Table 9-3 | MPLS L3VPN Internet Access Design Options Comparison | 249 |
| List | PE-CE routing design questions | 250 |
| Table 9-5 | Comparison of PE-CE Routing Protocols | 265 |

## Complete Tables and Lists from Memory

Print a copy of Appendix D, "Memory Tables" (found on the companion website), or at least the section for this chapter, and complete the tables and lists from memory. Appendix E, "Memory Tables Answer Key," also on the companion website, includes completed tables and lists to check your work.

## Define Key Terms

Define the following key terms from this chapter and check your answers in the glossary:

Virtual Routing and Forwarding (VRF), route distinguisher (RD), route target (RT), VPN label, network manager, controller, orchestrator, edge routers, virtual networks (VNs), TLOC extensions, underlay network, overlay network

# CHAPTER 10

# Security

## This chapter covers the following topics:

- **Infrastructure Security:** This section covers security at the infrastructure and device level, with the associated network design implications.

- **Perimeter Security and Intrusion Prevention:** This section covers protecting the perimeter and discusses the different firewall deployment models and network designs.

- **Network Control and Identity Management:** This section covers network control and identity management, with an emphasis on the corresponding 802.1X capabilities and their network design characteristics and requirements.

This chapter builds on what we covered in Chapter 4, "Security Is Pervasive," where we focused on security design (the "why" a modern network needs an overarching security architecture) and the current industry security methodologies, frameworks, and ideologies that network designers need to know. This chapter focuses on the specific security technologies, protocols, features, and capabilities, and how to properly leverage them within a network design.

This chapter covers the following "CCDE v3.0 Core Technology List" section:

- 5.0 Security

## "Do I Know This Already?" Quiz

The "Do I Know This Already?" quiz allows you to assess whether you should read this entire chapter thoroughly or jump to the "Exam Preparation Tasks" section. If you are in doubt about your answers to these questions or your own assessment of your knowledge of the topics, read the entire chapter. Table 10-1 lists the major headings in this chapter and their corresponding "Do I Know This Already?" quiz questions. You can find the answers in Appendix A, "Answers to the 'Do I Know This Already?' Quizzes."

**Table 10-1** "Do I Know This Already?" Section-to-Question Mapping

| Foundation Topics Section | Questions |
|---|---|
| Infrastructure Security | 1–7 |
| Perimeter Security and Intrusion Prevention | 8–10 |
| Network Control and Identity Management | 11, 12 |

1. Which of the following planes is responsible for controlling the fast-forwarding of traffic passing through a network device?
   a. Management plane
   b. Control plane
   c. Data plane

2. Which of the following planes is like the brain of the network device and usually controls and handles path selection functions?
   a. Management plane
   b. Control plane
   c. Data plane

3. Which of the following planes focuses on how a device is accessed and monitored?
   a. Management plane
   b. Control plane
   c. Data plane

4. For which type of traffic does a network device make a typical routing and forwarding decision regarding whether to send the traffic over its interfaces to directly attached nodes?
   a. Non-IP traffic
   b. Receive IP traffic
   c. Transit IP traffic
   d. Exception IP traffic

5. Which of the following types of traffic carries nonstandard attributes, such as a transit IP packet with an expired TTL?
   a. Non-IP traffic
   b. Receive IP traffic
   c. Transit IP traffic
   d. Exception IP traffic

6. Which of the following types of traffic is traffic destined to the network node itself?
   a. Non-IP traffic
   b. Receive IP traffic
   c. Transit IP traffic
   d. Exception IP traffic

**7.** Which of the following types of traffic would MPLS and CLNP be categorized into?

    **a.** Non-IP traffic

    **b.** Receive IP traffic

    **c.** Transit IP traffic

    **d.** Exception IP traffic

**8.** Which type of attack would a web application firewall (WAF) mitigate?

    **a.** Network direct access

    **b.** Layer 2 attack

    **c.** Network DoS attack

    **d.** Application targeted attack

**9.** Which of the following types of attack would an infrastructure ACL (iACL) mitigate? (Choose two.)

    **a.** Network direct access

    **b.** Layer 2 attack

    **c.** Network DoS attack

    **d.** Application targeted attack

**10.** Which type of attack would remotely triggered black hole (RTBH) mitigate?

    **a.** Network direct access

    **b.** Layer 2 attack

    **c.** Network DoS attack

    **d.** Application targeted attack

**11.** Which of the following authentication options would be the best design choice if a business wanted the highest level of security?

    **a.** EAP-MD5

    **b.** EAP-TLS

    **c.** LEAP

    **d.** PEAP + EAP-MSCHAP

**12.** Which of the following authentication options would be the best design choice if a business wanted the easiest deployment?

    **a.** EAP-MD5

    **b.** EAP-TLS

    **c.** LEAP

    **d.** PEAP + EAP-MSCHAP

# Foundation Topics

Designing a modern network that is secure is one of the most complicated tasks that network designers confront. Mobility and communication over the Internet are becoming one of the primary (de facto) methods of communications and are essential requirements (in most cases they are unstated requirements at the time of this writing. Businesses assume they will have them) for many businesses. Therefore, to address these trends and requirements, sophisticated security countermeasures are required. Typically, a good security design follows a structured approach that divides the network into domains and applies security in layers, with each layer focusing on certain types of security requirements and challenges. This is also known as *defense in depth*, where multiple layers of protection are strategically located across the network and where a failure of one layer to detect an attack or malicious traffic flow will not leave the network at risk. Instead, the multiple security layers work in a back-to-back manner to detect the attack or malicious flow in the network. Figure 10-1 summarizes the common security aspects that can be applied in a layered approach.

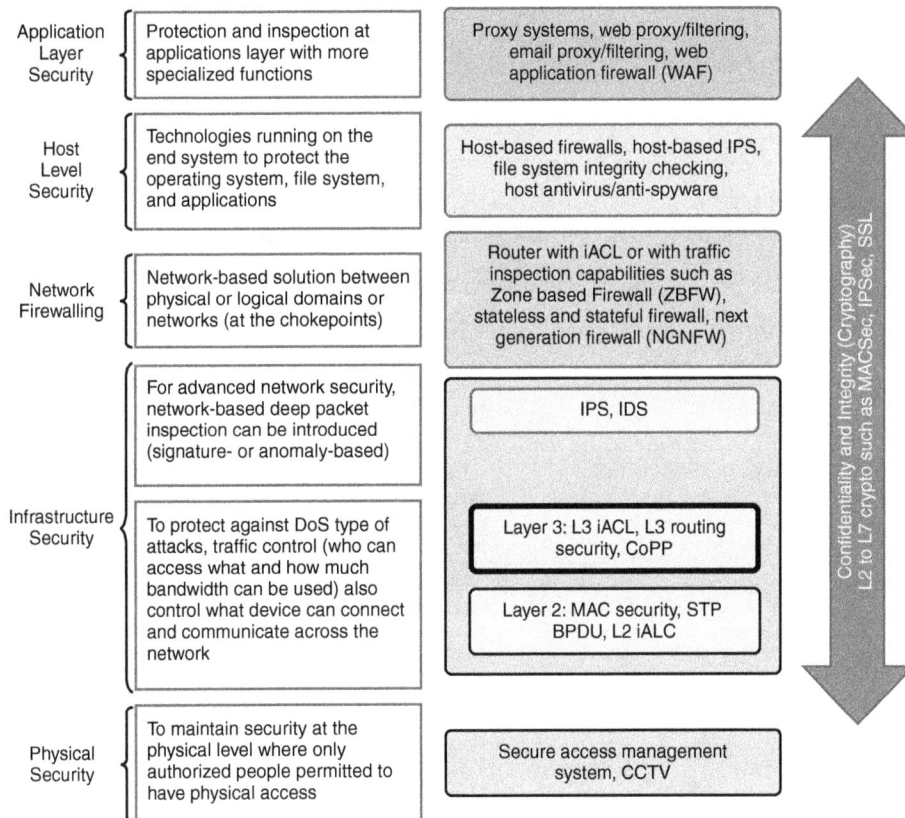

**Figure 10-1**  *Elements of Layered Network and Information Security*

# Infrastructure Security

This section covers infrastructure security and network firewall considerations, in brief, focusing on the integration and impact with regard to network design.

## Device Hardening Techniques

Securing network devices is the first and essential consideration to protect any network. For instance, even though you may have a very secure control plane and network edge design and policies, if one or two network devices are compromised, the network can be taken down easily by introducing black-holing routes, generating a large amount of malicious traffic that is technically sourced from an internal (trusted) network device, or the attacker can even take advantage and access other internal zones that are not accessible externally. Additionally, if someone can access a network device, it means all the traffic passing through that device can be sniffed and copied. This is dangerous for an organization that has sensitive information, such as credit card numbers or people's personal information. Therefore, protecting network devices is a fundamental requirement to achieving a secure network design.

**Key Topic**

To approach device-level protection in a more structured manner, as a network designer you need to understand the following types of traffic that network devices normally handle, as summarized in Figure 10-2. Understanding the different traffic types passing through any network will help to identify how each traffic type can impact the device and the possible countermeasures for each. The types of traffic are:

- **Transit IP traffic:** Traffic for which a network device makes a typical routing and forwarding decision regarding whether to send the traffic over its interfaces.

- **Receive IP traffic:** Traffic destined to the network node itself, such as toward a router's IP address, and requires CPU processing.

- **Exception IP traffic:** Any IP traffic carrying a nonstandard "exception" attribute.

- **Non-IP traffic:** Typically related to non-IP packets and almost always is not forwarded.

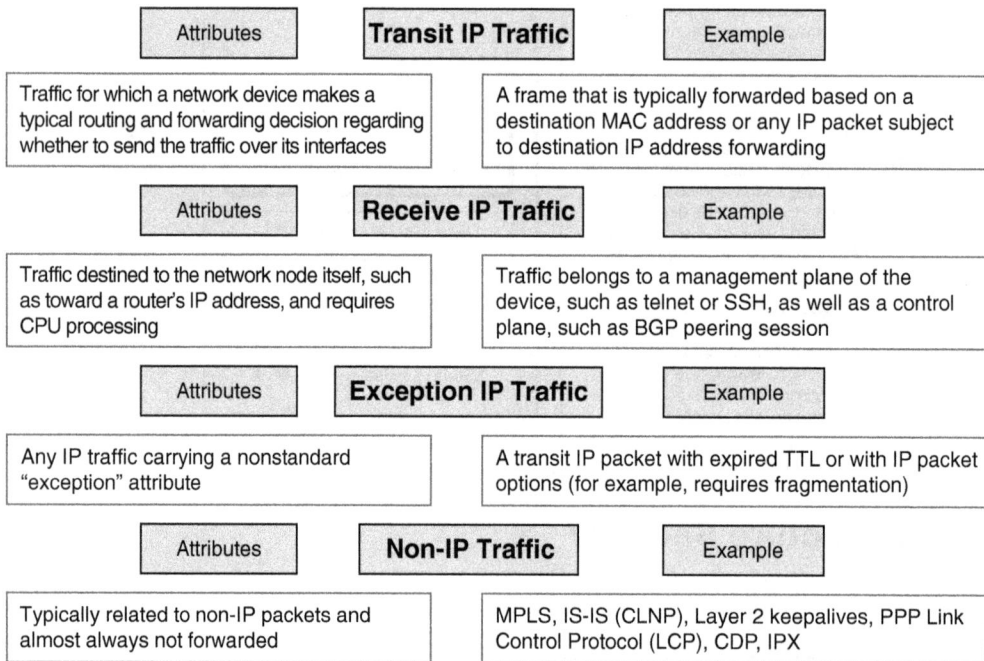

| Attributes | Transit IP Traffic | Example |
|---|---|---|
| Traffic for which a network device makes a typical routing and forwarding decision regarding whether to send the traffic over its interfaces | | A frame that is typically forwarded based on a destination MAC address or any IP packet subject to destination IP address forwarding |

| Attributes | Receive IP Traffic | Example |
|---|---|---|
| Traffic destined to the network node itself, such as toward a router's IP address, and requires CPU processing | | Traffic belongs to a management plane of the device, such as telnet or SSH, as well as a control plane, such as BGP peering session |

| Attributes | Exception IP Traffic | Example |
|---|---|---|
| Any IP traffic carrying a nonstandard "exception" attribute | | A transit IP packet with expired TTL or with IP packet options (for example, requires fragmentation) |

| Attributes | Non-IP Traffic | Example |
|---|---|---|
| Typically related to non-IP packets and almost always not forwarded | | MPLS, IS-IS (CLNP), Layer 2 keepalives, PPP Link Control Protocol (LCP), CDP, IPX |

**Figure 10-2** *Types of Traffic Handled by Network Devices*

The different traffic types listed in Figure 10-2 each have an impact in terms of overloading network devices with regard to the IP traffic planes covered in Table 10-2 and Figure 10-3, because each has its own attributes and security considerations.

**Table 10-2**  IP Traffic and Device Planes

| Device Plane | Security Concern Example |
|---|---|
| **Data plane** | Responsible for controlling fast-forwarding of traffic passing through a network device. However, this plane is normally limited to a certain number of packets per second (PPS) based on the hardware platform's throughput. Common mechanisms to protect this plane include infrastructure ACLs (iACLs), QoS toolset, remotely triggered black hole (RTBH), and Unicast Reverse Path Forwarding (uRPF). |
| **Control plane** | As discussed earlier in this book, the control plane is like the brain of the network node; it usually controls and handles all path selection functions. Therefore, any control plane–related issues, such as a flapping session with a BGP peer that advertises an extremely large number of prefixes, will impact not only the network stability but also the device itself (because of high CPU spikes in this case). Common mechanisms to protect this plane include iACLs, routing protection, and control plane policing (CoPP). |
| **Management plane** | As the name implies, this plane relates to the management traffic of the device, such as device access, configuration, troubleshooting, and monitoring. Therefore, its criticality is equal to the other two planes. Any unauthorized access can lead to a device- and network-wide crisis, such as injecting a black-holing route into the control plane or flooding the network with malicious traffic, which will ultimately impact all the hosts and users transiting through the network in general and this network device in particular. Common mechanisms to protect this plane include CPU and memory thresholding, AAA (authentication, authorization, and accounting), and CoPP. |

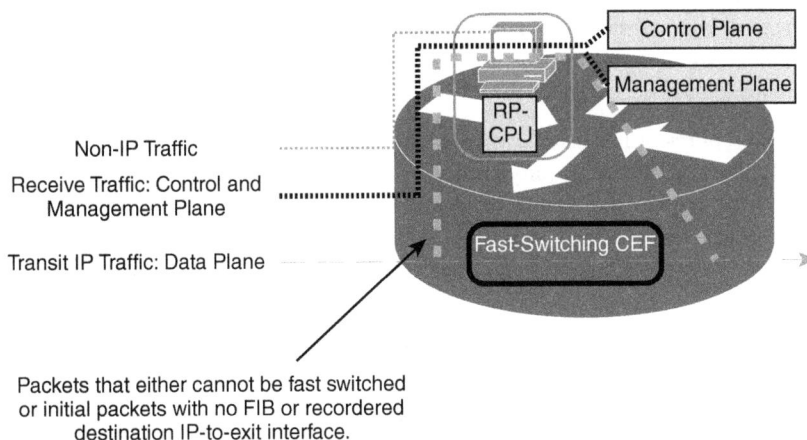

**Figure 10-3**  *Traffic Handling Within a Router*

Taking the different traffic types and how each is mapped into the relevant plane for processing in a structured manner, the following points should be considered as general and foundational guidance to achieving the desired level of device hardening and security level.

Consider that these points should not conflict or breach the security policy or company standards.

- **Physical security:** The first basic security consideration is that network devices be placed physically in a secure place, where only authorized people can access them.

- **Authentication, authorization, and accounting (AAA):** Network devices must be accessible only by authorized people. Ideally, the access privileges must be multilevel role-based access (authorization level "who can do what") combined with a reliable accounting (logging) to keep track of who has accessed the devices and what has been done.

- **Secure the device management plane:** Ideally, the management sessions and activities of the network devices must be performed over protected protocols such as Secure Shell (SSH), Secure Sockets Layer (SSL), Secure File Transfer Protocol (SFTP), or IPsec. Also, it is always advised that out-of-band management access to network devices be provisioned to offer secure, guaranteed direct access to critical network devices in situations where the network might be facing a DDoS attack (because in-band management access will be almost impossible).

- **Protect the hardware resources:** Maintaining a device hardware resource is the foundation to maintaining traffic flow through the device. Therefore, protecting hardware resources such as CPU is an essential countermeasure to be considered with regard to device hardening, such as CoPP to protect the CPU from spikes because of excessive control plane traffic. Control plane protection (CPP) also helps to protect from other types of traffic, such as exception and transit traffic.

- **Disable unused services:** A network node does not usually use every single service it has; normally, the device role and function within the network dictate which required services need to be turned on. In general, unused services should be disabled to limit the chances that an intruder or hacker can compromise one of these unused services to generate malicious traffic or DoS attacks.

- **Disable unused ports:** A network device with unused ports should be disabled. Furthermore, the port should be put into a VLAN that is dedicated to unused ports and only for that purpose (i.e., no traffic should be allowed on that VLAN). Another recommendation is to not use the default VLAN, VLAN 1, or to leverage native VLAN hopping. These simple configuration changes can limit the attack surface when properly applied.

- **Monitoring:** Network security is dynamic and always evolving and changing in response to the continuous technology and design developments and changes. As a result, there are always new system vulnerabilities, breaches, and intrusion developments. Therefore, to keep track of what is going on in terms of device utilization (bandwidth or hardware resource) and whether it is within the normal (baseline) limits, network and device monitoring helps network operators to always optimize device security either proactively or in a fast, reactive approach. Simple Network Management Protocol (SNMP) and NetFlow are the most common protocols used for this purpose, but not the only ones.

Although the device-level security considerations covered earlier are considered generic and standard foundational guidelines, the consideration of these guidelines will vary based on some other factors. Figure 10-4 identifies the primary influencing factors with regard to device-level security.

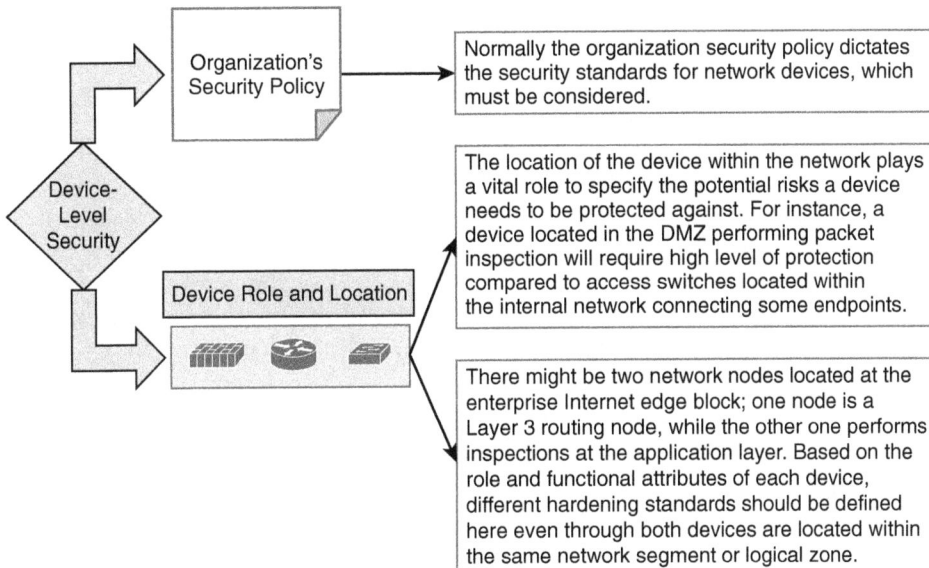

Normally the organization security policy dictates the security standards for network devices, which must be considered.

The location of the device within the network plays a vital role to specify the potential risks a device needs to be protected against. For instance, a device located in the DMZ performing packet inspection will require high level of protection compared to access switches located within the internal network connecting some endpoints.

There might be two network nodes located at the enterprise Internet edge block; one node is a Layer 3 routing node, while the other one performs inspections at the application layer. Based on the role and functional attributes of each device, different hardening standards should be defined here even through both devices are located within the same network segment or logical zone.

**Figure 10-4**   *Influencing Factors on Device-Level Security*

## Layer 2 Security Considerations

Securing Layer 2 or data link layer communication is as critical as securing higher layers in the network. One of the common practices is that network designers or operators focus more on securing higher layers across the network, such as the Layer 3 control plane, protocol, and session layers, without giving Layer 2 enough focus to secure it. An analogy mentioned in multiple sections in this book is that a building constructed with a weak foundation will not be able to deliver the promised value or end result and will be more vulnerable to damage. Similarly, a highly secure network on the higher layer, such as network and application layers, can be easily taken out of service by simple Layer 2 flooding across its switched network. Therefore, it is critical to secure the network in a bottom-up manner to achieve a reliable and self-healing network security design.

Layer 2, compared to other OSI layers, is the simplest with regard to security considerations and one of the most dangerous at the same time. For instance, if a hacker compromises a data center switch, the hacker can easily sniff all the traffic passing through this switch that might carry critical data. At the same time, this hacker can simply take the data center switched network down by flooding the ternary content-addressable memory (TCAM) table of the switch or sending malicious Layer 2 traffic storms. In addition, IPv6 raises a number of first-hop security (FHS) concerns that were not present in IPv4.

Layer 2 security can be categorized as the following to produce a structured and flexible Layer 2 security design:

10

■ **Authentication and authorization:** Normally, the Layer 2 switched network is the network edge or boundary where end users and hosts connect. Typically, it is the edge where rogue and unauthorized endpoints can be connected. Therefore, only authorized devices must be permitted to connect to the network at the switch port level. This normally can be achieved by using different mechanisms and features, such as authentication with IEEE 802.1X, which can integrate with a network admission control (NAC) system to authenticate users and their endpoints, port ACL (routed ACL, MAC ACL, VLAN ACL), dynamic ARP inspection (DAI), DHCP snooping, IP device tracking (IPDT), and other port security features. One or a combination of these mechanisms or features can be used based on the security goals, standards, and supported features. Also, IPv6 router advertisement (RA) guard, DHCPv6 guard, and IPv6 snooping help to mitigate IPv6 FHS concerns. Furthermore, the IEEE 802.1AE standard known as MACsec can be used to satisfy some security specifications that require protecting the data traversing an Ethernet LAN, such as between the Layer 2 DCI edge nodes of a secure data center in which MACsec offers the ability to authenticate and encrypt packets between the two MACsec-capable devices at the DCI edge of the interconnected data centers.

■ **VLAN design:** Although one of the primary goals of VLANs is to provide separation at Layer 2, the design of these VLANs can sometimes lead to security concerns. For example, if an extended VLAN across two data centers over a DCI is compromised, this may lead to a major impact and risk to both data centers. Therefore, it is always recommended that Layer 2 domains should be contained to a limited range within the network (smaller failure domain); this also avoids issues such as Address Resolution Protocol (ARP) or unicast flooding. However, practically this is not always achievable because the reality is that application requirements dictate how the network should be designed. Accordingly, protective features and policies have to be considered in scenarios like this, such as defining rate limiting along the path and mitigating the impact of the traffic flooding. In addition to the typical separation that VLANs provide to the design, private VLANs (PVLANs) offer a more advanced level of a controlled traffic separation between VLANs and between endpoints that reside within the same VLAN. Furthermore, the use of VLAN Trunking Protocol (VTP) can introduce a high risk to the switched network because inserting a switch that takes over the VTP server role will rewrite VLANs set up across the entire switched network under the same VTP domain. Therefore, securing VTP or considering VTP in transparent mode is always recommended to avoid its impact in situations like this.

■ **Layer 2 control plane:** As discussed earlier in this book, Layer 2 control protocols, such as Spanning Tree Protocol (STP), provide protective mechanisms against Layer 2 loops. However, STP is considered a double-edged sword because in the case of misconfiguration, with any security weakness any attacker can simply bring down the switched network or change its logical design. This is because no reliable authentication capability is built into STP. Therefore, if someone attaches a switch to the existing switched network that has better attributes than the other switch to make it the network root bridge, the network first will be unstable for a while during the election process of the new root bridge. Second, after this rogue switch is considered as the root, a disaster will probably occur in this network because of performance issues, and may introduce traffic black-holing at Layer 2. To further exacerbate the situation, most of the traffic will pass through this new root switch, in which case anyone can sniff the traffic or take it or copy it. Moreover, this behavior can be dangerous to Metro Ethernet providers if one of their customer switches becomes the root bridge. Therefore, protecting the Layer 2 control protocols used is an essential consideration to secure Layer 2 switched networks, such as protecting the root bridge by denying any other switch in the network from participating or sending a rogue message to be elected as a root bridge and filtering Layer 2 bridge protocol data unit (BPDU) messages at the network edge as well.

## Layer 3 Control Plane Security Considerations

The Layer 3 control plane is the intelligence over the network that steers traffic toward reaching its intended destination. This means any wrong information can lead to traffic black-holing and the eventual dropping of packets. In addition, in a network with unprotected routing, anyone can easily bring it down by either flooding a large number of prefixes (legitimate or not) by inserting a rogue router or redirecting traffic to a black-holing next hop or a sniffer. Therefore, securing the Layer 3 control plane is a must. However, the question here is this: Where is securing a control plane a must, and where is it just something beneficial to have? Although there is no one standard answer to this generic question, the following can be considered as general rules with regard to Layer 3 control plane security:

- Secure the control plane against rogue Layer 3 peers by authenticating routing sessions. This is a recommended practice to be considered across the entire routed domain and is strongly recommended when peering with an external routing domain.

- Device protection with regard to Layer 3 control plane functions is a primary element to be considered. For example, if a router that is not supposed to receive and process prefixes receives a large number of them, this node might run out of memory. In all likelihood, its CPU will be overloaded, and this will result in a malfunctioning node in the routing path, which will introduce instability (because the routing session will probably keep going up and down in this case). A large percentage of traffic will also be dropped. Therefore, it is important to limit the maximum number of routes to be accepted by the node. This is common in MPLS L3VPN Service Provider (SP) design, where normally the PE nodes limit the maximum number of prefixes per VRF/VPN.

- Accept only the routes that are expected to be received. This is common in MPLS L3VPN and ISPs, where the ISP, for example, will only permit the Provider Assigned "PA" or Provider Independent "PI" address space the customer has to advertise with the agreed subnet length, sometimes combined with uRPF check. For instance, a customer who owns /24 IPv4 public ranges will not normally be allowed to divide it and advertise it in the format of host routes (/32) unless it is agreed with the ISP. This is because the ISP always tries to protect its PE nodes from being overloaded with extra thousands (or even can reach hundreds of thousands) of extra routes.

- Advanced BGP techniques such as QoS policy propagation with BGP (QPPB) and remote black-holing trigger can be used in SP-style networks to offer a more dynamic, flexible, controlled, and secure control plane design that is almost impossible to be

10

achieved with the typical static filtering mechanisms, such as using ACLs or prefix lists.

■ QoS has powerful features that help to optimize Layer 3 control plane security design, usually by guaranteeing a minimum amount of bandwidth during congestion situations. This ensures the routing session will not be torn down, and it limits the amount of bandwidth to a predefined maximum allowed bandwidth. This protects the network from undesirable behavior (such as DoS attacks by a rogue router, for example) that generates a large amount of traffic using the same control plane marking or even TCP/UDP port. To take advantage of this QoS as a protective mechanism here, it ideally should be applied across the entire routing domain and not only at the edge of the network.

■ CoPP, as covered earlier, should be considered on every network node to relax the impact of control plane flooding and protect network nodes from CPU spikes by blocking unnecessary or DoS traffic.

**NOTE**   Although CoPP uses modular QoS CLI (MQoC), the QoS considerations earlier provide network-wide treatment on a per-hop basis for Layer 3 control plane traffic flows, whereas CoPP is focused only on a device level. As covered earlier, marking down the DSCP value of packets that are out of profile can impact how these packets will be treated by other nodes across the network.

## Remote-Access and Network Overlays (VPN) Security Considerations

The most common and proven mechanism for providing secure remote access for remote users and remote sites over the public Internet is the overlaid virtual private network (VPN). One common misconception is that the term VPN is always a synonym for IPsec. In fact, IPsec offers cryptography at the network layer, whereas VPN can take different forms, such as MPLS L3VPN, MPLS L2VPN, SSL VPN, DMVPN, and so on. Therefore, it is important that network architects and designers distinguish between VPN and IPsec. Having said that, secured VPN with IPsec is the most common and proven combination that is used to protect the various types of VPN solutions. Table 10-3 compares the most common overlay VPN solutions from different design aspects.

**Table 10-3**   Overlay VPN Solutions Comparison

|  | IPsec | GRE | DMVPN | GET VPN | Remote Access (Client Based) |
|---|---|---|---|---|---|
| Standard | Yes | Yes | No (Cisco proprietary) | No (Cisco proprietary) | Yes |
| Overlay transport | P2P IPsec tunnel | P2P GRE | mGRE | Tunnel-less | P2P IPsec tunnel |

| | IPsec | GRE | DMVPN | GETVPN | Remote Access (Client Based) |
|---|---|---|---|---|---|
| Protection model | Peer to peer | Peer to peer with IPsec | Peer to peer with IPsec | Group protection (RFC 3547) | Peer to peer |
| IP routing | Static routing | Tunneled dynamic routing | Tunneled dynamic routing | Standard IP WAN dynamic routing | Reverse-route injection |
| Offers a protected VPN over public Internet | Yes | Yes, with IPsec | Yes, with IPsec | No | Yes |
| Offers a protected VPN over the IP WAN transport | Yes | Yes, with IPsec | Yes, with IPsec | Yes | Yes |
| Requires some changes to the existing Layer 3 routing design when used over the WAN | Yes | Yes | Yes | No | Yes* |
| Requires additional network equipment | No | No | No | Yes, such as a key server | No |
| Supports NATing | Yes | Yes | Yes | No (NAT must be performed before encryption or after decryption) | Yes |

* This VPN is commonly used to provide connectivity over the Internet.

One main consideration with regard to an enterprise VPN is whether to enable the VPN across the enterprise WAN. The typical answer to this depends on the business requirements, security policy, and functional requirements. A network designer must consider the impact on the business-critical applications because of the increased packet overhead by the VPN tunneling, especially when VPN is combined with IPsec, as shown in Figure 10-5. Besides, it is common that security devices in the path, such as firewalls running NAT, add a layer of design and operational complexity to the VPN solution.

10

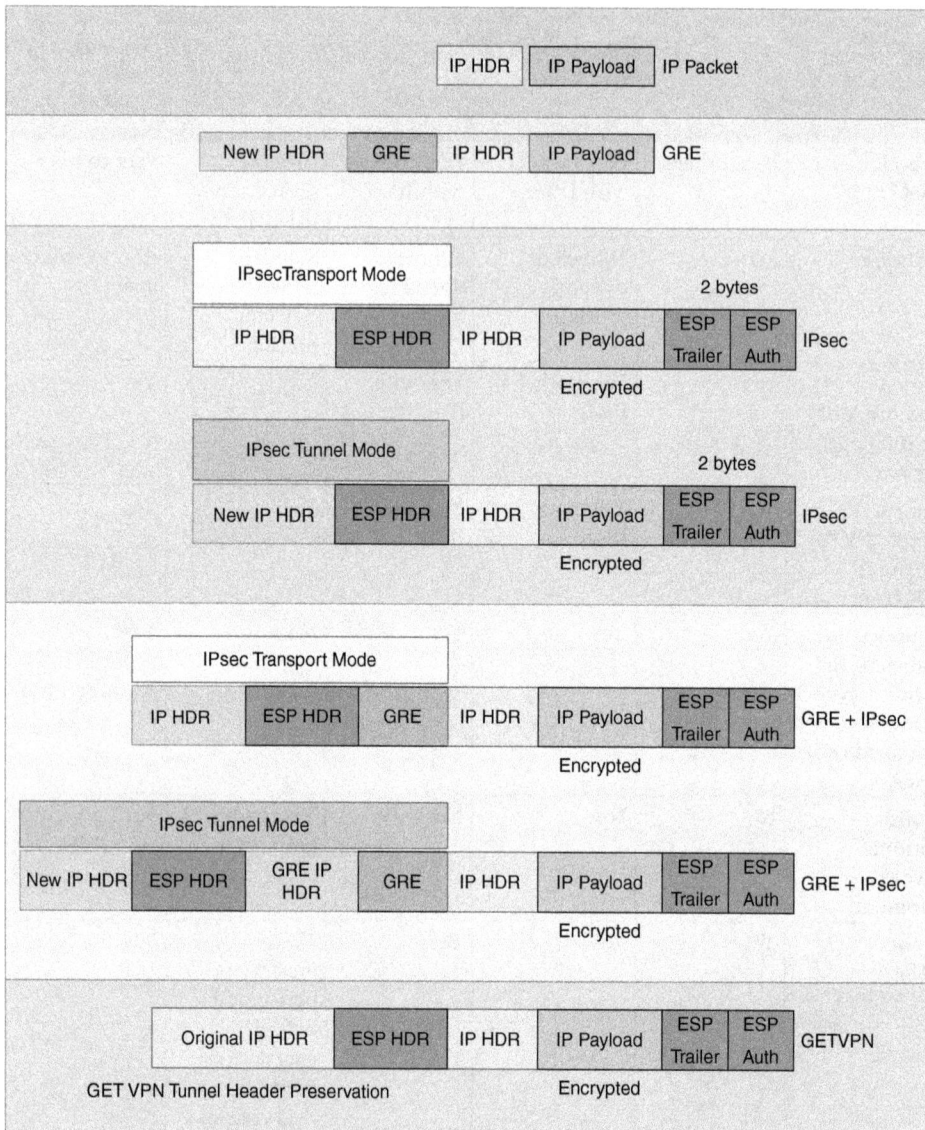

**Figure 10-5**  *IP Packet Overhead with VPN Solutions*

In addition, introducing a VPN across the WAN can increase the design and operational complexity in scenarios where dual WAN connectivity is required (either dual routers or links). In these cases, the routing redundancy design will be moved from the typical IGP or BGP to IPsec redundancy, which might not be a suitable option for some design requirements. Therefore, GETVPN is becoming the most common and desirable secure VPN solution of the WAN because GETVPN by nature offers tunnel header preservation, which facilitates routing encrypted packets using the underlying network routing infrastructure. There is no need to consider any redundancy design at the VPN level, such as dual hubs or stateful IPsec high availability.

Even though securing IP traffic across the WAN will introduce a new layer of complexity, secure IP communication is a top priority for some businesses, like those in financial services environments. Sending internal traffic across the WAN will be seen by the business as trusting SP setup, security, employees, and even contractors and third parties who work and integrate with this SP not to perform any malicious activity. Therefore, for this type of business, it is common that IPsec, such as secure DMVPN or GETVPN, is deployed across the WAN.

IPsec packet payloads can be encrypted, and IPsec receivers can authenticate packet origins. Internet Key Exchange (IKE) and Public Key Infrastructure (PKI) can also be used with IPsec. IKE is the protocol used to set up a security association (SA) with IPsec. PKI is an arrangement that provides for third-party verification of identities.

In other words, enabling secure IP communication over the WAN is not a common practice because it is seen as adding an unnecessary layer of complexity and may impact the performance of some business applications, whereas for those businesses that are concerned most about security, secure IP communication over the WAN is essential. Therefore, the decision of whether to enable secure IP communications across the WAN is almost always based on the business type, priorities, and impact on the overall network and application performance, in addition to operational complexity.

## Wireless Security Consideration

Wireless security has evolved over the years, starting with Wired Equivalent Privacy (WEP), which was quickly identified as a highly vulnerable wireless security protocol that was replaced by Wi-Fi Protected Access (WPA). The first version of WPA saw an increase in key size to 128 bits but still had the same limitations around encryption and cipher methods being leveraged like those of WEP. WPA2 was created as the new wireless security standard in 2004, with the most impactful change being the implementation of the Advanced Encryption Standard (AES). Leveraging AES in WPA2 provides an overall higher level of security and performance to the wireless network. WPA2 still had some limitations and security issues, the most prevalent being dictionary attacks. Thus, WPA3 was created, which provides the same level of authentication as WPA2 but also delivers an increased cryptographic strength for highly sensitive use cases. Table 10-4 compares these different wireless security protocols.

**Table 10-4**   Wireless Security Comparison

|  | WEP | WPA | WPA2 | WPA3 |
|---|---|---|---|---|
| Encryption | RC4 | TKIP with RC4 | CCMP and AES | AES |
| Key size | 40-bit | 128-bit | 128-bit | 128-bit (Personal) 192-bit (Enterprise) |
| Cipher method | Stream | Stream | Block | Block |
| Data integrity | CRC-32 | Message Integrity Code | CBC-MAC | Secure Hash Algorithm (SHA) |

**10**

|  | WEP | WPA | WPA2 | WPA3 |
|---|---|---|---|---|
| **Key management** | None | 4-way handshaking mechanism | 4-way handshaking mechanism | Simultaneous Authentication of Equals (SAE) handshake |
| **Client authentication** | WPE-Open WPE-Shared | PSK and 802.1X (EAP variant) | PSK and 802.1X (EAP variant) | SAE and 802.1X (EAP variant) |

From what is shown in Table 10-4, it's pretty straightforward that WPA3 is the best option, from both a security standpoint and a performance standpoint. With that said, at the time of writing not all wireless devices support WPA3, in which case WPA2 is the preferred option.

# Perimeter Security and Intrusion Prevention

Technically, a firewall is not a device built to perform routing or switching. Instead, it is a device that is primarily intended to perform security functions such as packet filtering, inspection, and VPN termination. Although almost all the firewalls nowadays support static and dynamic routing, the nature of these devices almost always impacts the network design. This impact varies based on different factors and attributes. The following are the primary and most common and influencing factors that network designers must always consider whenever there is a firewall in the network path:

- **Stateful versus stateless firewall:** Typically, stateless firewalls deal with traffic in a static manner where traffic flows are only monitored and filtered (allowed or blocked) based on static information such as IP source, IP destination, or port numbers. This behavior from a network design point of view can introduce limitations and operational complexities because every single traffic flow needs to be included in both directions (ingress and egress) as part of the firewall entry rules. Stateless firewalls are also not very friendly with applications that automatically negotiate ports, such as RTP (UDP) ports, which are normally automatically negotiated during the signaling and establishment of a VoIP or video call. Stateful firewalls, in contrast, offer traffic monitoring and filtering that is more aware of traffic flows to overcome the limitations of the stateless firewalls, especially with applications that require dynamic port negotiation to establish their session, such as FTP, VoIP, and video RTP media streams. The benefits of a stateless firewall are that they provide a high level of scalability and can easily support any asymmetric routing design required.

- **Stateful versus stateless firewall failover:** Even though some firewalls that work in active-standby mode are stateful firewalls from a traffic-handling perspective, the way these firewalls fail over from the active to the standby can take two forms, either stateless failover or stateful failover. With the stateless failover, the state table is not replicated to the standby peer firewall. This means in the event of active firewall failure, the active role will be moved to the standby firewall. The issue here is that after the failover happens, all the connections (active flows) have to be reinitiated. In contrast, stateful failover offers the ability to replicate the state table between the active and the standby firewalls to avoid reinitiation of the active session following a failover event. The impact of a session's reinitiation might not be an issue for some applications and businesses, but it can be a serious issue for others. Therefore, the decision whether to use stateless or stateful failover should ideally be driven by the design and application requirements.

■ **Routed versus transparent firewall modes:** The other important factor that has a direct impact on network design with regard to firewalls, apart from the operational attributes of the firewalls (stateful or stateless), is that firewalls can operate as a Layer 3 or Layer 2 node. Technically, from a network design point of view, introducing a Layer 3 node into a routed network means that there will be a need to change IP addressing or a need for new IPs in the area where this firewall will be added, in addition to considerations about whether the firewall supports dynamic routing. For instance, if a firewall is added to an Intermediate System-to-Intermediate System (IS-IS) routed domain and the firewall does not support IS-IS, in this case, static routing will be considered. This means a redesign of the routing will be required. Firewalls operating in Layer 2, however, will not introduce these challenges and can be inserted transparently into the routing domain. This is commonly known as *transparent mode* because there is no change required to the existing routing, IP addressing, or host default gateways. Furthermore, when a firewall operates in transparent mode, it often loses some of its functions or features, such as providing VPN termination. However, this is completely vendor- and device model–dependent and can vary from one to another to a large extent. Remember that the CCDE exam is not hardware- or vendor-dependent. Therefore, if something relates to nonstandard software or hardware specification, it should be provided as part of the design scenario.

■ **Virtualization:** With the adoption of virtualization, specifically hypervisors, a new model for firewalls is a virtual appliance that is deployed within a hypervisor. This is a shift from the physical firewall devices that networks have historically been accommodated with. The security goal with providing a firewall, with all of its associated capabilities, is to secure the virtualization environment from any east–west security propagation. This new model now also includes leveraging these virtual firewall appliances to provide security functionality between different layers of an application, like a three-tier application discussed in Chapter 3, "What's the Purpose of the Network?" These firewalls can now provide NAT, isolation, application inspection, policy-based routing, TLS inspection, and many more security capabilities within the application-specific traffic zones.

**NOTE** Hypervisors have taken this virtualized firewall appliance a step further; instead of leveraging a dedicated virtual firewall, the hypervisor has this functionality built-in, allowing these security capabilities to be instantiated at any location, at any layer, and at any virtual machine within the hypervisor's management control.

**10**

Detecting and mitigating common types of attacks is by far one of the most important security goals of all security devices. Based on the multiple sections and security layers discussed briefly in this section, Figure 10-6 provides a summarized list of some common network attacks and risks, along with the possible countermeasures and features that can be used to protect the network infrastructure and mitigate the impact of these attacks at different layers.

**Key Topic**

| Attack/Risk | Layer | Mitigation |
|---|---|---|
| Applications Targeted Attacks | Application | Web Proxy/Filtering, E-Mail Proxy/Filtering, Web Application Firewall (WAF) |
| Identity Spoofing, Unauthorized Access or Privilege Access | Device/Host | Strong Password Policy (Local Passwords, RADIU, TACACS+, OTP, PKI), Disabling Unused Services, CoPP Management Plane, Management ACLs, OOB Management Access |
| Virus/Worm/Trojan | Host/Application | IPS, Host-Based IPS/IDS, AV, Web/Mail Filtering |
| TCP SYN Flood | Network-L4 | Statefull Network Firewalls, Host-Based Firewalls/IPS, Network-Based IPS, Proxy Systems |
| IP Spoofing | Network-L3 | uRPF, Stateful FW, Signature-Based IDS/IPS, L3 Cryptography (for Example, Filtering Out RFC 2827 and 1918 at the Edge to External Networks) |
| Network Direct Access | Network - L2 - L4 | L2 iACL, L3 iACL, Firewalls |
| Routing Protocol Attacks | Network - L3 | Router Neighbor Authentication, Route Filtering, iACL, CoPP, System and Topological Redundancy |
| Layer 2 Attacks | Network - L2 | iACL, CoPP, System and Topological Redundancy |
| Network DoS Attack | Network - L2- L3 | L3 and L2 Network and Device Security Considerations + Remotely Triggered Black Hole (RTBH) Anomaly-Based IDS/IPS |
| Man-in-the-Middle, Sniffer | At Different Layers | Cryptography on Multiple Layers, Signature-Based IDS/IPS |

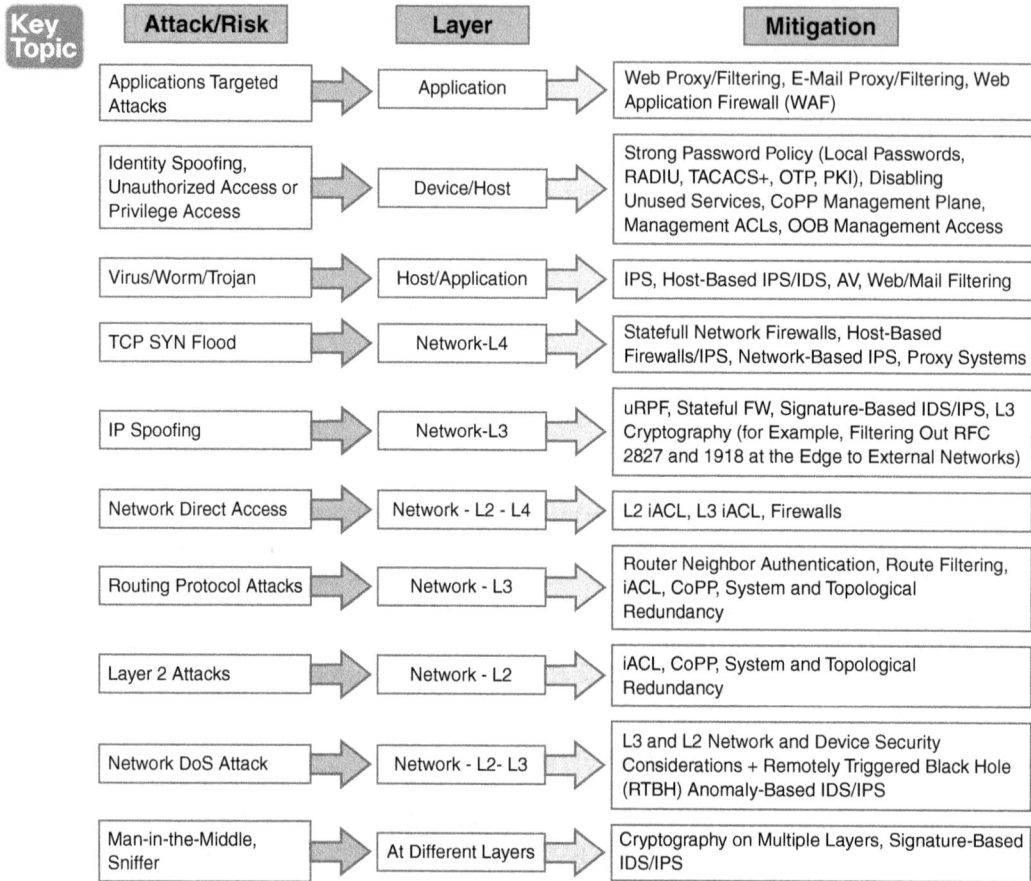

**Figure 10-6**  *Common Network Attacks and Mitigation Mechanisms*

## Network Control and Identity Management

Network access control (NAC) is a critical infrastructure service just like that of DNS, DHCP, NTP, and so forth, because without this service, user devices and users will not be able to authenticate to the network and get access to the resources they need to function. In addition, if the NAC capability is being leveraged for device administration (RADIUS/TACACS+), then the administrators lose the ability to make changes to the network devices until the NAC systems are back online.

NAC is also referred to as transaction-based or session-based security, as all the NAC solutions today provide the capabilities of authentication, authorization, and visibility, discussed next in order.

### Authentication

The first step in the NAC process is authentication, and every device and user gets authenticated. How each of these resources is authenticated depends on what protocols they support and the overall security architecture design being deployed.

There are three parts to 802.1X authentication:

1. **Supplicant:** A software client running on the end device that passes authentication information to the authentication server.

2. **Authenticator:** A network device that the client is connecting to. In a wired deployment, this could be a network switch, and in a wireless deployment, this could be the local access point the client is connecting to or the wireless controller that manages the access points.

3. **Authentication server:** A RADIUS server that contains an authentication database.

**Extensible Authentication Protocol (EAP)** is a protocol used to pass authentication information between the supplicant and the authentication server. There are a number of different types of EAP authentication options, with each one handling the authentication differently:

■ EAP-MD5: **EAP Message Digest Challenge (EAP-MD5)** is an EAP authentication type that provides base-level EAP support. EAP-MD5 is typically not recommended for implementation because it may allow the user's password to be derived. It provides for only one-way authentication—there's no mutual authentication of the client and network. It also doesn't provide a means to derive dynamic per session keys.

■ EAP-TLS: **EAP Transport Layer Security (EAP-TLS)** provides for certificate-based and mutual authentication of the client and the network. It relies on client-side and server-side certificates to perform authentication and can be used to dynamically generate user-based and session-based keys to secure future communications between the client and the network. One limitation of EAP-TLS is that certificates must be managed on both the client and server sides. For a large design, this could be a very cumbersome task. There are additional capabilities that can provision and deploy identify certificates (user and device certifications) during an initial login process, which can mitigate this overall limitation with EAP-TLS.

■ EAP-TTLS: **EAP Tunneled Transport Layer Security (EAP-TTLS)** is an extension of EAP-TLS. This security method provides for certificate-based, mutual authentication of the client and network through an encrypted tunnel, as well as a means to derive dynamic keys. Unlike EAP-TLS, EAP-TTLS requires only server-side certificates, which makes it easier to deploy than EAP-TLS but less secure.

■ PEAP: **Protected Extensible Authentication Protocol (PEAP)** provides a method to transport securely authentication data, including legacy password-based protocols. PEAP accomplishes this by using tunneling between PEAP clients and an authentication server. Like TTLS, PEAP authenticates clients using only server-side certificates, thus simplifying the implementation and administration of a secure network.

■ EAP-FAST: **EAP Flexible Authentication via Secure Tunneling (EAP-FAST)**, instead of using a certificate to achieve mutual authentication, authenticates by means of a PAC (Protected Access Credential), which can be managed dynamically by the authentication server. The PAC can be distributed to the client either manually or automatically. Manual delivery is via disk or a secured network distribution method. Automatic distribution is in-band, over the network.

**10**

■ **LEAP: Lightweight Extensible Authentication Protocol (LEAP)** is an EAP authentication type that encrypts data transmissions using dynamically generated key, and supports mutual authentication. This is a Cisco proprietary protocol that Cisco licenses to other manufacturers and vendors to use.

Table 10-5 summarizes and compares the EAP types listed.

**Key Topic**

**Table 10-5** 802.1X EAP Types Comparison

|  | EAP-MD5 | EAP-TLS | EAP-TTLS | PEAP | EAP-FAST | LEAP |
|---|---|---|---|---|---|---|
| Client-side certificate required | No | Yes | No | No | No (PAC) | No |
| Server-side certificate required | No | Yes | Yes | Yes | No (PAC) | No |
| WEP key management | No | Yes | Yes | Yes | Yes | Yes |
| Rogue AP detection | No | No | No | No | Yes | Yes |
| Deployment difficulty | Easy | Difficult (client certificate deployment) | Moderate | Moderate | Moderate | Moderate |
| Security | Poor | Very high | High | High | High | High when strong passwords are used. |

In some situations, there will be devices that do not support 802.1X authentication, in which case **MAC Authentication Bypass (MAB)** can be used. With MAB the authenticator (network device) sends the MAC address of the client device to the authentication server, which is likely a RADIUS server, to check if it permits the MAC address. MAB is a very insecure option, but for those devices that have yet to adopt 802.1X support, it might be the only option.

## Authorization

Once a device or user has properly completed authentication, they start the authorization process. This step is about what the device or user should have access to. What do they need to complete their job or role? These can be referred to as use cases. A good use case example is network printers, and here are some questions that should be answered:

■ Do network printers require Internet access?

■ Do network printers require full east–west (lateral) network access?

In most cases, the answer to these questions is no, they don't. For most network printer use cases, we can allow the print server to access them, and that's all that is needed. This process should be followed for all use cases; just remember to only allow what the use case needs to work, not what the use case wants to have access to.

There are a number of different authorization policies that can be applied to a use case, and the following are some of the most common:

- **Downloadable ACL (dACL):** Most NAC deployments have the capability of pushing a preconfigured ACL to the switchport on a switch and, in some cases, to the specific client session on a switchport that has multiple active sessions concurrently. The NAC solution permits the access of this session while also applying the preconfigured dACL that may limit what that session can access. In the network printer use case example previously introduced, the printer's session on the switch's switchport can have a dACL applied that only allows the printer to access the print server subnet.

- **VLAN override:** When designing and deploying a NAC architecture, some deployments find value in preconfiguring a dead VLAN (some use the 666 VLAN identifier) as the default access VLAN on all switchports. By default, this VLAN has either limited access or no access to anything on the network, other than what is required to properly authenticate. Once a client has properly authenticated, the NAC solution can dynamically override the preconfigured VLAN to a VLAN that has the authorized access the session requires.

- **Voice domain:** When authenticating and authorizing VoIP phones via a NAC solution, for voice traffic to properly function, get the appropriate QoS markings end to end, there has to be a way to identify the traffic of a device as voice traffic. This is where the voice domain option can be leveraged. Once a VoIP phone has been properly authenticated, via either 802.1X or MAB, the NAC solution can assign it to the voice domain that is configured on the switch.

From a network design perspective, how a resource is authenticated can also be an important characteristic that can be leveraged in the authorization process to potentially segment or validate that resource in a specific situation for that business. A production, real-world example of this is designing a NAC deployment where all owned and managed devices by the business are authenticated via EAP-TLS. Here these devices are in a specific internal identity source, such as Microsoft Active Directory, that validates these devices are owned and managed by the business. Because these devices are managed by the business, it is most often easier to deploy identity certificates, which are a requirement for EAP-TLS authentication. At this point, the NAC design can provide access based on the device and user information in the internal identity source. This could be limited access by leveraging a dACL that restricts specific protocols, changing the configured VLAN with the VLAN override capability, or in some cases providing full network access.

**10**

In addition, this business has a bring your own device (BYOD) policy that allows business users to connect their personal devices to the corporate network. With this real-world example, all business-owned resources would be authenticating via EAP-TLS, which is easy to identify from a network design perspective. For these BYOD devices, though, the business does not own or manage them, so EAP-TLS would in most cases not be the best option (it would be a management nightmare). Here is where the business could leverage PEAP-MSCHAPv2 for all BYOD devices and then limit the access provided to Internet-only access (guest access) or some limited subnet of full network access. Now, from a network design perspective, we can easily see what are business-owned devices (all EAP-TLS authenticated devices) and what are BYOD devices (all PEAP-MSCHAPv2 authenticated devices), while ensuring proper authentication and authorization of all devices and meeting business outcomes and business success.

## Visibility

In regard to network access control and identity management, when we are talking about visibility, we are not talking about real-time visibility, real-time traffic flows, or real-time analytics. When a NAC solution is properly designed and deployed, the business starts to see where devices, users, and resources live, when they are connecting to the network, from which locations they are connecting to the network, how they are authenticating, and what is being denied. For example, without a NAC solution deployed, a corporate business might inadvertently allow a gaming system to connect to its network. This just might be something that occurs regularly and has the potential to cause a business impact if that gaming system is allowed to use all available bandwidth. With a NAC solution, this access can be denied from the start. On the flip side, if this wasn't a corporate business but was a higher education campus environment, the college might want to allow the gaming system to connect to the network while also limiting the available bandwidth the gaming system can consume.

Another great example of this visibility is third-party network switches and hubs. In a corporate environment, these devices can run wild, uncontrollable at times, unless you deploy port security, which can also be a large management overhead. Instead, if properly leveraging a NAC solution, these third-party switches and hubs can be denied access to the network right from the start.

The visibility provided by a NAC deployment is really the first step toward real-time visibility.

# Summary

Security and the network design focus on security are extremely imperative. We've covered security in two dedicated chapters in this book now, the other being Chapter 4, where we covered security design. Security is interweaved and overlaid on top of every network design, solution, and architecture. Every security-focused network design decision will have its trade-offs to account for, and we can most assuredly take security too far where it causes the network to be unusable by the business users. We can also not take security far enough, leaving vulnerabilities, attack surfaces, and architecture points wide open. There has to be a happy middle ground where the network and the data on top of it are being properly secured to maintain its confidentiality, integrity, and availability while not limiting the ability of the business and its users to complete their approved functions.

The topics covered in this chapter focused primarily on infrastructure- and device-level security. If a device hasn't been properly hardened (locked down), it is much easier to compromise. The simple task of removing, or disabling, all services, protocols, and functions that are not being utilized can save the entire network from a catastrophic outage. In addition, leveraging different security-focused protocols to help secure not only the device but also the Layer 2 and Layer 3 control planes limits the overall attack surface of the infrastructure.

Having purpose-built security devices that provide specific security capabilities such as firewall, IDS, and NAC helps secure not only what is being connected to the LAN but also what is being allowed into the perimeter of the network. Understanding these components and their capabilities is a requirement for all network designers. Knowing the role a NAC solution plays, that it is an infrastructure service like DNS, DHCP, NTP, and so on, allows a network designer to ensure it is properly designed and the network it supports is properly designed for it.

# Reference

Al-shawi, Marwan, *CCDE Study Guide* (Cisco Press, 2015)

# Exam Preparation Tasks

As mentioned in the section "How to Use This Book" in the Introduction, you have a couple of choices for exam preparation: the exercises here, Chapter 18, "Final Preparation," and the exam simulation questions in the Pearson Test Prep Software Online.

# Review All Key Topics

Review the most important topics in this chapter, noted with the Key Topic icon in the outer margin of the page. Table 10-6 lists a reference of these key topics and the page numbers on which each is found.

**Key Topic**

**Table 10-6**   Key Topics for Chapter 10

| Key Topic Element | Description | Page Number |
|---|---|---|
| Paragraph | Device-level protection and types of traffic | 282 |
| Figure 10-6 | Common Network Attacks and Mitigation Mechanisms | 294 |
| Table 10-5 | 802.1X EAP Types Comparison | 296 |

# Complete Tables and Lists from Memory

Print a copy of Appendix D, "Memory Tables" (found on the companion website), or at least the section for this chapter, and complete the tables and lists from memory. Appendix E, "Memory Tables Answer Key," also on the companion website, includes completed tables and lists to check your work.

# Define Key Terms

Define the following key terms from this chapter and check your answers in the glossary:

transit IP traffic, receive IP traffic, exception IP traffic, non-IP traffic, data plane, control plane, management plane, supplicant, authenticator, authentication server, Extensible Authentication Protocol (EAP), EAP Message Digest Challenge (EAP-MD5), EAP Transport Layer Security (EAP-TLS), EAP Tunneled Transport Layer Security (EAP-TTLS), Protected Extensible Authentication Protocol (PEAP), EAP Flexible Authentication via Secure Tunneling (EAP-FAST), Lightweight Extensible Authentication Protocol (LEAP), MAC Authentication Bypass (MAB)

**10**

# CHAPTER 11

# Wireless

## This chapter covers the following topics:

- **IEEE 802.11 Standards and Protocols:** This section covers the different wireless standards and protocols, with the associated network design aspects.

- **Enterprise Wireless Network Design:** This section covers some of the most common enterprise wireless design use cases, their requirements, and associated implications.

In today's world of technology, wireless networks are more prevalent and all around us. If you are traveling, going through airports, staying at hotels, and riding trains, wireless networks are there. If you are visiting hospitals, malls, convention centers, or even attending concerts or sporting events, wireless networks are there.

With all of these wireless networks, the expectation or unstated requirement, as discussed in Chapter 1, "Network Design," is that the network must work. When you connect to the airport wireless network, you expect to be able to join a work call or watch that new show you've been dying to catch up on, on your favorite streaming service. When you are watching your favorite sports team in their first home game of the season, you expect the wireless network to work so you can post a social message on your favorite social media platform about your favorite player and the play they just made.

All of this is to say, when we think about network design and the CCDE today, we have to include wireless network design. Network designers need to know how to design a wireless network for high density, voice and video, high availability, and many more use cases. In this chapter, we are going to touch on the most common wireless network topics and capabilities, but understand that this is, as always, not an all-inclusive list. The topic of wireless network design, with all of its facets and capabilities, warrants its very own dedicated network design book.

This chapter covers the following "CCDE v3 Core Technology List" sections:

- 6.1 IEEE 802.11 Standards and Protocols
- 6.2 Enterprise wireless network

## "Do I Know This Already?" Quiz

The "Do I Know This Already?" quiz allows you to assess whether you should read this entire chapter thoroughly or jump to the "Exam Preparation Tasks" section. If you are in doubt about your answers to these questions or your own assessment of your knowledge of the topics, read the entire chapter. Table 11-1 lists the major headings in this chapter and their corresponding "Do I Know This Already?" quiz questions. You can find the answers in Appendix A, "Answers to the 'Do I Know This Already?' Quizzes."

**Table 11-1** "Do I Know This Already?" Section-to-Question Mapping

| Foundation Topics Section | Questions |
|---|---|
| IEEE 802.11 Standards and Protocols | 1–3 |
| Enterprise Wireless Network Design | 4–9 |

**CAUTION** The goal of self-assessment is to gauge your mastery of the topics in this chapter. If you do not know the answer to a question or are only partially sure of the answer, you should mark that question as wrong for purposes of the self-assessment. Giving yourself credit for an answer you correctly guess skews your self-assessment results and might provide you with a false sense of security.

1. Which of the following wireless client specifications is helpful in designing the size of an AP cell?

   a. The antenna gain

   b. The receiver sensitivity

   c. The number of spatial streams

   d. Client's software operating system

2. A high-density area in a wireless design is determined by which one of the following statements?

   a. More clients are using the 5-GHz band than the 2.4-GHz band.

   b. A small number of clients in an area are using high-bandwidth applications.

   c. A higher number of clients are associated with each AP in an area.

   d. A higher amount of RF coverage is needed in an area.

3. Suppose a customer wants users on their wireless network to authenticate with a username and password before being allowed wireless network access. Which of the following items could be leveraged to add this security requirement in your wireless network design to meet the customer's needs?

   a. RADIUS

   b. TACACS

   c. SMNP

   d. AES servers

4. In a data-only wireless deployment (non-real-time traffic), which one of the following statements is true?

   a. Strict jitter requirements must be met.

   b. Strict latency requirements must be met.

   c. Strict packet loss requirements must be met.

   d. No specific requirements must be met.

**5.** Suppose a customer wants to use a real-time application that requires jitter to be less than 30 milliseconds. Which one of the following wireless network deployment models should be leveraged for this wireless network design requirement?

    **a.** Data deployment model

    **b.** High-density deployment model

    **c.** Voice deployment model

    **d.** There is not enough information given to determine a deployment model.

**6.** Suppose you are working on a wireless network design that is required to support a high density of clients within a large lecture space at a college. You begin by adjusting the AP's transmit power level down to its lowest setting, but you find that the AP's cell size is still too large for your design. What is the next step to take for you to reduce the AP's cell size?

    **a.** Use an external omnidirectional antenna.

    **b.** Use an external patch antenna.

    **c.** Install a second AP next to the first one and use the same channel on each.

    **d.** Enable the lowest data rate to reduce the cell size.

**7.** Which of the following statements are valid design goals for wireless network design that will support voice over the wireless network for calls? (Choose all that apply.)

    **a.** Make 12 Mbps the lowest mandatory data rate.

    **b.** Design for call capacity per AP.

    **c.** Use every possible 5-GHz nonoverlapping channel.

    **d.** Consider avoiding 5-GHz DFS channels.

**8.** When you meet with a customer to gather information about an upcoming wireless project, which one of the following items would be the most helpful as you prepare to design the wireless solution?

    **a.** The scope of the project

    **b.** A list of the buildings and locations that need wireless service

    **c.** Floor plans of buildings that need wireless service

    **d.** Diagrams of the physical network infrastructure

**9.** Regarding a wireless client in relation to the AP it's connecting too, which of the following statements is incorrect?

    **a.** As the client moves away from the AP, the AP's signal strength decreases.

    **b.** As the client moves away from the AP, the usable data rate decreases.

    **c.** As the client moves toward the AP, the SNR increases.

    **d.** As the client moves toward the AP, the usable data rate decreases.

## Foundation Topics

# IEEE 802.11 Standards and Protocols

When talking about wireless networks and designing them, we as network designers must understand the different 802.11 standards and protocols, as each has its own capabilities, pros, and cons. Network designers need to be able to compare all of these standards and protocols to then align them with the end-user device requirements. Keep in mind that not all end-user devices support all of these standards and protocols, so you will be required to identify the standard and protocol that meets the needs of everything in the design, down to the least-common denominator.

## Device Wireless Capabilities

When you begin the wireless design process, you need to find out the associated capabilities of the devices within the customer's network. Again, not all devices will have the same functionality. There are also scenarios where you and the customer will not be able to identify what devices will be used, such as in the bring your own device (BYOD) model. The types of information you need to collect include 802.11 support, radio frequency (RF) capabilities, and supported security suites. The information provided will help you determine what the common standard protocol is and allow you to design a wireless network that can ensure all the customer's devices will be supported.

For example, suppose a SaaS customer has hired you to design its campus HQ wireless network. The customer's device inventory includes wireless 4K security cameras, wireless Voice over IP (VoIP) devices, standard laptops, and high-definition tablets, all with their own specific wireless capabilities. All of these devices are critical to the success of the SaaS customer and its overall business, so your wireless network design must support all of the devices completely. Table 11-2 summarizes the associated wireless capabilities that have been collected that these devices support.

**Table 11-2**  Example SaaS Device Wireless Capabilities

|  | 4K Security Camera | Wireless VoIP | Standard Laptops | High-Def Tablets |
|---|---|---|---|---|
| Bands supported | 2.4 GHz, 5 GHz | 2.4 GHz, 5 GHz | 2.4 GHz, 5 GHz | 2.4 GHz, 5 GHz |
| 802.11 | a/b/g/n/ac | a/b/g/n | a/b/g/n | a/b/g/n/ac |
| Channels: 2.4 GHz <br> Channels: 5 GHz | 1 to 13 <br> 36–48, 52–64, 100–140, 144, 149–161 | 11 channels <br> 20 channels | 1 to 13 <br> 36–48, 52–64, 100–140, 149–165 | 1 to 13 <br> All, but DFS not recommended |
| Channel width | 2.4 GHz: 20, 40 MHz <br> 5 GHz: 20, 40, 80 MHz | 2.4 GHz: 20 MHz <br> 5 GHz: 20, 40 MHz | 2.4 GHz: 20 MHz <br> 5 GHz: 20, 40, 80 MHz | Unspecified |

**11**

| | 4K Security Camera | Wireless VoIP | Standard Laptops | High-Def Tablets |
|---|---|---|---|---|
| Data rates<br>802.11b<br>802.11g<br>802.11a<br>802.11n<br>802.11ac | Unspecified | Unspecified | 1–11 Mbps<br>6–54 Mbps<br>6–54 Mbps<br>MCS 0–7<br>MCS 0–8 | 1–11 Mbps<br>6–54 Mbps<br>6–54 Mbps<br>MCS 0–6<br>— |
| 802.11k<br>802.11r(FT)<br>802.11w | Not supported<br>Not supported<br>Not supported | Supported<br>Supported<br>Supported | Unspecified<br>Supported<br>Unspecified | Unspecified<br>Unspecified<br>Unspecified |
| Spatial stream | 2×2 | Unspecified | Unspecified | 1×1 |
| Antenna gain | Unspecified | Unspecified | 2.4 GHz: 2.4 dBi<br>5 GHz: 3.0 dBi | Unspecified |
| Transmit power (max, U.S.) | Unspecified | 16 dBm | 2.4 GHz: 13 dBm<br>5 GHz: 12dBm | 2.4 GHz: 13 dBm<br>5 GHz: 12dBm |

Taking a look at the 802.11 support among these devices, you can easily see that all of these critical devices support both the 2.4-GHz and 5-GHz bands. Furthermore, you can see that two of the four devices support 802.11a/b/g/n, while the other two add support for 802.11ac. This is important information to ensure the access points (APs) that you select in the design provide the required data rates at the corresponding 802.11 level. Next you look at the different channels and data rates each device supports. The key here is that you want to leverage the highest data rate that all devices support, and in most cases disable the lower ones so they are not leveraged. If you leave the lower data rates available, it will limit the overall performance of the access point as devices connect to it on those rates as well as the higher rates. Just keep in mind that a lot of legacy devices only support the lower data rates, and if a design disables them, those devices will not be able to connect to the wireless network.

In some wireless designs it may be prudent to create different **service set identifiers (SSIDs)**, with different APs advertising them for these legacy devices that only operate at the lower data rates. An SSID is used as a wireless network name and can be made up of case-sensitive letters, numbers, and special characters. When designing wireless networks, we give each wireless network a name which is the SSID. This allows end users to distinguish from one wireless network to another.

Continuing with this example scenario, the channels offered may be different as some countries only allow specific channels, and thus those devices will only operate in that specific country. For example, if a device only supports 2.4-GHz channels 1 through 11 (the wireless VoIP devices in this scenario), it will only be able to operate in the United States. The Table 11-2 column for high-definition tablets refers to a feature called **Dynamic Frequency Selection (DFS)**. DFS enables an AP to dynamically scan for **Radio Frequency (RF)** channels and avoid those used by other radio devices in the area. RF is a wireless electromagnetic signal used as a form of data communication.

For best performance in a wireless environment, wireless devices should be able to distinguish received signals as legitimate information they should be listening to and ignore any background signals on the spectrum. **Signal-to-noise ratio (SNR)** ensures the best wireless functionality. SNR is the difference between the received wireless signal and the noise floor. The noise floor is erroneous background transmissions that are emitted from either other devices that are too far away for the signal to be intelligible, or by devices that are inadvertently creating interference on the same frequency. For example, if a client device's radio receives a signal at -75 dBm, and the noise floor is -90 dBm, then the effective SNR is 15 dB. This would then reflect as a signal strength of 15 dB for this wireless connection. Generally, an SNR value of 20 dB or more is recommended for data networks, whereas an SNR value of 25 dB or more is recommended for a wireless network supporting real-time application traffic.

Shifting capabilities to RF, the transmit power level of one device might be very different from that of another device, as shown in Table 11-2. This difference is directly related to the size of the battery in the device—the smaller the device, the smaller the battery, and less power the device can leverage to transmit a signal. The location of the device's antenna can also have an impact on the RF performance. Some devices have embedded antennas, while other devices have external antennas, which allows them to be extended as needed. Lastly, the placement of the device in relation to access points, with obstacles obstructing the path, can degrade and limit the connections.

## Wireless Security Capabilities

In the majority of situations, customers will require the most secure network. Wired networks have some level of inherent security because the data transmissions are contained within the physical wire. In contrast, wireless networks transport data over the air, potentially allowing other devices to eavesdrop or manipulate the traffic.

When conducting a discovery session to identify wireless network design requirements, a network designer should ask the following questions:

- How should access to the wireless network be secured?

- How should data traveling over a wireless LAN be secured?

- Do wireless users need to be protected from each other?

- How will wireless users be protected from everyone else?

The 802.11 standard defines two methods of authentication: open and WEP. All devices should support both methods. WEP has been deprecated due to its inherent security weakness. Open authentication is used in conjunction with 802.1X (network access control [NAC]) security and multiple EAP options to offer a wide range of authentication methods, which was discussed in Chapter 10, "Security." Table 11-3 shows the different security specifications for the four device types we have been reviewing.

11

**Table 11-3**  Example SaaS Device Wireless Security Capabilities

|  | 4K Security Camera | Wireless VoIP | Standard Laptops | High-Def Tablets |
|---|---|---|---|---|
| WEP/WPA/TKIP Security | WPA (AES), WPA2 (AES), WPA-Enterprise, WPA2-Enterprise | WEP, TKIP, AES-CCMP, WPA (AES), WPA2 (AES), WPA-Enterprise, WPA2-Enterprise | WPA (AES), WPA2 (AES), WPA-Enterprise, WPA2-Enterprise | WPA (AES), WPA2 (AES), WPA-Enterprise, WPA2-Enterprise |
| EAP (802.1X) Security | EAP-TLS, EAP-PEAP, EAP-TTLS (PAP, CHAP, MSCHAP, MSCHAPv2) | LEAP, PEAP, MSCHAPv2, EAP-FAST, EAP-TLS | EAP-FAST, PEAP-GTC, PEAP-MSCHAPv2, EAP-TLS | PEAP-MSCHAPv2, EAP-TLS |

Reviewing the device security capabilities in Table 11-3, each device can support a comprehensive list of wireless security features and protocols. What you should notice is that these features and protocols are not the same across all devices. Network designers must be aware that not all devices support all security features, protocols, or options, and identify the proper security mechanisms that all devices support. Securing access to a wireless network involves applying one of the authentication methods in Table 11-3 to screen the users and devices that try to join.

During the wireless design process, you should list the wireless networks that the customer wants, along with the security mechanisms that should be leveraged to ensure the wireless network is secured. For example, a wireless network meant for the standard corporate laptops might require WPA2-Enterprise with AES, 802.1X with EAP-TLS, and digital certificates. Another wireless network offered to the high-def tablets might require PEAP-MSCHAPv2. Keep in mind, with most wireless products, a wireless SSID cannot run multiple security mechanisms simultaneously (i.e., WPA2-Enterprise and 802.1X cannot be used together). This limitation adds a constraint to any wireless network design and has the potential impact to add a number of SSIDs as the security mechanisms required in a design increase.

## Wireless Traffic Flow

Network designers need to keep in mind that end-user device traffic flow on a wireless network can be very different compared to the wired side. In the context of a wireless network, the wired network is also known as a *distribution network*. The distribution network provides the means of connecting multiple APs together. In most commercial wireless network designs, the distribution system is already designed and implemented, enabling the wireless APs to leverage it. The way an end user device's traffic will flow highly depends on the configuration mode of the access point. The following are the two AP modes:

■ **Autonomous:** In this mode the AP acts as a direct link between the wireless and wired sides of the network.

■ **Centrally controlled:** In this mode all wireless client traffic is forwarded to the wireless controller via a CAPWAP tunnel, discussed next. CAPWAP is defined in RFC 5415.

**NOTE**   Some vendor solutions have additional features that allow wireless traffic to be locally switched at the access point. For example, Cisco has the FlexConnect option within the Wireless LAN Controller (WLC) configuration that allows for this exact forwarding behavior.

**NOTE**   The following CAPWAP discussion highlights the Cisco-specific implementation as an example case study.

**Control and Provisioning of Wireless Access Points (CAPWAP)** is a logical network connection between access points and a wireless LAN controller. CAPWAP is used to manage the behavior of the APs as well as tunnel encapsulated 802.11 traffic back to the controller. CAPWAP sessions are established between the AP's logical IP address (obtained through DHCP) and the controller's management interface. Whether in local mode or FlexConnect mode, CAPWAP sessions between the controller and AP are used to manage the behavior of the AP. When in local mode, CAPWAP is additionally used to encapsulate and tunnel all wireless client traffic so that it can be centrally processed by the controller. CAPWAP sessions use UDP for both the control and data channels, as follows:

■ **CAPWAP Control Channel:** Uses UDP port 5246

■ **CAPWAP Data Channel:** Uses UDP port 5247 and encapsulates (tunnels) the client's 802.11 frames

Figure 11-1 illustrates the different CAPWAP channels between an AP and a controller.

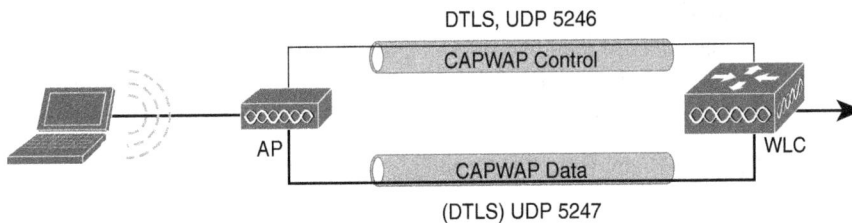

**Figure 11-1**   *Cisco CAPWAP Control and Data Plane Channels*

11

FlexConnect mode has options for local or central forwarding of traffic. A wireless design can use FlexConnect with central forwarding for a branch site AP that should tunnel traffic to the controller but should also be able to tolerate a WAN failure situation. APs are typically deployed in their own VLAN in a campus environment, or in multiple VLANs if leveraging a Layer 3 access design. Like most devices, it's a good idea to isolate APs.

If there is a security device that limits what traffic is allowed within the path between the access point and the wireless LAN control, the corresponding ports/protocols listed previously will need to be added to ensure proper communication of the CAPWAP tunnel.

As the number of APs grows, so does the number of CAPWAP tunnels terminating on the controller. Figure 11-2 illustrates the logical connection of multiple CAPWAP sessions over the physical infrastructure.

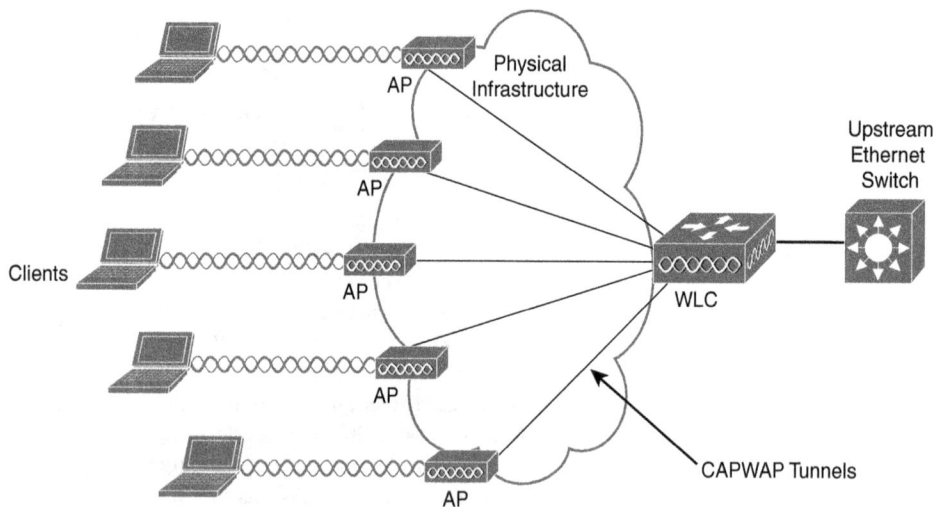

**Figure 11-2**   *Cisco CAPWAP Sessions Between the APs and the Controller*

After reviewing Figure 11-2, load and scalability concerns might arise. Considering that all APs in local mode use a CAPWAP tunnel to transport wireless data traffic back to the controller, designing both the physical environment and the logical environment to handle the load and scale is paramount. When designing a network, keep in mind that the CAPWAP tunnels are riding on top of a physical network that will have bandwidth and resource availability limitations. With this in mind, you can leverage your top-down design mindset to carefully analyze the performance requirements to ensure the wireless network can meet the needs of the business. This can include the following design aspects; while some of them are the physical underlay, they have a direct impact on the overlay wireless CAPWAP tunnel:

- The physical connection between the AP and the access switch

- An estimation of oversubscription of the uplink of the access switch to the network

- Backbone capacity of the core network

- WAN connection speeds if the controllers are centralized and APs are in local mode

- Network access speeds to the controller

- Performance capabilities of the controller

Another area where the logical path requires careful consideration is the path between the controller and the key infrastructure services, such as the AAA and DHCP servers. Additional infrastructure services, including AAA/NAC, DHCP, DNS, SDN controller, and many more, may be placed at locations throughout the network that have firewalls protecting them. Understanding the logical path between these services will often require opening of firewall rules for the service to interface with the controller.

As with CAPWAP, the wireless controller's management interface is used to communicate with AAA/NAC servers, as well as a host of other services, including directory servers, other controllers, and more.

For DHCP, a controller proxies communication to the DHCP server on behalf of clients using the controller's IP address in the VLAN associated to the WLAN of those clients.

Table 11-4 summarizes the ports that must be open to allow the controller to communicate with key services.

**Table 11-4**  Summary of Cisco Wireless LAN Controller Services Ports

| Service | Port |
| --- | --- |
| RADIUS authentication and authorization | UDP port 1812 (some older versions use UDP port 1645) |
| RADIUS accounting | UDP port 1813 (some older versions use UDP port 1646) |
| DHCP server | UDP port 67 |
| DHCP client | UDP port 68 |

# Enterprise Wireless Network Design

From an enterprise wireless network perspective, we have to shift our focus from what has been the norm for years to a more modern state. Most, if not all, wireless networks are high density. Historically, this used to be location specific—stadiums, convention centers, and higher education facilities being the most common high-density wireless networks. Today, the list includes, among many others, enterprise networks, airports, and hospitals, and that list is growing longer as more businesses embrace the wireless network. These days, nearly everything is high density!

## Designing for High Density

What does *high density* mean from a wireless perspective? What makes a high-density wireless network different from a standard wireless network? There are two main differentiators with a high-density wireless network: the number of users within a certain area (density) and coverage area. High-density networks ensure clients are able to have sufficient throughput regardless of the thousands of other clients within close proximity. This is one of the primary issues in venues such as stadiums and auditoriums where massive numbers of users and devices need wireless access within close proximity of each other.

11

In simple terms, *high density* means having more clients attached to the same AP, or set of APs, within a smaller physical location. A high-density wireless design is focused on RF coverage maximizing the effective **AP cell** size for the area in question. Let's break this down into the key RF relationships in high-density wireless design to help design a proper solution:

**Key Topic**

- **AP to client:** How clients hear APs

- **Client to AP:** How APs hear clients

- **AP to AP:** How APs hear each other

This all starts with Layer 1 of the OSI model, RF design. In a wireless world, this is all about the antenna: determining which type of antenna should be leveraged and where it should be placed. For antenna selection, we need to consider the density of clients that we want that AP to serve and how far away the AP can be placed and still enable clients to connect to it. For the antenna placement, we want to ensure clear line of sight to clients; that is, ensure there are no obstacles in the way. Identifying strategic locations can be the best tool for determining where to place antennas. Think of locations where users congregate and consider offering them direct frequencies that provide better throughput and signal strength.

If a wireless-enabled device can see the wireless network, it should be able to join it within the coverage area. Taking a look at Figure 11-3, the left side shows this example for two devices within the AP cell size as defined in the figure.

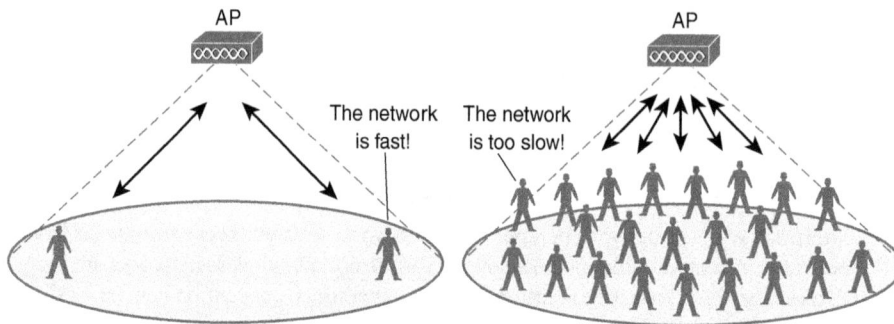

**Figure 11-3** *High-Density Wireless Example*

As more devices and clients move within the coverage area of the AP, increasing the density of clients, they will all compete for the available wireless network resources, including the physical infrastructure resources supporting the APs, such as bandwidth. This is shown in Figure 11-3 on the right side.

**Key Topic**

From a wireless design perspective, the best way to provide sustained performance to a high density of clients is to predetermine and distribute clients across multiple APs and channels. When doing this, we also have to limit the cell size of each AP, to limit the number of clients that can connect. Because we are limiting the cell size, we also need to increase the number of APs being leveraged. Figure 11-4 and Figure 11-5 represent the network design scenario of an auditorium, showing the number of APs and their respective cell sizes. The design in Figure 11-4 has only two APs, both with very large cell sizes. The design in Figure 11-5 has 13 APs, with very small cell sizes.

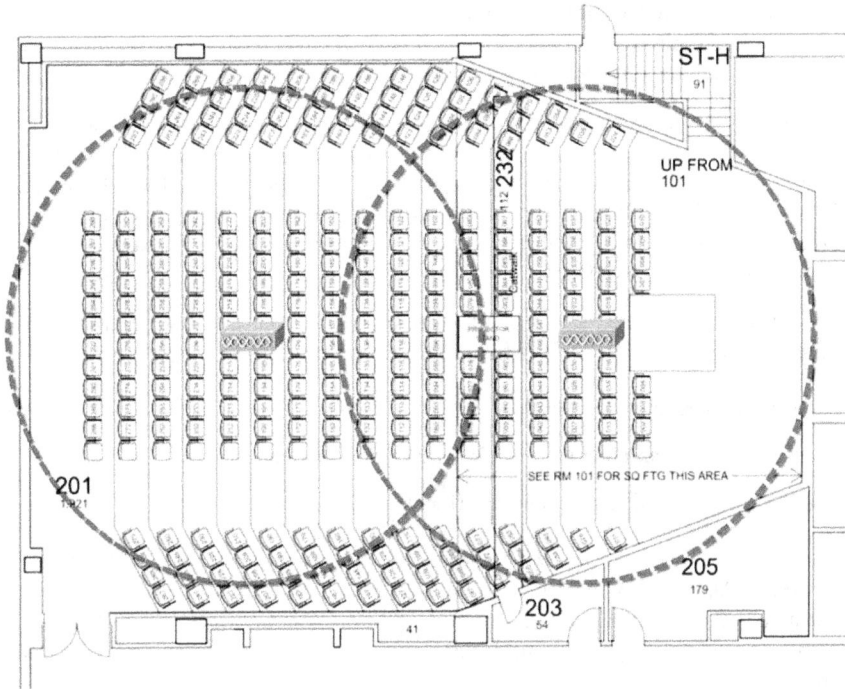

**Figure 11-4**  *High Density of Users in an RF Design Based on Coverage Only*

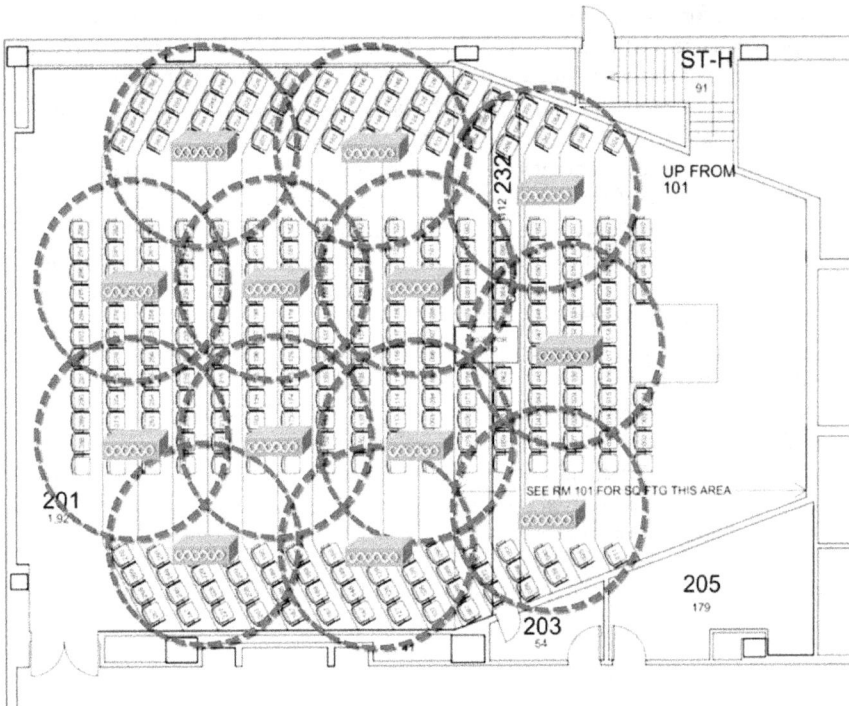

**Figure 11-5**  *High Density of Users in an RF Design Based on High Density*

In addition to the wireless footprint being added with a high-density deployment, we also have to keep in mind the physical network implications. Each AP needs to connect back to a switch to provide the connected clients access to the physical network.

From a design perspective, we can easily identify when and where we will need to add APs. Think of areas such as theaters, auditoriums, stadiums, and classrooms, with configured seating, which allows us to plan for the number of high-density clients properly. Identifying density in office spaces is also easy, and they usually have a lower client density because of their seating arrangements and layout. In addition, when designing any wireless network, we must consider that the average user carries three to five wireless devices.

Each vendor's hardware will have its own methods to reduce the cell size, but generally you can follow these three actions no matter which vendor's wireless APs are being leveraged:

■ Limit the AP's transmit power level.

■ Select an appropriate antenna for the AP, with the correct gain.

■ Reduce the size of a cell using omnidirectional antenna.

You should focus on leveraging the 5-GHz band in high-density areas, making sure that AP channels are nonoverlapping. You will need to take advantage of using more channels within a small dense area. AP cells should have 10 to 15 percent overlap. Leverage Cisco's Dynamic Transmit Power Control (DTPC) to automatically influence the transmit power levels of compatible clients.

We want to focus on maximizing the spectrum and limit unnecessary noise. This is why you should always aim for a single SSID in a high-density area. The more SSIDs in the area, the worse the performance for the end clients. This is because each SSID requires a separate beacon and will beacon at the minimum mandatory data rate. Each broadcast SSID will also respond to null probe requests. All of this noise is just exponential amounts of airtime wasted for real client traffic. Recall from earlier in this chapter that an SSID cannot leverage different authentication methods. This affects a network design decision that is very impactful to the overall performance of the design. If there is a requirement, or multiple requirements, to have different security mechanisms for different device types, you will be required to create multiple SSIDs, which will have limited performance in a high-density deployment.

## Designing for Voice and Video

Most user devices will make use of what is normally called "data" traffic that has no special requirements (QoS) or expectations other than basic responsive throughput. For the occasions where users are leveraging a real-time application like VoIP, videoconferencing, or collaboration, the network must handle the traffic differently. Real-time applications require special consideration as you design, configure, and operate a wireless network. The following effects need to be minimized so that voice and video sessions can be heard and seen consistently, without interruption or corruption:

**Key Topic**

■ **Latency:** The amount of time required to deliver a packet or frame from a transmitter to a receiver

■ **Jitter:** The variance of the end-to-end latency experienced as consecutive packets arrive at a receiver

■ **Packet loss:** The percentage of packets sent that do not arrive at the receiver

To keep these factors minimized, the adverse conditions in a wireless environment must be controlled. For example, a source of interference can cause packet errors that interrupt a voice or video stream. As packets are lost, they can be retransmitted and delayed, increasing latency and jitter. Other factors like poor RF coverage, high channel utilization, and excessive collisions can also impede good data throughput and integrity.

When designing for wireless voice and video, it is best to leverage the 5-GHz band as much as possible. The 5-GHz band includes more channels that do not overlap, which reduces the change of interference from non-wireless devices. AP cell efficiency should be at the forefront when designing for voice and video. Lower data rates require more time to transmit the same data than a higher data rate. When that lower data rate is being leveraged, it takes up the channel longer than the higher data rate. A better approach is to limit data transmissions to use the higher data rates so that stations transmit data quicker and get off the air sooner, allowing for other stations to leverage that same channel. This directly affects latency and jitter, which, as discussed previously, are critical to real-time applications.

An effective design should take call capacity into account. For example, voice calls use bidirectional RTP streams to transport audio. Each call uses two separate streams, but they cannot be transmitted simultaneously because of channel contention. At a 24-Mbps data rate, up to 27 simultaneous bidirectional RTP streams can exist, or up to 13 calls. A 6-Mbps data rate can support only 13 streams or 6 calls. Therefore, the maximum number of calls depends on the data rate used, as well as the channel utilization.

## Summary

This chapter covered various wireless network design topics, including IEEE 802.11 standards and protocols and enterprise wireless network use cases. To avoid wireless design defects, network designers need to always incorporate the wireless network design topics covered in this chapter in an integrated holistic approach rather than designing in isolation. Moreover, considering the top-down design approach is a fundamental requirement to achieving a successful business-driven design. Although this chapter focused on the wireless side, there are design implications within the physical infrastructure that also need to be kept in mind. For example, as you add more APs to solve wireless requirements (high density of users), each of these APs still requires physical network connectivity and power. These are all factors that must go into your wireless network design decision process.

## Reference

Al-shawi, Marwan, *CCDE Study Guide* (Cisco Press, 2015)

## Exam Preparation Tasks

As mentioned in the section "How to Use This Book" in the Introduction, you have a couple of choices for exam preparation: the exercises here, Chapter 18, "Final Preparation," and the exam simulation questions in the Pearson Test Prep Software Online.

11

## Review All Key Topics

Review the most important topics in this chapter, noted with the Key Topic icon in the outer margin of the page. Table 11-5 lists a reference of these key topics and the page numbers on which each is found.

**Key Topic**

**Table 11-5**  Key Topics for Chapter 11

| Key Topic Element | Description | Page Number |
|---|---|---|
| List | Wireless design questions | 305 |
| List | Wireless AP modes | 307 |
| Paragraph | CAPWAP tunnel description | 307 |
| List | High-density wireless design key RF relationships | 310 |
| Paragraph | High-density wireless design model | 310 |
| List | Wireless voice and video real-time application requirements | 312 |

## Complete Tables and Lists from Memory

There are no Memory Tables or Lists in this chapter.

## Define Key Terms

Define the following key terms from this chapter and check your answers in the glossary:

service set identifiers (SSIDs), Dynamic Frequency Selection (DFS), Radio Frequency (RF), signal-to-noise ratio (SNR), Control and Provisioning of Wireless Access Points (CAPWAP), AP cel

# CHAPTER 12

# Automation

## This chapter covers the following topics:

- **Zero-Touch Provisioning:** This section covers zero-touch provisioning and the network design elements around it.

- **Infrastructure as Code:** This section covers Infrastructure as Code (IaC) and the corresponding network design elements relevant to the inherent capabilities IaC provides.

- **CI/CD Pipelines:** This section covers CI/CD pipelines and the network design elements and capabilities that come with them.

I'm sure some of you are wondering why automation is a topic in network design and on the CCDE certification exam. This is a network design certification, after all. For network engineers, it can be hard to grasp the impacts of automation on the network and, more importantly, on the business. Networks and the corresponding network designs have never been more complex. Leveraging automation can have a large impact on the network and the business but it highly depends on the underlying network design. Assuming a network has been automated, how does that affect the design of the network?

How long would it take to build out 12 data centers worldwide manually? The architecture and overall network design could be very similar for each data center, but if we are manually configuring each device, at each location, this daunting task could very well take multiple years. Let's just say it takes 12 months to complete. That's one data center each month.

Now think about building these 12 data centers with automation. Every component, configuration, feature, capability, and functionality is templated with corresponding variables. Let's assume it takes a month to fully template all of these elements out, figure out the automation workflows and orchestration process, and we start building the data centers in month 2. By taking advantage of automation, the business can build multiple data centers at the same time, with the same resources. What took 12 months manually can be completed in 3 months using automation, in some cases even faster.

With a network built, automation can further be leveraged to complete operations and maintenance (O&M) tasks, troubleshoot network issues and resolve them, instantiate business intent end to end—and many more capabilities are being identified every day.

In addition to the increase in build time, automation limits user errors, reduces total cost of ownership, reduces network outages, and increases service agility. This is the business impact of automation, and why business leaders will require it within the design of their networking infrastructure.

This chapter focuses on the underlying impact of automation on a business and how a network designer can properly structure a network design for automation. This chapter does not cover how to build or write automation. Thus, this chapter will not cover programming languages, API calls, orchestration tools, data models, and so forth, but it will cover the specific capabilities of automation that network designers need to know to properly design a network.

This chapter covers the following "CCDE v3.0 Core Technology List" sections:

- 7.1 Zero-touch provisioning

- 7.2 Infrastructure as Code

- 7.3 CI/CD Pipeline

# "Do I Know This Already?" Quiz

The "Do I Know This Already?" quiz allows you to assess whether you should read this entire chapter thoroughly or jump to the "Exam Preparation Tasks" section. If you are in doubt about your answers to these questions or your own assessment of your knowledge of the topics, read the entire chapter. Table 12-1 lists the major headings in this chapter and their corresponding "Do I Know This Already?" quiz questions. You can find the answers in Appendix A, "Answers to the 'Do I Know This Already?' Quizzes."

**Table 12-1**  "Do I Know This Already?" Section-to-Question Mapping

| Foundation Topics Section | Questions |
|---|---|
| Zero-Touch Provisioning | 1, 2 |
| Infrastructure as Code | 3 |
| CI/CD Pipelines | 4, 5 |

**CAUTION**  The goal of self-assessment is to gauge your mastery of the topics in this chapter. If you do not know the answer to a question or are only partially sure of the answer, you should mark that question as wrong for purposes of the self-assessment. Giving yourself credit for an answer you correctly guess skews your self-assessment results and might provide you with a false sense of security.

1. Which of the following problems are resolved or reduced using zero-touch provisioning? (Choose two.)
   a. The amount of time needed to deploy new infrastructure
   b. Maintaining code across different versions
   c. Troubleshooting network outages caused by human error
   d. Increased time to market on new features, functionality, and capabilities.

   **2.** Which of the following are technical benefits of leveraging zero-touch provisioning? (Choose three.)

   **a.** All cabling is validated and correct.

   **b.** Day 1 configuration is loaded.

   **c.** ZTP-enabled devices upload their configuration when a save is issued.

   **d.** All devices are running the proper code version.

   **3.** Which of the following are problems resolved or reduced by leveraging infrastructure as code? (Choose two.)

   **a.** The amount of time needed to deploy new infrastructure

   **b.** Maintaining code across different versions

   **c.** Troubleshooting network outages caused by human error

   **d.** Increased time to market on new features, functionality, and capabilities.

   **4.** Which of the following are problems resolved or reduced by a CI/CD pipeline? (Choose three.)

   **a.** The amount of time needed to deploy new infrastructure

   **b.** Maintaining code across different versions

   **c.** Troubleshooting network outages caused by human error

   **d.** Increased time to market on new features, functionality, and capabilities.

   **5.** What are the four steps of a CI/CD pipeline?

   **a.** Source, build, test, deploy

   **b.** Build, test, code, deploy

   **c.** Build, test, deploy, monetize

   **d.** Source, test, build, monetize

# Foundation Topics

# Zero-Touch Provisioning

The overwhelming increased demand for automation has stressed the overall network design and architectures. Therefore, adding new services to an oversaturated and oversubscribed network fulfilling numerous tasks such as network management, quality of experience, and optimization is not feasible today. The answer to this problem comes in many forms, one of which is the automation of the most common daily tasks and responsibilities related to operations and maintenance of the network.

**Zero-touch provisioning (ZTP)** is based on software-defined network (SDN) solutions and network functions virtualization (NFV) concepts. The intent and outcome of a ZTP capability is to have any new network device fully configured automatically, in a plug-and-play situation. The benefits of ZTP are the following:

**Key Topic**

- All physical cabling is verified correct.

- All devices are running the proper code version.

- Pre-provision additional sites, pods, and devices ahead of time, before the business needs them.

■ All devices push their configurations back whenever a save is issued.

■ A ZTP controller can restore the latest configuration should a device return into the ZTP process.

In addition to reducing capital expenditure, ZTP is designed to reduce user errors and save time. Administrators can leverage automation tools to roll out the different aspects and components of the network devices, requiring no input from the network engineer once the process is started, which is normally when the device is first connected to the current production infrastructure. Figure 12-1 depicts a common ZTP provisioning process.

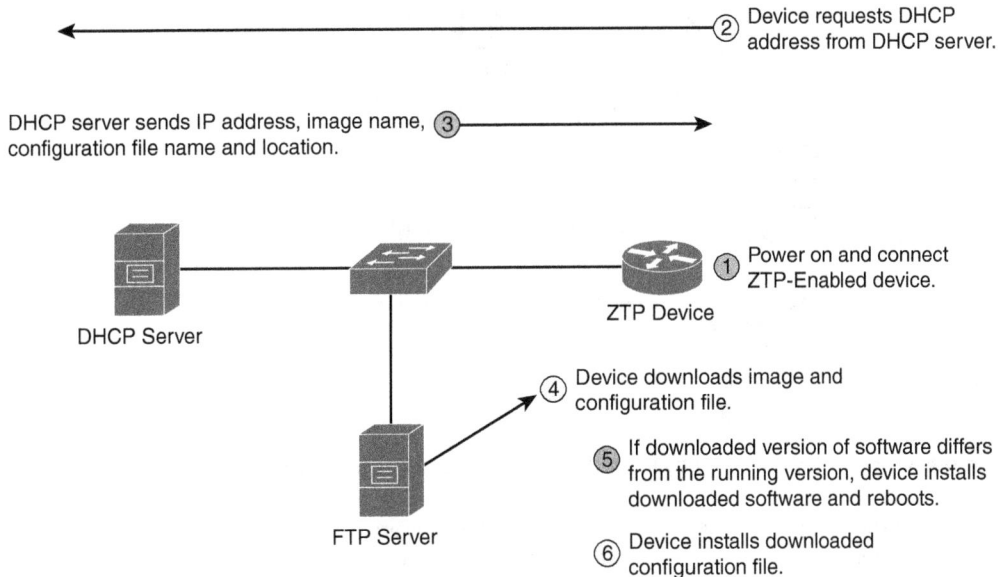

② Device requests DHCP address from DHCP server.

DHCP server sends IP address, image name, ③ configuration file name and location.

DHCP Server

ZTP Device

① Power on and connect ZTP-Enabled device.

④ Device downloads image and configuration file.

⑤ If downloaded version of software differs from the running version, device installs downloaded software and reboots.

FTP Server

⑥ Device installs downloaded configuration file.

**Figure 12-1**  *ZTP Provisioning Process*

**NOTE**   There must be a minimum amount of network infrastructure created to allow ZTP to communicate with the network devices it will configure. In most cases this means setting up a management network to allow all network devices to be automatically provisioned with ZTP via their out-of-band management interfaces.

## Infrastructure as Code

**Infrastructure as Code (IaC)** is a new approach to infrastructure automation that focuses on consistent, repeatable steps for provisioning, configuring, and managing infrastructure. Over the years, infrastructure has been provisioned manually. Deploying new capabilities, applications, and network devices used to take weeks, months, and even years to complete, which created a demand for a new process of provisioning and configuring devices that is more effective and more efficient. IaC fulfills that demand.

Infrastructure as Code is all about the representation of infrastructure through machine-readable files that can be reproduced for an unlimited amount of time. Figure 12-2 shows how leveraging IaC can streamline the building of multiple environments, one each for development, staging, and production.

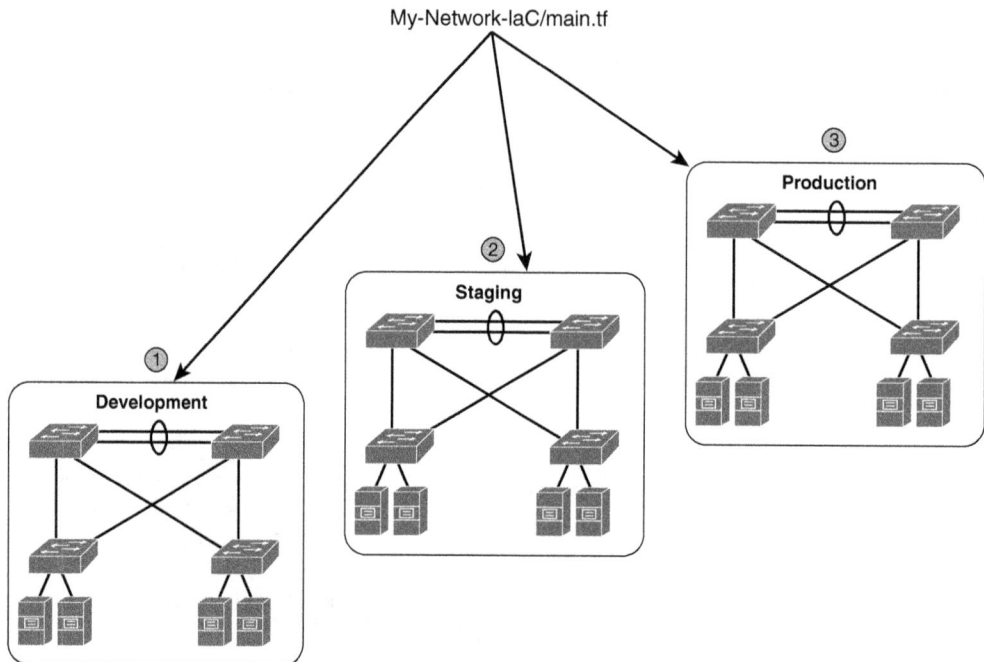

**Figure 12-2** *Leveraging IaC to Build Out Multiple Identical Environments*

Once these readable files that represent the infrastructure have been created, it is best to store them in a version control system like Git. Doing this allows these files to be updated and tracked, and a previous version can be easily rolled back if needed. Now, new infrastructure can be instantiated as needed, be it servers, virtual machines, or network devices, by pulling from Git or a similar version control system.

Being able to spin up new infrastructure in a repeatable, consistent fashion can be extremely useful for a business, such as in cases where an application is in large demand at a specific time. In such cases, all corresponding resources, servers, VMs, load balancers, firewalls, network devices, and so forth can be automatically instantiated to meet the high demand. Once the demand decreases, these resources can then be decommissioned and deprovisioned. This specific capability is called *elasticity* and is something all network designers should understand. Businesses will require this capability because it not only provides a very critical operational efficiency at the right time, but also reduces the long-term cost that manually adding resources would cause a business.

With Infrastructure as Code, we can read the code in the files to identify all the corresponding capabilities, characteristics, features, and functionality that will be instantiated when it gets provisioned. The entire team can review the code, make necessary changes, leverage version control to track changes, and ensure the code is compliant with security standards that the business is required to follow.

One of the main adoption concerns with IaC is that this approach welcomes, and even encourages, change. Historically, change was not something infrastructure teams and the

12

business could handle. The business would provide the infrastructure team a handful of change windows a year, and in most cases, each was very limited in time.

With IaC, change is a catalyst to improve the reliability and the performance of the infrastructure as a whole. Just as source code goes through multiple versions and becomes better with each new release, infrastructure becomes more resilient and reliable with each new version.

By using IaC, we can now have identical environments for development, staging, and production, as previously described. Each environment can be built the exact same way, leveraging the same automation and version control system in place. The different teams can properly test new features, functionality, and capabilities prior to rolling out in production, which leads to significantly fewer errors and troubleshooting steps.

Here are some high-level steps an organization should complete to properly adopt Infrastructure as Code:

**Key Topic**

- Organize configurations into objects that can be stored.

- Leverage a source of truth system (Git) to store these objects.

- Leverage current workflows and processes to instantiate configurations from these objects stored in the source of truth system.

- Complete full validation testing to ensure proper functionality as intended.

Businesses and their networks are far more complex than in the past. Businesses are requiring consistent uptime as they are relying on the network, specifically connectivity to the Internet, for most resources. Businesses need flexibility and elasticity to support their business needs. Infrastructure as Code can help support dynamic businesses while enhancing data and cybersecurity.

## CI/CD Pipelines

When making manual network changes via the CLI, it is suggested to make small, incremental changes versus large, wholesale changes. For example, copying into the CLI a hundred lines of configuration changes and then having to troubleshoot what part of the copy broke the network is much more difficult than leveraging automation tools to complete the same work. Figuring out what part of the change isn't working is always troublesome, and often could create service downtime, an outage, or a few hours of frustration (at a minimum).

Continuous integration/continuous delivery (CI/CD) is a common software development practice used by developers to merge code changes into a central repository multiple times a day, sometimes multiple times an hour, and automate the entire software release process. With **continuous integration (CI)**, each time the code has been changed, the build and test process is automatically executed for the given application, providing instant feedback to the different developers on what's working and, more importantly, what's not working in their code. With **continuous delivery (CD)**, the relevant resources are automatically provisioned and deployed, which can sometimes consist of multiple disparate stages for more complex projects. The important aspect of a CI/CD pipeline is that all of these processes are fully automated, documented, and visible to the entire project team.

The four steps in a CI/CD pipeline are source, build, test, and deploy. Figure 12-3 shows the CI/CD pipeline process workflow.

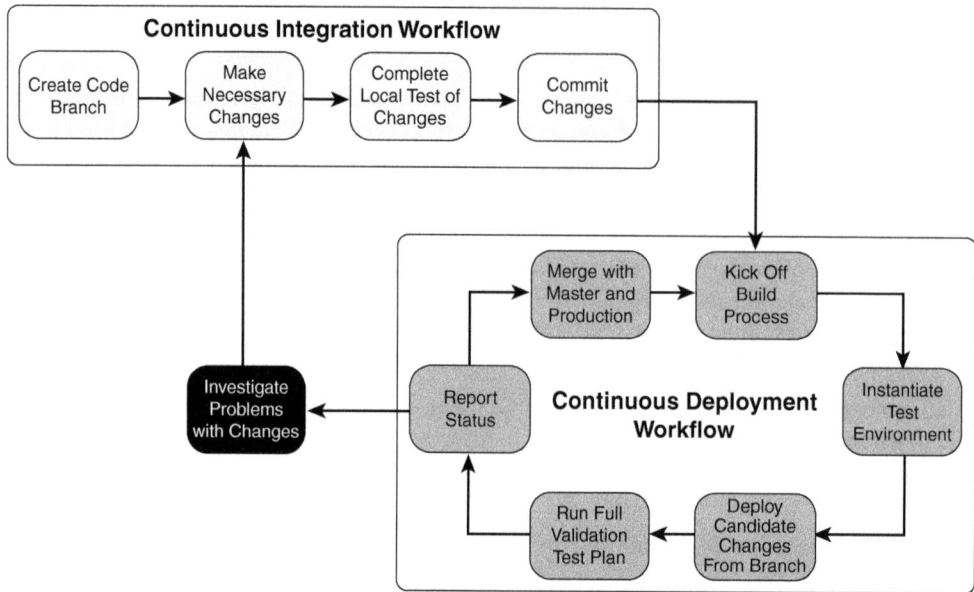

**Figure 12-3** *CI/CD Pipeline Process*

A network-focused CI/CD pipeline takes what has worked in software development and applies it to the network by automating the provisioning process, such as day one and day zero configuration builds, running automated functionality tests to verify configurations, and deploying changes to the production network. Automated pipelines remove manual errors and help implement standardized change processes.

Continuous integration leverages direct device communication. Network devices and automation tools talk directly sharing data and traffic telemetry. The network configurations are then treated as code—Infrastructure as Code (as discussed previously). The network configuration is stored in a version control system (e.g., Git). For a network engineer to make any change, they must first pull down the latest copy of the stored configuration, making their own branch of it. All changes are tested locally, and when testing is completed, the changes are committed and pushed back to the master copy in the version control system, kicking off the new build process.

The most common benefits of CI/CD are the following:

**Key Topic**

- The time required to enable new features and functionality is decreased.

- The process to make changes is streamlined.

- Changes are deployed quicker.

- Changes are simplified.

- Changes are small, steady, and frequent, decreasing the overall risk of the change.

- Removal of the human element from making changes reduces human errors.

- Intuitive test automation completed against proposed changes to validate those changes will not cause any unforeseen outages, thus minimizing network outages.

- Low-touch/automated change deployment to increase operational efficiency.

- Changes are more productive and lead to a more established better network.

Now that we have covered all of these automation capabilities, let's compare them to each other from two direct perspectives: a network design perspective and a business benefits perspective. Table 12-2 lists a number of the network design elements that should be followed to properly adopt these automation capabilities within a business, and Table 12-3 highlights the corresponding business benefit for each of these automation capabilities.

**Key Topic**

**Table 12-2**  Network Design Elements Required for Automation

| Network Design Element | Network Design | Automation Capability Affected |
|---|---|---|
| Each site, pod, and building block of the network design must be "templateable" and repeatable. This might include the specific subnets used, which ports on each device are used for what connections, the naming of the devices, and many more templateable items in the network design.<br><br>No one-off designs or arbitrary topologies should be leveraged.<br><br>No snowflake (Special designs / implementations against the norm) | Modularity | Zero-touch provisioning<br><br>Infrastructure as Code<br><br>Continuous integration/continuous delivery pipelines |
| Identify the smallest number of standard models and have the discipline to stick to them. For example, having small office, medium office, and large office models for all remote sites. Each remote site model would then have metrics and thresholds that would elevate the network design to the next level within the site models. Maybe a small office is for 10 or fewer users, a medium office is for 11 to 25 users, and a large office is for 26-plus users. Sometimes these thresholds are business-focused metrics (OPEX/CAPEX), while other times they are technology focused (e.g., number of ports, number of switches, number of line cards). | Hierarchy of design | Infrastructure as Code<br><br>Continuous integration/continuous delivery pipelines |
| Everything in the design should be modular building blocks so that if the need arises to add functionality, it can be easily added as a modular block. For example, if there is a need to add a second core block or a new Internet/perimeter block, it should be an easy addition to the overall network design. This can go all the way down to the role a single device is performing, such that of a CE-PE device. | Modularity<br>Hierarchy of design | Infrastructure as Code<br><br>Continuous integration/continuous delivery pipelines |

**Key Topic**

**Table 12-3** Business Benefits of Automation

| Automation Capability | Business Benefits |
|---|---|
| Zero-touch provisioning | Reduced CAPEX/OPEX. |
| | Decreased time to deployment. |
| | Reduced network outages with human errors. |
| Infrastructure as Code | Reduced CAPEX/OPEX. |
| | Reduced network outages with fewer human errors. |
| CI/CD pipelines | Increased time to market on new features, functionality, and capabilities. |
| | Increased revenue in some instances. |
| | Reduced CAPEX/OPEX. |
| | Reduced network outages with fewer human errors. |
| | Increased troop multiplier. |

## Summary

This chapter focused on the underlying impact of automation on a business and how a network designer can properly structure a network design for automation. This chapter specifically covered the automation capabilities of zero-touch provisioning, Infrastructure as Code, CI/CD pipelines, and the corresponding network design elements around them. After highlighting these capabilities individually, they were collectively compared to each other from a network design perspective and then from a business perspective.

## Reference

Al-shawi, Marwan, *CCDE Study Guide* (Cisco Press, 2015)

## Exam Preparation Tasks

As mentioned in the section "How to Use This Book" in the Introduction, you have a couple of choices for exam preparation: the exercises here, Chapter 18, "Final Preparation," and the exam simulation questions in the Pearson Test Prep Software Online.

## Review All Key Topics

Review the most important topics in this chapter, noted with the Key Topic icon in the outer margin of the page. Table 12-4 lists a reference of these key topics and the page numbers on which each is found.

**Table 12-4**   Key Topics for Chapter 12

| Key Topic Element | Description | Page Number |
|---|---|---|
| List | The benefits of ZTP | 318 |
| List | The high-level steps an organization should take to adopt Infrastructure as Code | 321 |
| List | Most common benefits of CI/CD pipelines | 322 |
| Table 12-2 | Network Design Elements Required for Automation | 323 |
| Table 12-3 | Business Benefits of Automation | 324 |

## Complete Tables and Lists from Memory

There are no Memory Tables or Lists for this chapter.

## Define Key Terms

Define the following key terms from this chapter and check your answers in the glossary:

zero-touch provisioning (ZTP), Infrastructure as Code (IaC), continuous integration (CI), continuous delivery (CD)

# CHAPTER 13

# Multicast Design

**This chapter covers the following topics:**

- **Multicast Switching:** This section covers the multicast switching protocols and network design elements for them.

- **Multicast Routing:** This section covers the multicast routing concepts and protocols, and the corresponding network design elements.

- **Multicast Design Consideration:** This section highlights the multicast design considerations that network designs should be aware of.

The typical behavior of unicast traffic is to send a copy of each packet to each receiver in any given network segment. As a result, the more receivers in the network of any given application, such as streaming video, the more bandwidth consumed. The impact of this behavior in small networks is almost always acceptable because of the limited number of receivers (end users). In contrast, in large-scale enterprise networks, where there might be thousands or even more receivers of certain applications such as IPTV streaming, this behavior wastes a large amount of bandwidth because of the unnecessary extra bandwidth required to transport unicast packets, which will ultimately lead to additional cost and overutilization of the available bandwidth.

Broadcast transmissions, however, lead to even worse situations, because they will forward data packets to all portions of the network, which can consume a lot of bandwidth and hardware resources across the network when there are actually only a few intended recipients.

Therefore, in today's networks, there are many applications and services developed to operate over multicast-enabled transport to help network operators overcome the aforementioned limitations. In general, with multicast, organizations can gain several advantages, such as the following:

- **Efficient and cost-effective transport:** More efficient network and bandwidth utilization can help to reduce the cost of network resources.

- **Optimized bandwidth and efficiency:** Enhancing the overall utilization over the network can reduce congestion caused by existing applications that are inefficiently transmitting unicast to groups of recipients, thereby allowing the network to support a larger number of recipients with simultaneous access to the application in a more efficient way.

- **Provide the business access to new applications:** Many applications and services nowadays are developed to operate over multicast-enabled networks that offer more efficient and scalable communication, specifically distributed multimedia business applications such as trading, market data, distance learning, and videoconferencing.

- **Open new business opportunities:** Leveraging these technologies opens new revenue-generating opportunities for both enterprises and service providers (SPs) by offering innovative services that were not possible over unicast transport, such as real time streaming services and IPTV.

**NOTE** Practically and ideally, the decision to enable multicast should be driven by business application requirements. However, as a network architect or designer, you can sometimes suggest that the business migrate certain applications from the unicast version to multicast version if the option exists and the transition is going to optimize the network and application performance. For instance, Moving Picture Experts Group (MPEG) high-bandwidth video applications usually consume a large amount of network bandwidth for each stream. Therefore, enabling IP multicast will enable you to send a single stream to multiple receivers simultaneously. This can be seen from the business point of view as a more cost-effective and bandwidth-efficient solution, especially if the increased bandwidth utilization translates to increased cost (such as WAN bandwidth).

This chapter covers the following "CCDE v3.0 Core Technology List" sections:

- 2.4 Multicast switching
- 3.9 Multicast routing concepts

# "Do I Know This Already?" Quiz

The "Do I Know This Already?" quiz allows you to assess whether you should read this entire chapter thoroughly or jump to the "Exam Preparation Tasks" section. If you are in doubt about your answers to these questions or your own assessment of your knowledge of the topics, read the entire chapter. Table 13-1 lists the major headings in this chapter and their corresponding "Do I Know This Already?" quiz questions. You can find the answers in Appendix A, "Answers to the 'Do I Know This Already?' Quizzes."

**Table 13-1** "Do I Know This Already?" Section-to-Question Mapping

| Foundation Topics Section | Questions |
|---|---|
| Multicast Switching | 1, 2 |
| Multicast Routing | 3–8 |
| Multicast Design Consideration | 9, 10 |

**CAUTION** The goal of self-assessment is to gauge your mastery of the topics in this chapter. If you do not know the answer to a question or are only partially sure of the answer, you should mark that question as wrong for purposes of the self-assessment. Giving yourself credit for an answer you correctly guess skews your self-assessment results and might provide you with a false sense of security.

1. Which multicast switching protocol should be used to limit the flooding of multicast IPv4 packets on a switch?

    a. MLD snooping

    b. PIM

    c. IGMP snooping

    d. Auto-RP

2. Which multicast switching protocol should be used to limit the flooding of multicast IPv6 packets on a switch?

    a. MLD snooping

    b. PIM

    c. IGMP snooping

    d. Auto-RP

3. Which multicast technique is used to avoid multicast loops?

    a. DF election (this is a most plausible distractor)

    b. RPF

    c. PIM-SSM

    d. PIM-BSR

4. Which multicast routing protocol would be best to use for an application that requires a many-to-many traffic pattern?

    a. PIM-BIDIR

    b. RPF

    c. PIM-SSM

    d. PIM-BSR

5. Which multicast routing protocol would be best to use if a network designer didn't want to include an RP in the network design?

    a. PIM-BIDIR

    b. RPF

    c. PIM-SSM

    d. PIM-BSR

6. Which protocol would be best to use if a network design required a standards-based way to determine an RP in a multicast network?

    a. PIM-BIDIR

    b. RPF

    c. PIM-SSM

    d. PIM-BSR

7. When using PIM-BIDIR where the RP will be in the data path, where should a network designer place the RP?

    a. Between the sources and receivers of the critical application

    b. Closest to the critical application

    c. Closest to the receivers

    d. Between the sources and receivers of the noncritical application

**8.** Which of the following are factors that should be considered when selecting the placement of a multicast RP? (Choose three.)

   **a.** Application multicast requirements

   **b.** Embedded-RP

   **c.** Multicast protocol

   **d.** Multicast tree

   **e.** RPF

   **f.** Phantom RP

**9.** Which of the following is an RP resiliency technique where two or more RPs are configured with the same IP address and subnet mask?

   **a.** Auto-RP

   **b.** Embedded-RP

   **c.** Anycast-RP

   **d.** Phantom RP

**10.** Which of the following is an RP resiliency technique specifically used for PIM-BIDIR where two RPs are configured with the same IP address but different subnet masks?

   **a.** Auto-RP

   **b.** Embedded-RP

   **c.** Anycast-RP

   **d.** Phantom RP

# Foundation Topics

# Multicast Switching

More and more Layer 2 networks are requiring multicast switching in their design because the applications being leveraged by the business require multicast support. This section covers the different multicast switching protocols—Internet Group Management Protocol (IGMP) and Multicast Listener Discovery (MLD)—and the corresponding attributes network designers need to know to properly design a Layer 2 network to run multicast properly.

## Multicast IP Address Mapping into Ethernet MAC Address

Normally, in any IP multicast–enabled environment, hosts (receivers) are required to receive a single traffic stream that has a common destination MAC address that maps to a multicast channel or group. Therefore, a mapping of IP to MAC is required here per multicast destination group IP. The range from 0100.5e00.0000 through 0100.5e7f.ffff is the available range of Ethernet MAC addresses for IPv4 multicast, and only half of these MAC addresses are available for use by IP multicast, where the IP to MAC mapping will place the lower 23 bits of the IPv4 multicast group address into these available 23 bits in the MAC address. As a result, MAC addressing will not always be unique in Layer 2 switched networks, because the upper 5 bits of the IP multicast address are dropped as part of the IP to MAC mapping, as shown in Figure 13-1.

In other words, as shown in Figure 13-1, the first 4 bits of the IP multicast 32 bits are always set to 1110 (Class D). Therefore, there will be only 28 unique bits left of the IP multicast address's information. Furthermore, because the remaining 28 unique bits that belong to the

Layer 3 IP multicast address information cannot be mapped in a 1:1 manner to the available 23 bits of Layer 2 MAC address space, this will lead to the loss of 5 bits of the address information in this mapping process. Consequently, it will result in what is known as 32:1 address mapping, which means that each Layer 2 multicast MAC address can potentially map to 32 IP multicast addressees. This mapping ratio, 32:1, is only true for IPv4, as IPv6 has a much higher ratio.

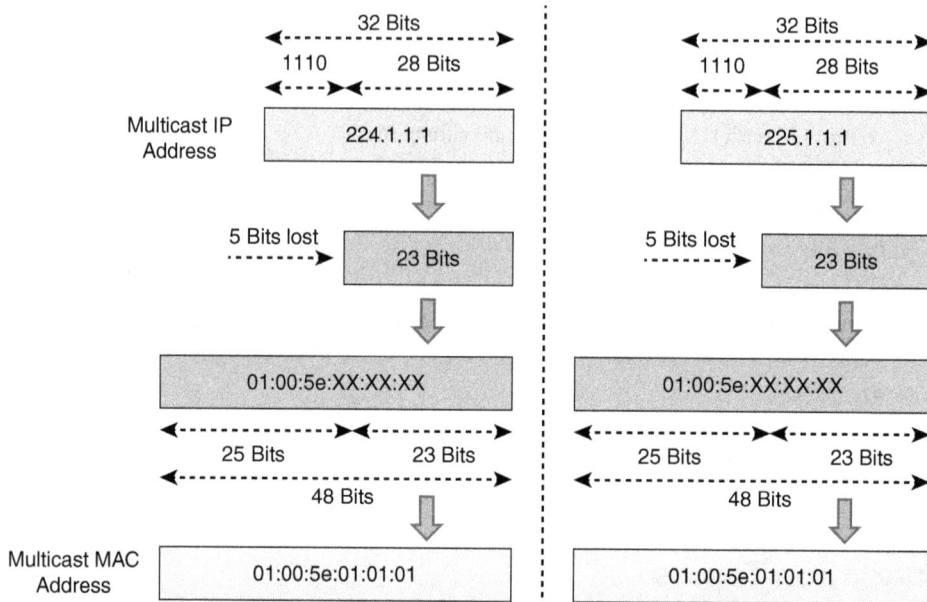

**Figure 13-1** *Multicast Group Layer 3 IP-to-Ethernet MAC Mapping*

To illustrate this point, consider a scenario of two IPTV channels streaming over different IP multicast group IDs (per channel), such as 239.1.1.1 and 239.2.2.2. Based on the mapping mechanism described earlier, hosts connected to a common Layer 2 switch and interested in receiving different IPTV channels will end up receiving both streams because both multicast groups' IPs will map to the same multicast MAC address on the Layer 2 switch. In this particular example, this will usually result in reduced efficiency in a large network with a large number of IPTV receivers connected to the same Layer 2 switched network.

Similarly, with IPv6, the mapping will be from 128-bit Layer 3 IPv6 multicast addresses to the 48-bit MAC address, where only the lower 32 bits of the Layer 3 address are preserved, while 80 bits of information are lost during the mapping process. This means, like IPv4, there is a possibility that more than one IPv6 multicast address might map to the same 48-bit MAC address.

Therefore, even though this might not always be an issue in LAN environments, it is still a critical point to be considered by network designers with regard to IP multicast design over a switched network. This can be even more critical if the switched network is connecting two different multicast domains (interdomain multicast scenario).

Internet Group Management Protocol (IGMP) snooping offers a simple solution for Layer 2 switched networks by eliminating the flooding to every single port (avoiding broadcast forwarding style). With IGMP snooping, the switch intercepts IGMP packets as they are being flooded and uses the information in the IGMP packets to determine which host/port should receive packets directed to a multicast group address. Similarly, Multicast Listener Discovery (MLD) snooping must be considered in a Layer 2 switched network when IPv6 multicast is used.

Nevertheless, even when multicast optimization protocols in a Layer 2 network are used, such as IGMP snooping, Layer 2 switches may flood all multicast traffic that belongs to the MAC address range of 0x0100.5E00.00xx, which belongs to the respective Layer 3 addresses in the link-local block (reserved for the use of routing protocols and topology discovery protocols) to all the ports within the same Layer 2 segment on the Layer 2 switch. Therefore, IP multicast addresses that map to the MAC address range 0x0100.5E00.00xx should be avoided whenever possible.

Moreover, in scenarios such as the one shown in Figure 13-2, when there are multiple Layer 3 routers connected over a Layer 2 switch using one shared segment/VLAN, IGMP snooping cannot optimize multicast traffic flooding within the Layer 2 switch. This is because these routers technically do not send IGMP membership reports for the desired multicast flows. Instead, the routers use routing protocol control messages among them, such as Protocol-Independent Multicast (PIM). As a result, there will be no IGMP messages for IGMP snooping to intercept and use for multicast traffic forwarding optimization, and the multicast video stream will be forwarded to all the routers connected to this switch within the same VLAN.

To optimize this design and overcome this issue, routers that are not supposed to receive the multicast stream simply can be placed in a different VLAN in this switch, such as the router connected to the WAN in Figure 13-2. Alternatively, the IGMP multicast route (mroute) proxy feature may be used to convert PIM join messages back into IGMP membership reports.

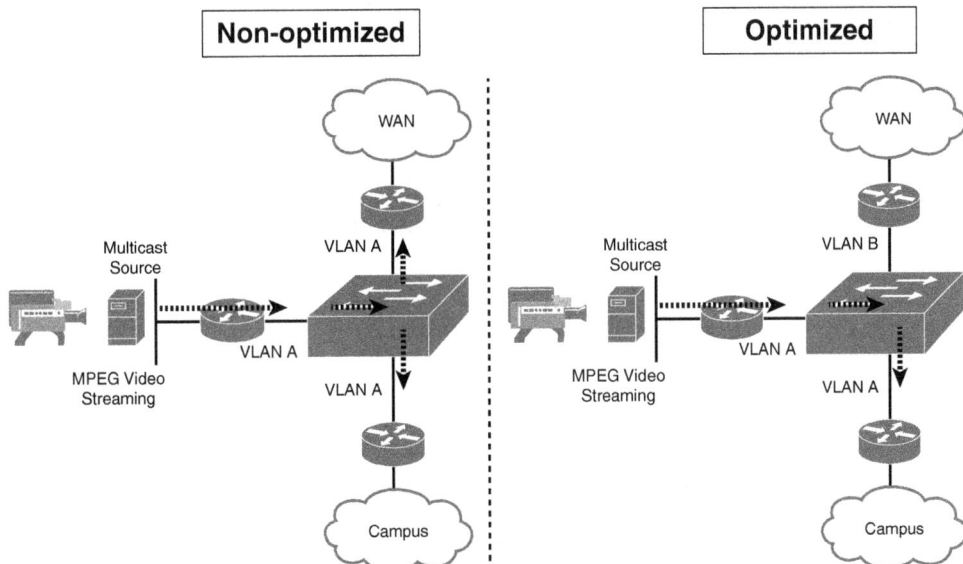

**Figure 13-2**  *Design Issue: IGMP Snooping*

**NOTE**  Network designers must consider maintaining the existing Layer 3 unicast communication between the WAN router and other routers in this network after placing the WAN router in a different VLAN, such as adding Layer 3 VLAN interfaces or adding additional interfaces/subinterfaces from other routers within the same VLAN.

**NOTE**  IGMP snooping may maintain forwarding tables based on either Ethernet MAC addresses or IP addresses. Because of the MAC-overlapping issues covered earlier with regard to mapping an IP multicast group address to Ethernet addresses, the forwarding based on IP address is desirable if a switch supports both types of forwarding mechanisms.

## Multicast Routing

This section starts by discussing the key consideration to achieve successful multicast forwarding, and then covers the most common protocols used to route multicast traffic within a single multicast domain. It also discusses multicast routing between different multicast domains.

### Reverse Path Forwarding

It is critical to understand the concept of reverse path forwarding and how it can impact multicast traffic forwarding and design. **Reverse path forwarding (RPF)** is the mechanism used by Layer 3 nodes in the network to optimally forward multicast datagrams. The RPF algorithm uses the rules shown in Figure 13-3 to decide whether to accept a multicast packet, forward it, or drop it.

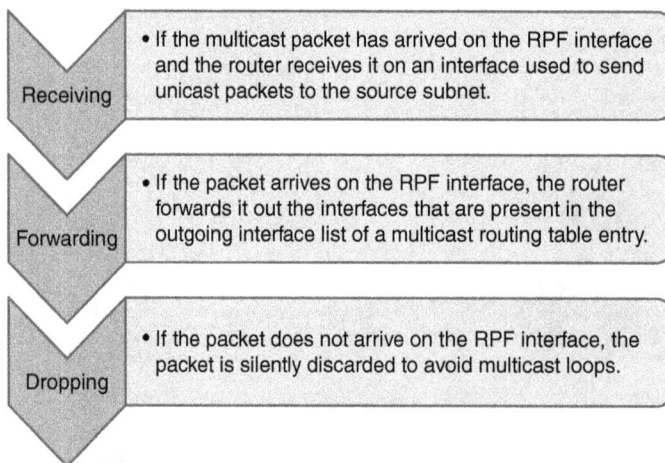

**Receiving**
- If the multicast packet has arrived on the RPF interface and the router receives it on an interface used to send unicast packets to the source subnet.

**Forwarding**
- If the packet arrives on the RPF interface, the router forwards it out the interfaces that are present in the outgoing interface list of a multicast routing table entry.

**Dropping**
- If the packet does not arrive on the RPF interface, the packet is silently discarded to avoid multicast loops.

**Figure 13-3**  *Multicast RPF Check Rules*

Therefore, a successful RPF check is a fundamental requirement to establishing multicast forwarding trees and passing multicast content successfully from sources to receivers. For instance, in the scenario shown in Figure 13-4, multicast was enabled only on interface G0/1, while unicast routing prefers interface G0/0 to reach the multicast sender (host IP 192.168.1.10). Hence, multicast traffic will not be received in this case because of RPF check failure. This issue can be fixed in different ways, such as aligning multicast-enabled interfaces with the unicast routing table, using static multicast route (mroute), or using Multicast Border Gateway Protocol (MP-BGP) to overcome the issue associated with the failure of RPF check. However, the primary goal here is first to be aware of the RPF impact. Then, as a network designer, you can decide which mechanism is more feasible to use to avoid the failure of RPF check. In other words, to achieve a successful IP multicast design, RPF check must be considered along with the possible mechanism that helps to avoid any forwarding failure while at the same time minimizing increased control plane complexity. For example, if BGP is used as the unicast control plane protocol, MP-BGP will avoid adding extra control plane complexity, because one routing protocol will be used to control both unicast and multicast routing in this case. Technically, MP-BGP here will only help to avoid RPF check failure over the desired interface to be used for multicast forwarding when the existing unicast routing prefers another path or interface. The network depicted in Figure 13-4 does not illustrate a real network and is conceptual only.

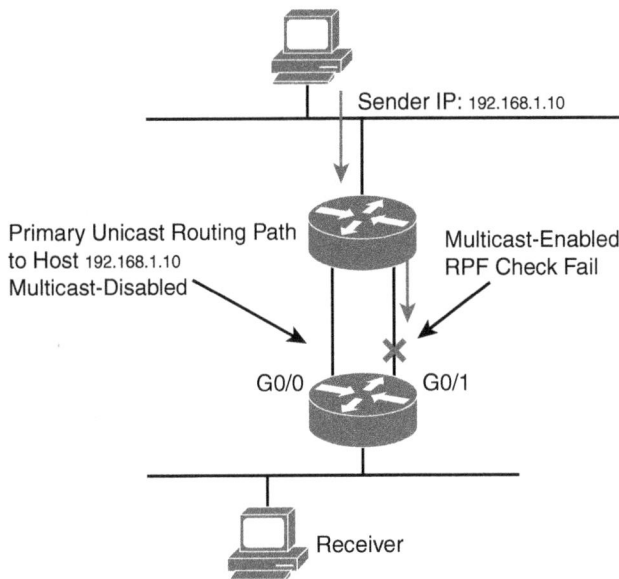

Sender IP: 192.168.1.10

Primary Unicast Routing Path to Host 192.168.1.10 Multicast-Disabled

Multicast-Enabled RPF Check Fail

G0/0    G0/1

Receiver

**Figure 13-4**  *RPF Check Failure Scenario*

## Multicast Routing Protocols

Multicast traffic flows are IP traffic, which almost always traverses the network using the existing unicast routing. However, for the multicast receivers and senders to locate each other across the network and communicate using the underlay IP network, a multicast routing protocol must use this function, and usually each multicast routing protocol has its own characteristics in terms of the discovery and forwarding style (multicast trees) of IP multicast traffic streams. Table 13-2 compares the most common multicast routing protocols.

**Key Topic**

**Table 13-2**  Comparison of Layer 3 Multicast Routing Protocols

| Multicast Protocol Type | IP Version | Supported Multicast Trees | Rendezvous Point | Suitable Applications | Supported IGMP/MLD Version | Scope |
|---|---|---|---|---|---|---|
| PIM-Dense Mode (PIM-DM) | 4 | Source tree only | Not required | Legacy | IGMP 1, 2, and 3 | Intradomain |
| PIM-Sparse Mode (PIM-SM) | 4/6 | Shared tree and source tree | Required | One to many | IGMP 1, 2, and 3; MLD 1 and 2 | Intradomain and interdomain |
| PIM-SSM | 4/6 | Forwarding only. Uses shortest-path tree (SPT) | Not required | One to many | IGMP 3, MLDv2 | Intradomain and interdomain |
| PIM-BIDIR | 4/6 | Shared tree only | Required | Many to many Many to one | IGMP 1, 2, and 3; MLD 1 and 2 | Intradomain |

**NOTE**  Although some of the PIM protocols covered in Table 13-2 technically support multicast interdomain routing, it is not common for them to be used to provide multicast interdomain routing without other protocols such as Multicast Source Discovery Protocol (MSDP).

Although having multiple flavors of PIM might be seen as an added complexity to the multicast network design, it can be seen as added flexibility for an environment with different types of multicast applications. For instance, **PIM Source-Specific Multicast (PIM-SSM)** can be deployed for certain enterprise communication applications, PIM-BIDIR for financial applications, and PIM-SM for other general IP multicast communications.

RP discovery, however, is one of the primary design aspects that must be considered during the planning and design phase of any IP multicast design task. Table 13-3 summarizes the common mechanisms used to locate or discover the intended multicast RP within a multicast domain.

**Key Topic**

**Table 13-3**  Multicast RP Discovery Mechanisms

| | Manual/Static | Automatic (Auto-RP) | BSR | Embedded-RP |
|---|---|---|---|---|
| Supported IP version | IPv4/v6 | IPv4 | IPv4/IPv6 | IPv6 |
| Scalability | Scalable, with a potential of increased operational overhead.* | Scalable | Scalable | Scalable |

| | Manual/Static | Automatic (Auto-RP) | BSR | Embedded-RP |
|---|---|---|---|---|
| Operational simplicity and flexibility | Inflexible with management overhead in large deployments. Changing RP's IP requires all the nodes to be updated. | Flexible with simplified operation.** Offers more protocol's filtering capabilities than BSR. | Flexible with simplified operation. Offers less traffic overhead (as RP information is encapsulated in PIM packets). | Flexible with simplified operation, dynamic. |

* If static RP is used with a redundant RP mechanism such as Anycast-RP with MSDP or anycast=RP PIM (RFC 4610), static RP can offer a scalable solution without significant increase in the operational overhead.

** Taking into account traffic and processing overhead because of the specific multicast groups used to propagate RP information, also special consideration in hub-and-spoke topology "NBMA" is required.

**Auto-RP** is a Cisco-proprietary protocol. Bootstrap Router (BSR), in contrast, is the standards-body method of electing an RP (RFC 5059). BSR operates completely based on PIMv2 and PIM-SM, which offer an optimized bandwidth during the flooding and discovery process as compared to Auto-RP. In contrast, Auto-RP can scope multiple RP addresses per domain and operate over either PIMv1 or PIMv2, with the ability to fall back to dense mode if required. As a general rule, network designers should avoid considering BSR and Auto-RP protocols within the same domain at the same time. That said, in some scenarios where a multicast PIMv2 domain that does not support Auto-RP (for example, non-Cisco network nodes) needs to interoperate with a multicast PIMv1 domain running Auto-RP, both Auto-RP and BSR functions are required. One simple solution to facilitate the interoperability between these two different multicast PIM domains is to deploy a network node at the multicast domain's boundary to perform the Auto-RP mapping agent and BSR functions. Also, this approach is applicable to PIM migration scenarios between Version 1 and Version 2.

**NOTE** The RP is required to initiate new multicast sessions with sources and receivers. During these sessions, the RP and the first-hop router (FHR) may experience some periodic increased overhead from processing. However, this overhead will vary based on the multicast protocol in use. For instance, the RP with PIM-SM Version 2 requires less processing than in PIM-SM Version 1, because the sources only periodically register with the RP to create state. In contrast, the location of the RP is more critical in network designs that rely on a shared tree, where the RP must be in the forwarding path, such as PIM-BIDIR. (The following section covers this point in more detail.)

## RP Placement

Normally, the placement of a multicast RP is influenced primarily by the following factors:

- The multicast protocol that is used
- The multicast tree model
- Application multicast requirements (for example, one to many versus many to many)
- Targeted network between the sources and receivers (LAN versus WAN)

As a rule of thumb, when the source tree along with the **shortest-path tree (SPT)** are considered, RP placement is not a big concern even though it is commonly recommended to be placed closer to the multicast sources, because in this case the RP is not necessarily required to be in the data forwarding path. However, in some multicast applications, such as many-to-many types of multicast applications, the receivers might operate as senders at the same time using different multicast groups for receiving and sending. Technically, even when SPT is enabled (where the last-hop router [LHR] cuts over to the SPT source tree), the source tree is always created between the source and the RP, in which an (S,G) state is created on all the nodes between each source and its RP before the switch over to SPT takes place, as shown in Figure 13-5. This may lead the nodes in the path (between the many receivers/senders and the RP) to hold a large number of (S,G) states. (In trading environments, this number can reach up to a few thousands of source feedback streams sourced from the receivers that operate as multicast senders as well.)

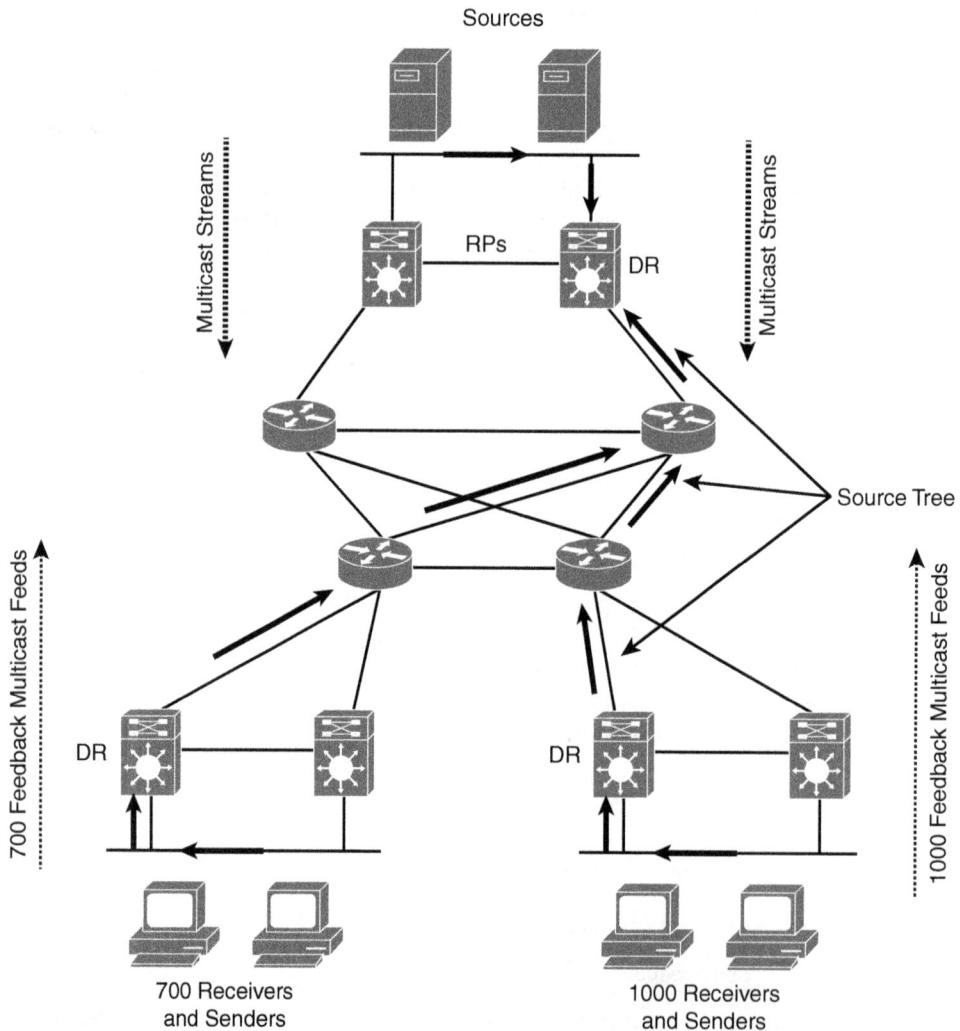

**Figure 13-5**  *Many-to-Many Multicast Applications Using PIM-SM RP*

Therefore, to reduce the number of (S,G) states on these nodes, you can add different RPs close to the receivers that require sending feeds to the feedback groups, as shown in Figure 13-6. Also, MSDP can be introduced in this design among the RPs to ensure that all RPs for any given group will be aware of other active sources. This design option with MSDP is limited to IPv4 only. This design is suitable in a multicast environment that requires the receivers to be able to send feedback/streams using a separate group from the actual data source group. Alternatively, PIM-SM can be migrated to PIM-BIDIR in this environment.

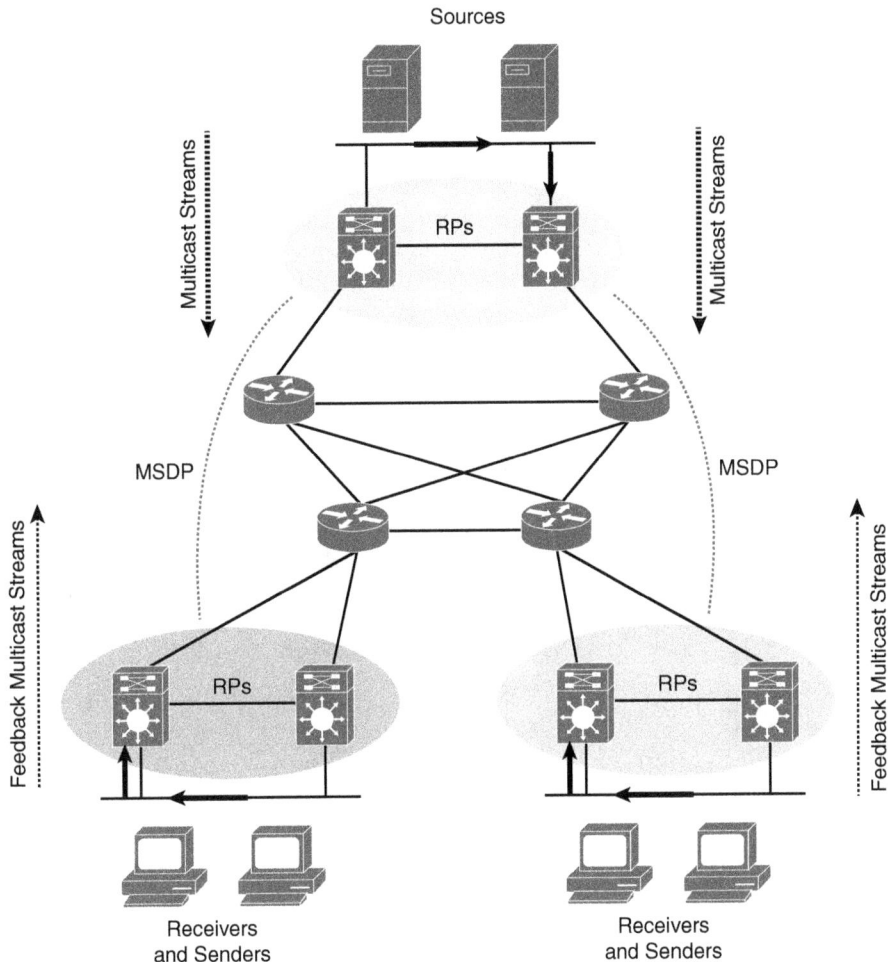

**Figure 13-6**   *Optimized: Many-to-Many Multicast Applications Using PIM-SM RPs*

However, when the **shared tree** is used as the forwarding multicast tree, such as using PIM-BIDIR where the RP will be in the data path, network designers must carefully and wisely consider the placement of the RP within the network because all the traffic will flow through the RP. For example, if multiple multicast streams are sourced from different senders distrusted at different locations across the network, identifying which stream should be given priority is a key to place the RP in the most optimal path between the sources and receivers based on the following:

**Key Topic**

- Application's criticality to the business

- Bandwidth consumption of each application's stream

- Number of sources and receivers of each application

- The underlying transport and topology of the network between the sources and receivers (LAN versus WAN, hierarchical versus hub and spoke)

For example, in the scenario shown in Figure 13-7, there are two hub nodes each connected to a different data center with multicast applications streaming over a shared tree. Although hub 1 is connected directly to the business-critical multicast application that requires high bandwidth, multicast streams will reach each of the remote site's receivers via hub 2. This is because hub 2 is defined as the RP for the multicast shared tree (such as PIM-BIDIR), where the RP must be in the data forwarding path, which can result in a congested data center interconnect (DCI) link and degraded application quality. Therefore, to optimize the path for the business-critical application with high-bandwidth requirements, the RP function must be moved to hub 1 to serve as an RP either for the entire multicast domain or at least for the multicast groups used by the applications located in DC-1. This can provide a more optimal path for multicast streams sourced from DC-1 toward the spokes/receivers.

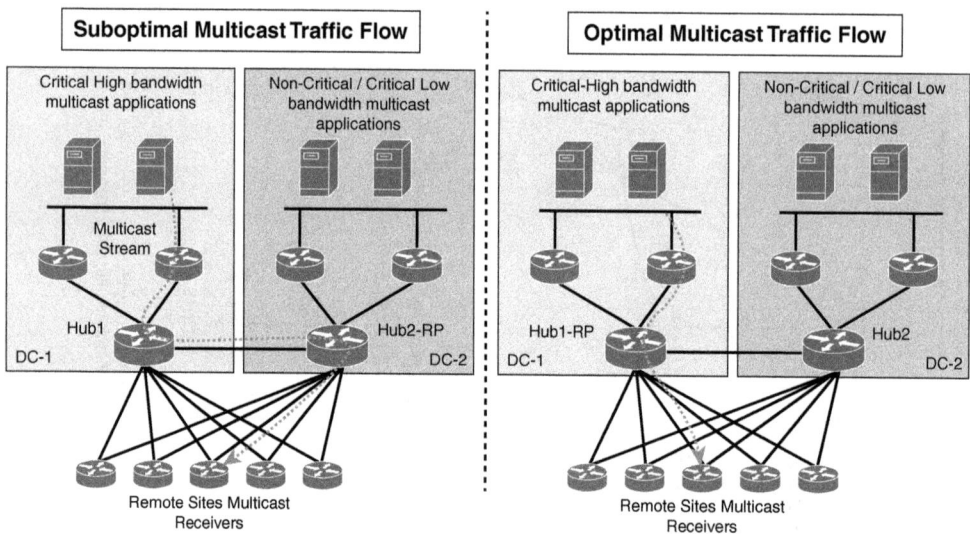

**Figure 13-7**  *RP Placement Consideration: Shared-Tree Multicast*

**NOTE**  In both scenarios, the assumption is that the RPF check is considered based on the utilized path for multicast traffic.

## Interconnecting Multicast Islands

One of the most common multicast design challenges occurs when two or more multicast-enabled networks need to communicate over a unicast-only network—for example, when a retail business is planning to roll out a digital media signage solution across all the remote sites, where the media server of this solution will stream the media contents centrally from the data center to the remote sites "acting as a multicast streaming source." This digital media signage solution requires multicast to be enabled end to end. However, the WAN provider in this scenario supports only unicast. In this case, this enterprise either needs to migrate to a new WAN provider that supports multicast or needs to consider a tunneling mechanism such as GRE in which IP multicast traffic can ride this tunnel and be transported over the IP unicast WAN network, as shown in Figure 13-8. Moreover, if the network in the middle is MPLS enabled, this may facilitate building a self-deployed L2VPN (such as point-to-point E-Line), on top of which multicast routing then can be enabled and transported.

### Multicast over GRE/LTPv3/DMVPN

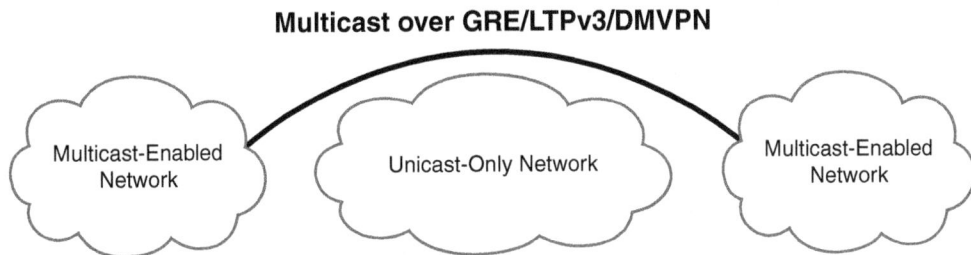

**Figure 13-8**   *Connecting Multicast Islands*

Which decision is better depends entirely on the situation and the requirements, such as timeframe to deploy the solution, cost considerations, and equipment support to the overlay technology. (For example, do the WAN edge nodes support GRE?) Furthermore, the most important influencing factor here is the characteristics of the multicast application used. For instance, IP tunneling such as GRE or DMVPN may introduce some limitations, such as limited scalability if a many-to-many multicast application is used. Therefore, in this case, it might be more feasible for large-scale organizations to consider a core/WAN transport that supports multicast routing rather than considering a solution that may potentially introduce limitations to the business in the future.

The other common scenario here is when there is a network device in the network path that does not support multicast, such as a firewall. This device will usually break the multicast network into two networks, where a mechanism such as IP tunneling is required to pass multicast traffic through this device, as shown in Figure 13-9, assuming this approach (tunneling traffic through the firewall) will not breach the organization's security policy standards.

**Single Multicast Domain**

Multicast over IP tunnel

Multicast sub-domain       Non-multicast       Multicast sub-domain
                              nodes

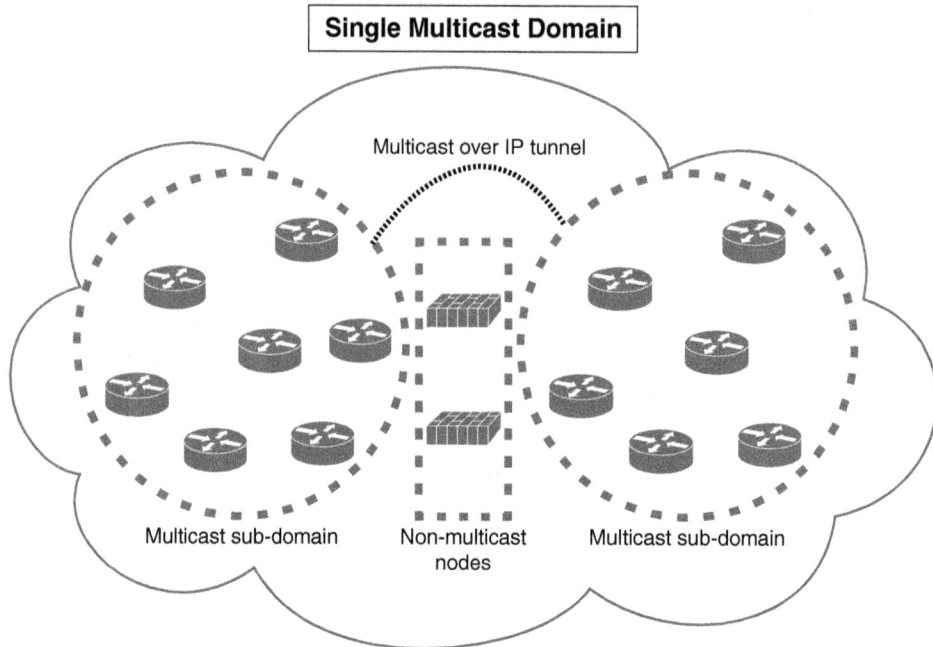

**Figure 13-9** *Connecting Multicast Islands over a Non-Multicast-Enabled Node*

## Interdomain Multicast

The multicast protocols discussed earlier focused mostly on handling multicast in one multicast domain. The term *multicast domain* can be defined as an interior gateway protocol (IGP) domain, one BGP autonomous system (AS) domain, or it can be based on the administrative domain of a given organization. For example, one organization might have multiple multicast domains, with each managed by a different department; for instance, one domain belongs to marketing, and another domain belongs to engineering. Other common scenarios of multiple multicast domains are between service providers and after a merger or acquisition between companies. Therefore, it is important sometimes to maintain the isolation between the different multicast domains by not sharing streams and RP feeds and at the same time offering the ability to share certain multicast feeds and RP information as required (in a controlled manner). This section covers the most common protocols that help to achieve this type of multicast connectivity.

## Multicast BGP

As discussed earlier, a successful RPF check is a fundamental requirement to establish multicast forwarding trees and pass multicast content successfully from sources to receivers. However, in certain situations, unicast might be required to use one link and multicast to use another, for some reason, such as bandwidth constraints on certain links. This situation is common in interdomain multicast scenarios. Multicast Border Gateway Protocol (MP-BGP; sometimes referred to as MBGP) is based on RFC 2283, "Multiprotocol Extensions for BGP-4." MP-BGP offers the ability to carry two sets of routes or network layer reachability information (NLRI) (sub-AFI), one set for unicast routing and one set (NLRI) for multicast routing. BGP multicast routes are used by the multicast routing protocols to build data distribution trees and influence the RPF check. Consequently, service providers and enterprise networks can control which path multicast can use and which path unicast can use using one control plane protocol (BGP) with the same path selection attributes and rules (such as AS-PATH, LOCAL_PREFERENCE, and so on).

## Multicast Source Discovery Protocol

**Multicast Source Discovery Protocol (MSDP)** is most commonly used to provide a mechanism of RP redundancy along with Anycast-RP (covered later in this chapter). MSDP is the most common protocol used to interconnect multiple IPv4 PIM domains because of its controlled and simplified approach of interconnecting PIM domains, in which it allows PIM domains to use an interdomain source tree instead of a common shared tree. With PIM, each RP is usually aware of the multicast sources and receivers within its PIM boundary (domain). MSDP peers use TCP sessions and they send Source Active (SA) messages to inform other MSDP peers about an active source within the local multicast domain, as shown in Figure 13-10.

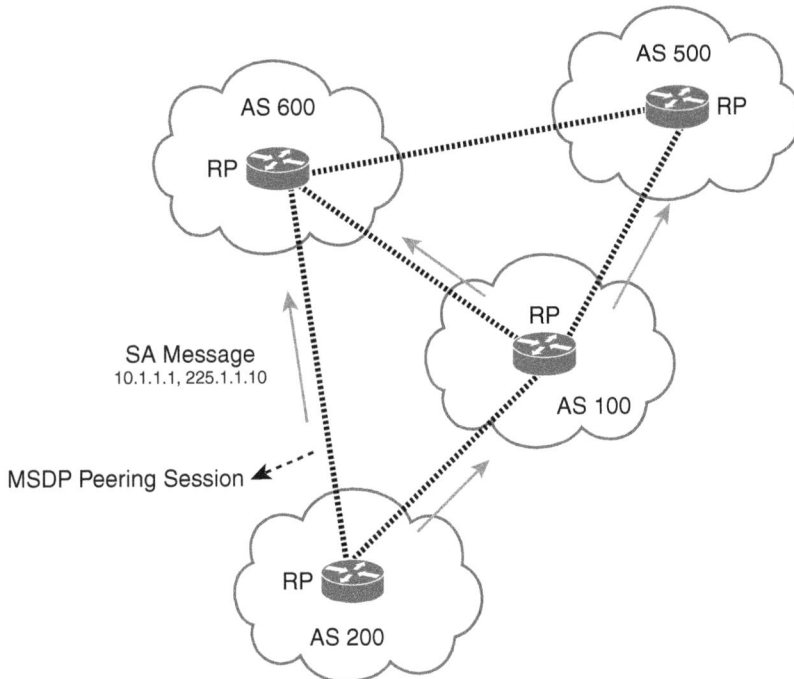

**Figure 13-10**  *Typical Interdomain Multicast Scenario with MSDP*

Consequently, all the MSDP peers (RPs) will be aware of all sources within the local domain and in other domains. RPF check is a fundamental consideration here, as well. However, with MSDP, some rules drive whether RPF check for SA messages is to be performed, as summarized in Table 13-4.

**Table 13-4** Cisco MSDP RPF Check Rules

| RPF Check Performed | RPF Check Not Performed |
|---|---|
| The sending MSDP peer is also an interior MP-BGP peer. | The sending MSDP peer is the only MSDP peer (for example, if only a single MSDP peer or a default MSDP peer is configured). |
| The sending MSDP peer is also an exterior MP-BGP peer. | The sending MSDP peer is a member of a mesh group. |
| The sending MSDP peer is not an MP-BGP peer. | The sending MSDP peer address is the RP address contained in the SA message.* |

* This table covers the Cisco-specific MSDP RPF check rules. There is a standardized list of rules in RFC 3618.

One common design issue with regard to multidomain multicast using MSDP and BGP that leads to RPF failure is that the IP address of the interior MP-BGP peer is different from the MSDP IP (for example, BGP using the physical IP, and MSDP using the loopback IP), as shown in Figure 13-11.

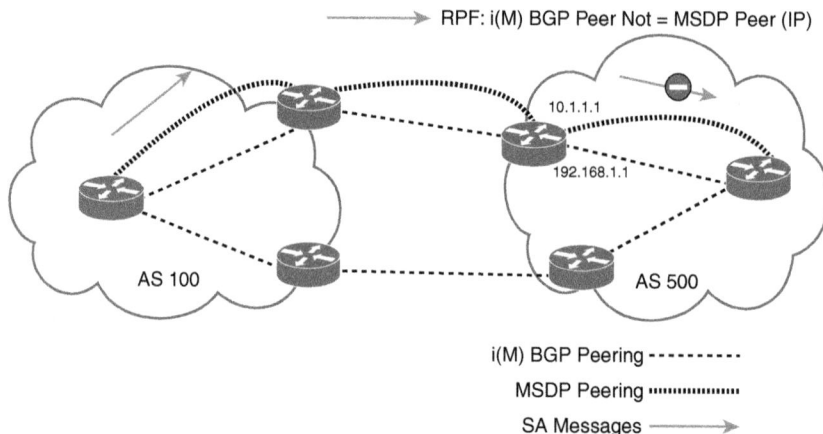

**Figure 13-11** *Common RPF Check Failure with MSDP and BGP*

Therefore, it is important that the address used for both MP-BGP and MSDP peer addresses is the same.

In the scenario shown in Figure 13-12, AS 500 is providing transit connectivity service to a content provider (AS 300) that offers IPTV streaming. End users need to connect to the streaming server IP 10.1.1.1 over AS 500. AS 500 has two inter-AS links, and they want to offer this transit service with AS 300 using the following traffic engineering requirements:

- Unicast traffic between AS 500 and AS 300 must use the link with 5 Gbps as the primary path and fail over to the 10-Gbps link in case of a failure.

- Multicast traffic must use the 10-Gbps link between AS 500 and AS 300. However, in case of a link or node failure, multicast traffic must not fail over to the other link (to avoid impacting the quality of other unicast traffic flowing over the 5-Gbps inter-AS link). Multicast group addresses that are in the range of 232/5 must not be shared between the two domains (AS 300 and AS 500).

■ Currently, the IPTV system needs to use only the range of 225.1.1.0/24. Therefore, only sources in AS 300 with this range must be accepted by AS 500.

To achieve these requirements, the following design aspects must be considered:

■ To ensure multicast traffic flow is over the 10-Gbps link only and without facing RPF check failure, MP-BGP will be used to advertise the multicast sources IPs (e.g., 10.1.1.1) and filter out these IPs from being advertised/received over the 5-Gbps link.

■ MSDP peering must be established between multicast RPs of AS 300 and AS 500 to exchange SA messages about the active source within the local domain in each AS.

■ PIM RP filtering and MSDP filtering is required to ensure that the RPs will only send/ accept sources within the multicast group IP range (225.1.1.0/24).

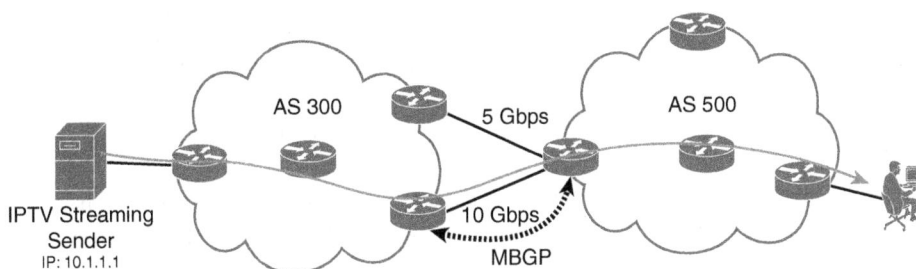

**Figure 13-12**  *Interdomain Multicast Design Scenario*

## Embedded-RP

Although PIM SSM offers the ability for IPv6 multicast to communicate over different multicast domains, PIM SSM still does not offer an efficient solution for some multicast deployments where many-to-few and many-to-many types of applications exist, such as videoconferencing and multiuser games applications. Also, in some scenarios, the multicast sources between domains may need to be discovered. Furthermore, MSDP cannot be used to facilitate interdomain multicast communication as with IPv4, because it has deliberately not been specified for IPv6. Therefore, the most common and proven solution (at the time of this writing) to facilitate interdomain IPv6 communication is the IPv6 **Embedded-RP** (described in RFC 3306) in which the address of the RP is encoded in the IPv6 multicast group address, and specifies a PIM-SM group-to-RP mapping to use the encoding, leveraging, and extending unicast-prefix-based addressing. The IPv6 Embedded-RP technique offers network designers a simple solution to facilitate interdomain and intradomain communication for IPv6 Any-Source Multicast (ASM) applications without MSDP.

However, network designers must consider that following an RP failure event, multicast state will be lost from the RP after the failover process because there is no means of synchronizing states between the RPs (unless it is synchronized via an out-of-band method, which is not common). In addition, with MSDP, network operators have more flexibility to filter based on multicast sources and groups between the domains. In contrast, with Embedded-RP, there is less flexibility with regard to protocol filtering capabilities, and if there is no other filtering mechanism, such as infrastructure access lists, to limit and control the use of the RP within the environment, a rogue RP can be introduced to host multicast groups, which may lead to a serious service outage or information security risk.

# Multicast Design Considerations

This section focuses on multicast design considerations on typical enterprise networks.

## Application Multicast Requirements

Designing IP multicast is just like all other IP network designs. To be a business-driven design and facilitate business goals, it has to follow the top-down approach by first focusing on the business goals and needs, taking into account the characteristics of the applications to be used. With regard to multicast applications, they can be categorized into three primary types, as listed in Table 13-5.

**Table 13-5**  Multicast Application Types

| Application Characteristic | Application Examples |
|---|---|
| One to many | Newsfeeds, push media, paging, announcements, some database updates |
| Many to many | Multimedia conferencing, multiplayer games, concurrent processing |
| Many to one | Data collection, polling, resource discovery |

Therefore, understanding at a high level the characteristics of the multicast applications is essential to determine which multicast protocol and design might be appropriate.

## Multicast Resiliency Design Considerations

In today's modern and converged networks, it is common that several business-critical applications run over multicast, where packet or connectivity loss is a highly undesirable experience, especially if the application is serving some core business applications, such as in financial services and trading applications, where minutes of disconnectivity can translate to the loss of hundreds of thousands of dollars.

Similarly, for content providers such as the IPTV provider, the quality of the multimedia and its availability is critical to maintaining its reputation and meeting the service-level agreement (SLA) requirements with the subscribers. Therefore, the design of IP multicast must incorporate the consideration of high availability and resiliency of the solution. However, with IP multicast, two levels can be considered to offer a reliable service. The first level is the underlying unicast IP infrastructure. The second level is the core component of the multicast network, such as RPs and other network-based solutions that are multicast specific, such as redundant sources, multicast with path diversity, and FRR. Specifically, the availability of RP in a PIM multicast domain is critical, especially for PIM-BIDIR and PIM-SM. Therefore, a design for a reliable multicast-enabled infrastructure must consider the availability of the RP as well. This section covers the common mechanisms and approaches used to address multicast resiliency design requirements.

## Anycast-RP

The concept of **Anycast-RP** is based on using two or more RPs configured with the same IP address on their loopback interfaces. Typically, the Anycast-RP loopback address is configured as a host IP address (32-bit mask). From the downstream router's point of view, the Anycast IP will be reachable via the unicast IGP routing. Because it is the same IP, IGP normally will select the topologically closest RP (Anycast IP) for each source and receiver. MSDP peering and information exchange is also required between the Anycast-RPs in this design, because it is common for some sources to register with one RP and receivers to join a different RP, as shown in Figure 13-13.

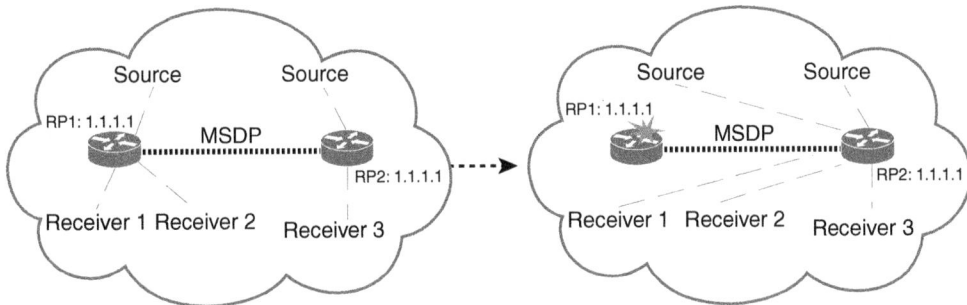

**Figure 13-13**  *Anycast-RP with MSDP*

In the event of any Anycast-RP failure, IGP will converge, and one of the other Anycast-RPs will become the active RP, and sources and receivers will fail over automatically to the new active RP, thereby maintaining connectivity.

**NOTE**  IPv6 does not support MSDP. Therefore, each RP has to define other RPs in the network as PIM RP set to maintain a full list of sources and receivers in the network. Alternatively, Anycast-RP using PIM, described in RFC 4610, can be used instead where the Anycast-RP functionality can be retained without using MSDP.

## Anycast-RP Using PIM

Anycast-RP using PIM (described in RFC 4610) enables both IPv4 and IPv6 to retain the same ultimate goal network designers aim to achieve using Anycast-RP with MSDP, and with fewer protocols. In other words, Anycast-RP can still be used without the need to establish MSDP peering sessions between the RPs. The primary difference is that each RP will send copies of the Register messages to the other (peering) RPs that usually received them from the registered sources or the designated router (DR) that is directly attached to the multicast source LAN. Consequently, each RP can maintain a list of the sources of each multicast group registered with different RPs (S,G). As a result, the multicast streams from the different sources distributed across the PIM multicast domain can reach all receivers even if they are associated with different RPs.

As shown in Figure 13-14:

1. The closest RP (anycast loopback IP) for each source and receiver will be selected by the underlying unicast routing protocol (IGP).

2. When RP-B receives the PIM Register message from multicast source S-B via R-1 (DR), it will decapsulate it and then forward it across the shared tree toward the interested (joined) receivers.

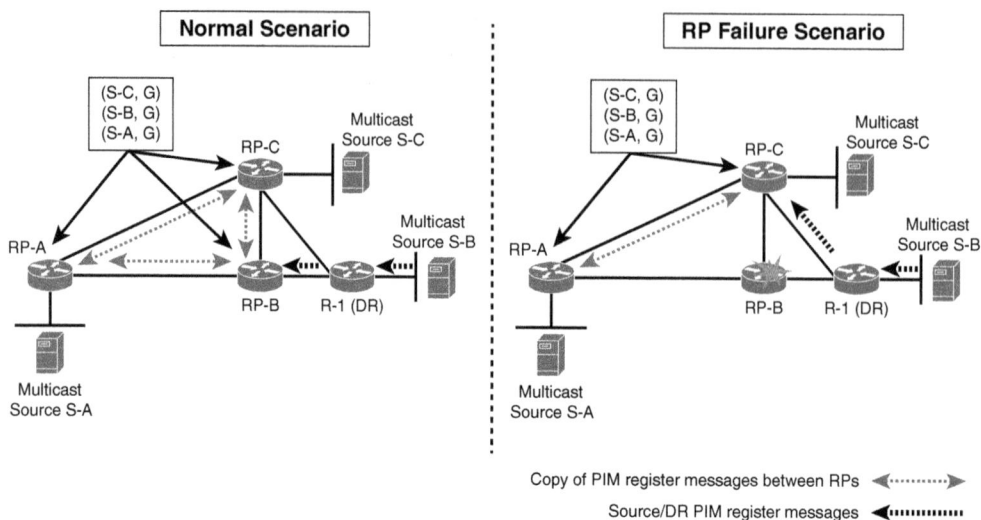

**Figure 13-14**  *Anycast-RP PIM*

3.  Assuming all RPs (RP-A, RP-B, and RP-C) are deployed as Anycast-RP PIM among each other, in this case RP-B will send a copy of this Register message to RP-A and RP-C.

4.  RP-A and RP-C will usually decapsulate the Register message sent from RP-B and in turn will forward it down to the shared tree (if there is any interested/joined receiver; otherwise, the RP can discard the packet).

5.  If RP-B fails, IGP will route any multicast messages from the multicast source behind R-1 (S-B) toward the closest Anycast-RP (in this example, RP-C).

However, MSDP may still be required in IPv4 interdomain multicast scenarios (for example, if the internal sources need to be advertised to multicast RPs outside of the local PIM domain, and similarly if learning of external multicast sources is required). In addition, MSDP rides on top of the Transmission Control Protocol (TCP), so it also offers guaranteed delivery. Also, from an operations and troubleshooting point of view, MSDP compared to Anycast-RP RFC 4610 offers higher flexibility because it has an SA cache and multiple **show** commands that help network operators to know in a simplified way where it learned stuff from and what is in the cache.

## Phantom RP

With PIM-BIDIR, all the packets technically flow over the shared tree. Therefore, redundancy considerations of the RP become a critical requirement. The concept of a **phantom RP** is specifically used for PIM-BIDIR, and the phantom RP does not necessarily need to be a physical RP/router. An IP subnet that is routable in the network can serve the purpose as well, where the shared tree can be rooted as shown in Figure 13-15.

Phantom RP 10.1.1.1

Loopback: 10.1.1.1/30            Loopback: 10.1.1.1/29

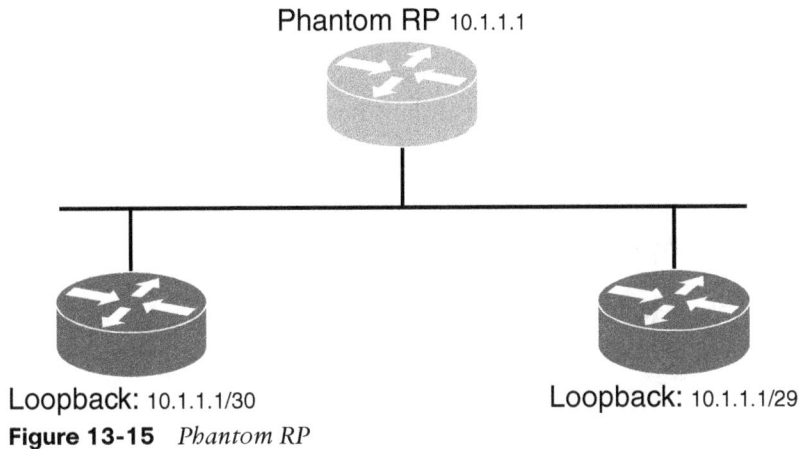

**Figure 13-15**  *Phantom RP*

As shown in Figure 13-15, if you configure two routers in a network with the same IP address and different subnet masks, IGP can control the preferred path for the root (phantom RP) of a multicast shared tree based on the longest match (longest subnet mask) where multicast traffic can flow through. The other router with the shorter mask can be used in the same manner if the primary router fails. This means the failover to the secondary shared tree path toward the phantom RP will rely on the unicast IGP convergence. Furthermore, if Auto-RP is used as a dynamic RP discovery mechanism to reduce operational overhead, a mapping agent is required for Auto-RP to work properly.

## Live-Live Streaming

The term *live-live* refers to the concept of using two live simultaneous multicast streams through the network using either a path separation technique or a dedicated infrastructure and RPs per stream. For instance, as shown in Figure 13-16 and Figure 13-17, the first stream (A) is sent to one set of multicast groups, and the second copy of the stream (B) is sent using a second set of multicast groups. Each of these groups will usually be delivered using separate infrastructure equipment to the end user with complete physical path separation, as shown in Figure 13-16. This design approach offers the ultimate level of resiliency that caters for any failure in a server or in network component along the path.

However, using single infrastructure such as MPLS-enabled core associated with MPLS Traffic Engineering (MPLS-TE) to steer the streams over different paths across the core infrastructure, as shown in Figure 13-17, offers resiliency for any failure on the server side; however, it may not offer full network resiliency (because both streams will use the same core infrastructure).

That said, if the MPLS provider caters for different failure scenarios (optical, node, link, and so on, along with switchover time that is fast enough to be performed without being noticed by the applications, such as using MPLS-TE FRR, and also avoids any shared risk link group [SRLG] along the path), it can offer a reliable and cost-effective solution.

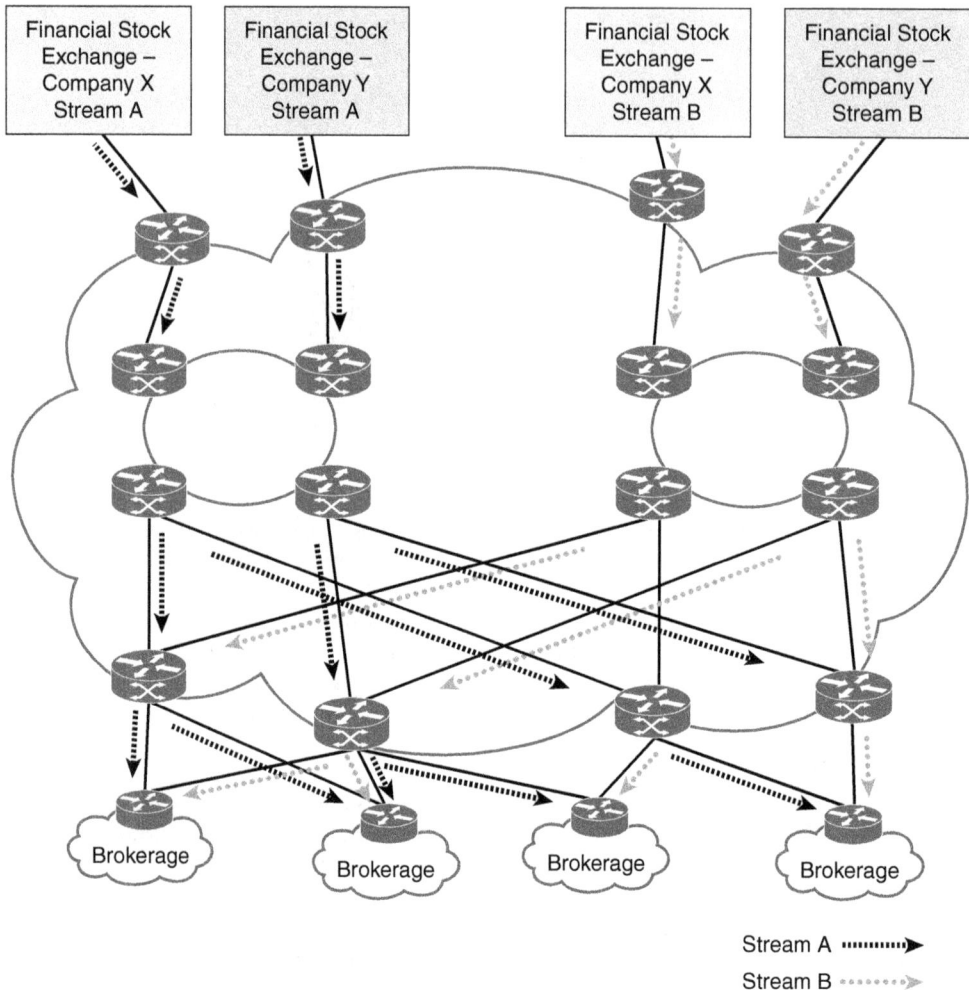

**Figure 13-16** *Live-Live Stream over Separate Core Networks*

One of the primary drivers to adopt such an expensive design approach (live-live) is the strict requirement of some businesses (commonly in the financial services industry because each millisecond is worth money) to aggressively minimize the loss of packets in the multicast data streams by adopting a reliable and low-latency multicast solution that does not introduce retransmissions.

## First Hop Redundancy Protocol-Aware PIM

In a shared LAN segment with a pair of Layer 3 gateways deployed with a First Hop Redundancy Protocol (FHRP) such as Hot Standby Routing Protocol (HSRP) to provide first-hop redundancy, the PIM designated router (DR) election is unaware of the FHRP redundancy deployment. As a result, the HSRP active router (AR) and the PIM DR of the same LAN segment will probably not be the same router because PIM has no inherent redundancy capabilities and normally operates independently of HSRP.

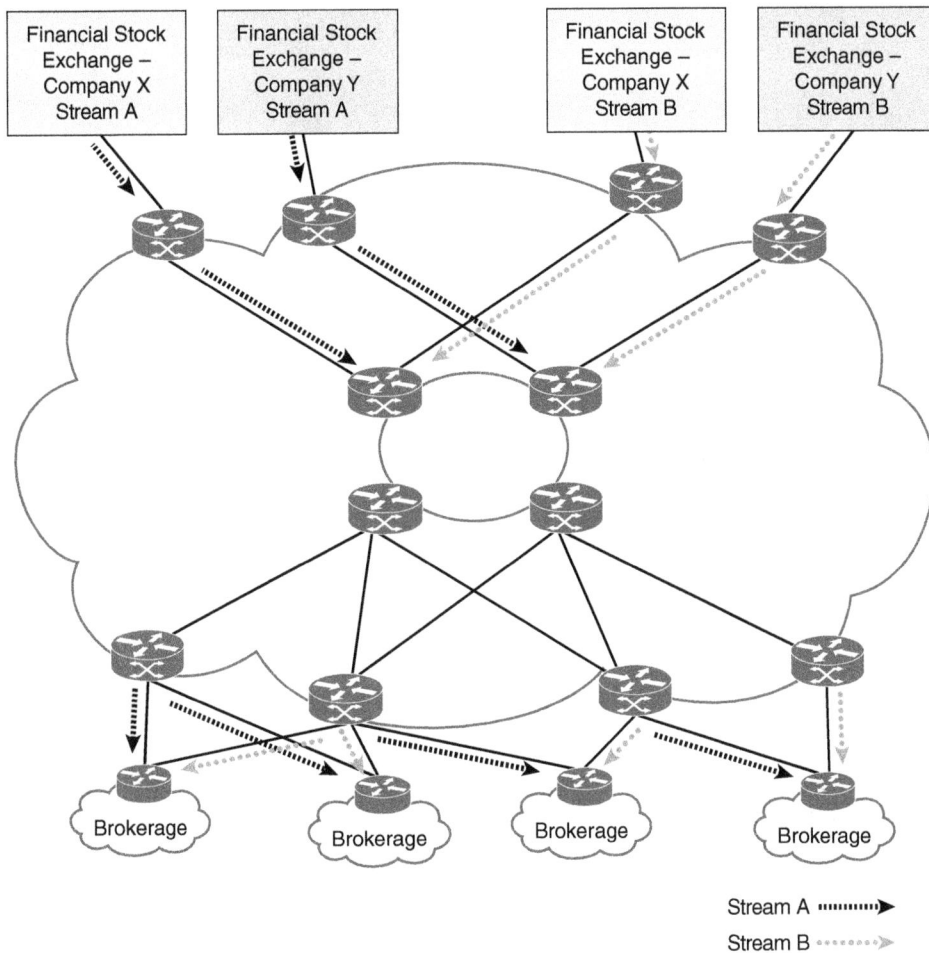

**Figure 13-17**  *Live-Live Stream over Single-Core Network*

HSRP-Aware PIM helps network designers to ensure that multicast traffic is forwarded via the HSRP (AR) that will be acting as the PIM (DR) for the LAN segment as well. This will allow PIM to leverage HSRP redundancy and to offer failover capability derived from the HSRP states in the device. With HSRP-Aware PIM enabled, PIM sends an additional PIM Hello message using the HSRP virtual IP addresses as the source address for each active HSRP group when a device becomes HSRP active. Therefore, downstream devices such as a Layer 3 node or a multicast streaming server can use either a static route or a default gateway that points to the FHRP virtual IP (VIP), which offers a more reliable and simplified operation with a simple configuration at the downstream node side, as shown in Figure 13-18.

From a resiliency point of view, following an HSRP AR failover or recovery event, the newly elected HSRP AR will assume the responsibility for the routing and forwarding of all the traffic sent to the HSRP VIP. At the same time, PIM will tune the priority of the DR based on HSRP state to maintain the alignment of the HSRP AR and the multicast DR functions on the same node.

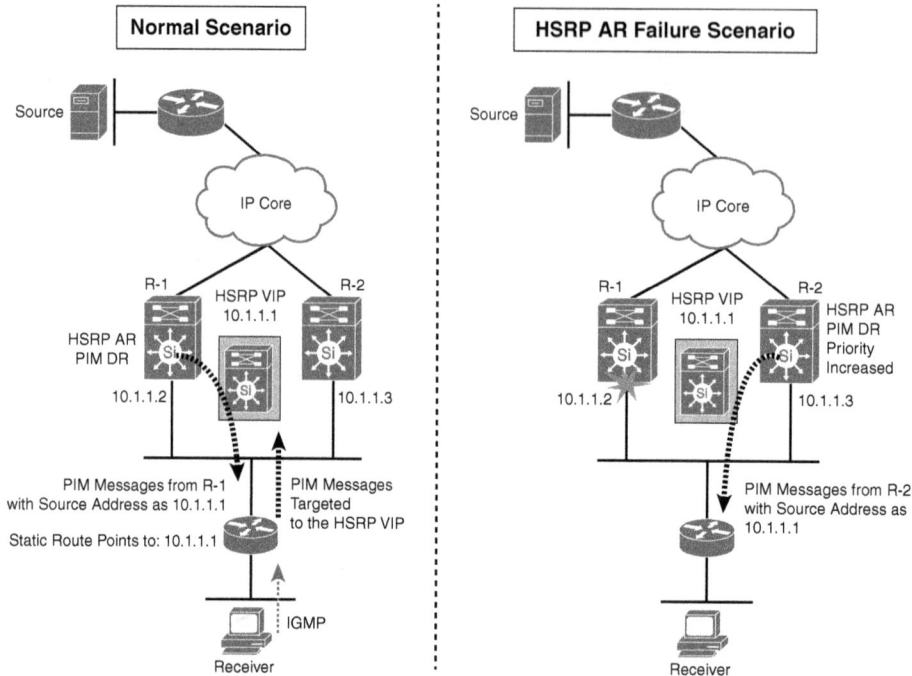

**Figure 13-18**  *HSRP-Aware PIM*

## Summary

In this chapter, we focused on multicast design, covering multicast switching, routing, and overall design considerations. Based on the multicast design options, considerations, and constraints covered in this chapter, network designers should always answer the following questions before considering any design recommendation or strategy:

■ What is the targeted business or environment (enterprise versus SP)?

■ Why is multicast required? Is it to run a business-critical application, to provide a transport network for customers to run their applications, or to provide a service, such as IPTV?

■ What are the characteristics of the multicast application? Many to many versus one to many, IPv4 or IPv6? This can help to decide the type of RP and where the RP should be positioned in the network.

■ Is the targeted network under single or multiple administrative domains? If multiple, is there any requirement to provide multicast communications between the different domains (interdomain)?

■ What is the Layer 3 routing protocol in use, and what does the logical design look like? This helps to determine RPF check failure considerations.

■ Is there any constraint related to enabling multicast across the network path? (For example, the WAN provider may not support multicast.)

- Does any network device in the path not support multicast, such as firewalls?

- Are there any staff knowledge limitations with regard to certain IP multicast protocols?

- Are there any high-availability requirements, such as redundant RPs?

## References

Al-shawi, Marwan, *CCDE Study Guide* (Cisco Press, 2015)

## Exam Preparation Tasks

As mentioned in the section "How to Use This Book" in the Introduction, you have a couple of choices for exam preparation: the exercises here, Chapter 18, "Final Preparation," and the exam simulation questions in the Pearson Test Prep Software Online.

## Review All Key Topics

Review the most important topics in this chapter, noted with the Key Topic icon in the outer margin of the page. Table 13-6 lists a reference of these key topics and the page numbers on which each is found.

**Table 13-6**   Key Topics for Chapter 13

| Key Topic Element | Description | Page Number |
|---|---|---|
| Figure 13-1 | Multicast Group Layer 3 IP-to-Ethernet MAC Mapping | 330 |
| Table 13-2 | Comparison of Layer 3 Multicast Routing Protocols | 334 |
| Table 13-3 | Multicast RP Discovery Mechanisms | 334 |
| List | RP placement | 335 |
| List | RP optimal path placement guidelines | 338 |
| Table 13-5 | Multicast Application Types | 344 |

## Complete Tables and Lists from Memory

There are no Memory Tables or Lists for this chapter.

## Define Key Terms

Define the following key terms from this chapter and check your answers in the glossary:

reverse path forwarding (RPF), PIM Source-Specific Multicast (PIM-SSM), Auto-RP, shortest-path tree (SPT), shared tree, Multicast Source Discovery Protocol (MSDP), Embedded-RP, Anycast-RP, phantom RP

# CHAPTER 14

# Network Services and Management

**This chapter covers the following topics:**

- **IPv6 Design Considerations:** This section covers critical IPv6 topics and network design elements for IPv6.

- **Quality of Service Design Considerations:** This section covers QoS models, concepts, migrations, and the corresponding network design elements.

- **Network Management:** This section covers the network management concepts, protocols, and the corresponding network design elements.

This chapter covers multiple networking and IP service design concepts and considerations that are considered as additional core topics for the purpose of the CCDE v3 written and practical exam at the time of this writing. The different topics discussed in this chapter might be presented as an application or service to be used to achieve a business need. For example, a business-critical application might require quality of service (QoS) to be enabled across the network to work properly.

Note, as well, that this chapter focuses on the design drivers, considerations, and approaches that network designers can consider based on the different design requirements, without covering any deep technical details.

**NOTE** IPv6-specific design considerations are covered in this chapter, but there are no specific CCDE blueprint line items for IPv6. This is because it is expected that IPv4 and IPv6 are inherently included throughout every CCDE blueprint domain and topic. The additional IPv6 section is included in this chapter because it contains critical topics that all network designers and CCDE candidates should know.

This chapter covers the following "CCDE v3.0 Core Technology List" sections:

- 4.6 QoS techniques and strategies

- 4.7 Network management techniques

- 4.8 Reference models and paradigms that are used in network management (FCAPS, ITIL, TOGAF, and DevOps)

**NOTE** For the Core Technology List item 4.8, only FCAPS is covered in this chapter. ITIL and TOGAF are covered in chapter 5 while DevOps (Automation) was covered in chapter 12.

# "Do I Know This Already?" Quiz

The "Do I Know This Already?" quiz allows you to assess whether you should read this entire chapter thoroughly or jump to the "Exam Preparation Tasks" section. If you are in doubt about your answers to these questions or your own assessment of your knowledge of the topics, read the entire chapter. Table 14-1 lists the major headings in this chapter and their corresponding "Do I Know This Already?" quiz questions. You can find the answers in Appendix A, "Answers to the 'Do I Know This Already?' Quizzes."

**Table 14-1** "Do I Know This Already?" Section-to-Question Mapping

| Foundation Topics Section | Questions |
|---|---|
| IPv6 Design Considerations | 1–4 |
| Quality of Service Design Considerations | 5–8 |
| Network Management | 9, 10 |

**CAUTION** The goal of self-assessment is to gauge your mastery of the topics in this chapter. If you do not know the answer to a question or are only partially sure of the answer, you should mark that question as wrong for purposes of the self-assessment. Giving yourself credit for an answer you correctly guess skews your self-assessment results and might provide you with a false sense of security.

**1.** What is the first step an enterprise should take when migrating to an IPv6-only network if it wants to ensure that no service interruption occurs?

 **a.** Leverage an overlay, like VXLAN, to support IPv6 hosts within the data center.

 **b.** Leverage a translation mechanism that is based on a load balancer, pure DNS, or classical NAT64.

 **c.** Migrate certain modules of the enterprise network first while leveraging a translation/tunneling mechanism to maintain communication between IPv6 and IPv4 islands.

 **d.** Migrate the core to be in dual-stack mode first, then migrate the other modules as time allows.

**2.** Which of the following IPv6 design approaches for migrating to IPv6 would allow a network to provide IPv6 access inbound or outbound at the enterprise Internet edge?

 **a.** Leveraging an overlay, like VXLAN, to support IPv6 hosts within the data center

 **b.** Leveraging a translation mechanism that is based on a load balancer, pure DNS, or classical NAT64

 **c.** Migrating certain modules of the enterprise network first while leveraging a translation/tunneling mechanism to maintain communication between IPv6 and IPv4 islands

 **d.** Migrating the core to be in dual-stack mode first, then migrating the other modules as time allows

**3.** Which of the following IPv6 mechanisms would allow for end-to-end IPv6 and IPv4 functionality?

   **a.** Dual stack

   **b.** ISATAP

   **c.** GRE

   **d.** mGRE

**4.** Which of the following IPv6 mechanisms would allow network-to-network connectivity to transit IPv6 over IPv4-only devices without any additional control plane protocols?

   **a.** Dual stack

   **b.** ISATAP

   **c.** GRE

   **d.** mGRE

**5.** Which of the following QoS queuing characteristics is an example of Weighted Fair Queuing?

   **a.** Supports real-time queueing and minimum bandwidth guarantee

   **b.** Offers a dynamic distribution based on DSCP values

   **c.** Four queues with associated levels of importance, with the most important being serviced first

   **d.** Suitable for large links that have a low delay with very minimal congestion, but has no requirement for priority or classes of traffic

**6.** Which of the following QoS queuing characteristics is an example of FIFO?

   **a.** Supports real-time queueing and minimum bandwidth guarantee

   **b.** Offers a dynamic distribution based on DSCP values

   **c.** Typically, four to six queues with associated levels of importance, with the most important being serviced first

   **d.** Suitable for large links that have a low delay with very minimal congestion, but has no requirement for priority or classes of traffic

**7.** Which of the following QoS queuing characteristics is an example of Priority Queuing?

   **a.** Supports real-time queueing and minimum bandwidth guarantee

   **b.** Offers a dynamic distribution based on DSCP values

   **c.** Typically, four to six queues with associated levels of importance, with the most important being serviced first

   **d.** Suitable for large links that have a low delay with very minimal congestion, but has no requirement for priority or classes of traffic

**8.** Which of the following QoS queuing characteristics is an example of LLQ?

   **a.** Supports real-time queueing and minimum bandwidth guarantee

   **b.** Offers a dynamic distribution based on DSCP values

   **c.** Typically, four to six queues with associated levels of importance, with the most important being serviced first

   **d.** Suitable for large links that have a low delay with very minimal congestion, but has no requirement for priority or classes of traffic

9. Which network management protocol would be used if you wanted XML encoding and messages encrypted with SSH?

   a. YAML

   b. NETCONF

   c. RESTCONF

   d. gNMI

10. Which network management protocol would be used for north-bound controller communication leveraging JSON encoding while also leveraging HTTP for transport?

   a. YAML

   b. NETCONF

   c. RESTCONF

   d. gNMI

## Foundation Topics

## IPv6 Design Considerations

IPv6 is a broad topic. Therefore, this section focuses on one of the most critical and important topics, which is the integration and coexistence of IPv4 and IPv6, by covering the different design and technical options and how to follow a business-driven life-cycle design approach.

### IPv6 Business and Technical Drivers

Although the initial IPv6 draft "The Case for IPv6" was published as an IETF draft more than two decades ago (1999) (draft-ietf-iab-case-for-ipv6-06.txt), there was no serious move or adoption of IPv6 by enterprises and service providers. However, during the past few years, billions of smartphones and various mobile devices have been connecting to the public Internet and private networks. The overall Internet needs to evolve and grow to support this continuous growth, not to mention that the increased adoption of Internet-based services and public cloud hosting has added more demand on public IP addresses (IPs).

Furthermore, the new trends and technologies, such as the Internet of Things (IoT) and smart connected cities, are leading to an increasingly growing number of Internet-connected users, applications, sensors, and services that require a huge number of IPs. As the Internet transitions, organizations must start to consider adopting IPv6 to support and facilitate achieving their business goals in terms of organic growth, merger and acquisition, and continuity in the market. With the exhaustion of IPv4 addresses and the increased demand on IPs, organizations face many challenges, such as the following:

■ The exhaustion and constraints of IPs (IPv4), which adds many challenges to managing and provisioning new services that require new addressing spaces

■ Added complexity to merger and acquisition scenarios, where Network Address Translation (NAT), with its limitations to overcoming the overlapping addressing issue, is considered a primary design option

■ New market and service trends, where mobility and numerous connected devices (smart cities, IoT, and so on) are key drivers for both enterprise and service providers that are trying to adopt innovation and the provision of new services, for which a large number of IPs are required to accommodate the various IP-enabled endpoints

Therefore, the adoption of IPv6 is becoming a necessity to overcome these challenges and to meet these new technology solution requirements and demands. Hypothetically, if an organization is encountering one or more of the following situations, it is time to migrate to IPv6, or at least start to consider enabling IPv6:

- The organization is unable to expand their business to other global regions because of the exhaustion of the public IPv4 addresses.

- Deploying innovative network environments, applications, and devices such as IoT with a huge number of connected sensors is required to support smart connected communities.

- A service provider or an enterprise needs to maintain seamless connectivity across fixed and mobile users when using collaborative applications, and NAT is no longer an option.

- Enabled IPv6-based 4G/5G/LTE mobile networks or connecting workforces to these networks.

- The organization primarily works as a supplier or partner with public sector or government entities and is required to comply with their standards. In these entities, IPv6 is fast becoming the standard.

- Encountering IPv6 need in the network driven by the deployments of end-user and host operating systems and applications such as Microsoft Windows 11, Windows Server 20xx, Apple macOS, system virtualization, and large-scale multitenancy.

## IPv6 Address Types (Review)

Before taking a look at the different types of IPv6, it is important that network architects and designers be aware of the high-level differences and similarities between the two IP versions. Table 14-2 summarizes some of the technical similarities and differences between IPv4 and IPv6.

**Table 14-2**  Summary of IPv4 Versus IPv6

|  | IPv4 | IPv6 |
|---|---|---|
| Address scope | 32 bit | 128 bit, multiple scopes |
| IP allocation | Manual, Dynamic Host Configuration Protocol (DHCP) | Manual, stateless address autoconfiguration (SLAAC), DHCP |
| QoS | Differentiated services, integrated services | Differentiated services, integrated services, flow label |
| Multicast | Internet Group Management Protocol (IGMP), Protocol-Independent Multicast (PIM), Multiprotocol Border Gateway Protocol (MP-BGP) | Multicast Listener Discovery (MLD), PIM, MP-BGP |
| Security | No built-in support | IPsec built in |

**NOTE** Although IPv6 supports built-in IPsec, it is considered a myth and misconception that IPv6 is more secure than IPv4. This assertion stems from the original mandated use of IPsec in host-to-host communications, as specified in RFC 2401. Consequently, if IPsec is implemented, it will provide confidentiality and integrity between two hosts, but it still will not address any link operation vulnerabilities, attacks, and most of the denial-of-service (DoS) attacks.

With regard to IPV6 addressing, IPv6 has three types of unicast addresses:

- **Link local:** This address is nonroutable and can exist only on a single Layer 2 domain (fe80::/64). As described in RFC 4291, even when one or more routable IPv6 addresses are assigned to a certain interface, an IPv6 link-local address is still required to be enabled on this interface.

- **Unique local address (ULA):** This address is routable within the administrative domain of a given network (fc00::/7). This address, in concept, is similar to the IPv4 private address range (RFC 1918).

- **Global:** This address is routable across the Internet (2000::/3). This address is similar in concept to the IPv4 public ranges.

Network designers can mix and match the different IPv6 addressing types when designing an enterprise network. However, there are some issues associated with each design model, as summarized in Table 14-3.

**Table 14-3** IPv6 Address Types

| IPv6 Addressing Model | Scope | Design Simplicity | Manageability | Scalability |
|---|---|---|---|---|
| ULA only | Internal | Moderate | Moderate | Low |
| ULA + global | Internal and external | Complex | Complex | Moderate |
| Global only* | Internet and external | Simple | Simple | High |

* Because NAT is primarily used with IPv4 to overcome IP address shortages, it is always arguable that this is a less-secure model because it exposes internal device IPs to external networks (no address hiding using NAT). In addition, it is commonly claimed that NAT offers a layer of secure communication; however, that argument fails to cover the impact of NAT on many applications at higher layers.

## Migration and Integration of IPv4 and IPv6

Considerations related to transitioning to IPv6 and planning and designing the integration of both IP versions to coexist are critical aspects since the introduction of IPv6. However, the most common critical and challenging question for network architects and designers with regard to migrating and enabling IPv6 is where to start. For network architects or designers to achieve a successful IPv6 migration or integration, they must follow a structured approach based on the top-down design methodology, which consists of network discovery, assessment, planning, design, deployment, monitoring, and optimization, as shown in Figure 14-1.

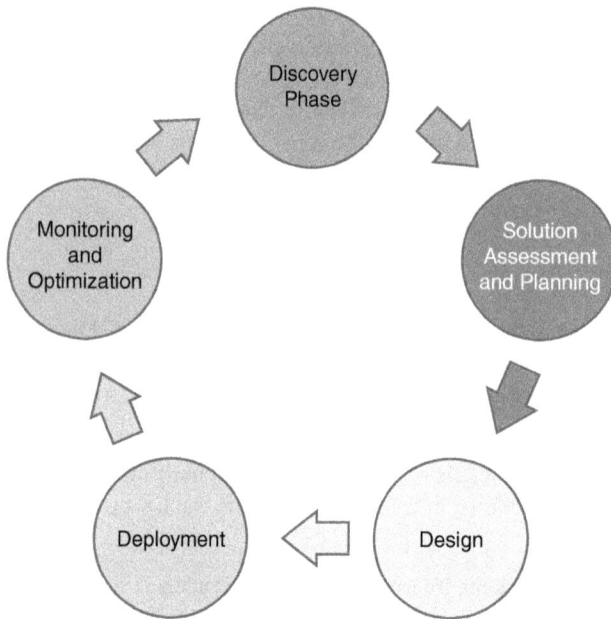

**Figure 14-1** *Migration to IPv6 Design Approach*

The sections that follow explain the phases shown in Figure 14-1.

## Discovery Phase

At this phase, network architects or designers usually focus on understanding and identifying the business goals and drivers toward the enablement of IPv6, in addition to other influencing factors such as project timeframe, government compliance, and the geographic distribution of the sites with regard to IP addressing availability. Information about other influencing factors from the network point of view at this stage also needs to be identified and gathered at a high level, such as whether the existing network infrastructures (LAN, WAN, security nodes, services, and applications) support IPv6 and whether the business is willing to invest and upgrade the nodes that do not support IPv6. Therefore, it is critical that the right and relevant information be gathered at this phase so that it can be analyzed and considered during the planning phase.

## Solution Assessment and Planning

Migrating or integrating IPv4 and IPv6 networks might seem as simple as enabling IPv6 routing across certain network areas; however, the reality is that several other factors must be considered during the planning phase to ensure that the design will deliver its intended goal and be prepared for the challenges and constraints that may arise during the deployment phase. After the completion of the high-level business and network discovery, network designers need to analyze each of the identified influencing factors and generate a migration or integration plan based on the gathered information. In general, the following considerations help at this phase to influence and drive the detailed design of the transition strategy and approaches at later stages:

■ **Goal:** Understanding the main purpose of the migration or integration (based on the information gathered during the previous phase) is important to driving the design in the most suitable direction. For instance, is the transition to IPv6 to access certain services in the data center? To comply with "regulatory compliance"? Or to enable IPv6 at the Internet edge only (for example, lack of public IPv4 pools)?

- **Infrastructure support:** Furthermore, the answers with regard to the goal become vital during the planning phase, because it is important to understand whether the entire infrastructure supports IPv6 or only the network edges (for example, provider edges [PEs] in service provider [SP] network and access or distribution in campus networks) and whether the business is willing to upgrade the devices that do not support IPv6. Based on this, the migration or integration plan guides the next phase (design) as you consider a technology solution to overcome any constraint with regard to IPv6 support across the network.

- **Existing services and applications:** Although many applications already support IPv6, many still do not, especially those that are developed in house. Therefore, understanding and assessing application support is critical because the upgrade, if it fails to consider how the IPv6 networks will reach these IPv4-only applications, will break the communication with these applications. Such a break may seriously impact the business.

In other words, the planning phase takes the output information from the discovery phase and analyzes it and uses it as a foundation to drive the selection of the appropriate migration/integration approach. Consider, for example, that an enterprise needs to migrate its network to be IPv6 enabled (end to end), but at present the core network components do not support IPv6 and the business is not allocating any budget to upgrade these components. In addition, access to some new IPv6-enabled applications hosted at the data center is an urgent requirement. This information should be collected during the discovery phase, and the network designer at the assessment and planning phase is expected to select the right approach to meet the requirements for this enterprise, taking into consideration the relevant constraints. In this case, the network designer may suggest either an IPv6 over IPv4 tunneling mechanism (either sourced from the workstation or using access/distribution switches), Domain Name System (DNS)-based translation, or 64NATing to facilitate accessing the new IPv6-based applications over the existing IPv4 core infrastructure.

The following subsections cover the approaches possible during the planning phase with regard to IPv6 based on the targeted environment (enterprise versus service provider). The selected approach should ideally be complemented with one or more of the technical mechanisms listed later in Table 14-6 at the design phase. Therefore, each phase (ideally) must take the outcomes of the previous phase as a foundational input to achieve a cohesive end-to-end business-driven design that avoids any "design in isolation" throughout all the phases.

### Transition Approaches to IPv6 for Enterprise Networks

After identifying the objectives of the migration or integration of IPv4 and IPv6, network designers must decide which approach is the most suitable, taking into consideration the objectives, the different requirements, and design constraints, such as timeframe, available budget, and whether IPv6 is supported by the network nodes. Table 14-4 summarizes the most common approaches, based on different design goals and priorities, to transition an enterprise network to be IPv6 enabled.

**Key Topic**

**Table 14-4** Approaches to Enable/Transition to IPv6 for the Enterprise

| Design Goal | Priorities | Timeframe | Design Approach | Design Considerations |
|---|---|---|---|---|
| Migrate the enterprise network to be pure IPv6 (IPv6-only) or dual stack | No service interruption | Flexible | Migrating the core to be in dual-stack mode first, then other enterprise modules can be gradually migrated to IPv6-only or dual stack, depending on the goals and requirements of the business. | Increased utilization of hardware resources. Increased control plane complexity in dual-stack mode. Core network components must support IPv6. Partially migrated blocks may require tunneling as an interim solution. |
| Migrate the enterprise network fully or partially to be pure IPv6-only or dual stack | Quickly migrating certain enterprise modules first, such as data centers | Limited | Migrating certain modules of the enterprise network first. A DNS translation or tunneling mechanism such as ISATAP is required to maintain the communications between IPv6 and IPv4 islands within the network. | This approach is suitable when the core device does not support IPv6 and requires either hardware or software upgrades. Increases design and control plane complexity. Increases operational complexity. |
| Migrate the data center network fully or partially to support communication with IPv6 hosts | Support virtualized and nonvirtualized IPv6 hosts | Flexible | Depending on the data center architecture, different approaches can be used. For example, a DC design can consider a typical dual stack, or an overlay can be considered using VXLAN to support IPv6 hosts; also, MPLS-enabled DC may consider 6PE/6VPE. | In all cases, dual stack increases hardware resource utilization, control plane complexity, and operational complexity. If the timeframe is limited, one quick and common solution is host-based tunneling (ISATAP, RFC 5214), assuming endpoints and hosts support this tunneling mechanism. |

| Design Goal | Priorities | Timeframe | Design Approach | Design Considerations |
|---|---|---|---|---|
| Provide IPv6 access either inbound or outbound at the enterprise Internet edge | Support translation between IPv4 and IPv6 at the Internet edge | Flexible | A translation mechanism is required that is either based on a load balancer, pure DNS, or classical NAT64. | Increases operational complexity at the Internet edge. Requires additional security considerations with regard to IPv6 enablement. If the timeframe is limited, it is always quicker to consider the current infrastructure support. For example, using NAT at the existing Internet edge router is quicker than provisioning a new DNS server to do the translation. |

14

### Transition Approaches to IPv6 for Service Provider Networks

In its overall approach and goal, enabling IPv6 in an SP network is slightly different from in enterprise networks. Typically, the SP networks enable IPv6 either to provide a transit path for other SPs (such as transit and peering Internet service providers) or to offer IPv6 connectivity to its customers. Which mechanism to use is mainly driven by the goal and the transport used by the SP, whether it is based on native IPv4 only or Multiprotocol Label Switching (MPLS) based (with or without MPLS VPN). Table 14-5 summarizes these approaches from the SP point of view.

> **NOTE**   Some of the transition approaches to IPv6 for Service Provider Networks include technical options that are out of scope for the CCDE v3 exam at the time of this writing. These options are still listed in Table 14-5 and Table 14-6 to provide the specifics for the transition, but this chapter does not go into the detail of explaining them. This includes the following topics: 6PE, 6VPE, and 6rd.

**Key Topic**

**Table 14-5**  Approaches to Enable/Transition to IPv6 for the Service Provider

| Goal | Transport | Possible Approaches |
|---|---|---|
| Provide IPv6 Internet transit | Native IPv4 | Dual stack, tunneling (manual RFC 2893, Generic Routing Encapsulation [GRE], Layer 2 Tunneling Protocol Version 3 [L2TPv3]) |
| Provide IPv6-based services and Internet access to residential clients | Native IPv4 | Dual stack, 6rd, tunneling such as IPv6 over L2TP |
| Provide IPv6 Internet access/transit | MPLS | 6PE, IPv6 over pseudowires |
| Provide IPv6 connectivity for MPLS L3VPN customers | MPLS | 6VPE |
| Provide IPv6 Internet access for MPLS L3VPN customers over a separate VPN or within the same customer VPN | MPLS | 6VPE |

Unlike with enterprise networks, enabling IPv6 in a service provider network can be more flexible and less interruptive. Service provider networks by nature are transport networks where no endpoints or hosts are directly connected. This means that there is no need to consider any directly connected host with this transition, except host and services used by the SPs to deliver specialized functions such as network management, DHCP, and authentication/authorization services. Besides, when MPLS is enabled (nowadays MPLS to a large extent is the de facto protocol for SP networks), the integration and enablement of IPv6 will be simpler with the MP-BGP overlay capability (MP-BGP-based 6PE, 6VPE). Furthermore, operators can consider the phased approach in which only the PE nodes that need to provide IPv6 transit connectivity need to be enabled with IPv6 first without introducing any change to the core (P) routers. However, the additional load and hardware resource limits must be calculated and taken into consideration.

**NOTE**  One of the primary considerations when migrating or integrating with IPv6 is to ensure that IPv6 is secured in the same manner that IPv4 is secured. Otherwise, the entire network will be vulnerable to network attackers and breaches. For instance, since the release of Microsoft Windows Server 2008, IPv6 has been native to the Windows OS, which supports transition technologies at the server/client level such as ISATAP. In this case, if one of these servers is compromised and the network security rules do not consider IPv6, malicious traffic can ride an IPv6 tunnel or packet without being blocked or contained by the security devices in the path.

As covered so far, the migration to IPv6 differs slightly in SP networks as compared to in enterprise networks, based on the fact that SPs mainly need to provide a transit path to their customers and other peering providers. However, today's SPs are not limited to just providing transit paths; they also offer many other services, such as hosted services and applications, Software as a Service (SaaS), cloud-based hosting data centers, and content services such as IPTV. Practically, service providers cannot simply migrate to native IPv6, because they normally have other services and customers who need IPv4. Therefore, the coexistence of IPv4 and IPv6 is inevitable nowadays for SPs. What is important to plan and design for is the strategy that can be used to incorporate IPv6 into the existing provider network in a way that is not interruptive and at the same time offers enough flexibility and reliability to optimize SPs' time to market with regard to IPv6-enabled services and transport while maintaining IPv4 support intact. One common approach that application and content providers use to support IPv6 is to enable IPv6 at the services level first. With this approach, providers that offer hosted services and applications, whether it is a cloud offering (such as SaaS) or any other IP service such as VoIP and Internet, usually enable or migrate these services to be IPv6 enabled first, and so customers that need to use these services must be IPv6 enabled, using a form of translation at the customer side such as NAT v4 to v6 at the enterprise Internet edge, or the provider might offer the translation service at a different cost or as a value-added service such as DNS-based translation. Also, the SP must ensure the underlay transport can carry IPv6 communications. In fact, this approach is similar (to a large extent) to enabling or rolling out IPv6 applications in an enterprise data center in which the enterprise needs to incorporate an interim solution to provide and maintain access to these applications to users with only IPv4 endpoints.

When the desired approach is decided based on the design requirements and constraints, network design can move to the next phase to start connecting the dots and generate a solution driven by the outcomes of the previous phases.

## Detailed Design

After selecting the suitable approach for the migration or integration between IPv4 and IPv6 networks, ideally based on the gathered and analyzed information during the planning phase, network designers at this stage can put together the details of the design, such as selecting the suitable integration mechanism, deployment details such as tunnels termination, IP addressing, routing design, network security details, and network virtualization considerations, if any are required. Typically, the outcome of the design phase will be used by the implementation engineers during the deployment phase to implement the designed solution. Therefore, if there is any point that is not doable or practically cannot be implemented, it will be reported back to this phase (to the network designer) to be revised and changed accordingly. There are various mechanisms and approaches with regard to integrating IPv6 and IPv4. For simplicity, these mechanisms can be classified as following:

- Dual stack
- Tunneling based
- Translation based
- MPLS environment solutions

Table 14-6 lists the various technical mechanisms that can be used to integrate and support the coexistence of both IPv4 and IPv6, taking into consideration some of the primary design aspects that influence the solution selection based on the design requirements.

**NOTE** Information in Table 14-6 is not a best practice or mandatory with regard to IPv6 integration and migration options. However, it is based on the most commonly considered technology solutions for certain scenarios. And network designers must always assess the different influencing factors before suggesting any approach or mechanism.

**Key Topic**

**Table 14-6** Mechanisms to Support Coexistence of IPv4 and IPv6

| Mechanism | Scenario | Targeted Environment | Design Concern |
|---|---|---|---|
| Dual stack | End-to-end IPv6 + IPv4. | Any environment that ultimately needs to move to end-to-end IPv6 | IPv6 support in all L3 ware platforms is required.<br><br>Increased control plane complexity.<br><br>May introduce scalability weaknesses when both IP versions are running together (depends on available hardware resources such as memory). |
| Tunneling: point to point (P2P) (L2TPv3, GRE RFC 2473) | Transit IPv6 over IPv4-only network. | A small number of IPv6 islands that need to interconnect over IPv4-network | Scalability and encapsulation overhead.<br><br>Increased control plane complexity. |
| Tunneling: ISATAP (RFC 5214) | Host-source tunnels that terminate at IPv6-enabled modules or services. | For trial IPv6 services or in case of IPv6 enabled partially (for example, only at the data center); mostly enterprise networks | Affects the overall network architecture. QoS, multicast, and NAT issues.<br><br>Adds control plane complexity.<br><br>Increases operational complexity. |
| Tunneling: mGRE | Interconnect IPv6 over IPv4 in a hub-and-spoke topology. | Interconnects IPv6-enabled remote sites in hub-and-spoke topology over IPv4 WAN.<br><br>Interconnects private IPv6 islands across public IPv4 clouds. | Multicast traffic has to go via the hub.<br><br>Adds control plane complexity.<br><br>Increases operational complexity. |
| Tunneling: 6rd (RFC 5969) | Used to extend IPv6 deployment to customer sites (usually residential gateway), with limited impact on existing IPv4 infrastructure. | Service provider networks that offer IPv6 services/Internet access over IPv4 service provider network to residential customers.<br><br>Simple, stateless, automatic IPv6-in-IPv4 encap/decap that offers fast IPv6 enablement. | Whether the network equipment supports 6rd.<br><br>Adds control plane complexity.<br><br>Increases operational complexity. |
| Tunneling: IPv6 over L2TP software | To offer IPv6 access for residential gateway. | Digital subscriber line (DSL)/residential service providers with limited investment.<br><br>Stateful architecture on L2TP network server (LNS). | Dual-stack IPv4/IPv6 service on residential gateway LAN side.<br><br>Increases operational complexity. |

| Mechanism | Scenario | Targeted Environment | Design Concern |
|---|---|---|---|
| Translation 64: NAT/SLB | Allows IPv6 handsets/ endpoints to access IPv4 Internet over Long Term Evolution (LTE)/4G/5G or IPv4 services. | Green-field IPv6 service providers or enterprise networks that need to be interconnected to legacy/ existing IPv4 network/ services. | Does not support every application type or protocol today. Performance may not match dual-stack design, depending on traffic load. |
| Translation: DNS | DNS in this scenario offers the translation between the v4 and v6 based on the source and targeted host. | Access applications or services by name. | Limited to services and applications that can be reached by name, not IP. NAT64 is usually required to leverage DNS64. |
| Translation: LISP | LISP encapsulation can facilitate the IPv6 communication over IPv4 transport. | Enterprise edge, data center, or WAN with a mix of IPv4 and IPv6 networks. | High operational complexity. Increased control plane complexity. Network devices must support LISP. |
| MPLS: 6PE | Facilitate enabling IPv6 over an existing MPLS and MP-BGP IPv4 network. | Large enterprises and service providers that want to provide IPv6 over their IPv4 infrastructure. | Does not provide traffic separation between different customer networks (no MPLS VPN support). Increases control plane complexity. |
| MPLS: 6VPE | Facilitate enabling IPv6 over an existing MPLS/MP-BGP IPv4 network for VPN customers. | MPLS VPN providers or enterprises with MPLS VPN networks. | Increases control plane complexity. May introduce scalability limitations because a separate routing information base (RIB) and forwarding information base (FIB) are required per customer. |

14

**NOTE**   Typically, adding any overlay or tunneling mechanism to the network will almost always increase the level of its operational complexity. However, this level varies based on several factors, such as network size, routing design, staff knowledge, and the nature of the selected technology itself.

### Deployment, Monitoring, and Optimization

These steps usually are more relevant to the implementation of the design followed by continuous monitoring for the implemented solution to ensure that the network is delivering the promised value and meeting the expectations. Ideally, the implementation should follow an implementation plan that specifies what the services are, and the futures that need to be enabled before proceeding with any step. It should also cover any potential risk associated with any change (for example, whether it is going to be disruptive to the existing production network). For instance, enabling IPv6 at the routing protocol level may lead to resetting the existing IPv4 peering sessions; normally this depends on the routing protocol, hardware platform, and software in use.

## Transition to IPv6 Scenario

ABC Corp. is an international real-estate company headquartered in Singapore, with 116 remote sites distributed across Asia, Australia, and Europe, as shown in Figure 14-2.

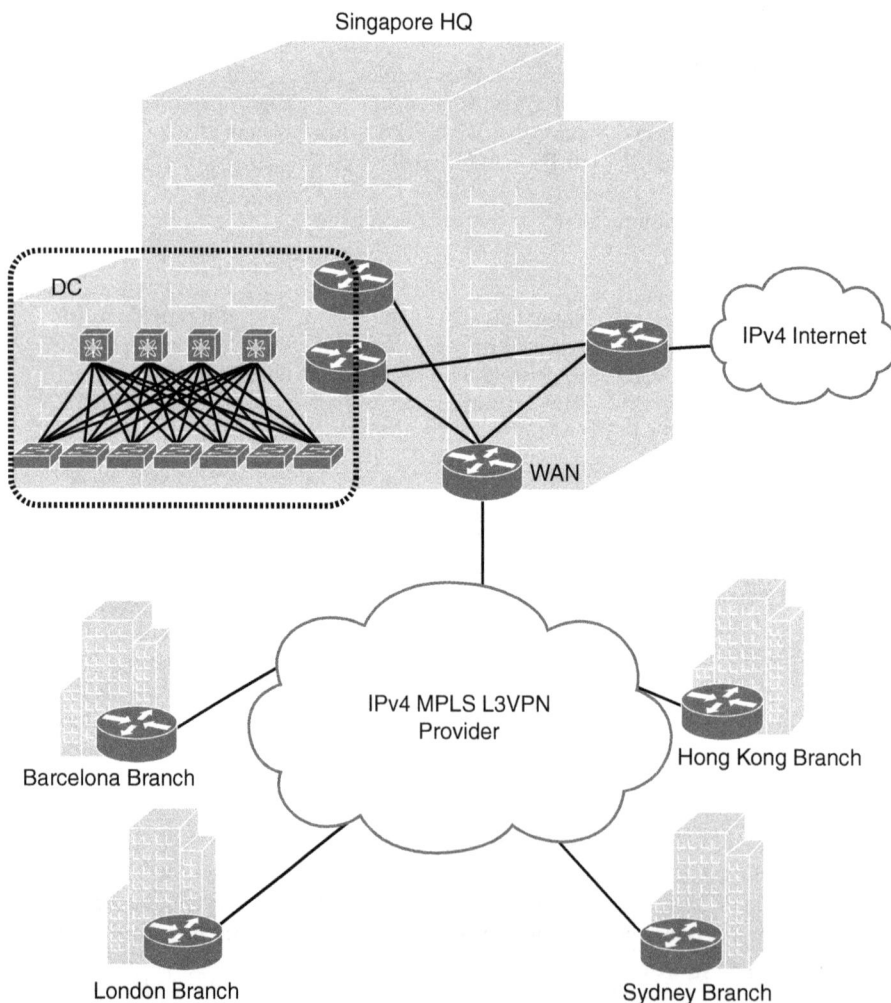

**Figure 14-2**  *ABC Corp. Network*

The CIO of ABC Corp. has decided to migrate the company's entire IP network infrastructure and applications to be primarily based on IPv6 to support some of the long-term business innovation and strategic plans. In addition, ABC Corp. wants to achieve the following with this transition project to maintain business continuity:

- Retain the ability for its internal users to access some legacy applications that do not support IPv6 and to access the IPv4 Internet

- Provide the ability for external users to access the ABC Corp. new IPv6 web-based services over the IPv4 Internet

Therefore, ABC Corp. has decided to add another Internet link dedicated to accessing IPv6 Internet services. Moreover, the security team of ABC Corp. requires accessing some (predefined) IPv4 Internet websites, "web-based services," by the internal IPv6-enabled users to appear as if they are accessible over the IPv6 Internet. For example, when an IPv6-enabled user accesses a website across the IPv4 Internet using a typical site, such as www.example.com, the resolved source IP address of the website's domain name by the DNS should appear to the user as an IPv6 source address instead of an IPv4 source address.

One of the ABC Corp. primary requirements is that the go-live of the IPv6 network project be within six weeks. Therefore, the company has hired a network consultant to provide a strategic approach that can help it to achieve this goal within this limited timeframe.

## Network Requirements Analysis

Based on the information provided, the following are the primary design constraints:

- ABC Corp. needs a quick (transition) solution to enable IPv6.

- ABC Corp. Internet and data center services are all located at the HQ/hub site (centralized model).

- The current MPLS VPN WAN provider does not support IPv6.

Design considerations include the following:

- The network must be end-to-end IPv6 enabled.

- Provide the ability to ABC Corp. end users to access the IPv4 Internet and IPv4-only legacy applications.

- Web services accessed over the IPv4 Internet by internal IPv6 clients must appear as if they are accessed over the IPv6 Internet.

- ABC Corp. IPv6 web services must be accessible over the IPv4 Internet.

The assumption is that network nodes across the network, end-user devices, applications, and hosts within the data center support IPv6.

## Design Approach

To meet the primary requirements of ABC Corp., taking into account the design considerations covered earlier, ABC Corp. can consider the following phased approach.

**Phase 1**

Provide fast IPv6 enablement across the network (see Figure 14-3):

- Enable IPv6 on all network nodes (dual stack), starting from the DC followed by other nodes such as WAN routers.

- Enable IPv6 routing on the network nodes (DC, WAN routers hub and spokes, and Internet edge).

- Enable stateful NAT64 at the IPv4 Internet edge gateway toward the IPv4 Internet to provide Internet access for the internal IPv6 devices.

- Introduce DNS64 functionality to satisfy the requirement of making Internet service IPv4 source addresses appear as if they are sourced from an IPv6 address (by synthesizing DNS A record into AAAA record).

- Enable static NAT64 at the DC edge nodes, where the IPv4 to IPv6 static mapping can provide internal IPv6-enabled users to access legacy IPv4-only applications.

- Enable static NAT64 at the IPv4 Internet gateway, where the IPv6 to IPv4 static mapping can enable external users to access ABC Corp. IPv6 web-based services over the IPv4 Internet.

- Interconnect the IPv6 network islands (spokes/remote sites) with the HQ/hub site using an IP overlay tunneling mechanism (preferably IPv6 over mGRE IP tunneling dynamic multipoint VPN [DMVPN]) over the IPv4 MPLS VPN WAN.

**Phase 2**

Design optimization (see Figure 14-3):

- Migrate the WAN to a provider that supports IPv6 MPLS L3VPN to be used instead of the overlaid IPv6 DMVPN over the WAN.

- Disable IPv4 routing from the network areas where no IPv4 clients/hosts exist, such as remote sites, which helps to reduce the load from the remote site network nodes (holding separate RIB/FIB tables for each IP version).

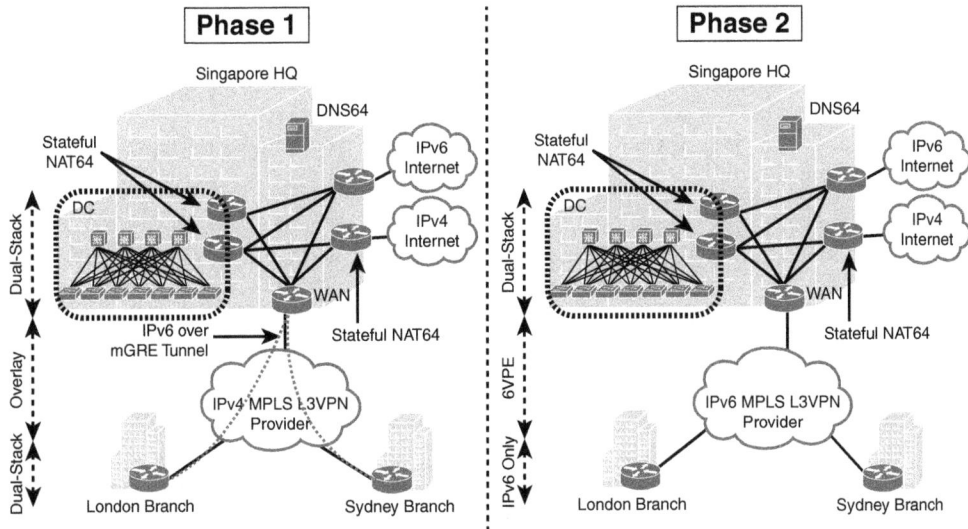

**Figure 14-3**  *ABC Corp. Transition Approach to IPv6: Phases 1 and 2*

> **NOTE**  For a sample WAN migration example, see Chapter 17, "Enterprise WAN Architecture Design."

## Quality of Service Design Considerations

In today's converged networks, there is an extremely high reliance on IT services and applications. In particular, there is increased demand on real-time multimedia applications and collaboration services (also known as media applications) over one unified IP network that carries various types of traffic streams such as voice, video, and data applications. For instance, voice streams can be found across the network in different flavors, such as standard IP telephony, high-definition audio, and Internet Voice over IP (VoIP). In addition, video communications also have various types, where each has different network requirements, such as video-on-demand, low-definition interactive video (such as webcams), high-definition interactive video (such as telepresence), IP video surveillance, digital signage, and entertainment-oriented video applications. However, there can be an unlimited number of data applications.

Therefore, to deliver the desired quality of experience to the different end users, network designers and architects need to consider a mechanism that can offer the ability to prioritize traffic selectively (usually real-time and mission-critical applications) by providing dedicated bandwidth, controlled jitter and latency (based on the application requirements), and improved loss characteristics, and at the same time ensuring that providing priority for any traffic flow will not make other flows fail. This mechanism is referred to as quality of service (QoS). The following sections discuss the design approaches and considerations of QoS using a business-driven approach.

## QoS High-Level Design: Business-Driven Approach

To design and deploy QoS successfully across the network, in particular converged networks, network designers or architects need to consider the top-down design approach to initially understand the critical and important applications from the business point of view. Then, they can assess the optimal QoS design strategies to meet the business and application requirements.

As discussed earlier, the goal of the network is to become a business enabler. To achieve that, the network ideally should act as a service delivery mechanism that facilitates the achievement of the business's goals. Therefore, it is critical to align QoS design with business priorities, expectations, and requirements to deliver the desired level of quality of experience.

For instance, an organization may have a financial application that is sensitive to packet loss, and any loss of connectivity can cost the business a large amount of money. Accordingly, this application must be treated as a high-priority application. Similarly, if an SP needs to meet a very strict SLA with its customers to deliver their voice traffic with no more than 1 percent of end-to-end packet loss, this SP must consider a suitable QoS strategy and apply the right QoS design to meet these SLA requirements. Otherwise, it will lead to a business loss. This loss can be tangible, such as a penalty, intangible, such as reputation, or most likely both. Therefore, network designers must have the right QoS design approach and strategy to avoid the high volume of application complexities in today's networks and provide the desired level of quality of experience to the business and end users. Table 14-7 provides a summarized (top-down) QoS design approach and strategy (in order).

**Key Topic**

**Table 14-7** Top-Down QoS Design Approach Summary

| Strategic Goal | Approach | Design Considerations |
|---|---|---|
| Understand business requirements. | Understand business priorities and goals. | Identify primary business drivers. Highlight the constraints (for example, budget). |
| Identify the scope. | Understand the scope of the QoS design, such as campus, WAN, VPN, or service provider edge or end to end across different blocks. | Is the application used within the campus, across the WAN, or over VPN? Is there any network in the path that is not directly controlled, such as a WAN? |
| Identify mission-critical applications. | Identify which applications need to be treated differently. Identify which applications are not nonbusiness applications. | Identify the mission-critical applications or services (for example, SAP, FCoE, VoIP, TelePresence). |
| Understand application requirements. | Identify the characteristics of each application. | What sort of network delivery is required: TCP, UDP, unicast, multicast, and so on? Application sensitivity to packet loss, jitter, and delay. |

| Strategic Goal | Approach | Design Considerations |
|---|---|---|
| Select a design strategy and identify the technical constraints. | Clarify the end-to-end design strategy in terms of number of QoS classes, QoS toolset to be used, and so on. | What traffic classification strategy is to be used within the LAN (for example, 8 or 12 classes)? |
| | | What MPLS DiffServ tunneling mode is used? |
| | | Is the core/WAN underlay native IP or MPLS based? |
| | | What are the classes of service (CoS) supported over the WAN (for example, MPLS provider)? |
| | | Can the targeted network node support the required number of queues? Or does it support any priority queueing (technical constraints)? |

## QoS Architecture

In general, there are two fundamental QoS architecture models:

**Key Topic**

- **Integrated services (IntServ):** This model, specified in RFC 1633, offers an end-to-end QoS architecture based on application transport requirements (usually per flow) by explicitly controlling network resources and reserving the required amount of bandwidth (end to end along the path per network node) for each traffic flow. Resource reservation protocols, such as RSVP, and admission control mechanisms form the foundation of this process.

- **Differentiated services (DiffServ):** This model, specified in RFC 2475, offers a QoS architecture based on classifying traffic into multiple subclasses where packet flows are assigned different markings to receive different forwarding treatment (per-hop behavior [PHB]) per network node along the network path within each differentiated services domain (DS domain).

**NOTE**   The aforementioned QoS architectural models are applicable for both IPv4 and IPv6, because both IP versions include the same 8-bit field in their headers, which can be used, for example, for DiffServ (IPv4: Type of Service [ToS]; IPv6: Traffic Class). Therefore, the concepts and methodologies discussed in this section are intended for both IPv4 and IPv6 unless otherwise specified, such as if any application supports and uses the added 20-bit Flow Label field of the IPv6 header (RFC 8200, 6437). However, the larger IPv6 packet's header needs to be considered when calculating the aggregate bandwidth of traffic flows.

## QoS DiffServ Architecture and Toolset

A true and effective QoS design must cover traffic flows of applications and services end to end. This can sometimes be complicated because each single traffic flow traverses multiple networks and is architected with a different QoS philosophy. Therefore, for network architects and designers to create such a cohesive design model, the design must be divided into domains commonly called (and described in RFC 2475 as) differentiated services domains

(DS domains). Each DS domain usually consists of multiple interconnected network nodes (such as switches and routers) operating under a common service provisioning policy, along with a set of PHB groups enabled on each node that offers network designers a framework to facilitate the design of scalable and flexible DS domains.

**NOTE** Per-hop behavior (PHB) defines a forwarding behavior with regard to the treatment characteristics for the traffic class, such as its dropping probabilities and queueing priorities, to offer different levels of forwarding for IP packets (described in RFC 2474). DS domains described in this section refer to any domain with QoS policies and differentiated treatments, regardless if it is IP Precedence based (RFC 791), Assured Forwarding based (RFC 2597), or a mixture of both.

**NOTE** *Service provisioning policy* refers to the attributes of traffic conditioners (QoS policies) deployed at each DS–domain boundary and mapped across the DS domain.

Each single DS domain can be divided into two primary types of DS network nodes. The first type is internal, which usually refers to the nodes belonging to a single DS domain. As described earlier, these nodes should have the same QoS or DS provisioning policy. The second type of DS node is the DS boundary node, which faces other DS or non-DS-capable domains. Typically, these nodes are the ones that must be responsible for applying traffic policies (QoS policies) on traffic flows in both directions (ingress and egress) based on a predefined or agreed model between the DS domains that is ideally driven by traffic and application requirements, as shown in Figure 14-4.

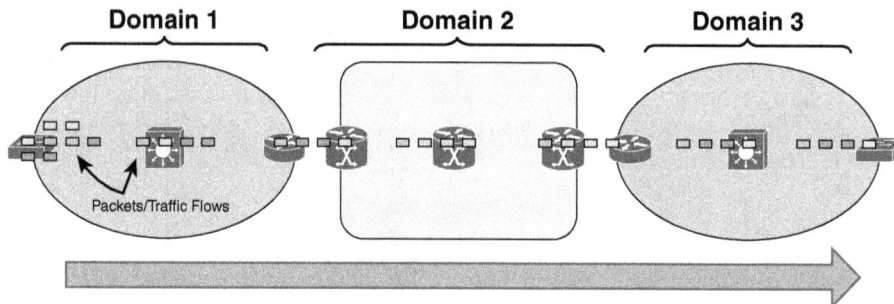

**Figure 14-4** *QoS DS Domains*

Practically, DS domains can take different forms, such as the following:

- An enterprise domain with a SP domain in the middle (WAN transport).

- Within an enterprise, there can be multiple DS domains: campus LAN, WAN, DC, and DMVPN over Internet.

The second scenario consists of multiple DS domains, which all belong to a single administrative authority and can be combined under one global DS domain or region. Each DS domain in that region has the flexibility to have its own QoS provisioning standards to offer a more structured and tiered QoS domains design for large-scale networks with a large number of distributed blocks, as shown in Figure 14-5.

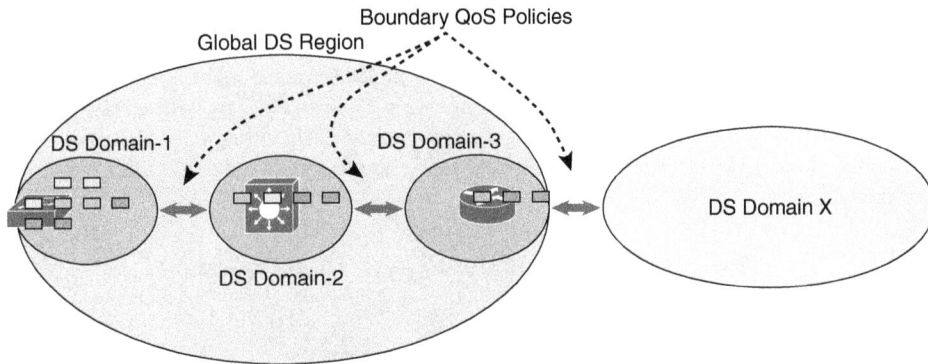

**Figure 14-5** *Multitier DS Domains*

Each domain is usually under a single administrative and control authority. Hypothetically, each authority defines cohesive end-to-end, measurable, and quantifiable attributes per DS subdomain and per global DS domain, which should be driven by the characteristics and requirements of the services or applications running across this domain (traffic flow aggregates). Typically, there can be multiple possible points across each QoS domain specifying where policies and traffic treatment can be enforced to influence the experience of packets as they enter, cross, and exit a DS domain. Traffic conditioning and QoS policies are the key QoS elements that serve this purpose within each domain and between the different domains using the following primary QoS toolset:

- Traffic classification and marking

- Traffic profiling and congestion management

- Congestion avoidance (active queue management)

- Admission control

The subsequent subsections cover these QoS toolsets in more detail.

## Traffic Classification and Marking

*Traffic classification* refers to the process of selecting frames or packets in a traffic stream based on the content of some portion of the frame or packet header to which different policies can then be applied. *Traffic marking*, on the other hand, writes a value into the packet header to be identified by QoS policies and placed in the desired class with the desired treatment at different stages during the end-to-end packet trip from the source to the intended destination (within and across DS domains).

However, classification of traffic does not always require marking to be applied. For instance, in some scenarios, traffic only needs to be selected based on the value of a combination of one or more IP header fields, such as source address, destination address, source port, destination port numbers, or an incoming interface then to be associated with a QoS policy action, such as placing it in a predefined QoS queue. Classification of traffic almost always should be performed at the point of network access (as close to the traffic source as possible), and then to be associated with the appropriate marking value (usually ToS header bits) in which network designers can have the flexibility to select this traffic and apply the desired QoS policies at any node across the network (usually within a single DS domain or region).

In addition, marking can establish trust boundaries at the edge of the network. A *trust boundary* refers to the point within the network where markings such as class of service (CoS) or differentiated services code point (DSCP) begin to be accepted as its set by the connected node or endpoint, such as an IP phone that set its voice traffic with a DSCP-PHB value of EF. However, not every endpoint can mark its traffic. Therefore, trust boundaries are commonly classified into three primary models: trusted, untrusted, and conditional trust, as shown in Figure 14-6.

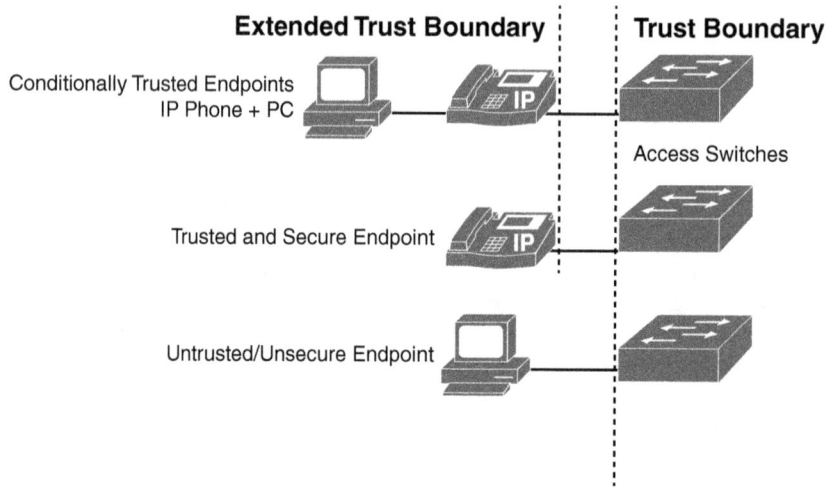

**Figure 14-6** *QoS Trust Boundaries*

**Key Topic**

- **Trusted model:** This model can be used with endpoints that can mark their traffic. At the same time, these endpoints have to be approved and trusted from a security point of view, such as IP phones, voice gateways, wireless access points, videoconferencing, and video surveillance endpoints. In addition, "ideally" these trusted endpoints should not be mobile (i.e., fixed endpoints) to be reflected at the switch port level in a more controlled manner.

- **Untrusted model:** This model usually considers using manual traffic classification and marking. The most common candidates of this model are PCs and servers because these endpoints are subject to attack and infection by worms and viruses that can flood the network with a high volume of malicious traffic. Even worse, this traffic might be marked with a CoS or DSCP value that has priority across the network, which usually leads to a true denial-of-service (DoS) situation. However, PCs and servers normally run business-critical applications that need to be given certain service differentiation across the network, such as a PC running a software-based phone or a server running business-critical applications such as SAP or CRM. With this model network, designers can selectively classify each of the desired application's traffic flows to be treated differently across the network and mark it with the desired CoS/DSCP value along with a policy that either limits each application class to a predefined maximum bandwidth or marks down the out-of-profile traffic to a CoS/DSCP value that has lower priority across the network. Furthermore, as a simple rule of thumb, any endpoint that is not under control by the enterprise should be considered untrusted, and the classification and marking of traffic flows can be controlled selectively and manually.

■ **Conditional trust model:** This model offers the ability to extend the trust boundary of the network edge to the device or endpoint it is connected to. This is based on an intelligent detection of a trusted endpoint, usually an IP phone. (In Cisco solutions, this is achieved by using Cisco Discovery Protocol [CDP].) However, the IP phone in this scenario has a PC connected to the back of the IP phone. Therefore, by extending the trust boundary to the IP phone, the IP phone can send its traffic in a trusted manner while overriding PC traffic to a DSCP value of 0. This model offers a simple and easy method to roll out large IP telephony deployments with minimal operational and configuration complexity. However, if the PCs have some applications that need to be marked with a certain DSCP value, such as a softphone or a business video application, in this case, manual traffic classification and marking are required at the edge port of the access switch to identify this traffic and mark it with the appropriate CoS/DSCP value and ideally associate it with a policer.

Furthermore, marking values can be changed (re-marked) based on the QoS design requirements at any location within a DS domain or between DS domains. The most common example is that between DS domains there might be a mismatch between the ToS values used to classify traffic. Therefore, at the egress or ingress, re-marking has to take place to maintain an end-to-end unified QoS model. In contrast, in a single DS domain, re-marking is commonly used to move traffic that is out of profile (out of policy limit) into a QoS class with lower bandwidth or priority as a protective countermeasure by re-marking the out-of-profile traffic flows with a different ToS value at any location within the DS domain. Table 14-8 summarizes the most common classification and marking options.

**Table 14-8** Summary of QoS Classification and Marking Options

| OSI Layer | Classification | Marking |
|---|---|---|
| Physical layer | Input interface | N/A |
| Layer 2 | VLAN ID, MAC, IEEE 802.1Q/p CoS | IEEE 802.1Q/p CoS |
| Layer 2.5 | MPLS label, MPLS EXP | MPLS EXP |
| Layer 3 | IP DSCP, IP based (such as IP source/destination) | IP (IPP, DSCP) |
| Layer 4 | Port-based source/destination | IPP, DSCP, EXP |
| Higher layers up to 7 | Application signature such as using a network-based application recognition (NBAR) framework | IPP, DSCP, EXP |

**NOTE** DSCP marking is more commonly used than IP Precedence (IPP) because of its higher flexibility and scalability to support a wider range of classes. However, in some scenarios, a mix of both may be required to be maintained, such as migration or integration between different domains (such as in merger and acquisition scenarios and WAN MPLS VPN service providers that offer their CoS classes based on IPP) where seamless QoS interoperability has to be maintained. In this case, class selector PHB is normally used to provide backward compatibility with ToS-based IP Precedence (RFC 4594, 2474).

Logically, after traffic flows are classified and marked (whether manually at the edge of the DS domain or automatically by being considered one of the trust boundary models discussed earlier), traffic flows have to be grouped in DS classes. Usually, application flows that share similar or the same traffic characteristics and network requirements such as delay, jitter, and packet loss can be placed under the same DS class to achieve a structured traffic grouping that helps network operators to assign the desired treatment at different locations across the DS domain (per class), such as assigning different queuing models per class to control traffic flows during periods of traffic congestion.

## Traffic Profiling and Congestion Management

During normal situations where traffic passing through the network is under or equal to the actual maximum available bandwidth capacity of its links, packets are usually sent out of the interface as soon as they arrive. In contrast, in traffic congestion situations, packets normally arrive faster than the outgoing interface can handle them. This will lead to undesirable outcomes to traffic flows that likely will impact the business-critical application's performance and users' quality of experience. In other words, technically, if the network is overprovisioned with bandwidth, there may be no need for QoS considerations, because it will add minimal value. However, it is a common practice for QoS to be enabled with a minimum number of classes to cater to critical applications in case of unpredicted congestion that might occur, such as a failure of the upstream network node, which may lead to overutilization of the secondary path if the capacity planning did not consider different failure scenarios.

Therefore, to cater to network congestion situations, a network designer must plan for an effective mechanism to be used to manage traffic during congestion periods. This is commonly referred to as *congestion management*. When congestion management is used across the network, nodes can queue the accumulating packets at the outbound interface until the interface (Tx-Ring) is free to send the queued packets. The transmission of the queued packets, however, is scheduled based on the assigned desired priority and a queuing mechanism configured at the interface level per traffic flow aggregate (predefined traffic profiling).

In fact, with congestion management, placing packets in a predefined transmission queue is one half of the job. The other half of it is how the different queues are serviced with respect to each other during a congestion period at the router/interface level. Table 14-9 summarizes the common queuing mechanisms that can be used for the purpose of congestion management, along with the characteristics of each.

**Table 14-9**  Common QoS Queuing

| | weighted fair queuing (WFQ) | Priority Queuing (PQ) | class-based weighted fair queuing (CBWFQ) |
|---|---|---|---|
| Characteristic | The WFQ algorithm offers a dynamic distribution among all traffic flows based on predefined values like that of DSCP. | Typically supports four queues with different priority levels, and the higher-priority queues are always services first. | Provides class-based queuing (user-defined classes) with a minimum bandwidth guarantee. Supports flow-based WFQ for undefined classes, such as class-default. Supports low-latency queuing (LLQ). |

There are other queuing techniques, but they are not covered in this section because they are less commonly used or offer basic queuing capabilities. However, this does not mean that they are not used or cannot be considered as an option. These other techniques include weighted round-robin (WRR) and custom queuing. **First-in, first-out (FIFO)** queuing is the default queuing when no other queuing is used. Although FIFO is considered suitable for large links that have a low delay with very minimal congestion, it has no priority or classes of traffic.

Although WFQ offers a simplified, automated, and fair flow distribution, it can impact some applications. For instance, a telepresence endpoint may require end-to-end 5 Mbps for the video Rapid Transport Protocol (RTP) media stream over a 10-Mbps WAN link, and there may be multiple flows passing over the same link, let's say ten flows in total. Typically, with WFQ fairness, telepresence video streams will probably get one-tenth of the total available bandwidth of the 10 Mbps, which will lead to degraded video quality. With CBWFQ, though, network designers can place the flows of the telepresence RTP media streams in their own class with a minimum bandwidth guarantee of 5 Mbps during congestion periods. In addition, interactive video traffic can be assigned to the LLQ to be prioritized and serviced first during an interface congestion situation, as shown in Figure 14-7.

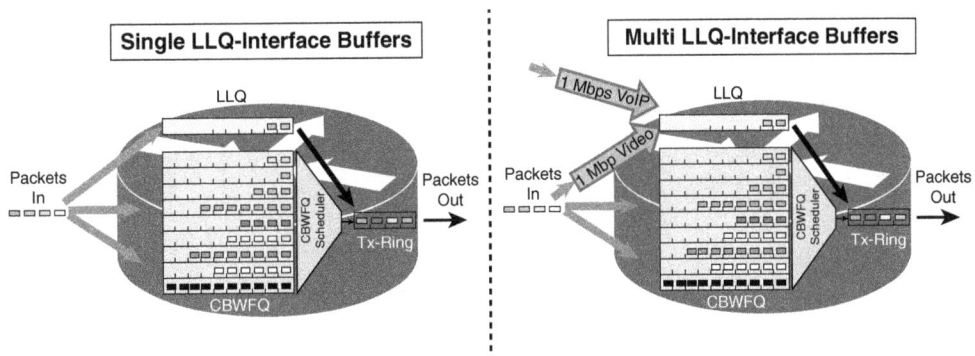

**Figure 14-7**  *CBWQ with LLQ*

As shown in Figure 14-7, CBWFQ supports two models of LLQ: single LLQ and multi-LLQ. Typically, with the multi-LLQ model, there will be a single aggregate LLQ from the software point of view, and multiple sub-LLQs can be enabled inside this strict priority queue. With this model network, operators can assign multiple traffic flow types to the LLQ, such as VoIP and video. However, inside the LLQ itself, the service is based on the FIFO concept. Therefore, within each parent LLQ admission control, it is required to protect LLQ from another LLQ, such as protecting a voice LLQ from a video LLQ.

**NOTE**  The different Cisco software versions, such as IOS, include a built-in policer (implicit policer) with the LLQ, which limits the available bandwidth of the LLQ (such as real-time traffic flows) to match the bandwidth allocated to the strict-priority queue, thus preventing bandwidth starvation of the non-real-time flows serviced by the CBWFQ sched-uler. However, this behavior (implicit LLQ policing) is applicable only during periods of interface congestion (full Tx-Ring). A similar concept is applicable to the multi-LLQ model, where a state implicit policer is enabled per LLQ.

### Hierarchical QoS

The most common scenario at the enterprise edge is that links are provisioned with sub-line rate. For instance, the WAN service provider may provide the physical connectivity over a 1-Gbps Ethernet copper link, whereas the actual provisioned bandwidth can be 10 Mbps, 50 Mbps, or any sub-line rate. The problem with this setup, even if QoS policies are applied on this interface, such as CBWFQ, is that it will not provide any value or optimization because QoS policies normally kick in when the interface experiences congestion. In other words, because the physical link line rate in this scenario is higher than the actual provisioned bandwidth, there will be no congestion detected even if the actual provisioned sub-line rate is experiencing congestion, which means that QoS has no impact here. Therefore, with hier-archical QoS (HQoS), the shaper at the parent policy (as shown in Figure 14-8) can simulate what is known as backpressure to inform the router that congestion has occurred in which QoS policies can take place.

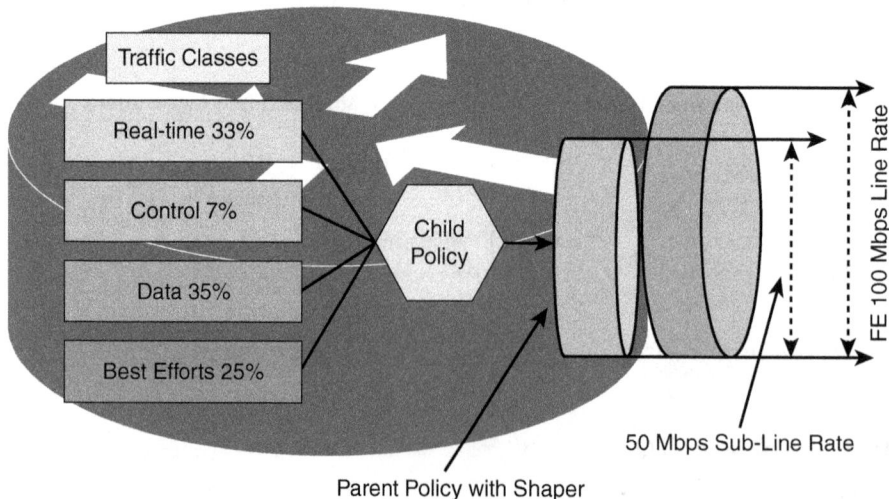

Figure 14-8  *HQoS*

## Congestion Avoidance (Active Queue Management)

The congestion management techniques discussed in the previous section help network designers and operators to manage the front of the queue in terms of which packets should be sent out first. Congestion avoidance algorithms, in contrast, manage the tail of the queue in terms of which packets should be dropped first when queuing buffers are full. One of the most commonly used and proven techniques here is the weighted random early detection (WRED), where packets are dropped based on their ToS markings, either IP Precedence based (where packets with lower IPP values are dropped more aggressively than higher IPP values) or DSCP based (where higher AF drop precedence values are dropped more aggressively). When the WRED algorithm starts selectively dropping packets, it will usually impact TCP windowing mechanisms to adjust the rate of flows to manageable rates. Therefore, WRED helps to optimize TCP-based applications. WRED is a member of a general family of technologies called Active Queue Management (AQM).

## Admission Control

Admission control is a common and essential mechanism used to keep traffic flows in compliance with the DS domain traffic conditioning standards, such as an SLA between a service provider and its customers that specifies the maximum allowed traffic rate per class and in total per link, where excess packets will be discarded to keep traffic flows within the agreed traffic profile (SLA). There are two primary ways that admission control can be performed: traffic policing and traffic shaping. With traffic policing, when traffic streams reach the predefined maximum contracted rate, excess traffic is either dropped or re-marked (marked down), as shown in Figure 14-9.

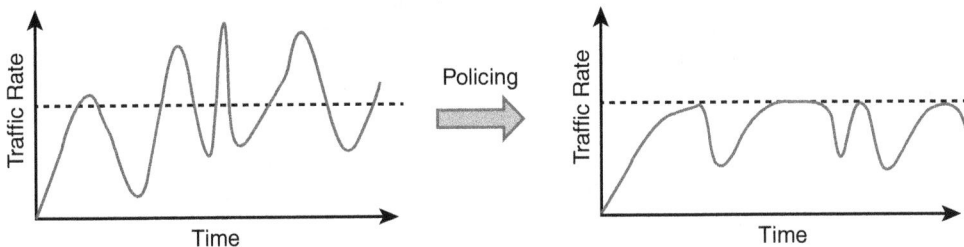

**Figure 14-9**   *Traffic Policing*

Traffic shaping, in contrast, keeps excess packets in a queue, buffered and delayed, and then schedules the excess for later transmission over increments of time. As a result, traffic shaping will smooth packet output rate and prevent unnecessary drops, as shown in Figure 14-10. However, the buffering of excess packets may introduce delay to traffic flows, especially with deep queues. Therefore, with real-time traffic it is sometimes preferred to police and drop excess packets rather than delay it and then transmit it, to avoid the degraded quality of experience.

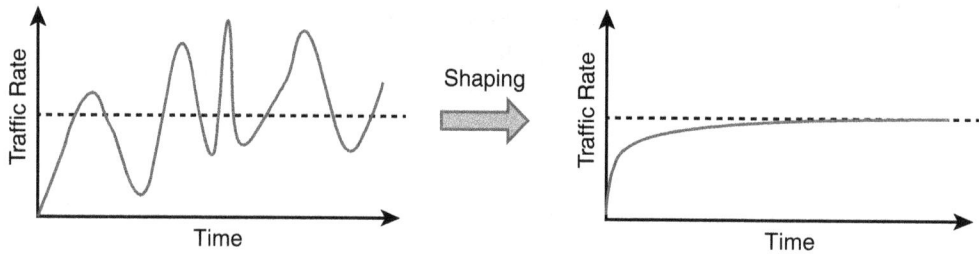

**Figure 14-10** *Traffic Shaping*

## QoS Design Strategy

Effective QoS design must always be measured end to end with regard to the packet's trip or traffic flow aggregates across the network (from the source to the intended destination). Otherwise, QoS will not deliver its value back to the business and end users. Therefore, it is important that network designers consider a consistent and unified QoS design model based on the different requirements, taking into consideration the underlying network infrastructures in terms of available bandwidth, traffic characteristics, and scope of the network, such as campus only versus WAN only or end to end across the entire enterprise (which translates to single DS domain versus multiple DS domains).

One of the critical elements of QoS design is traffic classification, marking, and profiling using a unified end-to-end approach. In particular, the strategy of marking IP traffic with a ToS value and mapping each type of traffic to a predefined QoS class will vary based on different factors. However, Figure 14-11 provides a generic summarized recommendation of a QoS design model (strategy) that can serve as a baseline. This 12-class model is based on both the Cisco QoS Baseline and the informational RFC 4594. One of the main benefits of this model is that it provides common and unified traffic marking and profiling characteristics across single and multiple DS domains, to a large extent.

**NOTE** While the IETF DiffServ RFCs provide a consistent set of PHBs for applications marked to specific DSCP values, they do not specify which application should be marked with which DSCP value. Therefore, considerable industry disparity exists in application-to-DSCP associations, which led Cisco to put forward a standards-based application marking recommendation in its strategic architectural QoS Baseline document (in 2002). Eleven different application classes were examined and extensively profiled and then matched to their optimal RFC-defined PHBs. More than four years after Cisco put forward its QoS Baseline document, RFC 4594 was formally accepted as an informational RFC (in August 2006). RFC 4594 puts forward 12 application classes and matches these to RFC-defined PHBs, as summarized in Figure 14-11.

| Application Class | Per-Hop Behavior | IETF RFC | Queuing & Dropping | Application Examples |
|---|---|---|---|---|
| VoIP Telephony | EF | 3246 | Priority Queue (PQ) | IP Telephony (IPT) |
| Broadcast Video | CS5 | 2474 | (Optional) PQ | IP Video Surveillance/IPTV |
| Real-time Interactive VC | CS4 | 2474 | (Optional) PQ | Telepresence |
| Multimedia Conferencing | AF4 | 2597 | BW Queue + DSCP WRED | IPT Video |
| Multimedia Streaming | AF3 | 2597 | BW Queue + DSCP WRED | Video on Demand (VoD), E-learning |
| Network Control | CS6 | 2474 | BW Queue | EIGRP, OSPF, BGP, HSRP, IKE |
| Call-Signaling | CS3 | 2474 | BW Queue | SCCP, SIP, H.323 |
| Mgmt (OAM) | CS2 | 2474 | BW Queue | SNMP, SSH, Syslog |
| Low-Latency Data | AF2 | 2597 | BW Queue + DSCP WRED | ERP Apps, CRM Apps, Database Apps |
| High-Throughput Data | AF1 | 2474 | BW Queue + DSCP WRED | E-mail, FTP, Backup Apps, Content Distribution |
| Best Effort | DF | | Default Queue + RED | Default Class |
| Low-Priority Data | CS1 | 3662 | Min BW Queue (Deferential) | YouTube, iTunes, BitTorent, Xbox Live |

**Figure 14-11**  *Twelve-Class QoS Baseline Model Based on Cisco and RFC 4594 Baselines*

> **NOTE**  The most significant of the differences between the Cisco QoS Baseline and RFC 4594 is the RFC 4594 recommendation to mark call signaling from AF31 to CS3 (as per the original QoS Baseline of 2002). It is important to remember that RFC 4594 is an informational RFC; in other words, it is only an industry best practice and not a standard.

The 12-class QoS model is a comprehensive and flexible model, which can be standardized and considered across the enterprise network. However, this model is not always viable or achievable for the following reasons:

- Not all enterprises or SPs are ready or need to deploy such a wide QoS design model.

- This 12-class QoS design can introduce a level of end-to-end QoS design and operational complexity, because most WAN providers offer either 4- or 6-class QoS models across the WAN.

> **NOTE**  The biggest concern with regard to the operational complexity is that it is prone to issues caused by human errors; this point is covered later in the "Network Management" section.

Ideally, what drives the number of classes (QoS model) is how many applications used across the network need special consideration and the level of service differentiation required for delivering the desired level of quality of experience. As shown in Figure 14-12, if the WAN provider is offering a 4-class QoS model, it is not an easy task to map from the 12-class model to the 4 classes (outbound), and from the 4-class model to the 12-class model (inbound) at each WAN edge, especially if there is a large number of sites. In contrast, if the enterprise is using a 4-, 6-, or 8-class QoS model, the operation and design complexities will be minimized with regard to QoS policies and configurations.

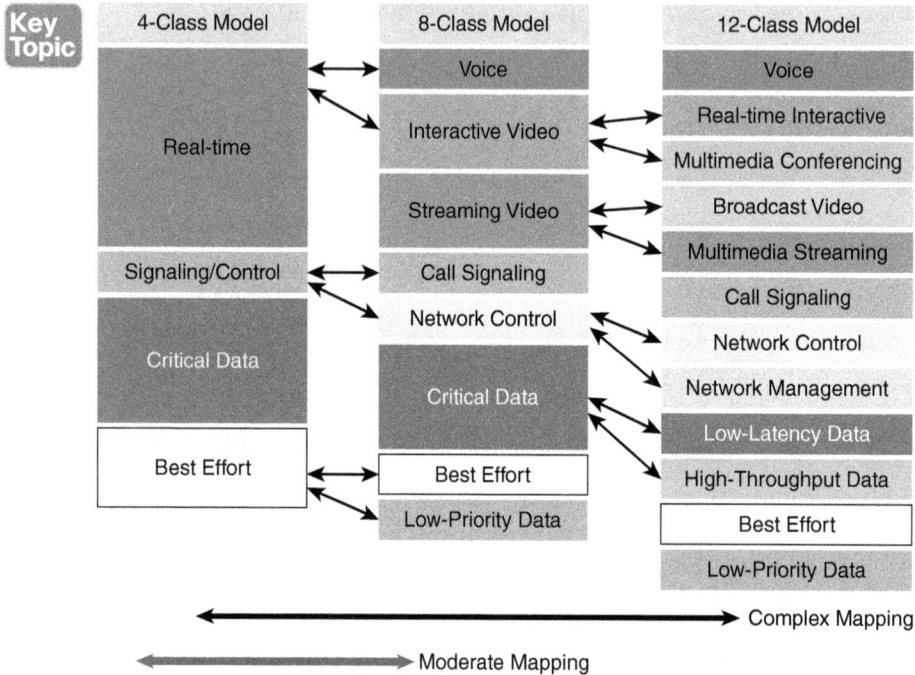

**Figure 14-12** *Mapping Between QoS Models with Different Classes*

However, both 4- and 6-class models include provisioning for only a single class for real-time (usually for voice), and if the video is added to the network, either a higher QoS class model (such as an 8-class model) is required to be considered or both voice and video traffic have to be provisioned under a single class, which may not be a desirable option for large deployments with a large number of IP telephony and video endpoints across multiple sites. Therefore, considering the top-down approach to identifying the business and user expectations, needs, and priorities in terms of the services and applications used, in addition to their level of criticality to the business, will help, to a large extent, drive the strategy of QoS design in the right direction (business-driven).

**NOTE** Network orchestration and automation tools may help eliminate the operational complexity discussed earlier. However, this is something that depends on the configuration and change management and on the platforms and architecture used. For instance, the level of automation in software-defined networks (SDNs) is always high, but with simplified manageability compared to other models.

Network designers must aim to balance between application requirements and QoS design with regard to the added design and operational complexity. As a general rule, network designers should always consider using the phased approach when possible, by starting with a simple QoS model (such as the 4-class model) as a baseline. In the future, if the requirements mandate more classes, it will be easier to cater to the new requirements by adding and incorporating new classes to the existing setup. This is based on the assumption that starting with a 4- or 6-class QoS model will be sufficient at that point of time with regard to the existing services and applications to be classified and profiled.

Although there are some standard recommended percentages of bandwidth allocation per class in multiple best practices design guides and white papers, such as real-time traffic (LLQ) to be assigned no more than 33 percent of the available or interface bandwidth and the best-effort class to be allocated 25 percent of the interface bandwidth, it is important to understand that these values represent a generic recommended bandwidth allocation per class based on common proven best practices. However, it is not a rule to consider these values as the standard bandwidth allocation ratios; instead, they should be considered as a foundation or a proven, tested reference when there is not enough information to build an accurate figure about bandwidth allocation in which these values can serve as a good baseline to start with. For example, Figure 14-13 illustrates a classic three-tier campus network with a WAN edge. In this network, there are 20 access switches dual-homed over 1-Gbps uplinks to the distribution switches. Similarly, the distribution switches are dual-homed to the core switch over 10-Gbps links.

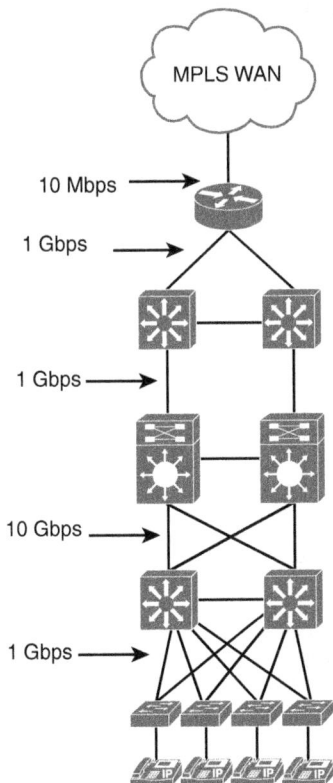

**Figure 14-13**  *QoS Bandwidth Allocation Example*

Each access switch has 30 IP phones connected to it, and VoIP calls are all based on the G7.11 codec (with approximately 80 kbps required per VoIP call). Typically, with the LAN, there is no restriction on how many calls can be placed at the same time. However, over the WAN, the voice system is designed to allow a maximum of 20 simultaneous calls.

Take these requirements and map them to the standard allocation of the LLQ, which is 33 percent. Within the campus LAN, allocating 33 percent of the links across the LAN is not an issue, because the network is already overprovisioned. However, provisioning the real-time (VoIP) LLQ with 33 percent of the 10-Gbps uplinks means that there will be 3.3-Gbps of bandwidth reserved during a period of congestion. This can be seen as a security risk by the security team because someone can flood the network with malicious traffic marked as DSCP EF, which will take precedence over other legitimate traffic and will consume the available 3.3-Gbps of bandwidth per uplink.

The WAN edge, however, has a 10-Mbps WAN link. Based on the voice system design, there will be only 20 simultaneous calls, meaning a maximum of (80 kbps × 20) 1.6 Mbps. Based on this, allocating 33 percent (= 3.3-Mbps) of the available WAN interface bandwidth to the LLQ (VoIP) means a waste of bandwidth in this case. Nevertheless, in this scenario, this organization still can allocate 25 percent of the links' available bandwidth to best effort (the default class) because there are no absolute requirements here on that part. Therefore, network designers need to consider the best practice recommended bandwidth or percentage allocation per class as a baseline and adjust the allocation based on the actual traffic flow requirements, taking into consideration security concerns, the available amount of bandwidth, and whether it is across the WAN, LAN, or a data center network.

Taking into consideration the different QoS design aspects, techniques, and strategies covered earlier in this section, network designers can consider the QoS design framework shown in Figure 14-14 as a foundational reference when approaching any QoS design task during the planning phase.

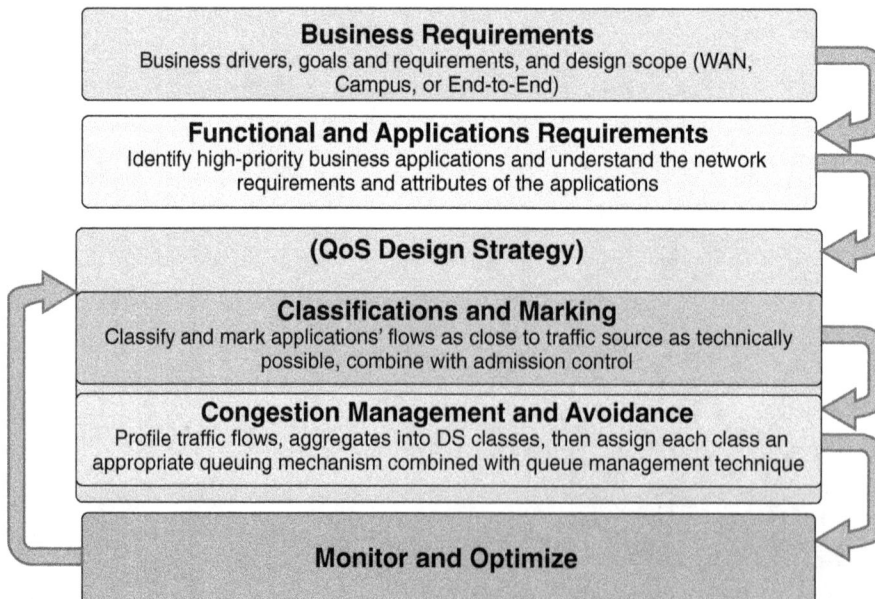

**Figure 14-14** *QoS Design Framework*

## Enterprise QoS Design Considerations

This section discusses QoS considerations that are applicable to enterprise-grade networks.

### Enterprise Campus

Today's campus networks are almost all provisioned with Gigabit/10 Gigabit of bandwidth across the LAN, in which QoS might be seen as an unnecessary service to be considered because the need for queuing is minimal or almost not required as compared to the WAN and Internet edge, where queuing is a primary function to enhance the service quality. Although this statement is valid to some extent, the need for QoS is not only limited to perform queuing functions.

The unified marking and accurate traffic classification (as close to the source as possible) that QoS can offer, when enabled across the campus LAN, also enables policing across the campus LAN to provide the flexibility and control to network operators to manage traffic usage based on different fields of traffic flows, such as ToS values. It can also be used as a protective mechanism in situations like DoS attacks, to mitigate their impact. Therefore, it is recommended that QoS be enabled across the campus LAN to maintain a seamless DS domain design in which classification and marking policies establish and impose trust boundaries, while policers help to protect against undesired flows at the access edge and across the LAN.

### Enterprise Edge

The enterprise edge (WAN, extranet, or Internet) is the most common place that traffic flow aggregation occurs (where more or a larger number of traffic flows usually arrive from the LAN side and need to exit the enterprise edge, which is usually provisioned with lower capacity). For instance, the LAN side might be provisioned with Gigabit/10 Gigabit, whereas the WAN has only 10 Mbps of actual available bandwidth. Therefore, QoS is always one of the primary functions considered at the enterprise edge to achieve more efficient bandwidth optimization, especially for enterprises that are converging their voice, video, and data traffic. Logically, the enterprise edge represents the DS domain edge where mapping and profiling of traffic flow to align with the adjacent DS domain is required to maintain a seamless QoS design. For instance, Figure 14-15 shows an example of a 12-class to 4-class QoS model mapping at the enterprise WAN edge router (CE) toward the SP edge (PE) to achieve end-to-end consistent QoS policies.

14

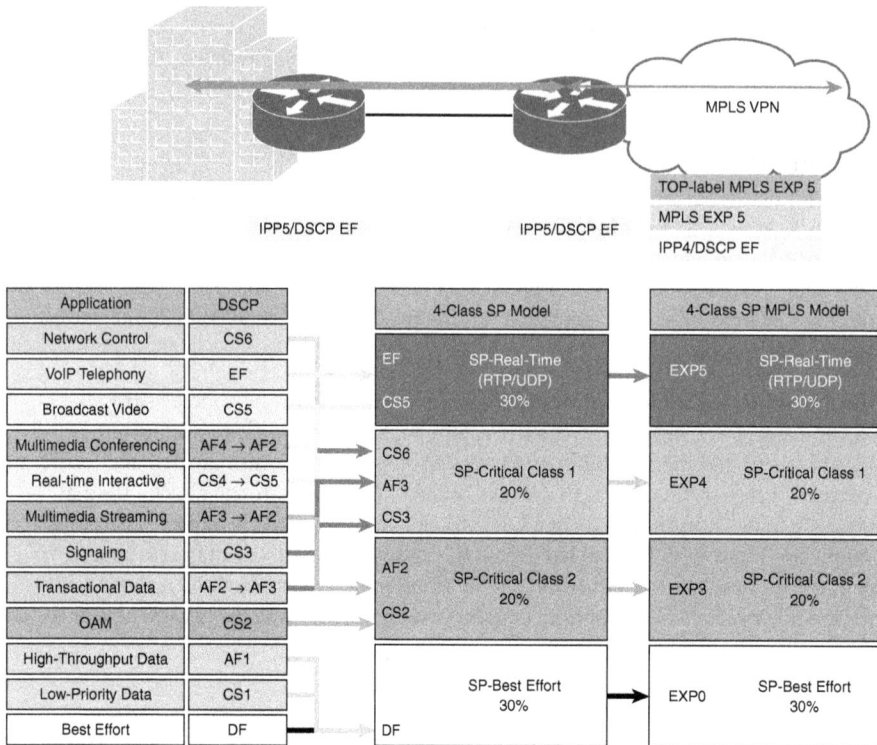

**Figure 14-15**  *QoS Mapping: Enterprise WAN Edge*

**NOTE**  The allocated bandwidth percentage in Figure 14-15 is only for the purpose of the example and based on the best practices recommendations. However, these values technically must be allocated based on the SLA between the service provider and enterprise customer.

**NOTE**  It is common that service providers offer their CoS based on IP Precedence marking only. As a result, the enterprise marking based on DSCP and deployed with the 8- or 12-class model may encounter inconsistent end-to-end markings. For instance, if there is a video RTP stream sent out from one site and its packets are marked with a DSCP value of DSCP 34 or AF41 (in binary, 100010), it will convert to IPP 4 (in binary, 100). In turn, it will come back as DSCP 32 (binary, 100000) at the other remote site. Therefore, a re-marking is required in this case at the other side (receiving) in the ingress direction to maintain a unified QoS marking end to end. Also, the used MPLS DiffServ tunneling mode and its impact on the original DSCP marking across the service provider network are covered later in this chapter.

## IP Tunneling QoS Design Considerations

As discussed earlier in this book, the attractive pricing drives many enterprises to consider deploying different VPN solutions as either alternative solutions to the private WAN transport or as redundant paths to the private WANs. In addition, some enterprises have to comply with their security policy that dictates considering a secure IP transport across the WAN/MAN, even if it is private, in which case the enterprise may consider IPsec, secure DMVPN, or GETVPN to protect the communication over the WAN or MAN transport.

One primary concern is the impact of increased packet overhead on the existing applications, because each VPN solution will add additional IP and ESP headers, as shown in Figure 14-16. In addition, VPN brings serious concern about bandwidth consumption. For instance, at Layer 3, a G.711 VoIP RTP stream needs 80 kbps. When this stream is sent over a GRE tunnel encrypted with IPsec, the total required bandwidth increases to ~112 kbps, which is an almost 40-percent bandwidth increase to transport the encrypted VoIP call. Therefore, bandwidth consumption with VPN is an essential point network designers need to consider, because it will usually impact the overall supported number of simultaneous voice calls or application sessions based on the actual available bandwidth.

14

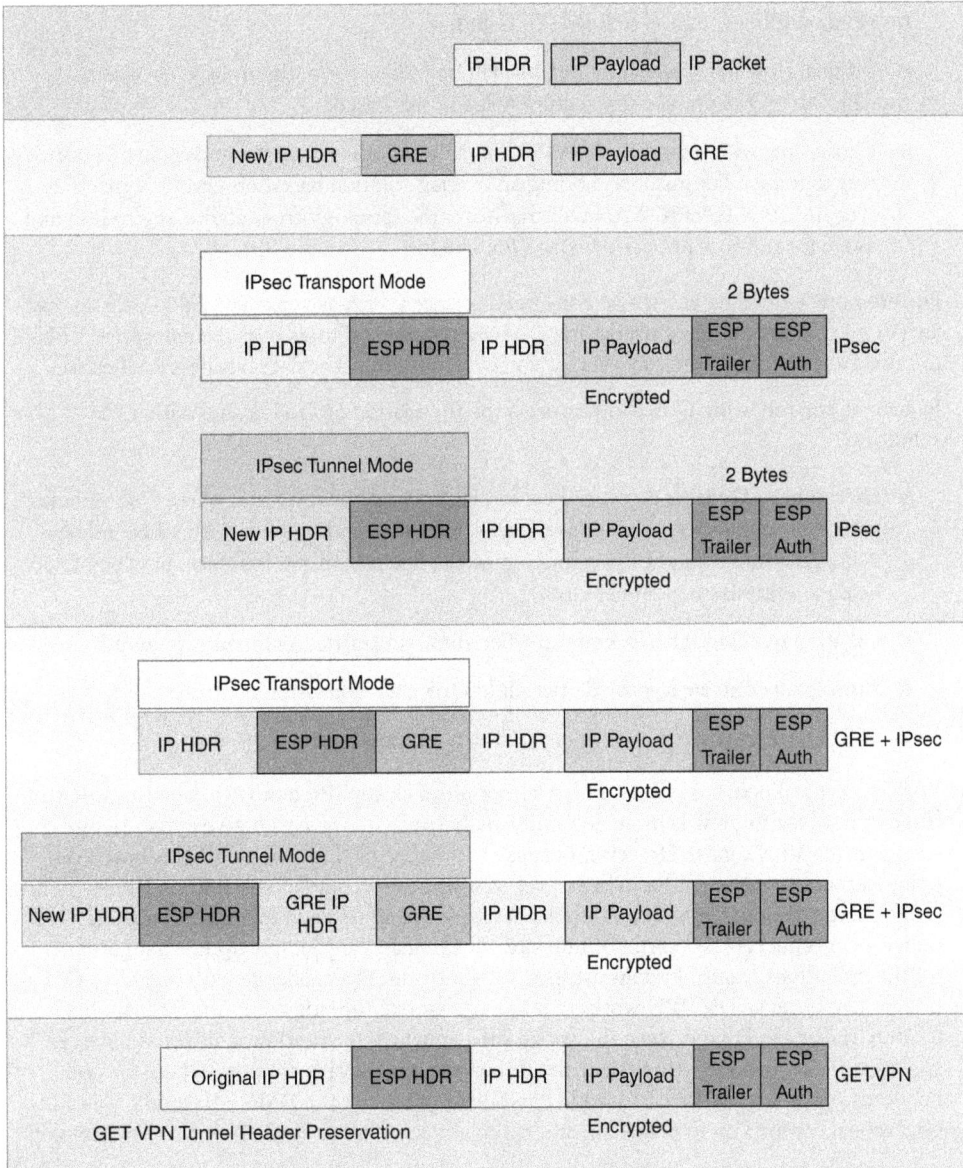

**Figure 14-16**  *IP Packet Overhead with VPN Solutions*

In addition, the following factors with regard to QoS design should also be taken into account by network designers when considering any VPN solution:

- Additional packet delay caused by the encryption and decryption functions.

- Pre-crypto queuing.

- The unforeseen impact of QoS packet reordering (prioritization) may lead the IPsec anti-replay mechanism to drop legitimate but out-of-sequence packets.

- ToS value preservation.

- Maximum transmission unit (MTU) issues.

- Original flow information by default will be hidden for outbound QoS policies. (The Cisco QoS pre-classify feature helps in this case.)

- Traffic flow with regard to the VPN logical topology versus the underlying IP transport topology. For instance, adding an overlay solution based on DMVPN phase 1 over an MPLS L3VPN WAN will transform the topology from any to any to hub and spoke for traffic sent over the DMVPN cloud.

Furthermore, extending enterprise converged services, such as voice and video over an overlay (VPN) solution, requires overlaying QoS techniques and tools over the enterprise VPNs to satisfy the different service levels for voice, video, and other data business applications.

In general, the following QoS toolset forms the foundation of QoS design with VPN solutions:

- **Hierarchical QoS:** As discussed earlier, HQoS helps in scenarios where QoS policies have to be applied on an interface with a sub-line rate. Similarly, with VPN and tunneling, the tunnel might use a fraction of the line rate of the interface, in which HQoS helps to optimize per-tunnel QoS.

- **CBWFQ queuing:** This is required when different traffic treatment is required.

- **Admission control:** To provide per-child QoS class admission control.

- **WRED:** Helps to further optimize TCP-based applications.

With the typical point-to-point VPN, whether it is a classic IPsec or GRE (combined with IPsec or not), traffic flow is more deterministic because it is always between two points or sites over the VPN tunnel. However, because the nature of this type of VPN is based on point-to-point tunneling, it has a high level of operational complexities and scalability limitations in large deployments. (For example, network operators need to maintain manual QoS policies per tunnel at the central or hub site, which makes any simple update a nightmare for the operations teams.) In contrast, DMVPN introduces a challenge with regard to QoS design here because DMVPN works in a hub-and-spoke topology and offers the ability to route traffic directly between the spoke sites, in which traffic flow is directed between spokes as well (not to mention the scale of remote site DMVPN support, where tens of hundreds of spokes can be easily added and integrated with the DMVPN cloud). This adds operational complexity to managing and maintaining seamless QoS where the hub normally has a single big pipe with higher bandwidth that can easily congest the spokes. Adding QoS policy per spoke at the hub site is a nonscalable and inflexible option.

The per-tunnel QoS feature with DMVPN will promote a zero-touch hub design that supports NHRP groups to dynamically provision QoS policies on a per-spoke basis at the DMVPN hub node facilitated by the NHRP group feature. Remote sites (spokes) can be profiled based on the provisioned WAN or Internet bandwidth and then automatically assigned QoS policy per spoke at the hub mGRE tunnel during the spoke tunnel registration (as shown in Figure 14-17), in which it will shape the traffic from the hub (higher-bandwidth side) toward the spoke (lower-bandwidth side) to be always aligned with the spoke's maximum download rate.

**HQ**

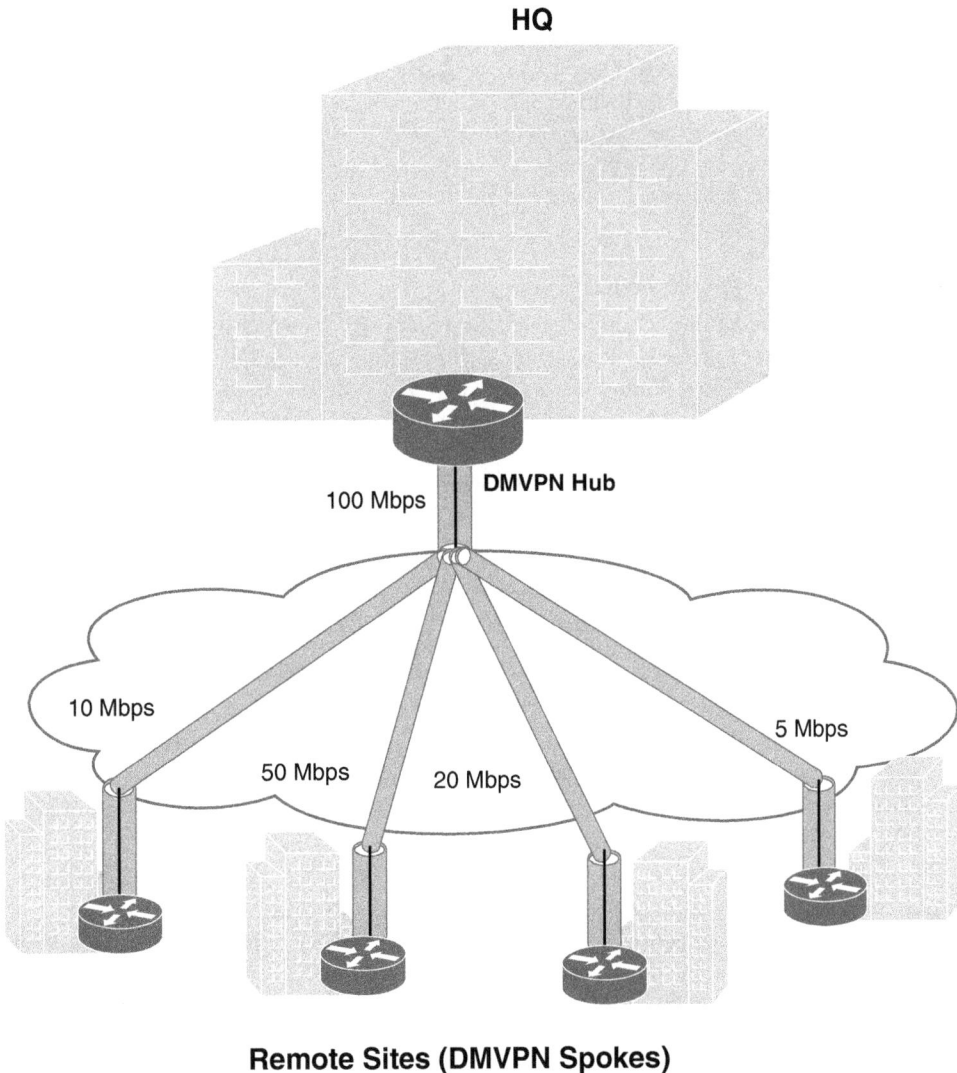

**Figure 14-17**  *IP Packet Overhead with VPN Solutions*

**NOTE**  At the hub site, network operators still need to define a policy to shape the traffic of the interface to match the actual provisioned sub-line rate of bandwidth. Spokes, however, should follow the typical HQoS deployment per site, where the bandwidth of the Internet/ WAN link must be shaped to the maximum upload provisioned capacity. This means direct spoke-to-spoke traffic streams will be controlled by the HQoS policies defined at the spokes level.

One key issue network designers must be aware of when considering per-tunnel QoS for a DMVPN solution is that GRE, IPsec, and the L2 overhead must be considered when calculating the required bandwidth for QoS shaping and queuing, because these headers will be included as part of the packet size. This is because queuing and shaping, technically, are executed at the outbound physical interface of the DMVPN mGRE tunnel. GETVPN, in contrast, preserves the entire original IP packet header (source and destination IP addresses, TCP and UDP port numbers, ToS byte, and DF bit) simply because QoS is applied at each GETVPN group member (because no tunnels are used). This makes GETVPN design and deployment simpler, and the same standard WAN QoS design can be applied, with the exception that packet size will be increased and therefore must be considered as part of QoS bandwidth calculations.

## Network Management

As discussed earlier in this book, today's modern networks carry multiple business-critical applications over one unified infrastructure such as voice, video, and data. In addition, in today's competitive telecommunications market, service providers always aim to satisfy their customers by meeting strict SLA requirements. As discussed earlier, various technologies, protocols, architectures, and constraints all collectively construct an operational network. However, designing and deploying a network using a business-driven approach will not guarantee the quality of the solution as well (that is, if the network is really operating as expected or not).

Moreover, traffic requirements in terms of pattern and volume can change over time as a natural result of business organic growth, merger and acquisition, and the introduction of new applications. This means that the network may end up handling traffic flows that it was not designed for. The question here is this: How can IT leaders and network operators know about what is going on? If the network is facing performance issues that need to be taken care of, how can the change and alteration be performed in a tracked and structured manner? In fact, configurations and changes are more of a concern compared to other aspects because any error can lead to either downtime or degraded performance, and generally, most network downtimes are caused by human error.

Consequently, there must be a set of procedures and mechanisms capable of measuring and providing real-time and historical information about every single activity across the network to help the IT team take action in a more proactive manner instead of relying on only the reactive approach to being able to effectively, and in a timely manner, identify and fix issues or abnormal behaviors in which the mean time to repair (MTTR) needs to be kept as short as technically possible. Furthermore, the action taken by IT should also be performed in a controlled and structured manner to be tracked and recorded, and also combined with some automation with regard to configuration and changes to help reduce the percentage of human errors. For the IT team to achieve this, they need a network management solution that controls the operation, administration, maintenance, and provisioning.

There are several industry standards and frameworks in the area of network management. This section discusses the ITU-T standard (FCAPS).

## Fault, Configuration, Accounting, Performance, and Security

FCAPS is a network management framework defined by the International Organization for Standardization (ISO) to help organizations classify the objectives of network management into the following five distinct layers or categories:

- **Fault management:** This management layer aims to minimize network outages by employing a set of procedures and activities to detect and isolate network issues, along with the appropriate corrective actions to overcome current issues and prevent them from occurring again. Alarms, fault isolation, testing, and troubleshooting are examples of fault management functions.

- **Configuration management:** This management layer aims to maintain a current inventory of network equipment, with its configurations to be used for planning, installation, and provisioning of new services and network equipment.

- **Accounting management:** It is implied by the name that this layer aims to ensure that each user or entity is billed or allocated an appropriate cost reference on the activities and utilization performed across the network. This can be measured based on various elements, such as a given service usage or bandwidth utilization. Usage management, pricing, auditing, and profitability analysis are examples of accounting management functions.

- **Performance management:** This layer aims to monitor and keep track of any performance issues, such as network bottlenecks, by continuously collecting and analyzing statistical information to monitor, correct, and optimize any reduced responsiveness across the network. This will potentially lead to enhanced capacity planning and quality measurement. Quality assurance, performance analysis, and monitoring are examples of the functions of this management layer.

- **Security management:** Typically, this layer focuses on the security of the management solutions (all the different layers above) in terms of access control, data confidentiality, and integrity with regard to network alarms, events, and reports. In addition, this layer sometimes refers to the monitoring and management of the network with regard to the security aspects such as unauthorized access, traffic spikes (DoS attacks), and targeted applications attacks.

## Network Management High-Level Design Considerations

Network management is an extensive and large topic. In addition to that, network management is defined and approached differently based on the entity or organization that creates the standards and the aim of the standards or framework, such as FCAPS versus ITIL. Therefore, it is impossible to cover such a large and extensive topic in a single section or chapter. This section covers the primary points and considerations at a high level to help network designers to drive the considerations around network management in the most suitable direction. Typically, there can be more than one right direction or approach.

The answers to the following questions help to form the foundation of the network management solution:

- What is the targeted environment (such as enterprise, MPLS VPN SP, application SP, cloud-hosting SP)?

- Is there any existing network management solution? If yes, does the solution follow any standard approach or framework such as FCAPS?

- Is the solution to be added to overcome an existing challenge or for enhancement purposes?

- Are there any business-related constraints, such as budget?

- What is the goal of the solution (for example, monitoring and fault management, capacity planning, billing, monitoring for security purposes, or a combination of these goals)?

- Are there any security constraints with regard to enabling network management, such as only out-of-band management, or can secure in-band management protocols be used?

After all or most of these questions have been answered, network designers should have a good understanding about the high-level targeted network management solution and should be able to start specifying its detailed design, which should answer at least the following questions:

- What information or events do network management solutions need to collect or monitor?

- Where is the best place to gather the intended information or report the relevant events?

- Where should this information or these events be sent after the collection process?

- What is the degree of detail required and is full or partial data collection required?

- Is the underlying transport network secure (internal) or untrusted (public Internet)?

- How is confidentiality and integrity of the polled or exported information maintained?

- What are the supported protocols and versions by the elements to be monitored and managed for the purpose of network management, such as SNMP or NetFlow?

Taking these questions into consideration, network designers can drive the solution selection and can specify which features and protocols are required and where they should be enabled. For example, an MPLS service provider decided to offer VoIP service to its customers to make voice calls to the public switched telephone network (PTSN) numbers across the service provider IP backbone, in addition to Internet Session Initiation Protocol (SIP) VoIP applications using Cisco Unified Border Element (CUBE) as a SIP gateway, as shown in Figure 14-18.

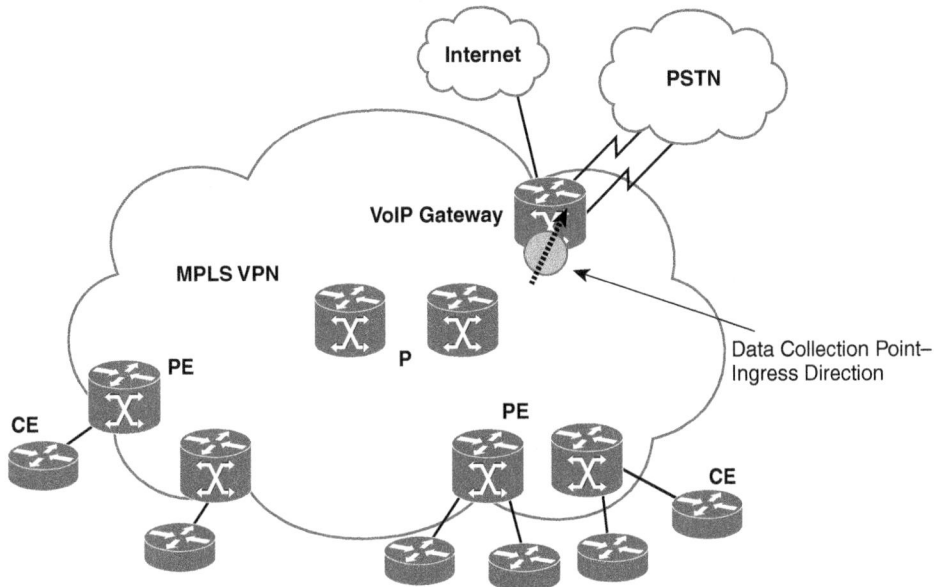

**Figure 14-18**  *MPLS VPN SP Sample Network*

According to the nature of this network, solution goals, and architecture, accounting management needs to be considered here to measure the utilization of the service. In this example, VoIP calls usually require a collection of the actual usage in terms of duration and destination number, which is normally an endpoint IP or another voice gateway IP or CUBE (such as to another SIP URI or PSTN number, local or international), and then a correlation between the collected information and pricing needs to be performed to generate a billing per VPN customer/user per call. By enabling NetFlow at the centralized CUBE, this service provider can export VoIP call usage to the NetFlow collector to obtain the desired reports for billing purposes.

In this particular example, the main question is where to enable data collection. Is it better at the service provider edge (PE), service provider core (P), or at the CUBE (voice gateway)? Based on the nature of the service and the traffic flow described, voice calls will be from customer sites toward the service's edge gateway, and no VoIP calls are made: neither directly between the different VPN customers nor between sites of the same VPN customer across the MPLS VPN provider backbone. This means that every VoIP traffic flow will pass through the voice gateway, where the service provider can perform ingress accounting per VoIP call flow in a centralized manner.

Another example is an enterprise with hundreds of remote sites connected over a single WAN network that carries different types of traffic, including VoIP, video, and data applications, as shown in Figure 14-19. This enterprise needs a network management solution that monitors WAN link utilization and reports any link that exceeds the available WAN bandwidth with regard to the traffic sourced from each remote site LAN. Based on these simple requirements, it is obvious that a performance management solution is required for the WAN links, where SNMP, for example, can be enabled per WAN link for the network management system (NMS) to collect the relevant Management Information Bases (MIBs) with regard to the utilization of each remote site WAN link. SNMP MIB polling also offers the ability to perform the polling based on a predefined time interval to retrieve the required information only when needed, to reduce the impact on the nodes' CPU and WAN performance.

However, if this enterprise requested that the utilization reports have to specify the top five protocols or applications using the WAN link during peak hours and the sources and destinations of these top five traffic flows during peak hours, in this case, egress accounting for the WAN links with NetFlow (IPFIX) is required to provide this level of visibility (per application, per session, or flow utilization).

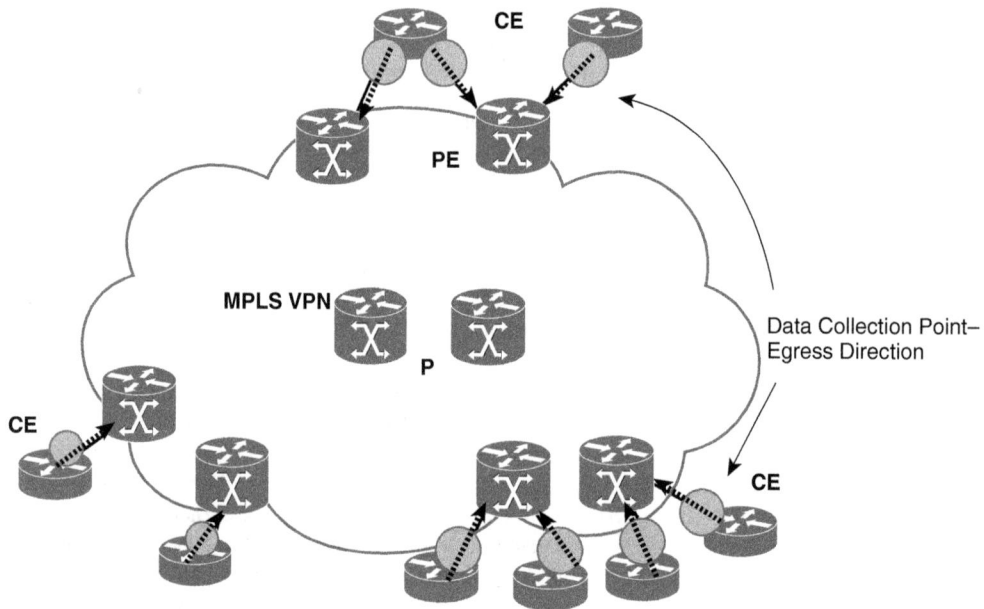

**Figure 14-19**  *Enterprise Network Sample for Network Management*

In this example, there are two primary design considerations worth analyzing. The first is the location of the data collection. Based on this scenario's requirements, a performance management solution is required to measure remote site WAN link utilization. This means a distributed data collection model is required here to measure and report the performance of the WAN link per remote site.

The second design consideration is in what direction the data or flows should be metered (ingress versus egress). Technically, in this particular example, to measure a CE router WAN link utilization for traffic sourced from the LAN side, in addition to identifying traffic types and top talkers over this link using different network management protocols (such as SNMP interface counters or NetFlow), measuring traffic in the egress direction will be an optimal choice. In addition, if the link measurement is considered at the egress direction in this example, it is recommended not to consider ingress data collection at the same node for the same purpose, to ensure that there will be no data duplication reported to the NMS.

From these two scenarios, it is obvious that the top-down approach is the most appropriate approach to developing and proposing a network management solution, starting from business goals and working toward the network management protocols.

## Multitier Network Management Design

In general, it is proven that integrating and structuring multiple management systems and tools in a hierarchical manner offers a more flexible and efficient network management solution when multiple elements and managements are present. For instance, this layered approach helps to reduce the number of alerts seen by network operations support staff, only presenting filtered and relevant information and alerts. However, to achieve this in a large-scale network, a considerable amount of structured integrations is required between multiple management systems and tools in a layered approach. As shown in Figure 14-20, this approach is recommended by Cisco systems because it offers the following benefits:

- Proactively identifies and corrects potential network issues before they become problems

- Offers optimized IT solution productivity by reducing and eliminating network connectivity loss to a minimum

- Focuses on the solution instead of the problem, which helps to reduce downtime duration (MTTR)

This multitier network management approach is based on bottom-up communication flow between the different management systems and tools using various protocols, including NetFlow, syslog, and SNMP. In fact, in a large network, it is almost always impossible to cope with the number of events reported from each network element to the NMS at the higher layers. In addition, in certain situations, a failure or fault in specific areas within the network can impact multiple devices. Typically, each device will independently alert the NMS. As a result, there will be duplicate instances of the same problem in this case.

With this architecture, the network management tier (NMT) receives the input from multiple elements and applications and then performs root-cause analysis by correlating the original information received from multiple sources and identifying the event that has occurred. This level of abstraction for event correlation provided by the NMT offers a simplified and efficient network management and operation solution. Network operators will be presented with the event deemed to be the most relevant and important, associated with the deduplication capabilities of network events, to reduce the number of unnecessary messages presented to the operations personnel. The service management tier in this approach provides an added intelligence and automation to the filtered events by the NMT and event correlation for more optimization, which will help network operators to move from complicated element management (per alert) to managing network events and identified problems. The level of automation and intelligence provided by this approach helps network operators avoid the classic network management approach also known as *box by box*, where the network operator or administrator needs to visit every single network node and manually configure, manage, and monitor each device separately.

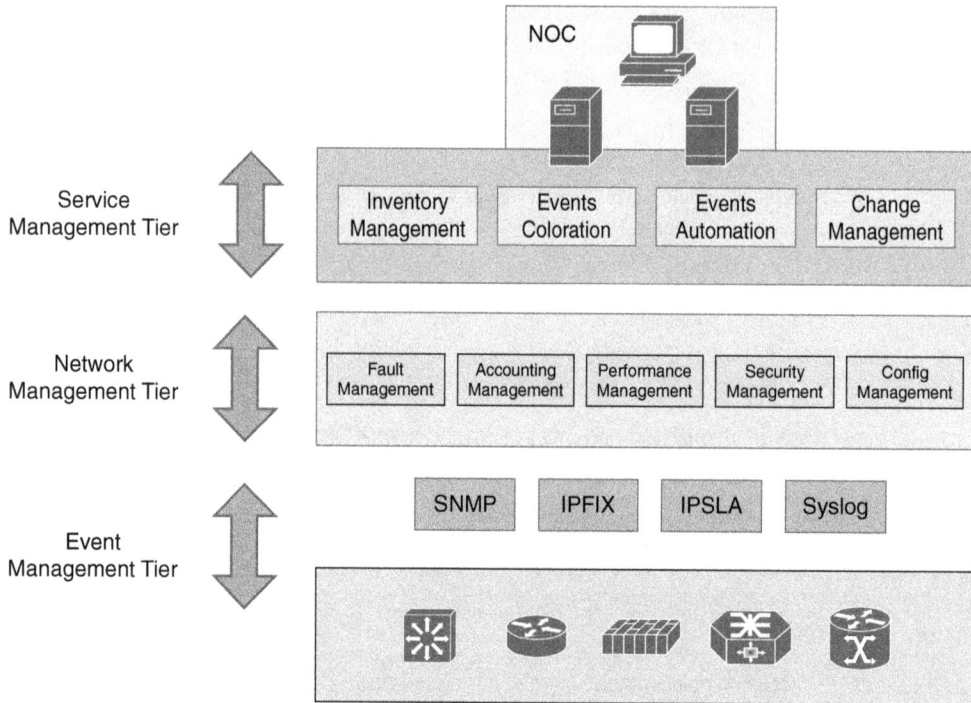

**Figure 14-20**  *Multitiered Network Management Solution*

## Model-Driven Network Management

With the advent of the different automation capabilities, which was discussed in Chapter 12, "Automation," there are new options when it comes to incorporating network management in an automatic way. The intent of this section is not to explain in full detail these concepts nor give a deep dive, but rather to show how these new automation capabilities have an impact on the network design for network management. With that said, this section covers YANG, NETCONF, RESTCONF, gNMI, and telemetry to give network designers a basic understanding to be able to make proper network design decisions when it comes to automated network management. In addition to the basic concepts listed, this section provides a couple of real-world examples, one focused on artificial intelligence for IT operations (AIOps) and the second focused on full-stack observability (FSO).

### YANG

**Yet Another Next Generation (YANG)** is an IETF standard (RFC 6020) data modeling language used to describe the data for network configuration protocols such as NETCONF and RESTCONF. YANG is extensible through augmentation, allowing new content to be added as needed to the YANG language. YANG has a hierarchical configuration structure within data models, which makes it very easy to read and reuse as needed. Figure 14-21 provides an example of a YANG data model. YANG is a full, formal contract language with rich syntax and semantics to build applications on.

```
Module my-interface {
}
namespace "com.my-interface";

    container interface {
        list interface {
            key name;
            leaf name { type string; }
            leaf admin-status { type enum; }
    rpc flap-interface {
        input {
            leaf name { type string; }
        }
        output {
            leaf result { type boolean; }
}
```

**Figure 14-21**  *Example YANG Data Model*

## NETCONF

**Network Configuration Protocol (NETCONF)** is a network management protocol defined by the IETF in RFC 6241. NETCONF provides rich functionality for managing configuration and state data. The protocol operations are defined as remote procedure calls (RPCs) for requests and replies in XML-based representation. NETCONF supports running, candidate, and startup configuration datastores. The NETCONF capabilities are exchanged during session initiation. Transaction support is also a key NETCONF feature. NETCONF is a client/server protocol and is connection-oriented over TCP. All NETCONF messages are encrypted with SSH and encoded with XML. A NETCONF manager is a client, and a NETCONF device is a server. The initial contents of the <hello> message define the NETCONF capabilities that each side supports. The YANG data model defines capabilities for the supported devices. In addition, other standards bodies and proprietary specifications define capabilities. Figure 14-22 highlights the different NETCONF operations and datastore capabilities.

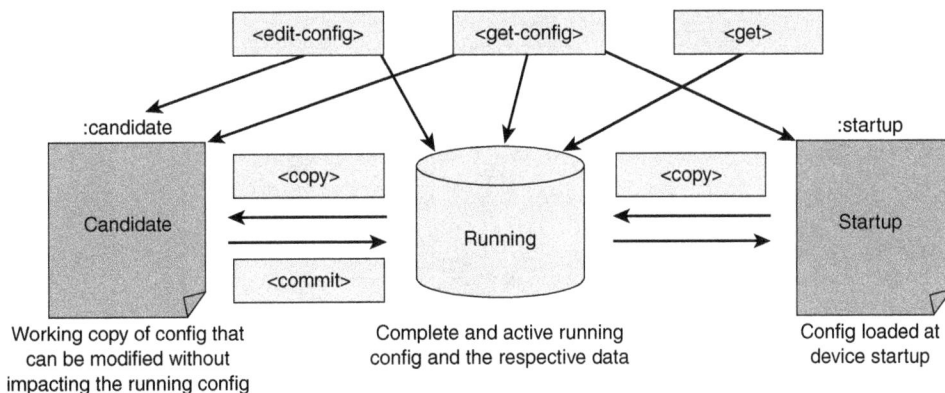

**Figure 14-22**  *NETCONF Operations and Datastore Capabilities*

## RESTCONF

**RESTCONF**, which is defined in RFC 8040, is an HTTP-based protocol that provides a programmatic interface for accessing YANG modeled data. RESTCONF uses HTTP operations to provide create, retrieve, update, and delete (CRUD) operations on a NETCONF datastore containing YANG data. RESTCONF is tightly coupled to the YANG data model definitions. It supports HTTP-based tools and programming libraries. RESTCONF can be encoded in either XML or JSON.

When comparing RESTCONF with NETCONF, RESTCONF has

- No notion of transaction

- No notion of lock

- No notion of candidate config and commit

- No notion of two-phase commit

- No <copy-config>

- XML or JSON, while NETCONF is only XML

In most design situations, it will be best to leverage NETCONF for routers and switches and RESTCONF for controller north-bound communication. Table 14-10 represents the different layers, highlighting NETCONF and RESTCONF characteristics at each layer.

**Table 14-10** Layering Model for NETCONF vs RESTCONF

| Layer | NETCONF | | RESTCONF | |
|---|---|---|---|---|
| Content | Configuration data | Notification data | Configuration data | Notification data |
| Operations | <get> <get-config> | | GET, POST, PATCH | |
| Messages | <rpc> | <notification> | HTTP payload | W3C Server-Sent Events within the HTTP payload |
| Secure Transport | SSH | | HTTPS | |

The operations that are listed in Table 14-10 are not an all-inclusive list of operations for both NETCONF and RESTCONF. Figure 14-23 takes the YANG example from Figure 14-21 and adds the corresponding RESTCONF HTTP operations on the right.

```
Module my-interface {
}
namespace "com.my-interface";                  GET: Gets a resource
                                                GET /restconf/data/my-interfaces:interfaces
        container interface {                   GET /restconf/data/my-interfaces:interfaces/interface/<some-name>
            list interface {
                key name;                       POST: Creates a resource or invoke operation
                leaf name { type string; }      POST /restconf/operations/my-interfaces:flap-interface + Data
                leaf admin-status { type enum; }
        rpc flap-interface {                    PUT: Replaces a resource
            input {                             PUT /restconf/data/my-interfaces:interfaces/interface/<some-name> + Data
                leaf name { type string; }
            }                                   DELETE: Removes a resource
            output {                            DELETE /restconf/data/my-interfaces:interfaces/interface/<some-name>
                leaf result { type boolean; }
}
```

**Figure 14-23** *Example YANG Data Model with RESTCONF HTTP Operations*

### gRPC Network Management Interface

**gRPC Network Management Interface (gNMI)**, developed by Google, provides the mechanism to install, manipulate, and delete the configuration of network devices and also to view operational data. The content provided through gNMI can be modeled using YANG. gRPC is a remote procedure call developed by Google for low-latency, scalable distributions with mobile clients communicating to a cloud server. gRPC carries gNMI and provides the means to formulate and transmit data and operation requests. When a gNMI service failure occurs, the gNMI broker (GNMIB) will indicate an operational change of state from up to down, and all RPCs will return a service unavailable message until the database is up and running. Upon recovery, the GNMIB will indicate a change of operation state from down to up, and resume normal handling of RPCs.

**14**

### Fully Automated Network Management

**Key Topic**

Today, IT systems create thousands of events per second. Having humans monitor all these events would be prohibitively expensive, and yet they still would not react quickly enough. This has paved the way for automated solutions to help solve this problem. At the time of writing, there are three prominent concepts in this area: artificial intelligence (AI) for IT operations (AIOps), closed-loop automation, and full-stack observability (FSO).

### Artificial Intelligence for IT Operations

AIOps helps in separating events with a business impact (from the noise) and resolving them autonomously. So, in the future, managing IT operations without AIOps will be challenging because of the rapid growth in data volumes and the rate of change. A modern-day AIOps platform detects anomalies, suppresses noise with correlation/de-duplication, helps in triaging and performing root cause analyses (RCA), and suggests/applies fixes.

### Closed-Loop Automation

Closed-loop automation (CLA) is a continuous process that monitors, measures, and assesses real-time network traffic and then automatically acts to optimize end-user quality of experience.

### Full-Stack Observability

Full-stack observability (FSO) is defined by metrics, events, logs, and traces. Modern applications span multiple environments. Today, a typical mobile application comprises hundreds of services communicating with each other over a zero-trust multi-cloud landscape, all of which have to work flawlessly. The level of complexity of these applications is tremendously higher than in decades past. We can no longer manage or optimize them because it is too much data with too little context and correlation. Traditional monitoring only gives visibility at the domain level, whether it be the network, infrastructure level, cloud, or database. The combined full-picture view is becoming more critical for the best user experience. This is where FSO comes into the forefront. Organizations require complete visibility and insights to properly take relevant action at the right time. To achieve this, there has to be a capability to measure the inner state of these applications based on the data generated by them, such as logs, metrics, and traces, which is also known as *observability*.

## Summary

This chapter covered various advanced IP topics and services that are part of any network design. To avoid design defects, network designers need to always incorporate these services in an integrated holistic approach rather than designing in isolation. Moreover, considering the top-down design approach is a fundamental requirement to achieving a successful business-driven design (for example, ensuring that the design complies with the organization's security policy standards). This chapter also emphasized the importance of considering the business priorities and design constraints in which network design ideally must adopt the "first things first" approach, which takes into consideration existing limitations, which may include staff knowledge, budget, or supported features and technologies.

## References

Al-shawi, Marwan, *CCDE Study Guide* (Cisco Press, 2015)

## Exam Preparation Tasks

As mentioned in the section "How to Use This Book" in the Introduction, you have a couple of choices for exam preparation: the exercises here, Chapter 18, "Final Preparation," and the exam simulation questions in the Pearson Test Prep Software Online.

## Review All Key Topics

Review the most important topics in this chapter, noted with the Key Topic icon in the outer margin of the page. Table 14-11 lists a reference of these key topics and the page numbers on which each is found.

**Key Topic**

**Table 14-11**  Key Topics for Chapter 14

| Key Topic Element | Description | Page Number |
|---|---|---|
| Table 14-4 | Approaches to Enable/Transition to IPv6 for the Enterprise | 360 |
| Table 14-5 | Approaches to Enable/Transition to IPv6 for the Service Provider | 362 |
| Table 14-6 | Mechanisms to Support Coexistence of IPv4 and IPv6 | 364 |
| Table 14-7 | Top-Down QoS Design Approach Summary | 370 |
| List | Fundamental QoS architecture models | 371 |
| List | QoS trust boundary models | 374 |
| Table 14-8 | Summary of QoS Classification and Marking Options | 375 |
| Figure 14-11 | Twelve-Class QoS Baseline Model Based on Cisco and RFC 4594 Baselines | 381 |
| Figure 14-12 | Mapping Between QoS Models with Different Classes | 382 |
| List | FCAPS layers | 391 |
| Section | Fully Automated Network Management | 399 |

## Complete Tables and Lists from Memory

Print a copy of Appendix D, "Memory Tables" (found on the companion website), or at least the section for this chapter, and complete the tables and lists from memory. Appendix E, "Memory Tables Answer Key," also on the companion website, includes completed tables and lists to check your work.

## Define Key Terms

Define the following key terms from this chapter and check your answers in the glossary:

weighted fair queuing (WFQ), Priority Queuing (PQ), class-based weighted fair queuing (CBWFQ), first-in, first-out (FIFO), Yet Another Next Generation (YANG), Network Configuration Protocol (NETCONF), RESTCONF, gRPC Network Management Interface (gNMI)

**14**

# CHAPTER 15

# Scalable Enterprise Campus Architecture Design

## This chapter covers the following topics:

**Campus Hierarchical Design Models:** This section covers the campus hierarchical models and the corresponding network design considerations.

**Campus Layer 3 Routing Design Considerations:** This section covers campus Layer 3 routing and the corresponding network design considerations.

**Campus Network Virtualization Design Considerations:** This section covers campus network virtualization and the corresponding network design considerations.

A campus network is generally the portion of the enterprise network infrastructure that provides access to network communication services and resources to end users and devices that are spread over a single geographic location. It may be a single building or a group of buildings spread over an extended geographic area. Normally, the enterprise that owns the campus network usually owns the physical wires deployed in the campus. Therefore, network designers typically tend to design the campus portion of the enterprise network to be optimized for the fastest functional architecture that runs on high-speed physical infrastructure (1/10/40/100 Gbps). Moreover, enterprises can also have more than one campus block within the same geographic location, depending on the number of users within the location, business goals, and business nature. When possible, the design of modern converged enterprise campus networks should leverage the following common set of engineering and architectural principles:

- Hierarchy

- Modularity

- Resiliency

This chapter covers the following "CCDE v3.0 Core Technology List" sections and provides design recommendations from an enterprise campus architecture standpoint:

- 2.0 Layer 2 Control Plane

- 3.0 Layer 3 Control Plane

- 4.0 Network Virtualization

## "Do I Know This Already?" Quiz

The "Do I Know This Already?" quiz allows you to assess whether you should read this entire chapter thoroughly or jump to the "Exam Preparation Tasks" section. If you are in doubt about your answers to these questions or your own assessment of your knowledge of the topics, read the entire chapter. Table 15-1 lists the major headings in this chapter and their corresponding "Do I Know This Already?" quiz questions. You can find the answers in Appendix A, "Answers to the 'Do I Know This Already?' Quizzes."

**Table 15-1** "Do I Know This Already?" Section-to-Question Mapping

| Foundation Topics Section | Questions |
| --- | --- |
| Campus Hierarchical Design Models | 1–5 |
| Campus Layer 3 Routing Design Considerations | 6, 7 |
| Campus Network Virtualization Design Considerations | 8, 9 |

**CAUTION**   The goal of self-assessment is to gauge your mastery of the topics in this chapter. If you do not know the answer to a question or are only partially sure of the answer, you should mark that question as wrong for purposes of the self-assessment. Giving yourself credit for an answer you correctly guess skews your self-assessment results and might provide you with a false sense of security.

1. Which of the following are the benefits of a routed access design model over a classical STP-based access design model? (Choose two.)

    a. Easier troubleshooting

    b. Flexibility to span Layer 2 natively

    c. Faster convergence time

    d. Potentially lower costs based on licensed features

2. Which of the following is a technical network design limitation of a routed access design model?

    a. More difficult troubleshooting

    b. Inability to span Layer 2 natively

    c. Slower convergence time

    d. Potentially higher costs based on licensed features

3. Which of the following is a business network design limitation of a routed access design model?

    a. More difficult troubleshooting

    b. Inability to span Layer 2 natively

    c. Slower convergence time

    d. Potentially higher costs based on licensed features

4. Which of the following is the best access-distribution connectivity model if a network design requires the most flexible design?

    a. Multitier STP based

    b. Switch clustering

    c. Routed access

    d. Layer 2 access

**5.** Which of the following is the best access-distribution connectivity model if a network design requires the most scalable design, both scale up and scale out?

   **a.** Multitier STP based

   **b.** Switch clustering

   **c.** Routed access

   **d.** Layer 2 access

**6.** Which of the following routing protocols for the campus has the highest architecture flexibility, proper route summarization, and no limitation to the number of tiers?

   **a.** EIGRP

   **b.** OSPF

   **c.** RIP

   **d.** IS-IS

**7.** Which of the following routing protocols for the campus supports MPLS-TE?

   **a.** EIGRP

   **b.** OSPF

   **c.** RIP

   **d.** BGP

**8.** Which of the following network virtualization techniques would be the proper fit if the network design required a high level of scalability?

   **a.** VLANs + 802.1Q + VRF

   **b.** VLANs + VRFs + GRE tunnels

   **c.** VLANs + VRFs + mGRE tunnels

   **d.** MPLS with MP-BGP

**9.** Which of the following designs would suit customers with a basic level of routing expertise?

   **a.** VLAN + 802.1Q + VRF

   **b.** VLANs + VRFs + GRE tunnels

   **c.** VLANs + VRFs + mGRE tunnels

   **d.** MPLS with MP-BGP

# Foundation Topics

## Campus Hierarchical Design Models

The hierarchical network design model breaks the complex flat network into multiple smaller and more manageable networks. Each level or tier in the hierarchy is focused on a specific set of roles. This design approach offers network designers a high degree of flexibility to optimize and select the right network hardware, software, and features to perform specific roles for the different network layers.

**Key Topic**

A typical hierarchical enterprise campus network design includes the following three layers:

- **Core layer:** Provides optimal transport between sites and high-performance routing. Due to the criticality of the core layer, the design principles of the core should provide an appropriate level of resilience that offers the ability to recover quickly and smoothly after any network failure event with the core block.

- **Distribution layer:** Provides policy-based connectivity and boundary control between the access and core layers.

- **Access layer:** Provides workgroup/user access to the network.

The two primary and common hierarchical design architectures of enterprise campus networks are the three-tier and two-tier models.

## Three-Tier Model

A **three-tier model**, illustrated in Figure 15-1, is typically used in large enterprise campus networks, which are constructed of multiple functional distribution layer blocks.

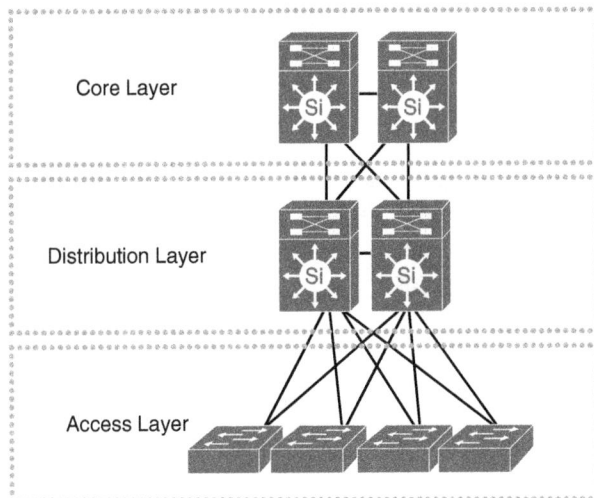

**Figure 15-1**  *Three-Tier Network Design Model*

## Two-Tier Model

A **two-tier model**, illustrated in Figure 15-2, is more suitable for small to medium-size campus networks (ideally not more than three functional disruption blocks to be interconnected), where the core and distribution functions can be combined into one layer, also known as *collapsed core-distribution architecture.*

**NOTE**  The term *functional distribution block* refers to any block in the campus network that has its own distribution layer such as a user access block, WAN block, or data center block.

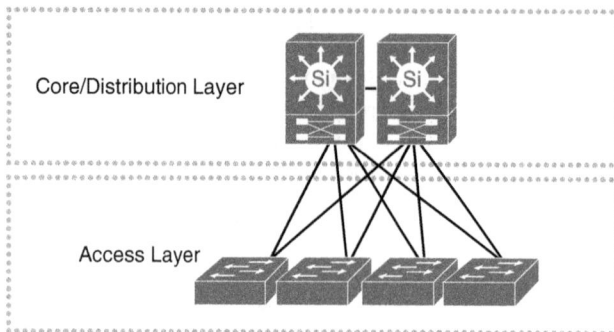

**Figure 15-2** *Two-Tier Network Design Model*

## Campus Modularity

By applying the hierarchical design model across the multiple functional blocks of the enterprise campus network, a more scalable and modular campus architecture (commonly referred to as *building blocks*) can be achieved. This modular enterprise campus architecture offers a high level of design flexibility that makes it more responsive to evolving business needs. As highlighted earlier in this book, modular design makes the network more scalable and manageable by promoting fault domain isolation and more deterministic traffic patterns. As a result, network changes and upgrades can be performed in a controlled and staged manner, allowing greater stability and flexibility in the maintenance and operation of the campus network. Figure 15-3 depicts a typical campus network along with the different functional modules as part of the modular enterprise architecture design.

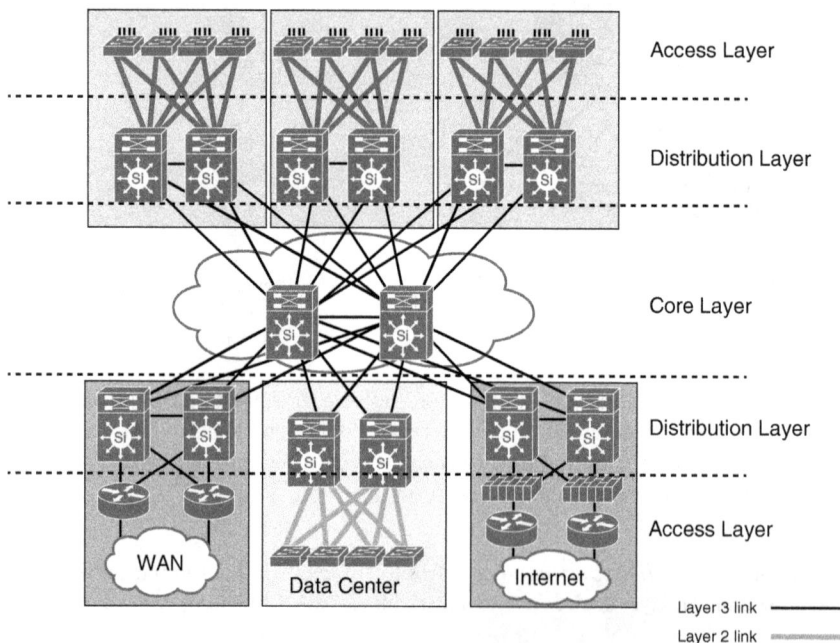

**Figure 15-3** *Typical Modular Enterprise Campus Architecture*

NOTE   Within each functional block of the modular enterprise architecture, to achieve the optimal structured design, you should apply the same hierarchical network design principle.

## When Is the Core Block Required?

A separate core provides the capability to scale the size of the enterprise campus network in a structured fashion that minimizes overall complexity when the size of the network grows (multiple campus distribution blocks) and the number of interconnections tying the multiple enterprise campus functional blocks increases significantly (typically leads to physical and control plane complexities), as exemplified in Figure 15-4. In other words, not every design requires a separate core.

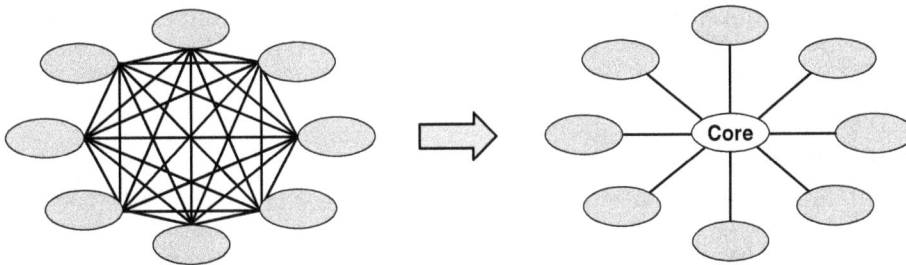

**Figure 15-4**   *Network Connectivity Without Core Versus with Core*

Besides the previously mentioned technical considerations, you should always aim to provide a business-driven network design with a future vision based on the principle "build today with tomorrow in mind." Taking this principle into account, one of the primary influencing factors with regard to selecting two-tier versus three-tier network architecture is the type of site or network (remote branch, regional HQ, secondary or main campus), which will help you, to a certain extent, identify the nature of the site and its potential future scale (from a network design point of view). For instance, it is rare that a typical (small to medium-size) remote site requires a three-tier architecture even when future growth is considered. In contrast, a regional HQ site or a secondary campus network of an enterprise can have a high potential to grow significantly in size (number of users and number of distribution blocks).

Therefore, a core layer or three-tier architecture can be a feasible option here. This is from a hypothetical design point of view; the actual answer must always align with the business goals and plans (for example, if the enterprise is planning to merge or acquire any new business); it can also derive from the projected percentage of the yearly organic business growth. Again, as a network designer, you can decide based on the current size and the projected growth, taking into account the type of the targeted site, business nature, priorities, and design constraints such as cost. For example, if the business priority is to expand without spending extra on buying additional network hardware platforms (reduce capital expenditure [CAPEX]), in this case, the cost savings is going to be a design constraint and a business priority, and the network designer in this type of scenario must find an alternative design solution such as the collapsed architecture (two-tier model) even though technically it might not be the optimal solution.

**NOTE**  Following the principle "build today with tomorrow in mind" can lead CCDE candidates into a gold-plating situation. For the CCDE exam, candidates should solve the problem at hand, and no more, unless the scenario states otherwise.

That being said, sometimes (when possible) you need to gain the support from the business first to drive the design in the right direction. To gain the support of IT leaders of the organization, you need to highlight and explain to them the extra cost and challenges of operating a network that either was not designed optimally with regard to their projected business expansion plans or was designed for yesterday's requirements and is incapable of handling today's requirements. Consequently, this may help to influence the business decision as the additional cost needed to consider the three-tier architecture will be justified to the business in this case (long-term operating expenditure [OPEX] versus short-term CAPEX). In other words, sometimes businesses focus only on the reduction of CAPEX without considering that OPEX can probably cost them more in the long run if the solution was not architected and designed properly to meet their current and future requirements.

## Access-Distribution Design Model

Chapter 7, "Layer 2 Technologies," discussed different Layer 2 design models that are applicable to the campus LAN design, in particular to the access-distribution model. Technically, each design model has different design attributes. Therefore, network designers must understand the characteristics of each design model to be able to choose and apply the most feasible model based on the design requirements.

The list that follows describes the three primary and common design models for the access layer to distribution layer connectivity. The main difference between these design models is where the Layer 2 and Layer 3 boundary is placed and how and where Layer 3 gateway services are handled.

- **Multitier STP based:** This model is the classical or traditional way of connecting access to the distribution layer in the campus network. In this model, the access layer switches usually operate in Layer 2 mode only, and the distribution layer switches operate in Layer 2 and Layer 3 modes. As discussed earlier in this book, the primary limitation of this design model is the reliance on Spanning Tree Protocol (STP) and First Hop Redundancy Protocol (FHRP). For more information, see Chapter 7.

- **Routed access:** In this design model, access layer switches act as Layer 3 routing nodes, providing both Layer 2 and Layer 3 forwarding. In other words, the demarcation point between Layer 2 and Layer 3 is moved from the distribution layer to the access layer. Based on that, the Layer 2 trunk links from access to distribution are replaced with Layer 3 point-to-point routed links, as illustrated in Figure 15-5.

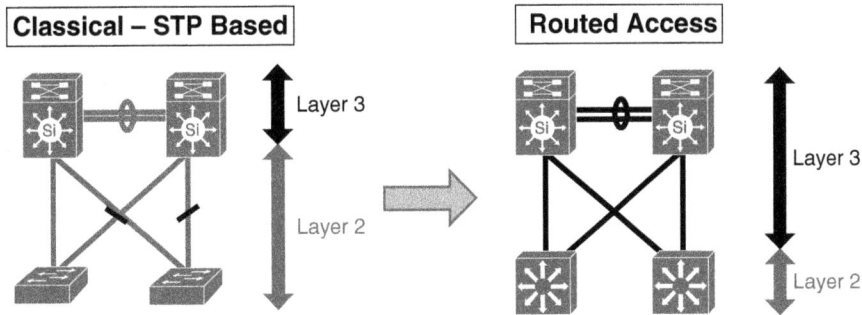

**Figure 15-5**  *Routed Access Layer*

The routed access design model has several advantages compared to the multitier classical STP-based access-distribution design model, including the following:

- Simpler and easier to troubleshoot, you can use standard routing troubleshooting techniques, and you will have fewer protocols to manage and troubleshoot across the network.

- Eliminate the reliance on STP and FHRP and rely on the equal-cost multipath (ECMP) of the used routing protocol to utilize all the available uplinks, which can increase the overall network performance.

- Minimize convergence time during a link or node failure.

**NOTE**  The routed access design model does not support spanning Layer 2 VLANs across multiple access switches, and this might not be a good choice for some networks. Although expanding Layer 2 over routed infrastructure is achievable using other different overlay technologies, this might add complexity to the design, or the required features may not be supported with the existing platforms for the access or distribution layer switches.

- **Switch clustering:** As discussed in Chapter 7, this design model provides the simplest and most flexible design compared to the other models discussed already. As illustrated in Figure 15-6, by introducing the switch clustering concept across the different functional modules of the enterprise campus architecture, network designers can simplify and enhance the design to a large degree. This offers a higher level of node and path resiliency, along with significantly optimized network convergence time. If you must extend Layer 2, this is the cleanest way to do it without adding STP or FHRP into the design.

The left side of Figure 15-6 represents the physical connectivity, and the right side shows the logical view of this architecture, which is based on the switch clustering design model across the entire modular campus network.

| **Physical Layout** | | | | **Logical Layout** | |
|---|---|---|---|---|---|

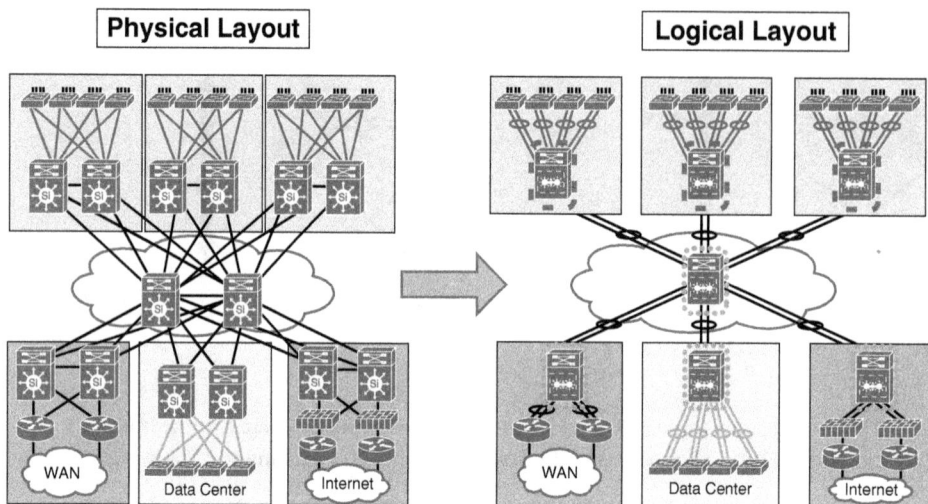

**Figure 15-6** *Switch Clustering Concept*

Table 15-2 compares the different access-distribution connectivity design models from different design angles.

**Table 15-2** Comparing Access-Distribution Connectivity Models

| | **Multitier STP Based** | **Routed Access** | **Switch Clustering** |
|---|---|---|---|
| **Design flexibility** | Limited (topology dependent) | Limited (for example, spanning Layer 2 over different access switches requires an overlay technology) | Flexible |
| **Scalability** | Supports scale up and limited scale out (topology dependent) | Supports both scale up and scale out | Scale up and limited scale out (typically limited to two distribution switches per cluster) |
| **Layer 3 gateway services** | Distribution layer (FHRP based) | Access layer (Layer 3 routing based) | Distribution layer (may or may not require FHRP*) |
| **Multichassis link aggregation (mLAG)** | Not supported | Not supported; instead relies on Layer 3 equal-cost multipath (ECMP) | Supported |
| **Access-to-distribution convergence time** | Dependent on STP and FHRP timers (relatively slow) | Interior Gateway Protocol (IGP) dependent, commonly fast | Fast |
| **Operational complexity** | Complex (multiple control protocols to deal with [for example, STP, FHRP]) | Moderate (advanced routing design expertise may be required) | Simple |

* Some switch clustering technologies, such as Cisco Nexus vPC, use FHRP (Hot Standby Router Protocol [HSRP]). However, from a forwarding plane point of view, both upstream switches (vPC peers) do forward traffic, unlike the classical behavior, which is based on active-standby.

> **NOTE**   All the design models discussed in this section are valid design options. However, the design choice must be driven by the requirements and design constraints, such as cost, which can influence which option you can select. For example, an access switch with Layer 3 capabilities is more expensive than a switch with Layer 2 capabilities only. This factor will be a valid tiebreaker if cost is a concern from the perspective of business requirements.

## Campus Layer 3 Routing Design Considerations

The hierarchal enterprise campus architecture can facilitate achieving more structured hierarchal Layer 3 routing design, which is the key to achieving routing scalability in large networks. This reduces, to a large extent, the number of Layer 3 nodes and adjacencies in any given routing domain within each tier of the hierarchal enterprise campus network.

In a typical hierarchal enterprise campus network, the distribution block (layer) is considered the demarcation point between Layer 2 and Layer 3 domains. This is where Layer 3 uplinks participate in the campus core routing, using either an interior gateway protocol (IGP) or Border Gateway Protocol (BGP), which can help to interconnect multiple campus distribution blocks together for end-to-end IP connectivity.

By contrast, with the routed access design model, Layer 3 routing is extended to the access layer switches. Consequently, the selection of the routing protocol is important for a redundant and reliable IP/routing reachability within the campus, considering scalability and the ability of the network to grow with minimal changes and impact to the network and routing design. All the Layer 3 routing design considerations discussed in previous chapters must be considered when applying any routing protocol to a campus LAN. Figure 15-7 illustrates a typical ideal routing design that aligns the IGP design (Open Shortest Path First [OSPF]) with the enterprise campus hierarchal architecture, along with the different functional modules. With route summarization to the core with OSPF, when links fail, there could be suboptimal routing situations, which is the case in any design with summaries at multiple area border routers (ABRs).

In the preceding example, the data center and other modules coexist within the enterprise campus to illustrate a generic IGP design over multiple modules interconnected over one core infrastructure (typical large campus design). However, in some designs, the data center is interconnected over a WAN or dedicated fiber links with the campus network. Also, the other blocks, such as Internet and WAN blocks, might be co-resident at the data center's physical location as well. In this case, you can use the same IGP design concept, and you can treat the WAN interconnect as an external link, as illustrated in Figure 15-8. Also, the enterprise core routing (that is, the OSPF backbone area) can be extended over the WAN. In other words, all the concepts of IGP and BGP design discussed earlier in this book have to be considered to make the right design decision. (There is no single standard design that you can use in different scenarios.)

15

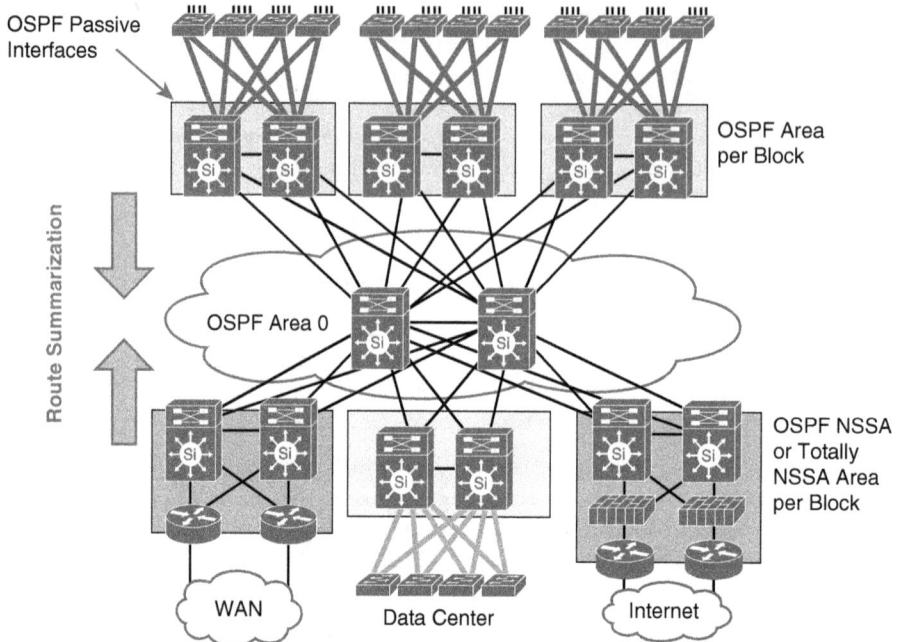

**Figure 15-7** *Campus Network: Layer 3 Routing*

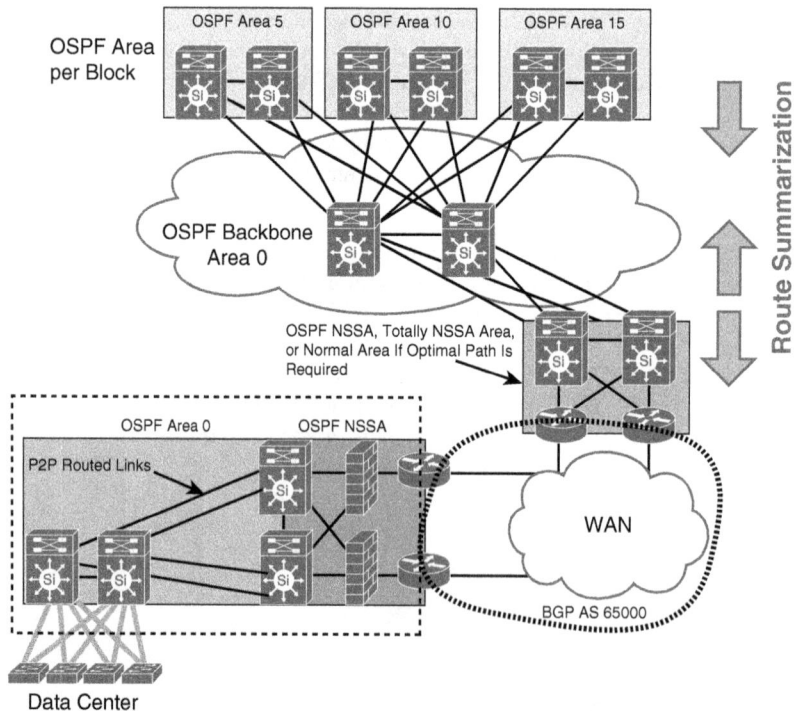

**Figure 15-8** *Campus Network: Layer 3 Design with WAN Core*

As discussed in Chapter 8, "Layer 3 Technologies," each protocol has its own characteristics, especially when applied to different network topologies. For example, the Enhanced Interior Gateway Routing Protocol (EIGRP) offers a more flexible, scalable, and easier-to-control design over a "hub-and-spoke" topology compared to a link-state routing protocol. In addition, although EIGRP is considered more flexible on multitiered network topologies such as three-tier campus architecture, link-state routing protocols have still proven to be powerful, scalable, and reliable protocols in this type of network, especially OSPF, which is one of the most commonly implemented protocols used in campus networks. Furthermore, in large-scale campus networks, if EIGRP is not designed properly with regard to information hiding and EIGRP query scope containment (discussed in Chapter 8), any topology change may lead to a large flood of EIGRP queries. In addition, the network will be more prone to EIGRP stuck-in-active (SIA) impacts, such as a long time to converge following a failure event and as an SIA timer puts an upper boundary on convergence times.

Consequently, each design has its own requirements, priorities, and constraints; and network designers must evaluate the design scenario and balance between the technical (protocol characteristics) and nontechnical (business priorities, future plans, staff knowledge, and so on) aspects when making design decisions.

Table 15-3 provides a summarized comparison between the two common and primary IGPs (algorithms) used in large-scale hierarchal enterprise campus networks.

**Table 15-3**   Link State Versus EIGRP in the Campus

| Design Consideration | EIGRP (DUAL) | Link State (Dijkstra) |
|---|---|---|
| Architecture flexibility | High (natively supports multitier architectures with route summarization) | High, with limitations (the more tiers the network has, the less flexible the design can be) |
| Scalability | High with proper query domain containment via EIGRP stubs and summarization | High with proper OSPF area design and area type selection |
| Convergence time (protocol level) | Fast (ideally with route summarization) | Fast (ideally with topology hiding, route summarization, and timers tuning) |
| MPLS-TE support | No | Yes |

## Campus Network Virtualization Design Considerations

Virtualization in IT generally refers to the concept of having two or more instances of a system component or function such as operating system, network services, control plane, or applications. Typically, these instances are represented in a logical virtualized manner instead of being physical.

Virtualization can generally be classified into two primary models:

- **Many-to-one virtualization model:** In this model, multiple physical resources appear as a single logical unit. The classical example of many-to-one virtualization is the switch clustering concept discussed earlier. Other examples include firewall clustering, and FHRP with a single virtual IP (VIP) that front ends a pair of physical upstream network nodes (switches or routers).

■ **One-to-many virtualization model:** In this model, a single physical resource can appear as many logical units, such as virtualizing an x86 server, where the software (hypervisor) hosts multiple virtual machines (VMs) to run on the same physical server. The concept of network function virtualization (NFV) can also be considered as a one-to-many system virtualization model.

## Drivers to Consider Network Virtualization

To meet the current expectations of business and IT leaders, a more responsive IT infrastructure is required. Therefore, network infrastructures need to move from the classical architecture (that is, based on providing basic interconnectivity between different siloed departments within the enterprise network) into a more flexible, resilient, and adaptive architecture that can support and accelerate business initiatives and remove inefficiencies. The IT and network infrastructure will become like a service delivery business unit that can quickly adapt and deliver services. In other words, it will become a "business enabler." This is why network virtualization is considered one of the primary principles that enable IT infrastructures to become more dynamic and responsive to the new and rapidly changing requirements of today's enterprises.

The following are the primary drivers of modern enterprise networks, which can motivate enterprise businesses to adopt the concept of network virtualization:

**Key Topic**

■ **Cost efficiency and design flexibility:** Network virtualization provides a level of abstraction from the physical network infrastructure that can offer cost-effective network designs along with a higher degree of design flexibility, where multiple logical networks can be provisioned over one common physical infrastructure. This ultimately will lead to lower CAPEX because of the reduction in device complexity and the number of devices. Similarly, it will lower OPEX because the operations team will have fewer devices to manage.

■ **Support a simplified and flexible integrated security:** Network virtualization also promotes flexible security designs by allowing the use of separate security policies per logical or vitalized entity, where users' groups and services can be logically separated.

■ **Design and operational simplicity:** Network virtualization simplifies the design and provision of the path and traffic isolation per application, group, service, and various other logical instances that require end-to-end path isolation.

**NOTE** It is important that network designers understand the drivers toward adopting network virtualization from a business point of view, along with the strengths and weaknesses of each design model. This ensures that when a network virtualization concept is considered in a given area within the network or across the entire network, it will deliver the promised value (and not be used only because it is easy to implement or it is an innovative approach). As discussed earlier in this book, a design that does not address the business's functional requirements is considered a poor design; consider the design principle "no gold plating" discussed in Chapter 1, "Network Design."

**NOTE** One of the main concerns about network virtualization is the concept of *fate sharing* (aka *shared failure state*), because any failure in the physical network can lead to a failure of multiple virtual networks running over the same physical infrastructure. Therefore, when the network virtualization concept is used, ideally a reliable and highly available network design should be considered as well. Besides the constraints about virtual network availability, there is always a concern about network virtualization (multitenant environment) where multiple virtual networks (VNs) operate over a single physical network infrastructure and each VN probably has different traffic requirements (different applications and utilization patterns). Therefore, there is a higher potential of having traffic congestion and degraded application quality and user experience if there is no efficient planning with regard to the available bandwidth, number of VNs, traffic volume per VN, applications in use, and the characteristics of the applications. In other words, if there is no adequate bandwidth available and the quality of service (QoS) policies to optimize and control traffic behaviors, one VN may overutilize the available bandwidth of the underlying physical network infrastructure. This will usually lead to traffic congestion because other VNs are using the same underlying physical network infrastructure, resulting in fate sharing.

15

This section covers the primary network virtualization technologies and techniques that you can use to serve different requirements by highlighting the pros and cons of each technology and design approach. This can help network designers (CCDE candidates) to select the best suitable design after identifying and evaluating the different design requirements (business and functional requirements). This section primarily focuses on network virtualization over the enterprise campus network. Chapter 17, "Enterprise WAN Architecture Design," expands on this topic to cover network virtualization design options and considerations over the WAN.

## Network Virtualization Design Elements

As illustrated in Figure 15-9, the main elements in an end-to-end network virtualization design are as follows:

**Key Topic**

- **Edge control:** This element represents the network access point. Typically, it is a host or end-user access (wired, wireless, or virtual private network [VPN]) to the network where the identification (authentication) for physical to logical network mapping can occur. For example, a contracting employee might be assigned to VLAN X, whereas internal staff are assigned to VLAN Y.

- **Transport virtualization:** This element represents the transport path that will carry different virtualized networks over one common physical infrastructure, such as an overlay technology like a generic routing encapsulation (GRE) tunnel. The terms *path isolation* and *path separation* are commonly used to refer to transport virtualization. Therefore, these terms are used interchangeably throughout this book.

- **Services virtualization:** This element represents the extension of the network virtualization concept to the services edge, which can be shared services among different logically isolated groups, such as an Internet link or a file server located in the data center that must be accessed by only one logical group (business unit).

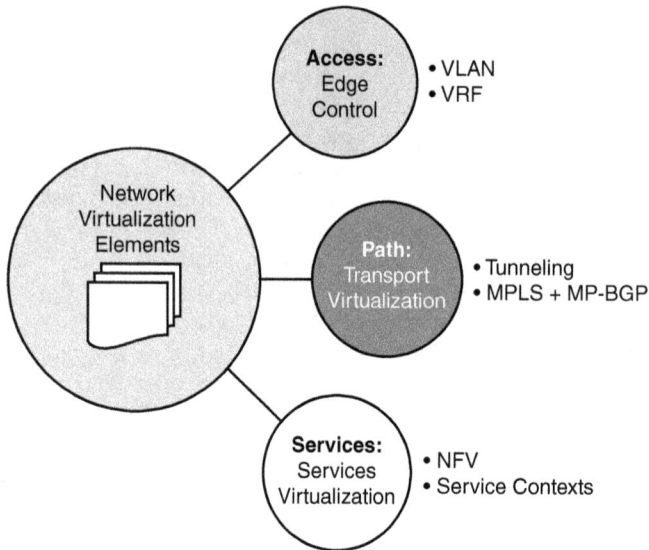

**Figure 15-9**  *Network Virtualization Elements*

## Enterprise Network Virtualization Deployment Models

Now that you know the different elements that, individually or collectively, can be considered as the foundational elements to create network virtualization within the enterprise network architecture, this section covers how you can use these elements with different design techniques and approaches to deploying network virtualization across the enterprise campus. This section also compares these different design techniques and approaches.

Network virtualization can be categorized into the following three primary models, each of which has different techniques that can serve different requirements:

■ Device virtualization

■ Path isolation

■ Services virtualization

Moreover, you can use the techniques of the different models individually to serve certain requirements or combine them to achieve one cohesive end-to-end network virtualization solution. Therefore, network designers must have a good understanding of the different techniques and approaches, along with their attributes, to select the most suitable virtualization technologies and design approaches for delivering value to the business.

## Device Virtualization

Also known as *device partitioning*, device virtualization represents the ability to virtualize the data plane, control plane, or both, in a certain network node, such as a switch or a router. Using device-level virtualization by itself will help to achieve separation at Layer 2, Layer 3, or both, on a local device level. The following are the primary techniques used to achieve device-level network virtualization:

- **Virtual LAN (VLAN):** VLAN is the most common Layer 2 network virtualization technique. It is used in every network where one single switch can be divided into multiple logical Layer 2 broadcast domains that are virtually separated from other VLANs. You can use VLANs at the network edge to place an endpoint into a certain virtual network. Each VLAN has its own MAC forwarding table and spanning-tree instance (Per-VLAN Spanning Tree [PVST]).

- **Virtual routing and forwarding (VRF):** A VRF is conceptually similar to a VLAN but from a control plane and forwarding perspective on a Layer 3 device. A VRF can be combined with a VLAN to provide a virtualized Layer 3 gateway service. As illustrated in Figure 15-10, each VLAN over an 802.1Q trunk can be mapped to a different subinterface that is assigned to a unique VRF, where each VRF maintains its own forwarding and routing instance and potentially leverages different VRF-aware routing protocols (for example, OSPF or EIGRP instance per VRF).

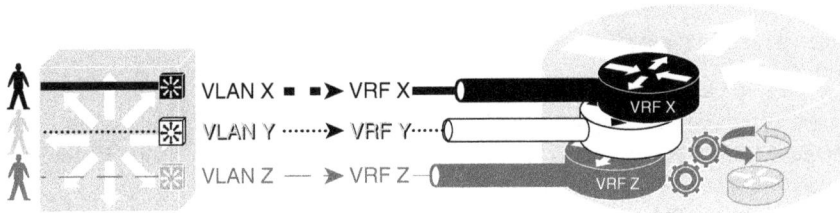

**Figure 15-10**  *Virtual Routing and Forwarding*

## Path Isolation

*Path isolation* refers to the concept of maintaining end-to-end logical path transport separation across the network. The end-to-end path separation can be achieved using the following main design approaches:

- **Hop by hop:** This design approach, as illustrated in Figure 15-11, is based on deploying end-to-end (VLANs + 802.1Q trunk links + VRFs) per device in the traffic path. This design approach offers a simple and reliable path separation solution. However, for large-scale dynamic networks (a large number of virtualized networks), it will be a complicated solution to manage. This complexity is associated with design scalability limitations.

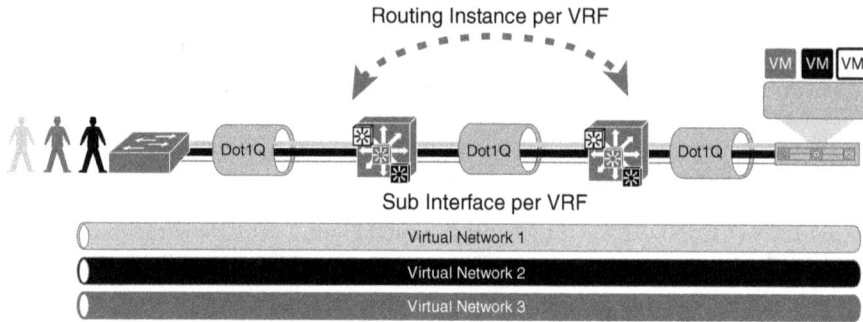

**Figure 15-11** *Hop-by-Hop Path Virtualization*

- **Multihop:** This approach is based on using tunneling and other overlay technologies to provide end-to-end path isolation and carry the virtualized traffic across the network. The most common proven methods include the following:

  - **Tunneling:** Tunneling, such as GRE or multipoint GRE (mGRE) (dynamic multipoint VPN [DMVPN]), will eliminate the reliance on deploying end-to-end VRFs and 802.1Q trunks across the enterprise network because the virtualized traffic will be carried over the tunnel. This method offers a higher level of scalability as compared to the previous option and with a simpler operation to some extent. This design is ideally suitable for scenarios where only a part of the network needs to have path isolation across the network.

    However, for large-scale networks with multiple logical groups or business units to be separated across the enterprise, the tunneling approach can add complexity to the design and operations. For example, if the design requires path isolation for a group of users across two "distribution blocks," tunneling can be a good fit, combined with VRFs. However, mGRE can provide the same transport and path isolation goal for larger networks with lower design and operational complexities. (See the section "WAN Virtualization" in Chapter 17 for a detailed comparison between the different path separation approaches over different types of tunneling mechanisms.)

  - **MPLS VPN:** By converting the enterprise to be like a service provider type of network, where the core is Multiprotocol Label Switching (MPLS) enabled and the distribution layer switches act as provider edge (PE) devices, each PE (distribution block) will exchange VPN routing over MP-BGP sessions, as shown in Figure 15-12. (The route reflector [RR] concept can be introduced, as well, to reduce the complexity of full-mesh MP-BGP peering sessions.)

    Furthermore, L2VPN capabilities can be introduced in this architecture, such as Ethernet over MPLS (EoMPLS), to provide extended Layer 2 communications across different distribution blocks if required. With this design approach, the end-to-end virtualization and traffic separation can be simplified to a very large extent with a high degree of scalability.

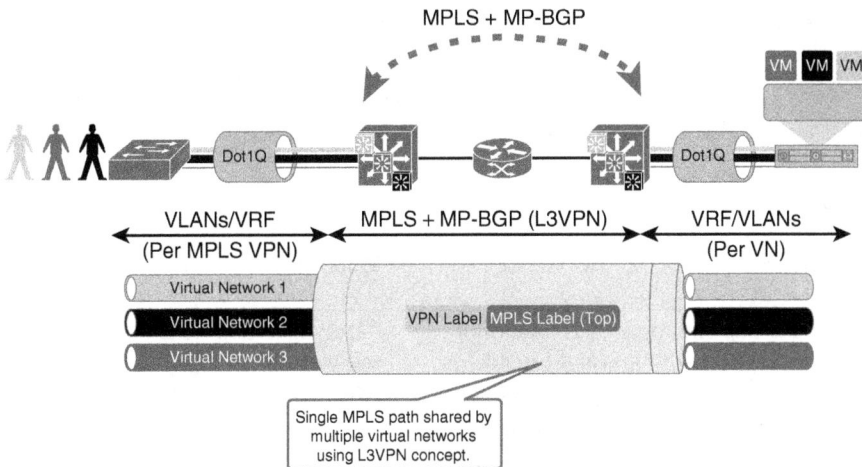

**Figure 15-12**  *MPLS VPN-Based Path Virtualization*

Figure 15-13 illustrates a summary of the different enterprise campus network virtualization design techniques.

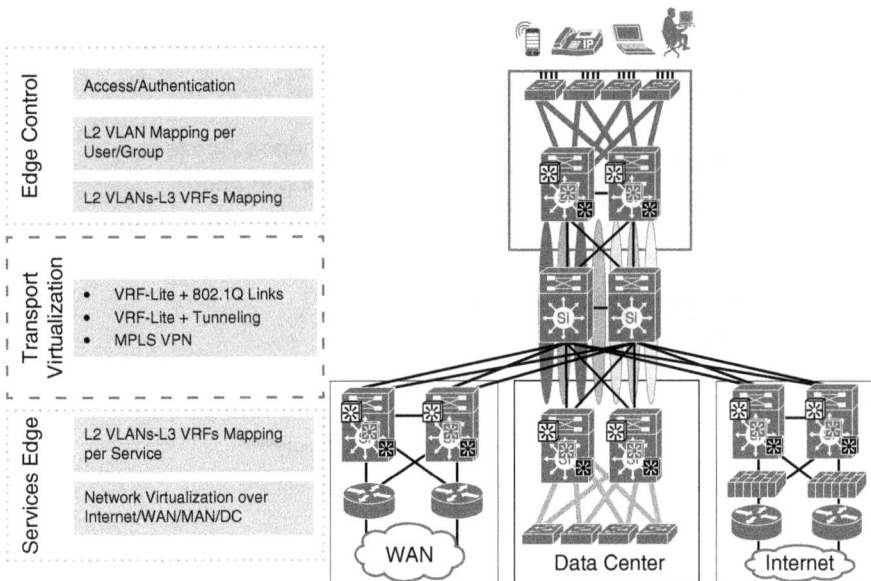

**Figure 15-13**  *Enterprise Campus Network Virtualization Techniques*

As mentioned earlier in this section, it is important for network designers to understand the differences between the various network virtualization techniques. Table 15-4 compares these different techniques in a summarized way from different design angles.

**Key Topic**

**Table 15-4** Network Virtualization Techniques Comparison

| | End to End (VLAN + 802.1Q + VRF) | VLANs + VRFs + GRE Tunnels | VLANs + VRFs + mGRE Tunnels | MPLS with MP-BGP |
|---|---|---|---|---|
| Scalability | Low | Low | Moderate | High |
| Operational complexity | High | Moderate | Moderate | Moderate to high |
| Design flexibility | Low | Moderate | Moderate | High |
| Architecture | Per hop end-to-end virtualization | P2P (multihop end-to-end virtualization) | P2MP (multihop end-to-end virtualization) | MPLS L3VPN-based virtualization |
| Operation staff routing expertise | Basic | Medium | Medium | Advanced |

## Service Virtualization

One of the main goals of virtualization is to separate services access into different logical groups, such as user groups or departments. However, in some scenarios, there may be a mix of these services in terms of service access, in which some of these services must only be accessed by a certain group and others are to be shared among different groups, such as a file server in the data center or Internet access, as shown in Figure 15-14.

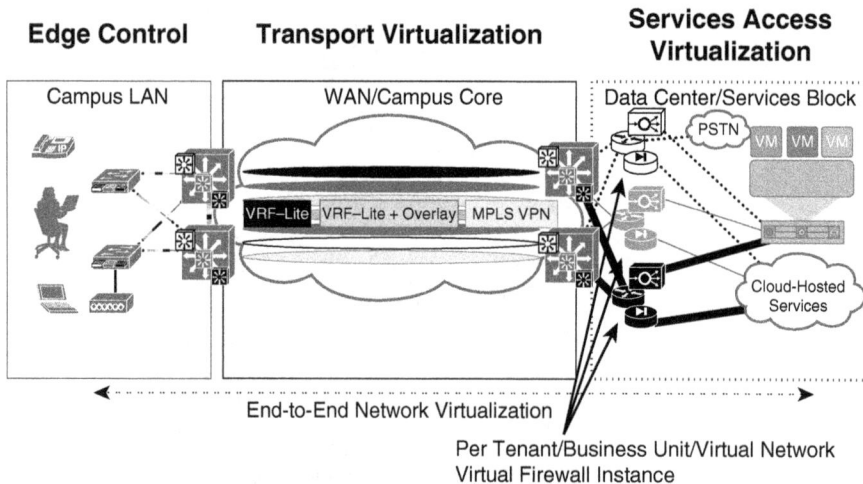

**Figure 15-14** *End-to-End Path and Services Virtualization*

Therefore, in scenarios like this where service access must be separated per virtual network or group, the concept of network virtualization must be extended to the services access edge, such as a server with multiple VMs or an Internet edge router with single or multiple Internet links.

The virtualization of a network can be extended to other network service appliances, such as firewalls. For instance, you can have a separate virtual firewall per virtual network, to facilitate access control between the virtual user network and the virtualized services and workload, as shown in Figure 15-15. The virtualization of network services can be considered as a "one-to-many" network device level virtualization.

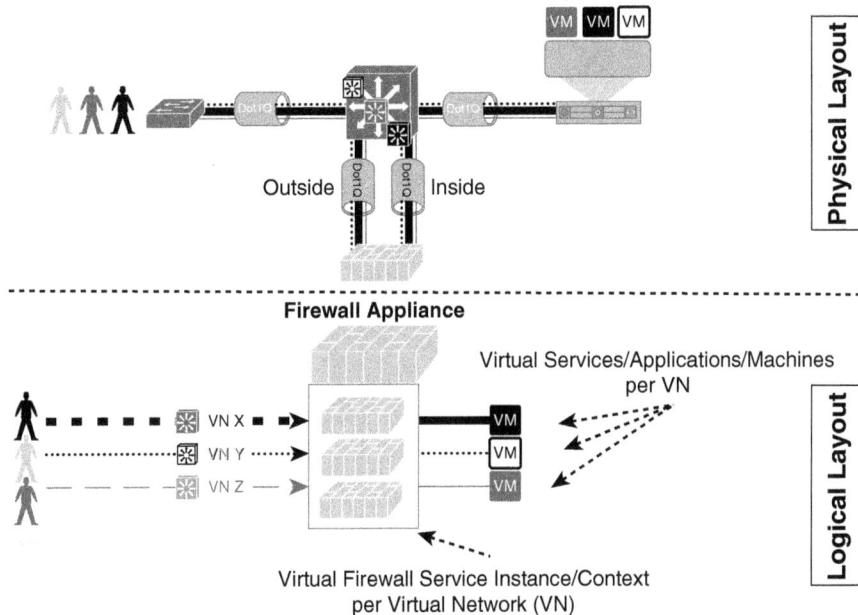

**Figure 15-15**  *Firewall Virtual Instances*

Furthermore, in multitenant network environments, multiple security contexts offer a flexible and cost-effective solution for enterprises (and for service providers). This approach enables network operators to partition a single pair of redundant firewalls or a single firewall cluster into multiple virtual firewall instances per business unit or tenant. Each tenant can then deploy and manage its own security policies and service access, which are virtually separated. This approach also allows controlled inter-tenant communication. For example, in a typical multitenant enterprise campus network environment with MPLS VPN (L3VPN) enabled at the core, traffic between different tenants (VPNs) is normally routed via a firewalling service for security and control (who can access what), as illustrated in Figure 15-16.

Figure 15-17 zooms in on the firewall services contexts to show a more detailed view (logical/virtualized view) of the traffic flow between the different tenants/VPNs (A and B), where each tenant has its own virtual firewall service instance located at the services block (or at the data center) of the enterprise campus network.

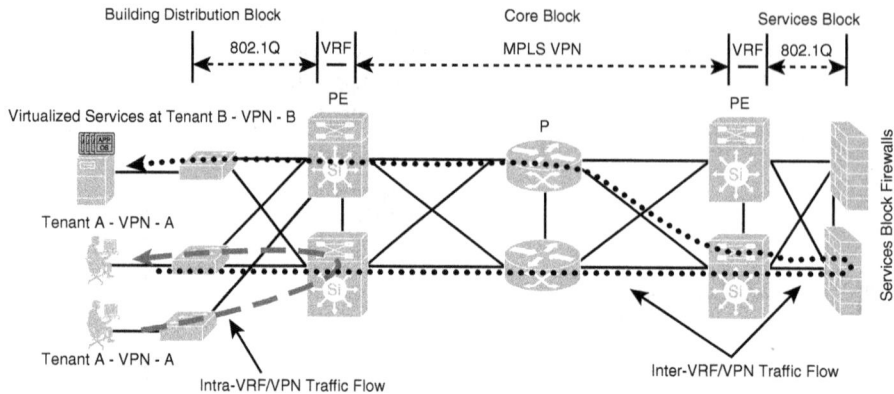

**Figure 15-16**  *Inter-Tenant Services Access Traffic Flow*

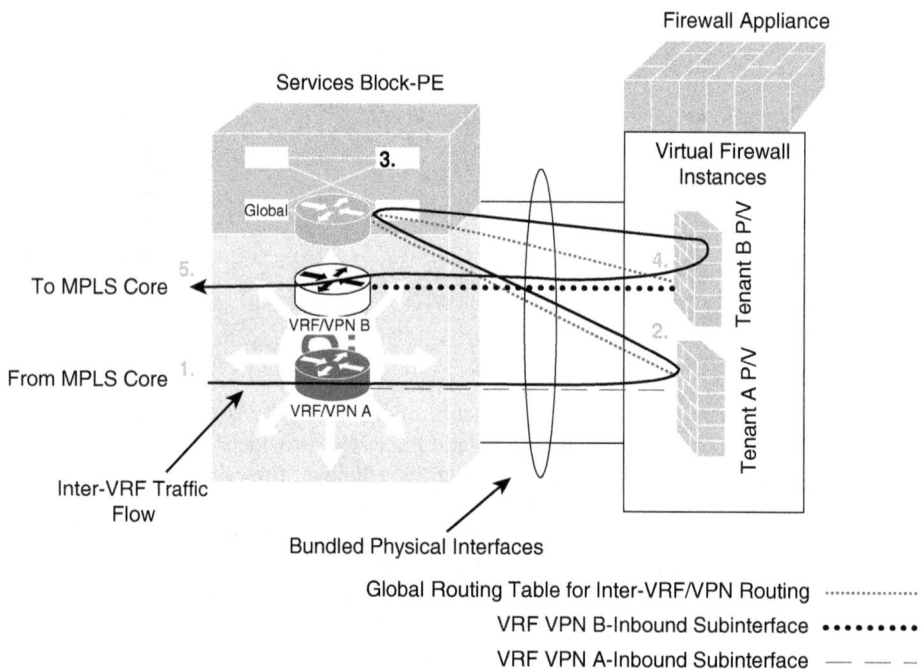

**Figure 15-17**  *Inter-Tenant Services Access Traffic Flow with Virtual Firewall Instances*

In addition, the following are the common techniques that facilitate accessing shared applications and network services in multitenant environments:

- **VRF-Aware Network Address Translation (NAT):** One of the common requirements in today's multitenant environments with network and service virtualization enabled is to provide each virtual (tenant) network the ability to access certain services (shared services) either hosted on premises (such as at the enterprise data center or services block) or hosted externally (in a public cloud). Also, providing Internet access to the different tenants (virtual) networks is a common example of today's multitenant network requirements. To maintain traffic separation between the different tenants (virtual networks) where private IP address overlapping is a common attribute in this type of environment, NAT is considered one of the common and cost-effective solutions to provide NAT per tenant without compromising path separation requirements between the different tenants' networks (virtual networks). When NAT is combined with different virtual network instances (VRFs), it is commonly referred to as *VRF-Aware NAT*, as shown in Figure 15-18. When multiple routing instances are combined into a single instance on a common device, with or without NAT, that is called *fusion routing* and the common device is called a *fusion router*.

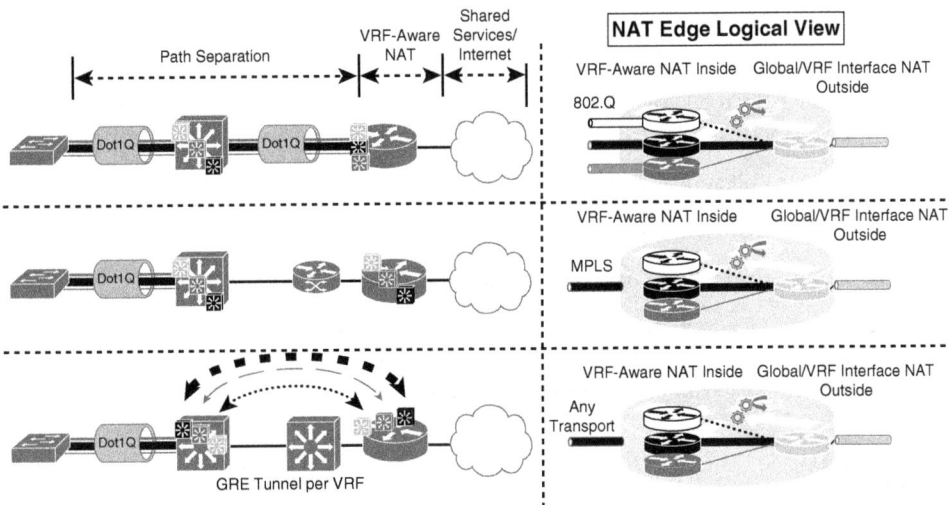

**Figure 15-18**  *VRF-Aware NAT*

*VRF-aware service infrastructure (VASI)* refers to the ability of an infrastructure or a network node, such as a router, to facilitate the application of features and management services (such as encryption and NAT) between VRFs internally within the same node, using virtual interfaces. For two VRFs to communicate internally within a network node (router), a VASI virtual interface pair can be configured. Each interface in this pair must be associated with a different VRF so that those two virtual interfaces can be logically wired, as illustrated in Figure 15-19. This capability is available in some high-end Cisco routers.

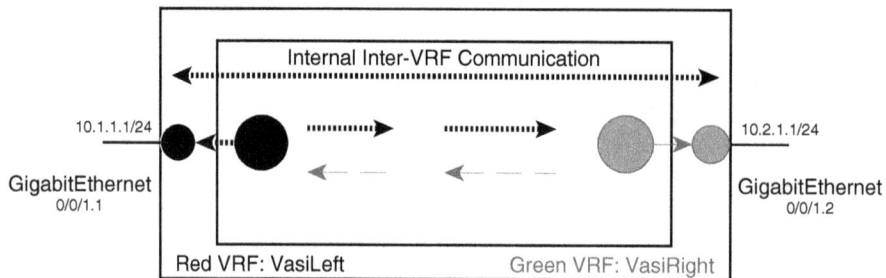

**Figure 15-19** *VRF-Aware Services Infrastructure*

- **Network functions virtualization (NFV):** The concept of NFV is based on virtualizing network functions that typically require a dedicated physical node, appliances, or interfaces. In other words, NFV can potentially take any network function typically residing in purpose-built hardware and abstract it from that hardware. As depicted in Figure 15-20, this concept offers businesses several benefits, including the following:

  - Reduce the total cost of ownership (TCO) by reducing the required number and diversity of specialized appliances

  - Reduce operational cost (for example, less power and space)

  - Offer a cost-effective capital investment

  - Reduce the level of complexity of integration and network operations

  - Reduce time to market for the business by offering the ability to enable specialized network services (especially in multitenant where a separate network function/ service per tenant can be provisioned faster)

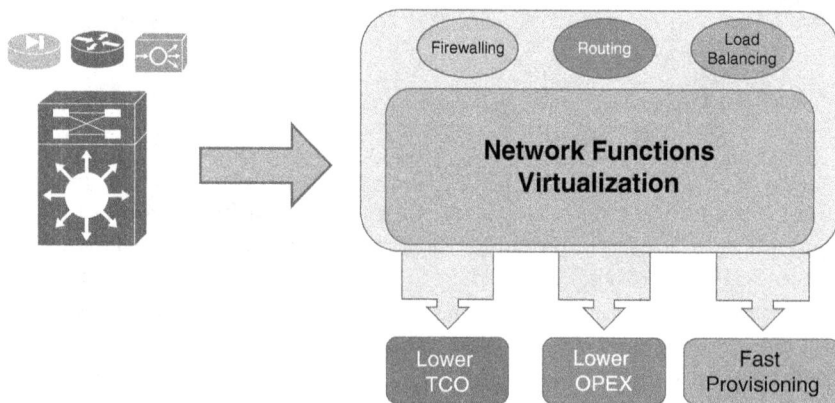

**Figure 15-20** *NFV Benefits*

The NFV concept helps businesses to adopt and deploy new services quickly (faster time to market) and is consequently considered a business innovation enabler. This is simply because purpose-built hardware functionalities have now been virtualized, and it is a matter of service enablement rather than relying on new hardware (along with infrastructure integration complexities).

**NOTE**   The concept of NFV is commonly adopted by service provider networks nowadays. Nonetheless, this concept is applicable and usable in enterprise networks and enterprise data center networks that want to gain its benefits and flexibility.

**NOTE**   In large-scale networks with a very high volume of traffic (typically carrier-grade), hardware resource utilization and limits must be considered.

## Summary

The enterprise campus is one of the vital parts of the modular enterprise network. It is the medium that connects the end users and the different types of endpoints such as printers, video endpoints, and wireless access points to the enterprise network. Therefore, having the right structure and design layout that meets current and future requirements is critical, including the physical infrastructure layout, Layer 2, and Layer 3 designs. To achieve a scalable and flexible campus design, you should ideally base it on hierarchal and modular design principles that optimize the overall design architecture in terms of fault isolation, simplicity, and network convergence time. It should also offer a desirable level of flexibility to integrate other networks and new services and grow in size.

However, the concept of network virtualization helps enterprises to utilize the same underlying physical infrastructure while maintaining access, and path and services access isolation, to meet certain business goals or functional security requirements. As a result, enterprises can lower CAPEX and OPEX and reduce the time and effort required to provision a new service or a new logical network. However, the network designer must consider the different network virtualization design options, along with the strengths and weaknesses of each, to deploy the suitable network virtualization technique that meets current and future needs. These needs must take into account the different variables and constraints, such as staff knowledge and the hardware platform-supported features and capabilities.

## References

Al-shawi, Marwan, *CCDE Study Guide* (Cisco Press, 2015)

## Exam Preparation Tasks

As mentioned in the section "How to Use This Book" in the Introduction, you have a couple of choices for exam preparation: the exercises here, Chapter 18, "Final Preparation," and the exam simulation questions in the Pearson Test Prep Software Online.

## Review All Key Topics

Review the most important topics in this chapter, noted with the Key Topic icon in the outer margin of the page. Table 15-5 lists a reference of these key topics and the page numbers on which each is found.

**Key Topic**

**Table 15-5**  Key Topics for Chapter 15

| Key Topic Element | Description | Page Number |
|---|---|---|
| List | Hierarchical enterprise campus network design layers | 405 |
| Section | When Is the Core Block Required? | 407 |
| List | Business drivers for network virtualization | 414 |
| List | Network virtualization design elements | 415 |
| Table 15-4 | Network Virtualization Techniques Comparison | 420 |

## Complete Tables and Lists from Memory

Print a copy of Appendix D, "Memory Tables" (found on the companion website), or at least the section for this chapter, and complete the tables and lists from memory. Appendix E, "Memory Tables Answer Key," also on the companion website, includes completed tables and lists to check your work.

## Define Key Terms

Define the following key terms from this chapter and check your answers in the glossary:

three-tier model, two-tier model, multitier STP based, routed access, switch clustering, many-to-one virtualization model, one-to-many virtualization model, edge control, transport virtualization, services virtualization

# CHAPTER 16

# Enterprise Internet Edge Architecture Design

## This chapter covers the following topics:

> **Enterprise Internet Edge Design Considerations:** This section provides design recommendations and considerations from an enterprise Internet edge architecture standpoint.

The *enterprise Internet edge* refers to the various enterprise network modules and components that facilitate efficient and secure communication to the Internet. This chapter covers different design options and considerations for the Internet modules.

This chapter covers the following "CCDE v3.0 Core Technology List" sections and provides design recommendations from an enterprise Internet edge architecture standpoint:

- 2.0 Layer 2 Control Plane

- 3.0 Layer 3 Control Plane

- 4.0 Network Virtualization

## "Do I Know This Already?" Quiz

The "Do I Know This Already?" quiz allows you to assess whether you should read this entire chapter thoroughly or jump to the "Exam Preparation Tasks" section. If you are in doubt about your answers to these questions or your own assessment of your knowledge of the topics, read the entire chapter. Table 16-1 lists the major heading in this chapter and the corresponding "Do I Know This Already?" quiz questions. You can find the answers in Appendix A, "Answers to the 'Do I Know This Already?' Quizzes."

**Table 16-1** "Do I Know This Already?" Section-to-Question Mapping

| Foundation Topics Section | Questions |
|---|---|
| Enterprise Internet Edge Design Considerations | 1–5 |

> **CAUTION** The goal of self-assessment is to gauge your mastery of the topics in this chapter. If you do not know the answer to a question or are only partially sure of the answer, you should mark that question as wrong for purposes of the self-assessment. Giving yourself credit for an answer you correctly guess skews your self-assessment results and might provide you with a false sense of security.

1. Which BGP attribute is applied inbound to influence traffic outbound and is carried throughout the AS in a multihoming design?
   a. LOCAL_PREFERENCE (LP)
   b. AS-PATH prepend
   c. Community values + (LP, AS-PATH, or weight)
   d. BGP weight
2. Which BGP attribute is applied outbound to influence inbound traffic flows in a multihoming design?
   a. LOCAL_PREFERENCE (LP)
   b. AS-PATH prepend
   c. MED
   d. BGP weight
3. Which of the following Internet multihoming options should be chosen if a business wants the most flexible option with limited monetary spending?
   a. Equal and unequal load sharing
   b. Active/standby
   c. Equal and unequal load sharing with two edge routers
   d. Equal and unequal load sharing with four edge routers
4. Which of the following Internet multihoming options should be chosen if a business just wants a backup option in case its primary Internet link suffers an outage?
   a. Equal and unequal load sharing
   b. Active/standby
   c. Equal and unequal load sharing with two edge routers
   d. Equal and unequal load sharing with four edge routers
5. Which of the following Internet multihoming options should be chosen if a business has two geographically disperse data centers with a large campus environment connected via a wide area network and the business wants to make Internet routing decisions based on its internal routing information?
   a. Equal and unequal load sharing
   b. Active/standby
   c. Equal and unequal load sharing with two edge routers
   d. Equal and unequal load sharing with four edge routers

# Foundation Topics

# Enterprise Internet Edge Design Considerations

The Internet edge is another module or block of the modular enterprise architecture and part of the enterprise edge module, which provides external connectivity to the other places in the network (PINs) across the enterprise. The Internet block in particular acts as a gateway to the Internet for the enterprise network. From a design point of view, the Internet edge design may significantly vary between organizations, because it is typically driven by the security policy of the business. In addition, large enterprises typically put the Internet edge or module either co-located within the campus network or within its data center network. Nevertheless, in both cases, the location will not impact the module design itself; it will only change the overall enterprise architecture and traffic flow. Therefore, the design concepts discussed in this section apply to both scenarios (location neutral, whether the Internet block is located within the campus or within the data center). With the advent of SD-WAN, the Internet edge could be at each branch location by leveraging direct Internet access circuits to provide this connectivity.

## Internet Edge Architecture Overview

To a large extent, the Internet edge design will vary between different networks based on the security policy of the organization and industry type. For instance, financial services organizations tend to have sophisticated multilayer Internet edge designs. In contrast, retail businesses usually have a less-complicated Internet edge design. Consequently, this can lead to a significant difference in the design, which makes it impractical to provide specific design recommendations for this block. However, Figure 16-1 illustrates a typical (most common) Internet block foundational architecture. It highlights the main layers and network components that form this module as part of the overall modular enterprise architecture.

This block should ideally follow the same principle of the hierarchical design model as part of the modular enterprise building block architecture. This hierarchical model offers a high level of flexibility to this block by having different layers, each focused on different functions. The distribution layer in this block aggregates all the communications within the block, and between the enterprise core and the Internet block, by providing a more structured design and deterministic traffic flow. This flexible design can be considered foundational architecture. It can then be changed or expanded as needed. For example, the demilitarized zone (DMZ) in Figure 16-1 can be replicated into multiple DMZs to host different services that require different security policies and physical separation. Some designs also place the VPN termination point in a separate DMZ for an additional layer of security and control.

Figure 16-2 highlights the typical primary functions and features at each layer of the Internet block architecture.

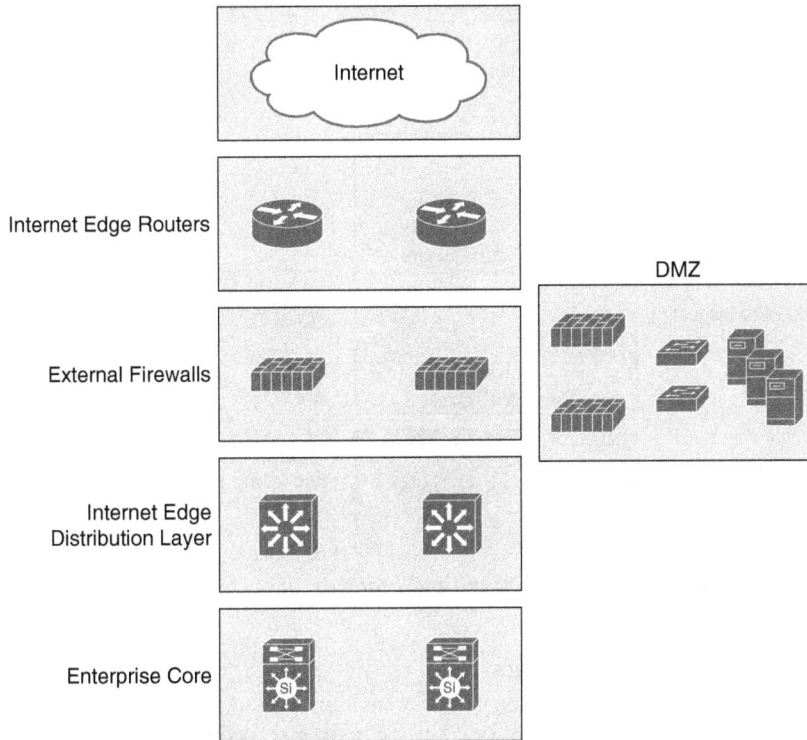

**Figure 16-1**  *Enterprise Internet Block Foundational Architecture*

**NOTE**  These functions can vary from design to design to some extent. Moreover, all of them are not necessarily required to be applied. For example, NAT can be applied at only one layer instead of doing multiple layers of NAT.

Consider, for instance, a scenario where the security requirements specify that all traffic passing through the external Internet block firewalls must be inspected, not tunneled. This slightly influences the design. Technically, it can be achieved in different ways, depending on the available options and platform capabilities. (For example, some next-generation firewalls can perform deep packet inspection, even for tunneled traffic.) Alternatively, the VPN tunnels can be terminated at the Internet edge router or a separate DMZ can be created for the VPN, where VPN tunnels can terminate (using a VPN concentrator or dedicated firewalls/routes). The decapsulated VPN traffic will then be sent back to the Internet edge firewalls for traffic inspection before reaching the internal network.

## Enterprise Multihomed Internet Design Considerations

The design of the Internet edge will vary to a large extent based on different variables and requirements. The major influencing factor is the organization's security policy. Similarly, the multihoming to the Internet follows the same concept, because it is part of this block. Because of this high degree of variability, this section covers the most common scenarios with multihoming to the Internet.

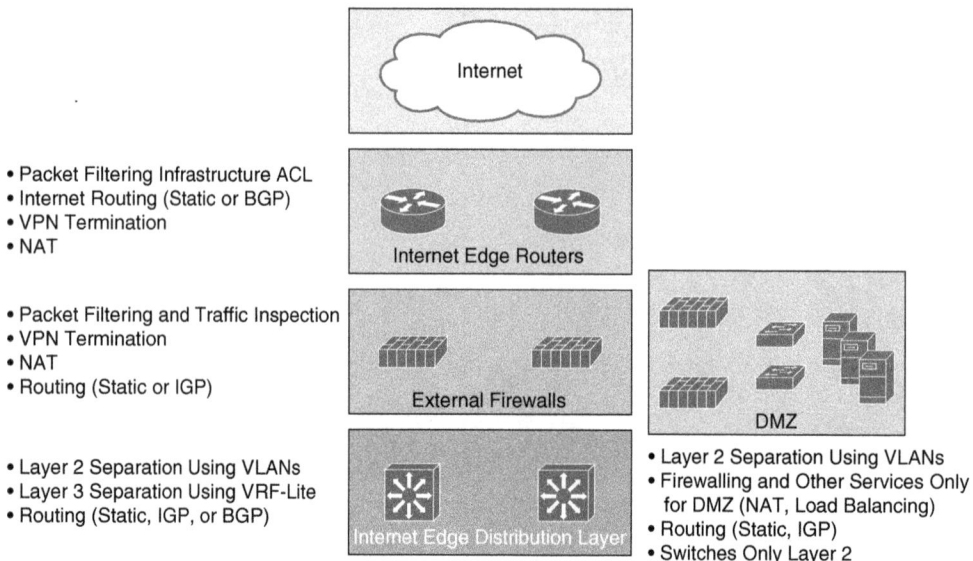

- Packet Filtering Infrastructure ACL
- Internet Routing (Static or BGP)
- VPN Termination
- NAT

- Packet Filtering and Traffic Inspection
- VPN Termination
- NAT
- Routing (Static or IGP)

- Layer 2 Separation Using VLANs
- Layer 3 Separation Using VRF-Lite
- Routing (Static, IGP, or BGP)

- Layer 2 Separation Using VLANs
- Firewalling and Other Services Only for DMZ (NAT, Load Balancing)
- Routing (Static, IGP)
- Switches Only Layer 2

**Figure 16-2**  *Internet Block per Layer Functions and Features*

## Multihoming Design Concept and Drivers

*Multihoming* refers to the concept of having two or more links to external networks, using either one or more edge nodes. This concept or connectivity model is common in large enterprises; however, the actual drivers toward adopting this connectivity model vary, and ideally the decision to do so should be driven by the business and functional requirements. In general, the most common drivers for enterprises include the following:

- Higher level of path and link redundancy.

- Increased service reliability.

- Offer the ability for the business to optimize the return on investment (ROI) of the external links through traffic load-balancing and load-sharing techniques.

- Cost control, where expensive links can be dedicated for a certain type of traffic only.

- Flexibility and efficiency:

  - This design approach increases the overall bandwidth capacity to and from the Internet (by using load-balancing or load-sharing techniques).

  - Provides the ability to support end-to-end network and path separation with service differentiation by having different Internet links and diverse paths end to end from the ISP to the end users (for example, to serve different entities within the enterprise using different Internet links based on business demand or a security policy).

From a design point of view, network designers need to consider several questions to produce a more business-driven multihoming Internet design. These questions can be divided into two main categories:

- Path requirements

    - Is the business goal high availability only?

    - Is the business goal to optimize the ROI of the existing external links?

    - Should available bandwidth be increased?

    - Should there be path and traffic isolation?

- Traffic flows characteristics

    - Is the business goal to host services within the enterprise and to be accessible from the Internet?

    - Is the business goal to host some of its services in the cloud or to access external services over the Internet?

    - Or both (hybrid)?

The reason behind considering these questions is to generate a design (typically BGP policies) that aligns with the business and functional requirements. In other words, designing in isolation without a good understanding of the different requirements and drivers (for example, business goals and functional and application requirements) will make it impossible to produce an effective business-driven multihoming design.

**NOTE** This information can be obtained in different ways, based on the gathered requirements. For example, it may be shown as functional requirements through utilization reports that show 75 percent of the traffic is outbound and 25 percent is inbound, in which it is clear that the traffic pattern is included toward accessing content over the Internet.

**NOTE** BGP is the most flexible protocol that handles routing policies and the only protocol that has powerful capabilities that can reliably handle multiple peering with multiple autonomous systems (interdomain routing). Therefore, this section only considers BGP as the protocol of choice for Internet multihoming design; however, some designs may use an interior gateway protocol (IGP) or static routing with multihoming. Typically, these designs eliminate all the flexibilities that you can gain from BGP multihoming scenarios.

## BGP over Multihomed Internet Edge Planning Recommendations

Designing a reliable business-driven multihoming connectivity model is one of the most complex design projects because of the various variables that influence the design direction. Therefore, good planning and an understanding of the multiple angles of the design are prerequisites to generating a successful multihoming design.

The following are the primary considerations that network designers must take into account when planning a multihomed Internet edge design with BGP as the interdomain routing protocol:

- Does the business have a public BGP autonomous system number (ASN) or should they leverage a private ASN. Typically, public ASNs offer more flexibility to the design, especially with multihoming to different ISPs.

- Provider-independent (PI) public IP addresses offer more flexibility and availability options to the design compared to provider-assigned (PA) IP addresses.

- A PI address combined with a public ASN obtained from the Regional Internet Registry (RIR), for example, provides the most flexible choice with Internet multihoming, especially for enterprises that host services to be accessed from the Internet across different ISPs.

- Receiving the full Internet routing table can help to achieve a more detailed traffic engineering policies. (For example, you can specify outbound traffic to use a certain link on a per geographical region basis, based on the IP prefixes along with its assigned BGP community values.)

**NOTE**  BGP community values can provide flexible traffic engineering control within the enterprise and across the ISP. Internet providers can match community values and predefined application policies per community value. For example, you can influence the path LOCAL_PREFERENCE value of your advertised route within the ISP cloud by assigning an *x* BGP community value to the route. Refer to RFC 1998.

## BGP Policy Control Attributes for Multihoming

As discussed in Chapter 8, "Layer 3 Technologies," several BGP attributes influence BGP path selection. Table 16-2 lists the most common simple and powerful BGP attributes that you can use to control route advertisements and influence the path selection in BGP multihoming design.

**Key Topic**

**Table 16-2**  Common BGP Attributes for Internet Multihoming

| Attribute | Usage Description |
|---|---|
| LOCAL_PREFERENCE (LP) | Influence outbound traffic flows |
| AS-PATH prepend | Influence inbound and outbound traffic flows |
| Community values + (LP, AS-PATH, or weight) | Influence inbound and outbound traffic flows within the customer AS and across the ISP's autonomous systems |
| BGP weight | Influence local router decision for outbound traffic flows (Cisco proprietary attribute) |

BGP community values technically can be seen like a "route tag," which can be contained within the one AS or be propagated across multiple autonomous systems to be used as a "matching value" to influence BGP path selection. For instance, one of the common scenarios with global ISPs is that each ISP can share the standard BGP community values used with its customers to distinguish IP prefixes based on its geographic location (for example, by region or continent). This offers enterprises the flexibility to match the relevant community value that represents a certain geographic location and associate it with a BGP policy such as AS-PATH prepending to achieve a certain goal. For example, an enterprise may want all traffic going to IP prefixes within Europe (outbound) to use Internet link 1, while all other traffic should use the second link. As illustrated in Figure 16-3, BGP community values can simplify achieving this goal to a large extent in a more dynamic manner.

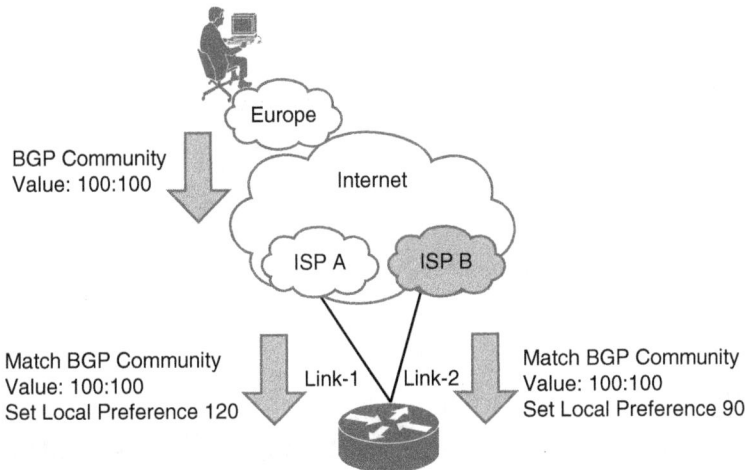

**Figure 16-3**  *BGP Community Value Usage Example*

## Common Internet Multihoming Traffic Engineering Techniques over BGP

This section covers the primary traffic engineering models and techniques that you can use with multihomed Internet connectivity over BGP.

### Scenario 1: Active-Standby

This design scenario (any of the connectivity models depicted in Figure 16-4) is typically based on using one active link for both inbound and outbound traffic, with a second link used as a backup. This design scenario is the simplest design option and is commonly used in situations where the backup link is a low-speed link and is only required to survive during any outage of the primary link.

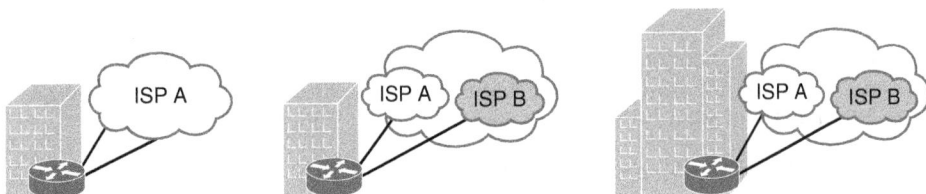

**Figure 16-4**  *Internet Multihoming Active-Standby Connectivity Models*

Figure 16-5 shows an active-standby scenario where ISP A must be used as the primary and active path for both ingress and egress traffic flows.

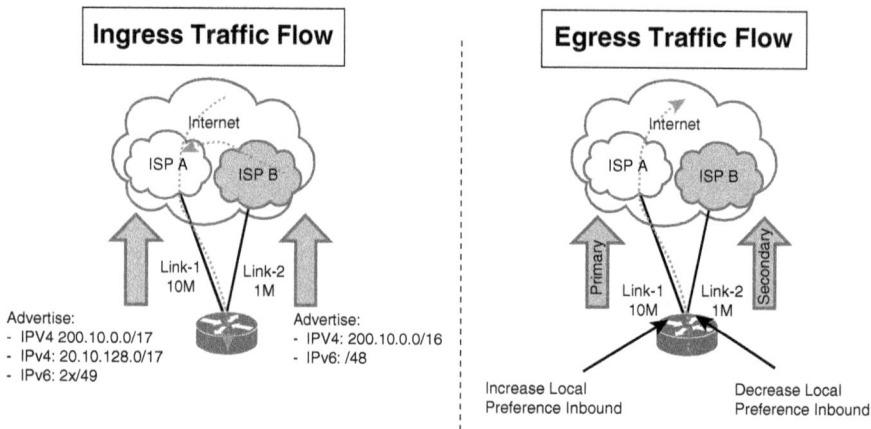

**Figure 16-5**  *Internet Multihoming Active-Standby Traffic Flow*

Table 16-3 outlines a possible solution for this scenario.

**Table 16-3**  Internet Multihoming Active-Standby

| Traffic Direction | BGP Policy |
|---|---|
| Ingress | Longest match over the preferred path, by dividing the prefix into two halves (for instance, advertise /16 as 2 × /17 over the preferred ingress path toward ISP A in the scenario in Figure 16-5) |
| Egress | Use LOCAL_PREFERENCE (set the preferred route/path with a higher value) |

Although AS-PATH prepending is one of the most common and obvious techniques that you can use in this scenario to influence ingress traffic, ISPs usually allocate a higher LOCAL_PREFERENCE value to prefixes learned by their customers than to the ones learned through other peering ISPs. In other words, even though the prefix 200.10.0./16 is advertised to both ISP A and B with AS-PATH prepending applied toward ISP B, ISP B's customers will always use the path over ISP B because it will be assigned a higher LOCAL_PREFERENCE (within ISP B cloud) than the one learned via ISP A when traffic is passing through ISP B. This means that the targeted goal will not be achieved optimally in this case because any traffic flow coming from any of ISP B's customers will use the low-bandwidth link (Link-2 in this case) to reach 200.10.0.0/16. Therefore, by dividing the /16 network into two /17s and advertising it over ISP A, if there is a customer of ISP B that wants to reach your prefix (one of the 2 × /17s advertised by ISP A), ISP B will never get to the /16 sent over ISP B link because the longer match over ISP A will always win in this case, as illustrated in Figure 16-5.

This behavior (where ISPs allocate a higher LOCAL_PREFERENCE value to prefixes advertised within their cloud) is altered sometimes when the ISP enables its customers to control what BGP attribute and value they need to assign to their prefixes when injected into the ISP cloud. This is normally achieved by using standard BGP community values to influence traffic routing within the ISP cloud (most commonly to change AS-PATH or LOCAL_PREF-ERENCE values within the ISP cloud). For example, in the scenario illustrated in Figure 16-6, if you assign a BGP community value 300:110 to prefix 200.10.0.0/16 toward ISP A (AS 300), its BGP LOCAL_PREFERENCE value within the SP cloud will be set to 110, which will be given preference across this SP cloud and its directly connected customers. Similarly, if you assign a BGP community value 500:70 to prefix 200.10.0.0/16 toward ISP B (AS 500), along with BGP AS-PATH prepending 3 × ASN, (100 100 100), this will make this prefix carry a low LOCAL_PREFERENCE value within AS 500 (assuming, 70 is less than the common default BGP LOCAL_PREFERENCE "100"). In addition, it will be seen (by AS 500) as a longer AS-PATH (100 100 100) compared to the prefix learned via ISP A (AS 300). Technically, the prefix 200.10.0.0/16 within AS 500 will be seen as follows:

Via AS 100: LOCAL_PREFERENCE = 70, AS-PATH = (100 100 100)

Via AS 300: LOCAL_PREFERENCE = 100, AS-PATH = (300 100) (preferred path)

With this approach, network designers can achieve a well-established optimal active-standby Internet connectivity model. However, two limitations apply to this approach. First, not every ISP provides this flexibility to its clients. Second, when customers are multihoming to different ISPs, the advertised prefixes must be PI to avoid having the upstream ISP aggregate the advertised prefix and break the entire design. For instance, if ISP A advertises the prefix 200.10.0.0/16 toward ISP B as 200.0.0.0/8, the path via R2 over ISP B will always be preferred by ISP B customers (longest match) regardless of which BGP attribute has been used.

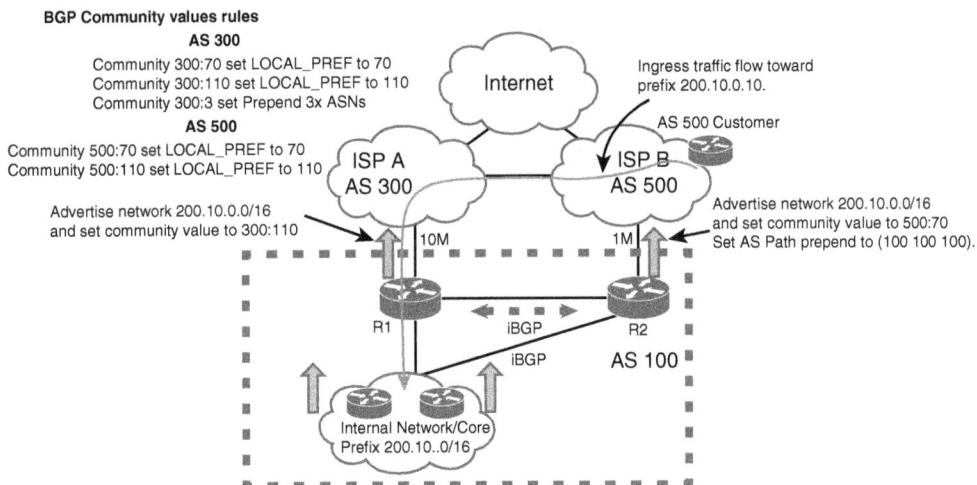

**Figure 16-6**  *Altering BGP Attributes Within the ISP Cloud*

### Scenario 2: Equal and Unequal Load Sharing

These types of design scenarios are commonly used either when there are two Internet links with the same bandwidth value and traffic needs to be distributed evenly across the two links to increase the overall bandwidth and reduce traffic congestion (equal load sharing) or when the Internet links have a different amount of bandwidth where traffic can be distributed in a weighted manner. Ideally, the link with higher bandwidth should handle more traffic than the link with lower bandwidth, as summarized in Figure 16-7.

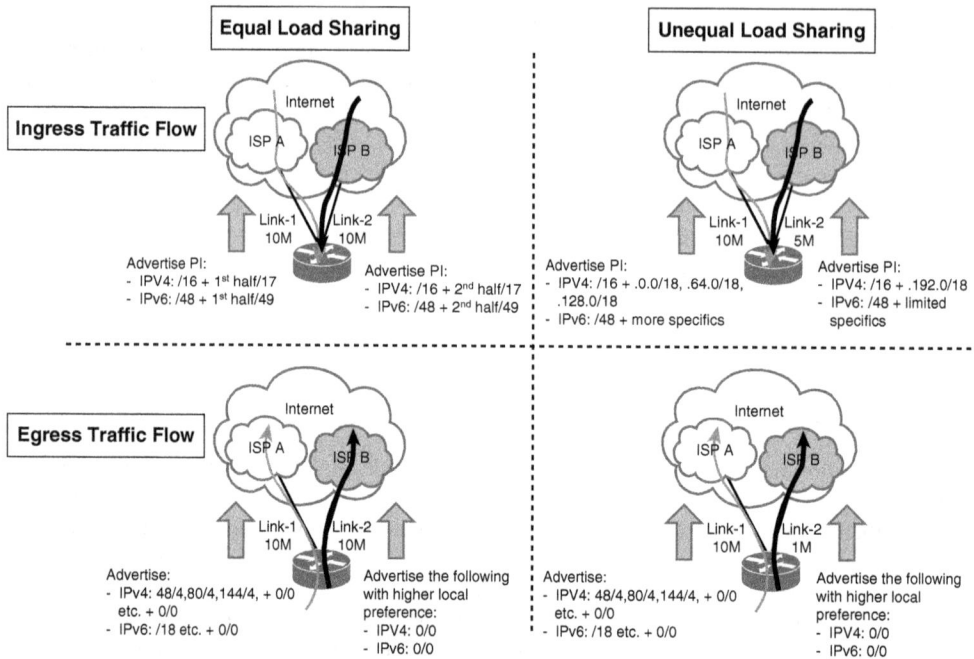

**Equal Load Sharing**

**Unequal Load Sharing**

**Ingress Traffic Flow**

Internet

ISP A    ISP B

Link-1    Link-2
10M    10M

Advertise PI:
- IPV4: /16 + 1st half/17
- IPv6: /48 + 1st half/49

Advertise PI:
- IPV4: /16 + 2nd half/17
- IPv6: /48 + 2nd half/49

Internet

ISP A    ISP B

Link-1    Link-2
10M    5M

Advertise PI:
- IPV4: /16 + .0.0/18, .64.0/18,
  .128.0/18
- IPv6: /48 + more specifics

Advertise PI:
- IPV4: /16 + .192.0/18
- IPv6: /48 + limited
  specifics

**Egress Traffic Flow**

Internet

ISP A    ISP B

Link-1    Link-2
10M    10M

Advertise:
- IPv4: 48/4,80/4,144/4, + 0/0
  etc. + 0/0
- IPv6: /18 etc. + 0/0

Advertise the following
with higher local
preference:
- IPV4: 0/0
- IPv6: 0/0

Internet

ISP A    ISP B

Link-1    Link-2
10M    1M

Advertise:
- IPv4: 48/4,80/4,144/4, + 0/0
  etc. + 0/0
- IPv6: /18 etc. + 0/0

Advertise the following
with higher local
preference:
- IPV4: 0/0
- IPv6: 0/0

**Figure 16-7** *Internet Multihoming Equal and Unequal Load Sharing*

Table 16-4 shows a typical solution.

**Table 16-4** Internet Multihoming Equal/Unequal Load Sharing

| Traffic Direction | BGP Policy |
|---|---|
| Ingress | The typical mechanism to use here is to divide the PI address into two halves. For example, an IPv4 /16 subnet can be divided into two /17 summary subnets, similarly to an IPv6 /48 subnet can be divided into two /49 summary subnets. Then advertise each half over a different link along with the aggregate (IPv4 /16 or IPv6 /48 in this example) over both links to be used in case of link failure. For unequal load sharing, you can use the same concept with more small subnets to be advertised over the path with higher capacity. |
| Egress | For the outbound traffic direction, you need to receive the full Internet route from one of the ISPs along with the default route from both. Accept with filtering only every other /4 for IPv4 (for example, 0/4, 32/4). IPv6 can use the same concept (IPv6 either selectively or the same concept). From the other link, increase the LOCAL_PREFERENCE for the default route.

In this case, the more specific route (permitted in the filtering) will be used over one link. Every other route that was filtered out will go over the second link using the default route. For unequal load sharing, more subnets can be accepted/allowed from the link with higher capacity. |

NOTE  ISPs usually deploy route filtering policies with their customers and with other peering ISPs, where only certain subnet lengths are accepted, to avoid propagating a large number of small networks such as v4/24 or v6/64. However, the subnets presented in this section are hypothetical to simplify the explanation of the discussed points.

Although the typical mechanisms used to influence path selection equally (to a certain extent) are the ones described in Table 16-4, this cannot guarantee fair or equal load distribution across both links (because traffic load cannot be measured based only on its network or subnet size). For instance, if a /24 subnet is divided into two halves (.0/25 and .128/25), advertising each half (/25) over two separate Internet links will (in theory) provide equal load distribution across both Internet links in the inbound direction. In fact, what is hypothetically correct is not always practically achievable; the reason is that if there are a few hosts such as an FTP or web server as part of the subnet 0/25, with high traffic volume destined to these servers, the link advertising 0/25 will be used more than the second Internet link. This same concept applies to the outbound traffic direction. In other words, achieving true equal load distribution must not be derived only by IP subnetting (dividing and advertising the available IP range equally across the available links). Instead, it also must be based on actual service utilization reports. Alternatively, you can use a specialized load-balancing appliance that can distribute traffic flow equally based on real-time link utilization.

Similarly, if unequal load sharing is required (for example, 70/30 load distribution, which means 70 percent of traffic should use the link with higher bandwidth and 30 percent should use the other link with lower bandwidth), the same principle and concerns discussed earlier with regard to equal load sharing apply here. In other words, proper planning and a study of utilization reports must be performed before designing and applying BGP policies to influence which network or host should be reachable over which link. Also, the same applies to outbound traffic. (What are the most targeted services in terms of traffic volume and number of traffic flows, such as cloud-hosted applications like Cisco Webex?) After these facts are identified, network designers can provide more practical multihoming with the desired proportional load distribution model (whether it is equal or unequal load distribution).

### Scenario 3: Equal and Unequal Load Sharing Using Two Edge Routers

The design techniques described in the preceding scenarios all apply to this scenario for the ingress traffic flow direction. However, the egress traffic flow direction depends on the LAN-side design behind the routers:

- **Using a firewall with a Layer 2 switch between the firewall and routers:** Use multiple groups of Hot Standby Routing Protocol (HSRP) or Virtual Router Redundancy Protocol (VRRP) for load sharing on the routers' side.

- **Using Layer 3 network node, such as a switch connecting directly to the edge routers:** You can use IGP with equal-cost multipath (ECMP) in this case.

**NOTE**  In both cases, you need to make sure that there is a link between the two Internet edge routers (physical or tunnel) with internal BGP (iBGP) peering over this link, to avoid traffic black-holing in some failure scenarios.

### Asymmetrical Routing with Multihoming (Issue and Solution)

The scenario depicted in Figure 16-8 demonstrates a typical design scenario with a potential for asymmetrical routing. This scenario is applicable to two sites or data centers with a direct (backdoor) link, along with a layer of firewalling behind the Internet edge routers. In addition, these firewalls are site-specific, where no state information is exchanged between the firewalls of each site. In addition, both sites are advertising the same address range toward the Internet (PI or PA) range. Therefore, a possibility exists that return traffic (of outbound traffic) originated from Site-1 going to the Internet using the local Internet link will come back over the Site-2 Internet link. The major issue here is that the firewall of Site-2 has no "state information" about this session. Therefore, the firewall will simply block this traffic. This can be a serious issue if the design did not consider how the network design will handle situations like this, especially during failure scenarios.

To optimize this design to overcome this undesirable behavior, network designers need to consider the following:

- **Control plane peering between edge routers in each site:** The first important point here is to make sure that both Internet edge routers are connected directly and use iBGP peering between them. This link can be physical or over a tunnel interface, such as a GRE tunnel. A network designer can optimize this design and mitigate its impact by adding this link along with associated BGP policies (make Site-1 internal prefixes more preferred over the Site-1 Internet link), which will help to avoid the blocking by the Site-2 firewall.

- **Organized IP addressing advertisement:** To make the preceding point work smoothly, a network designer needs to make sure that each site is advertising its own route prefixes as more specific (longest match principle), in addition to the aggregate of the entire PI subnet, as discussed earlier. For example, /16 might be divided into two /17s per site. (You can use the same concept with IPv6.)

- **NAT consideration:** The other point to be noted here is that if these prefixes are translated by the edge firewalls, one of the common and proven ways to deal with this type of scenario is by forming a direct route peering between the distribution/core layer nodes and the Internet edge routers. Consequently, the edge firewalls can perform NAT for traffic passing through them (as long as the two earlier considerations mentioned in this list are in place).

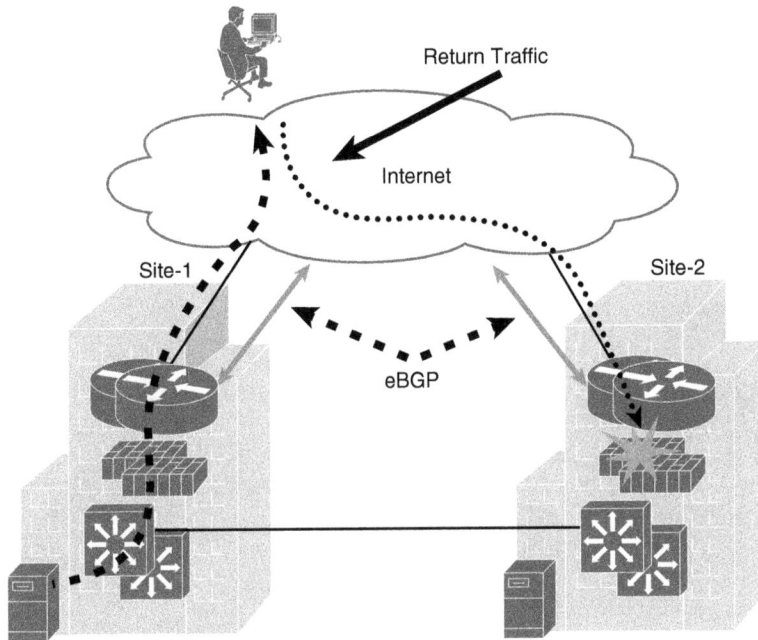

**Figure 16-8**  *Asymmetrical Routing*

**NOTE**  It's worth mentioning that asymmetric routing will completely break traffic flows with stateful firewalls in between, but even in networks without firewalls, asymmetric routing can be burdensome because it consumes valuable bandwidth from unrelated sites (transit sites). These can be even harder to troubleshoot because traffic flows continue to work despite the lurking problem.

For instance, in Figure 16-9 the Internet edge distribution is peering with the Internet edge router using multihop eBGP with private AS and advertising the PI prefixes over BGP (using static routes with a BGP network statement to advertise the routes). The firewall is using a default route toward the Internet edge router, along with NAT.

By incorporating the design recommendations to optimize the preceding scenario, a network designer can make the design more agile in its response to failures. For example, in the topology in Figure 16-8, if the link between Site-1 and the Internet were to go down for some reason, traffic destined for Site-1 prefixes (part of the first half of the /17) would typically go to the Site-2 Internet link (because we advertise the full /16 from both sites). Then traffic would traverse the inter-site link to reach Site-1. As a result, this design will eliminate the firewall blocking issue, even after an Internet link failure event, as illustrated in Figure 16-10.

Table 16-5 compares the three different Internet Multihoming Models from a design decision perspective.

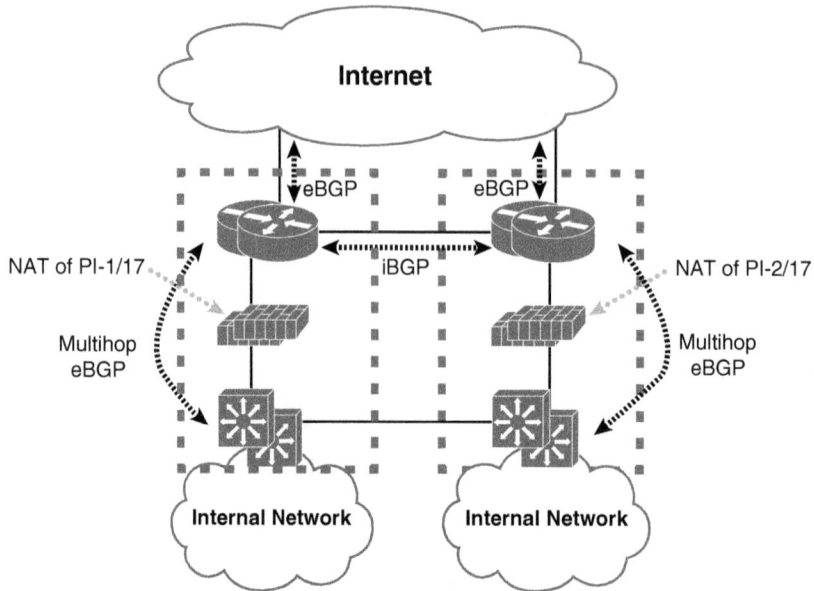

**Figure 16-9** *Optimized Multihoming Design with Firewalls*

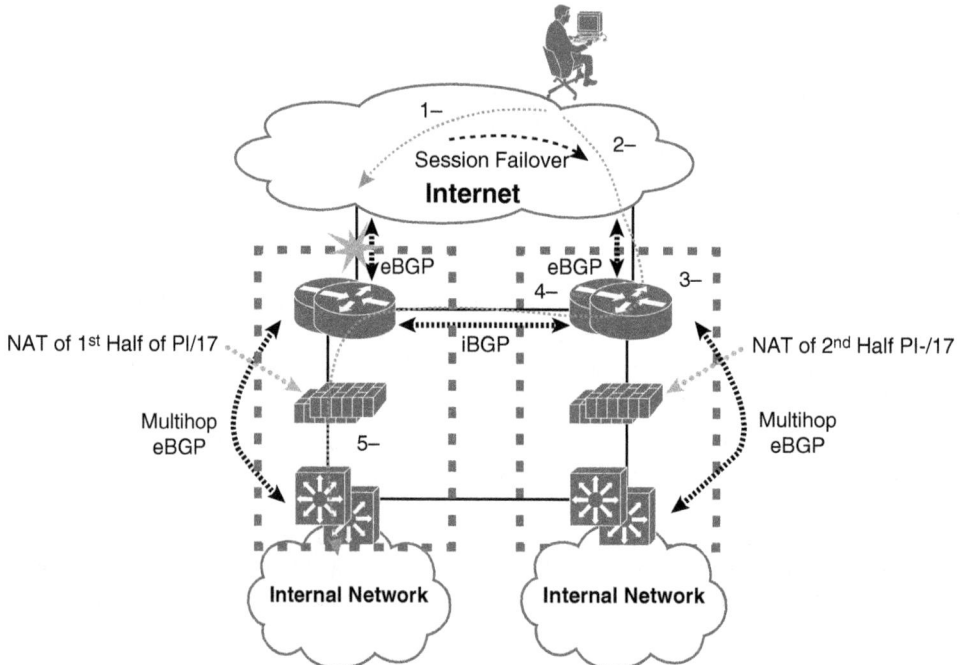

**Figure 16-10** *Optimized Multihoming Design: Failure Scenario*

**Key Topic**

**Table 16-5**    Comparing Internet Multihoming Models

| | Active/Standby | Equal and Unequal Load Sharing | Equal and Unequal Load Sharing with Two Edge Routers |
|---|---|---|---|
| Design flexibility | Least flexible | Flexible | Most flexible |
| Scalability | Least scalable | Scalable | Most scalable |
| Monetary cost | Low | Moderate | Very high |
| Operational efficiency | Least efficient | Efficient | Most efficient |
| Operational complexity | Simplest | Complex | Most complex |
| Design situation | The backup link is a low-speed link and is only required to survive during any outage of the primary link. | Traffic needs to be distributed evenly across multiple links to increase the overall bandwidth and reduce traffic congestion, or when the links have a different amount of bandwidth where traffic can be distributed in a weighted manner. For example, one link is 50-Mbps while another link is 100-Mbps, and the requirement is to use all of the available bandwidth efficiently. | Multiple data centers or multiple sites with a large campus environment hanging off the inside of the Internet edge architecture with the need to leverage the internal campus routing information to determine the best Internet exit point. |

16

## Summary

Today's enterprise businesses, in particular multinational and global organizations, primarily rely on technology services and applications to achieve their business goals. Therefore, the enterprise Internet edge architecture is one of the most vital and critical modules within the modern modular enterprise architecture. It represents the gateway of the enterprise network to the Internet, and today the Internet is an unstated requirement. As a network designer you must keep this in mind. Customers, businesses, and end users expect the Internet to always work, Network designers are responsible for making sure the Internet always works. Therefore, network designers must consider designs that can provide a common resource access experience to the remote sites and users without compromising any of the enterprise security requirements, such as end-to-end path separation between certain user groups. In addition, optimizing Internet edge design with business-driven multihoming designs can play a vital role in enhancing the overall Internet edge performance and design flexibility and can maximize the total ROI of the available links.

## References

Al-shawi, Marwan, *CCDE Study Guide* (Cisco Press, 2015)

## Exam Preparation Tasks

As mentioned in the section "How to Use This Book" in the Introduction, you have a couple of choices for exam preparation: the exercises here, Chapter 18, "Final Preparation," and the exam simulation questions in the Pearson Test Prep Software Online.

## Review All Key Topics

Review the most important topics in this chapter, noted with the Key Topic icon in the outer margin of the page. Table 16-6 lists a reference of these key topics and the page numbers on which each is found.

**Key Topic**

**Table 16-6**  Key Topics for Chapter 16

| Key Topic Element | Description | Page Number |
|---|---|---|
| Table 16-2 | Common BGP Attributes for Internet Multihoming | 434 |
| Section | Scenario 1: Active-Standby | 435 |
| Section | Scenario 2: Equal and Unequal Load Sharing | 437 |
| Section | Scenario 3: Equal and Unequal Load Sharing Using Two Edge Routers | 439 |
| Table 16-5 | Comparing Internet Multihoming Models | 443 |

## Complete Tables and Lists from Memory

Print a copy of Appendix D, "Memory Tables" (found on the companion website), or at least the section for this chapter, and complete the tables and lists from memory. Appendix E, "Memory Tables Answer Key," also on the companion website, includes completed tables and lists to check your work.

## Define Key Terms

No new key terms were introduced in this chapter.

# CHAPTER 17

# Enterprise WAN Architecture Design

## This chapter covers the following topics:

**Enterprise WAN Module:** This section covers critical WAN connectivity topics and network design elements.

**WAN Virtualization and Overlays Design Considerations and Techniques:** This section covers how the enterprise can extend network virtualization across the WAN.

**Enterprise WAN Migration to MPLS VPN Considerations:** This section covers example migration steps for WAN migration to MPLS VPNs.

The enterprise wide-area network (WAN) refers to the various enterprise WAN modules and components that facilitate efficient and secure communication between the different enterprise locations, including campuses, data centers, and remote sites. This chapter covers different design options and considerations for the WAN modules.

This chapter covers the following "CCDE v3.0 Core Technology List" sections and provides design recommendations from an enterprise WAN architecture standpoint:

- 2.0 Layer 2 Control Plane

- 3.0 Layer 3 Control Plane

- 4.0 Network Virtualization

## "Do I Know This Already?" Quiz

The "Do I Know This Already?" quiz allows you to assess whether you should read this entire chapter thoroughly or jump to the "Exam Preparation Tasks" section. If you are in doubt about your answers to these questions or your own assessment of your knowledge of the topics, read the entire chapter. Table 17-1 lists the major headings in this chapter and their corresponding "Do I Know This Already?" quiz questions. You can find the answers in Appendix A, "Answers to the 'Do I Know This Already?' Quizzes."

**Table 17-1** "Do I Know This Already?" Section-to-Question Mapping

| Foundation Topics Section | Questions |
|---|---|
| Enterprise WAN Module | 1–4 |
| WAN Virtualization and Overlays Design Considerations and Techniques | 5–7 |
| Enterprise WAN Migration to MPLS VPN Considerations | 8 |

1. Which topology would allow customers to roll out new transport services such as IPv6 or multicast most rapidly?

    a.   MPLS L2VPN WAN

    b.   MPLS L3VPN WAN

    c.   Internet as WAN

    d.   Internet as transport

2. Which of the following WAN transport models should be chosen if the design decision calls for controlling the number of routing neighborships deterministically?

    a.   MPLS L2VPN WAN

    b.   MPLS L3VPN WAN

    c.   Internet as WAN

    d.   Internet as transport

3. Which of the following WAN transport models should be chosen if the design decision calls for the cheapest WAN solution?

    a.   MPLS L2VPN WAN

    b.   MPLS L3VPN WAN

    c.   Internet as WAN

    d.   Internet as transport

4. Which of the following WAN transport models should be chosen if the design decision requires remote site scalability for a very large number of remote sites?

    a.   MPLS L2VPN WAN

    b.   MPLS L3VPN WAN

    c.   Internet as WAN

    d.   Internet as transport

5. Which of the following network overlay technologies should be chosen if the design decision requires the highest level of scalability, spoke registration combined with optional security, and spoke-to-spoke communication dynamically?

    a.   Remote access VPN

    b.   DMVPN

    c.   IPsec

    d.   GRE

6. Which of the following network overlay technologies provides support for a point-to-point network topology? (Choose two.)
    a. Remote access VPN
    b. DMVPN
    c. IPsec
    d. GRE

7. Which of the following network overlay technologies allows for a spoke-to-spoke traffic pattern?
    a. Remote access VPN
    b. DMVPN
    c. IPsec
    d. GRE

8. Which of the following should be taken into consideration when migrating from one WAN architecture to another? (Choose two.)
    a. Remote access VPN options
    b. Where the default route is generated
    c. The application specifications
    d. The logical and physical site architectures

# Foundation Topics

# Enterprise WAN Module

The WAN module is the gateway of the enterprise network to the other remote sites (typical remote branches) and regional sites. As part of the modular enterprise network architecture, this module aggregates and houses all the WAN or MAN edge devices that extend the enterprise network to the remote sites using different transport media types and technologies. Enterprises require a WAN design that offers a common resource access experience to the remote sites, along with sufficient performance and reliability.

As organizations move into multinational or global business markets, they require a flexible network design that reduces the time needed to add new remote sites. They also require the deployment of new technologies and capabilities that support emerging business applications and communications. In addition, it is becoming a common business requirement for users to have a consistent experience, *quality of experience (QoE)*, when connecting to the enterprise's online resources (such as applications and files), whether they are at the company headquarters or at a remote site; therefore, the WAN design should be flexible enough to accommodate different business application requirements. It should also offer the ability to scale bandwidth or to add new sites or resilient links without any major change to the overall design architecture.

From an enterprise architectural point of view, the primary WAN module can be either coresident physically within the data center block of the enterprise or at the primary enterprise campus. In both cases, there is "no difference" for the WAN module architecture itself, because the aggregation layer of the WAN module will be connected to the core layer of either the enterprise campus or data center.

This chapter covers and compares the different design options that you can use to provide the most suitable WAN design to meet various business, functional, and technical requirements.

## WAN Transports: Overview

One of the primary concerns of enterprise IT leaders is to manage costs and maintain reliable WAN infrastructures to meet their business goals. Furthermore, businesses are realizing that avoiding the complexities around the WAN transport between the different enterprise locations is becoming key to success in today's high-tech competitive market where technology solutions have become the primary business enabler and facilitator. Most commonly, enterprise IT leaders are always concerned about adopting a WAN solution that is capable and agile enough to address the following considerations:

- Interconnect the different enterprise locations and remote sites that are geographically dispersed

- Meet enterprise security policy requirements by protecting enterprise traffic over the WAN (secure transport), to offer the desired end-to-end level of protection and privacy across the enterprise network

- Cost-effective and reliable WAN by providing flexible and reliable transport that meets the primary business objectives and critical enterprise application requirements and that supports the convergence of voice, data, and video to satisfy the minimum requirements of today's converged enterprise networks

- Support business evolution, change, and growth by offering the desired level of agility and scalability to meet the current and projected growth of remote sites with flexible bandwidth rates

**17**

**NOTE**   The preceding factors are not necessarily the standard or minimum requirements for every enterprise network. Even so, these factors are the generic and common concerns that most IT enterprises have. These concerns can vary from business to business. For instance, many businesses have no concern about having unsecured IP communications over their private WAN.

Furthermore, today's Internet service providers (ISPs) can offer a dramatically enhanced Internet bandwidth and price performance with significantly improved service reliability. From a business perspective, this can be seen as high-bandwidth capacity at a cheaper cost; however, the level of end-to-end service efficiency and the level of reliability of the Internet might not be suitable for many businesses. The main point is that nowadays the Internet with secure virtual private network (VPN) overlay is adopted by many businesses as either their primary WAN transport or as a redundant WAN transport. The subsequent sections in this chapter cover it in more detail.

Consequently, there are multiple WAN topologies and transport models an enterprise can choose from, such as point to point, hub and spoke, and any to any. In addition, traffic over each of these topologies can be carried over Layer 2 or Layer 3 WAN transports, either over a private WAN network such as a Multiprotocol Label Switching (MPLS) provider or overlaid over the Internet. Typically, each model has its strengths and weaknesses in some areas.

Therefore, enterprise network designers must understand all the different aspects of each WAN transport and the supported topologies to select the right solution that is the best fit for the enterprise business, application, and functional requirements.

## Modern WAN Transports (Layer 2 Versus Layer 3)

The decision to select L2 or L3 for the enterprise WAN transport is completely a design decision, and ideally it has to be a business-driven decision. Therefore, it is hard to make a general recommendation that an L3 WAN is better than an L2 WAN for an enterprise. However, some factors can drive the decision in the right direction based on business, functional, and application requirements, along with the enterprise WAN layout and design constraints. This section highlights the advantages and disadvantages of both modern WAN transports (L2 and L3 based), and then compares them from different design angles. Once the differences between these two WAN transport options are identified, the job of selecting the right WAN technology or transport will be easier. It will simply be a matter of identifying the requirements and mapping them to the most suitable WAN transport model. In most cases, Layer 3 may be a better starting place and then the decision can be verified or changed to Layer 2 based on the requirements identified.

Layer 3 MPLS VPN as a WAN transport enables true any-to-any connectivity between any number of sites without the need for a full mesh of circuits or routing adjacencies. This can offer significant scalability improvements for enterprises with a large number of remote locations. With L3VPN, the provider typically exchanges routing information with enterprise WAN edge routers and forwards packets based on Layer 3 information (IP). In addition, L3VPN service providers (SPs) offer a complete control plane and forwarding plane separation per enterprise (customer), with each enterprise IP service assigned to its own virtual IP network (VPN) within the SP MPLS network. With regard to Layer 2 VPN services, however, the providers typically have no participation in the enterprise Layer 3 WAN control plane. This is because the provider forwards the traffic of any given enterprise based on its Layer 2 information, such as Ethernet MAC addresses.

**NOTE**   For the purpose of the CCDE exam, the scenario and requirements always determine the right choice. There might be situations where both WAN options seem to be equally valid, but typically one of them should be more suitable than the other because of a constraint or a requirement given to you on the exam. However, in some cases, there might be more than one right answer or optimal and suboptimal answers. Therefore, you need to have the right justification to support your design decision (using the logic of why, as discussed in Chapter 1, "Network Design").

### Layer 2 MPLS-Based WAN

Metro Ethernet (ME) services offered by ME SPs (Carrier Ethernet) are one of the most common Layer 2 WANs used by today's large enterprises. Layer 2 WAN (ME-based) offers two primary connectivity models for enterprises:

- **Ethernet line service (E-Line):** Also known as *Ethernet virtual private line (EVPL)*, provides a point-to-point service. With EVPL, the typical physical link is Ethernet (Fast Ethernet or Gigabit Ethernet), and the multiple circuits under one physical link are wired virtually over the ME provider cloud using VLANs as a service identifier.

■ **Ethernet LAN service (E-LAN):** Provides multipoint or any-to-any connectivity, also known as *Virtual Private LAN Services (VPLS)*, and offers any-to-any connectivity with high flexibility for the enterprise WAN.

Both E-line and E-LAN were covered in detail in Chapter 6, "Transport Technologies." From an enterprise perspective, the Layer 2 (ME) services provided by today's SPs over their MPLS core infrastructure appear either like a LAN switch for multipoint L2VPN services or like a simple passthrough link for the point-to-point L2VPN sites, as depicted in Figure 17-1.

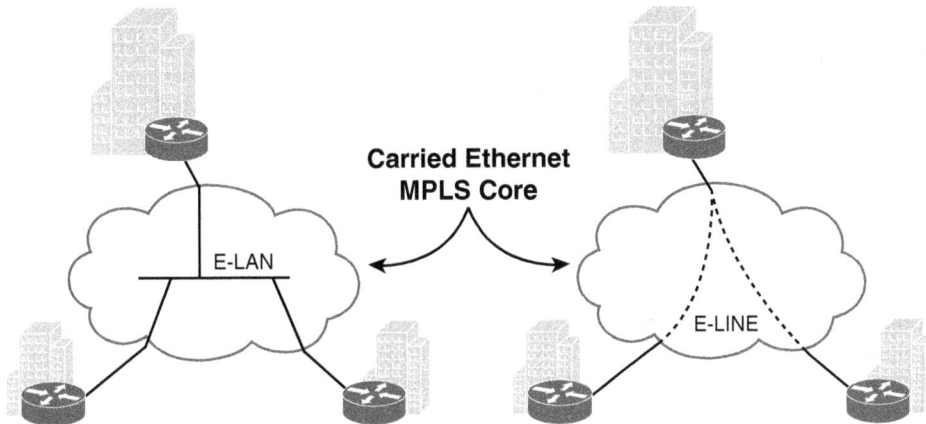

**Figure 17-1** *Layer 2 WAN MPLS Based*

**NOTE**    Although E-Tree is another type of ME connectivity model, it is a variation from E-LAN to provide a hub-and-spoke connectivity model.

**NOTE**    This section discusses these technologies from the enterprise point of view, as an L2 or L3 WAN solution. The service provider point of view is not covered in this book because it is beyond the scope of the CCDE v3 exam.

**NOTE**    L2VPN SPs can preserve the access media using the legacy access media type (such as ATM and Frame Relay) if this is required by the business or if there is a lack of ME coverage by the SP in certain remote areas. In addition, this type of connectivity (mixed) is a common scenario during the migration phases from legacy to modern L2 WAN services.

### Layer 2 MPLS-Based WAN Advantages and Limitations

Table 17-2 highlights the primary advantages and limitations of a Layer 2 MPLS-based WAN.

**Key Topic**

**Table 17-2** Layer 2 MPLS-Based WAN Advantages and Limitations

| Advantages | Limitations |
|---|---|
| **Bandwidth scalability:** At the time of this writing, L2 WAN services can scale from 1 Mbps to 100 Gbps, which makes this transport highly scalable in terms of bandwidth requirements. | **Limited access coverage:** There might be limited service access coverage of the L2VPN (ME) in some locations. |
| **Performance and quality of service (QoS):** By increasing bandwidth on a WAN circuit with L2WAN technologies, enterprises can have a low-latency and low-jitter WAN transport, which can be a good fit for converged networks (video, voice, and data). Furthermore, it can support end-to-end QoS classes as part of the service-level agreement (SLA) offered by the SP (class of service [CoS] or differentiated services code point [DSCP] based). | **Scalability concerns:** For large-scale networks with many remote sites connected over a common E-LAN service, there will be design limitations and issues on higher levels (control plane), such as a large number of routing adjacencies. |
| **Routing control:** With an L2 WAN, the enterprise will have full control over the design, implementation, and operations of routing across the WAN. This can also be considered added flexibility. For instance, with this model, enterprises have the freedom to choose the desired WAN routing protocol and can deploy enterprise-controlled WAN network virtualization. | **Routing topology limitations:** Routing protocol limitations on certain topologies can introduce limitations here, as well, because not every enterprise can afford to change its WAN routing protocol (for example, a full-mesh VPLS-based WAN with Open Shortest Path First [OSPF], where special care has to be taken for OSPF). |
| **Service offering availability:** SPs globally are moving toward adapting networks to provide ME services; therefore, it is available in a large number of places. | **Staff knowledge:** Enterprise staff knowledge and expertise can be considered a limiting factor as well. For instance, the staff might not be able to design, deploy, or operate large-scale WAN routing (for example, using Enhanced Interior Gateway Routing Protocol [EIGRP] or Border Gateway Protocol [BGP]). |
| **Topology flexibility:** With an L2 WAN, enterprises can have different topology layouts as required, such as point to point, point to multipoint, and multipoint to multipoint. Extending L2 effectively extends a fault domain. | |
| **Service flexibility:** With an L2 WAN, the enterprise can run any advanced IP/non-IP services without any reliance on the SP's support, such as IPv6 or multicast. | |
| **Cost-effective:** An L2 WAN, such as ME, replicates the cost model of Ethernet to the WAN. | |

**NOTE**   There are other variations of ME services, such as Ethernet private line (EPL), that support Ethernet over $x$WDM (dense wavelength-division multiplexing [DWDM], coarse wavelength-division multiplexing [CWDM]), SONET, or dedicated Ethernet interconnects over fiber. However, this type of service is more expensive than the other ones that are offered by MPLS SPs, such as VPLS or EVPL (Ethernet Virtual Circuit [EVC]) as an L2VPN ME service.

## Layer 3 MPLS-Based WAN

MPLS L3VPN enables enterprise customers to route traffic across the SP cloud as a transit L3 WAN network, with a simplified "one-hop" single routing session per link between the enterprise WAN edge router and provider edge routers. This means that the SP will typically offload all the enterprise WAN control plane complexities in terms of design and operations of the core WAN routing. As a result, enterprises will gain a significant savings in terms of operational expenses and speed up the time to add new remote sites, especially if there are hundreds or thousands of remote sites that need to be interconnected. In addition, this model will help to simplify and optimize end-to-end WAN QoS design and bandwidth volume planning of the enterprise WAN connectivity. Figure 17-2 illustrates a Layer 3 MPLS-based WAN. (The MPLS L3VPN cloud appears as single router [peer] from the remote site's point of view.)

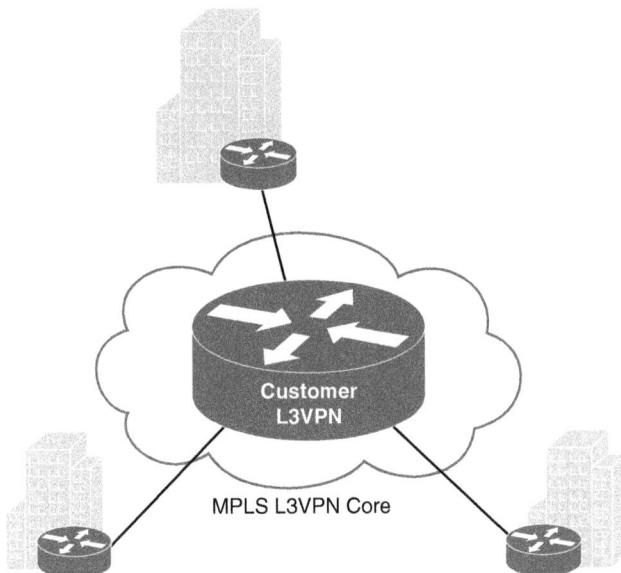

**Figure 17-2**   *Layer 3 MPLS-Based WAN*

## Layer 3 MPLS-Based WAN Advantages and Limitations

Table 17-3 highlights the primary advantages and limitations of a Layer 3 MPLS-based WAN.

**Key Topic**

**Table 17-3** Layer 3 MPLS-Based WAN Advantages and Limitations

| Advantages | Limitations |
|---|---|
| **Bandwidth scalability:** An MPLS L3VPN WAN can provide flexible bandwidth capacity and smooth upgrades as compared to other WAN transports. | **Cost of bandwidth:** Although L3WAN (L3VPN) offers a scalable and high speed of bandwidth, L2VPN (such as ME services) offers a higher bandwidth scale (for example, 10-Gbps wire rates) with a lower cost. |
| **Performance and QoS:** MPLS L3VPN can support consistent end-to-end QoS (DSCP driven) that enterprises can take advantage of, especially for real-time delay-sensitive traffic such as Voice over IP (VoIP).<br><br>QoS across L3 VPNs can be difficult because it may require QoS re-marking to comply with the different service carrier policies. | **IP addressing:** When enterprises move from legacy L2 WAN technologies such as Frame Relay, re-addressing of the WAN interfaces is required. |
| **Cost-effective:** L3VPN can reduce the operational cost because there will be no need to operate and maintain a large and complex routed WAN core infrastructure. | **SP dependencies:** With L3VPN, there is always dependency on SP support to deploy new network services, such as using IPv6 or multicast. |
| **Service offering availability:** Most of the SPs globally offer MPLS L3VPN. Therefore, it is available in a large number of places. | **Flexible topology:** For nonstandard layouts, such as hub and spoke, there might be an additional cost involved for the additional VPNs to be provisioned by the SP. |
| **Topology flexibility:** L3 WAN supports different topology layouts as required, such as point to point, point to multipoint, and any to any. | **LAN extension:** L3 WAN is typically a Layer 3 transit network, which by nature does not allow Layer 2 extensions. Therefore, if a LAN extension is required, it must be achieved using additional technologies. This will typically be an overlay technology, such as self-deployed L2VPN over MPLS over Generic Routing Encapsulation (GRE) over L3VPN WAN, which can add an additional layer of complexity to the network with regard to operations and troubleshooting. It also might not be a desirable solution for the requirements of some applications, due to the increased packets' header that may lead to fragmentation or serialization delay along the WAN path. |
| **Access flexibility:** With L3 WAN, the enterprise can be provisioned with any type of access media depending on the access availability of that location (for example, Ethernet, WiMAX, and VPN, over the Internet or 4G/5G). | |
| **Routing simplicity:** With L3VPN, enterprises will typically offload the core routing to the SP. From the enterprise point of view, only one routing peer/ session per link needs to be maintained. | |

### Internet as WAN Transport

Despite the fact that the nature of the Internet is a "best effort transport" that lacks end-to-end QoS support, the modern Internet can offer relatively high reliability and high-speed connectivity between various locations at a low cost. In addition, in today's modern businesses, many enterprises are increasingly hosting their services in the cloud and embracing many cloud-based services and applications offered as Software as a Service (SaaS), such as Cisco WebEx and Microsoft Office 365, which is changing the traffic pattern to be more toward the Internet.

Furthermore, the Internet can be a reasonable and cost-effective choice for remote sites as a primary transport when it is not feasible to connect over other WAN transport options or when there is a lack of WAN access coverage in certain remote areas. This design primarily relies on using VPN tunneling (overlay) techniques to connect remote sites to the hub site (enterprise WAN module located at the enterprise campus or data center) over the Internet. Ideally, this design is based on a hub-and-spoke connectivity using dynamic multipoint VPN (DMVPN) as the VPN overlay technology for the "Internet as WAN transport" design model, which offers the flexibility to provide any-to-any connectivity as well (direct spoke to spoke). However, point-to-point tunneling mechanisms such as the classical IPsec and GRE are still viable overlay options to be considered (considering typical scalability limitations with peer-to-peer [P2P] tunnels—see Table 17-5 later in the chapter for a detailed comparison of the different VPN mechanisms). Figure 17-3 depicts the different typical connectivity options available with the Internet as a WAN transport. In addition, GRE does have challenges, especially in a one-to-many NAT situation.

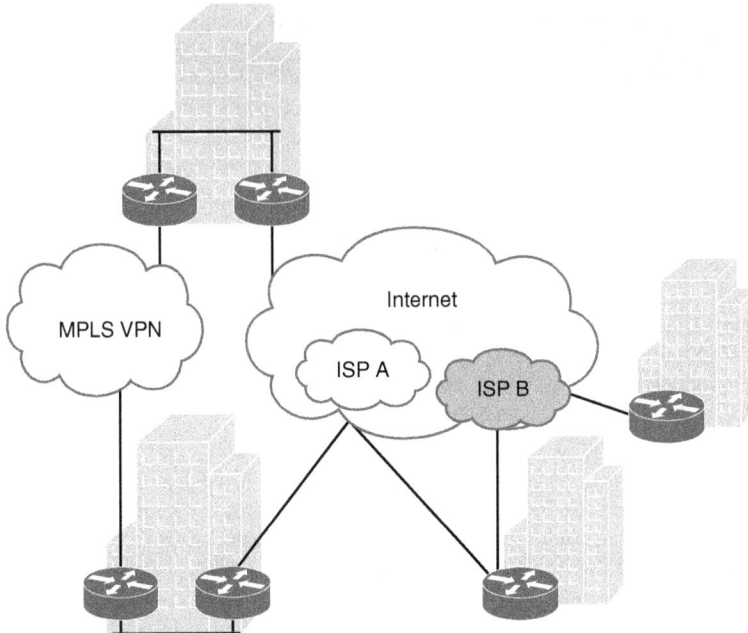

**Figure 17-3**  *Internet as a WAN Transport*

**NOTE** The decision of when to use the Internet as a WAN transport and how to use it in terms of level of redundancy and whether to use it as a primary versus backup path depends on the different design requirements, design constraints, and business priorities (see Table 17-5).

Furthermore, the Gartner Inc. report "Hybrid Will Be the New Normal for Next Generation Enterprise WAN" analyzes and demonstrates the importance of the integration of the Internet and MPLS WAN to deliver a cohesive hybrid WAN model to meet today's modern businesses, and applications' trends and requirements such as cloud-based services. As the report's Summary states, "Network planners must establish a unified WAN with strong integration between these two networks to avoid application performance problems."

**NOTE** The connectivity to the Internet can be either directly via the enterprise WAN module or through the Internet module (edge), as highlighted in Chapter 16, "Enterprise Internet Edge Architecture Design." This decision is usually driven by the enterprise security policy, to determine where the actual tunnel termination must happen. For instance, there might be a dedicated DMZ for VPN tunnel termination at the enterprise Internet edge that has a backdoor link to the WAN distribution block to route the decapsulated DMVPN traffic.

Table 17-4 highlights the primary advantages and limitations of the Internet as a WAN transport model.

**Table 17-4**  Internet as a WAN Transport Model Advantages and Limitations

| Advantages | Limitations |
|---|---|
| **Low-cost:** Offers low-cost WAN connectivity with relatively high bandwidth. | **Reliability:** Although today's ISP can offer a relatively high level of Internet service reliability, it still cannot satisfy the strict level of service reliability required by some businesses. Some business service level Internet offerings are available today, but the tradeoff is a higher monetary cost. |
| **Split tunneling:** Offload traffic from traversing any private WAN path that is destined to the Internet, such as accessing services hosted in a public cloud provider via direct Internet access (split tunneling). | **Consistent QoS:** Because the Internet by nature is best-effort IP transport, enterprises cannot maintain true end-to-end service differentiation and consistent QoS. |
| **Ubiquitous connectivity:** Flexible access because of the Internet connectivity can be provisioned over various media types such as wireless, Long-Term Evolution (LTE), 4G/5G, digital subscriber line (DSL), or Ethernet. | **Operations complexity:** Typically, using the Internet as a WAN transport requires an overlay mechanism such as GRE and DMVPN combined with protection (IPsec). These multiple technologies may add a layer of complexity when there is an issue to troubleshoot or a new site to be added, especially if other advanced IP services are enabled, such as IP multicast. |

| Advantages | Limitations |
|---|---|
| **Faster time of install:** Provisioning an Internet service is usually relatively quicker than provisioning other WAN services. This enables large enterprises that have many small remote sites they want to add quickly with flexible media access types and wide geographic coverage to accelerate their time to market; this is a common scenario in retail business. | |

## WAN Transport Models Comparison

Table 17-5 summarizes the selection criteria of different WAN transports discussed earlier in this chapter (from an enterprise point of view), considering various design aspects.

**Key Topic**

**Table 17-5**   Comparison of WAN Transport Models

|  | MPLS L2VPN WAN | MPLS L3VPN WAN | Internet as WAN |
|---|---|---|---|
| **Bandwidth** | Very flexible (can vary between 1 Mbps to 100 Gbps) | Flexible (less than L2 MPLS-based WAN [ME]) | Flexible with limitations, depending on the site location and connectivity provisioning type (DSL versus 4G versus 5G) |
| **WAN core routing control** | Enterprise managed and controlled | SP managed and controlled | Enterprise managed and controlled |
| **Cost** | Moderate | Usually more expensive than L2, especially when high bandwidth is required | Cheap |
| **CoS** | Depending on the SP, but can support CoS based on L2 marking and DSCP | End-to-end Layer 3 CoS (DSCP based) | End-to-end QoS guarantee not supported (only at the network edge) |
| **Staff experience** | Requires experienced staff to design and manage the core WAN routing | High level of routing expertise not required for the WAN | Requires experienced staff to design and manage the WAN routing and the overlay VPN setup |
| **Remote site scalability** | Introduces some routing/adjacency issues and limitations with a large number of sites | Can support very large scale of remote sites | Scalable to some extent (limited to the WAN router hardware capability [for example, supported number of VPN sessions]) |
| **Site physical connectivity** | Limited options (for example, legacy Frame Relay, ME) | Very flexible, can be any type of access (legacy, Ethernet, VPN over Internet to SP MPLS) | Flexible (DSL, Ethernet, 4G/5G) |

**17**

By understanding the differences between each WAN transport and the capabilities and limitations of each transport, network designers should be able to make a more business-driven design decision by mapping the suitable WAN transport to the different design requirements (for example, business, functional, and application). In addition, network designers ideally must consider the answers to the following questions during the planning phase of the WAN transport selection:

**Key Topic**

- Who is responsible for the core WAN routing management?

- Who manages the customer edge (CE) WAN devices?

- How critical is the WAN connectivity to the business? What is the impact of an outage on the WAN connectivity to the business in terms of cost and functions?

- What is the number of remote sites and what is the percentage of the projected growth, if any?

- Are there any budget constraints?

- What are the required WAN capabilities to transport business applications over the WAN with the desired experience (such as QoS, IP multicast, or IPv6)?

## WAN Module Design Options and Considerations

This section highlights the design considerations and the different WAN connectivity design options that pertain to the enterprise WAN module and remote site WAN connectivity as well.

### Design Hierarchy of the Enterprise WAN Module

The main goal of the WAN module is to aggregate the connectivity and traffic of the enterprise WAN that connects the enterprise with various types of remote locations. Therefore, this module provides the traffic and connectivity aggregation of the extended remote sites. Applying the hierarchical design principle can maximize the flexibility and scalability of this enterprise module. In addition, this structured approach will simplify adding, removing, and integrating different network nodes and services such as WAN routers, firewalls, and WAN acceleration appliances. As illustrated in Figure 17-4, applying hierarchy will enable each layer to perform certain functions using different features and tools in a more structured manner. Furthermore, the level of flexibility and adaptability offered by the hierarchal structure makes the WAN module design highly scalable and resilient.

### WAN Module Access to Aggregation Layer Design Options

The aggregation layer of the WAN module aggregates traffic and connectivity of access layer nodes, which are typically the WAN edge routers and other WAN services, such as firewalls and WAN acceleration appliances.

There are three common design options to interconnect WAN edge routers to the aggregation layer switches of the WAN module, illustrated in Figure 17-5.

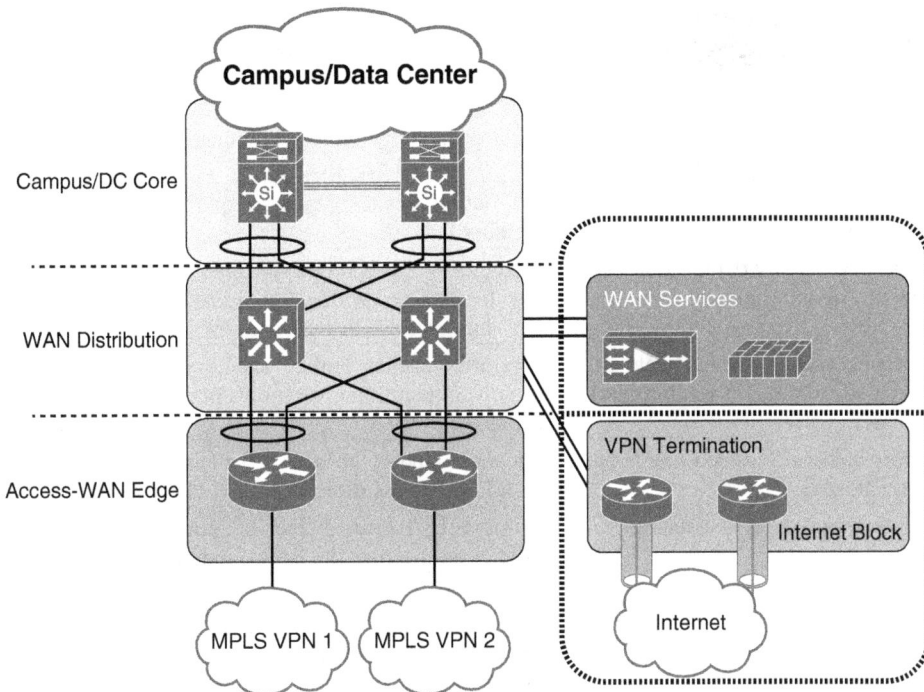

**Figure 17-4**    *WAN Module Hierarchal Design*

**17**

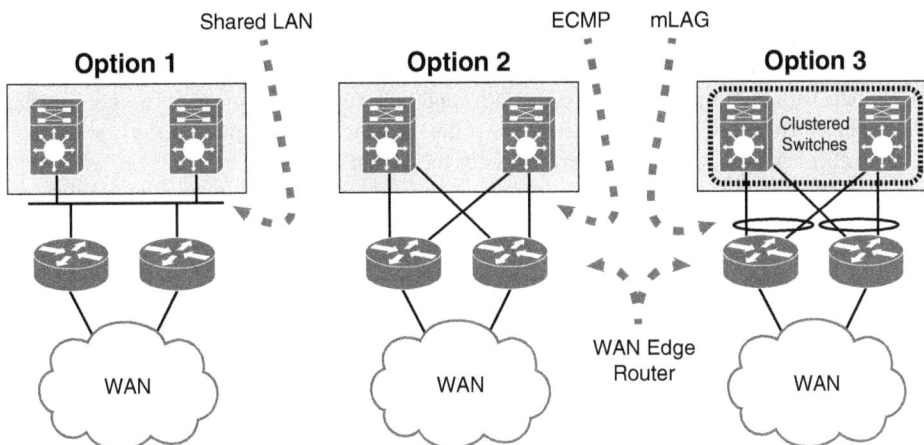

**Figure 17-5**    *WAN Access-Distribution Connectivity Options*

Design option 2 (equal-cost multipath [ECMP]) and option 3 (multichassis link aggregation [mLAG]) both offer a more flexible design compared to option 1; however, there are some technical differences between these two design options. Table 17-6 summarizes these differences. Although both design options (2 and 3) can achieve the same goal to a large extent, network designers should select the most suitable option based on the environment and required capabilities and features (optimal versus suboptimal).

**Table 17-6** ECMP Versus mLAG

| Option 2 | Option 3 |
|---|---|
| ECMP based. | mLAG based. |
| Link redundancy based on redundant L3 links (more routing peers). | Link redundancy based on redundant L2 mLAG links (fewer routing peers). |
| Routing reconvergence is required when one uplink fails. | No routing reconvergence is required if one mLAG member link fails. |
| Convergence time by default relies on routing protocol design and timers. | With mLAG, each flow typically utilizes one member link and will be limited to the capacity of that link (unless "flowlet" concept is used). |
| Supports both scale out and scale up. You can use more than two aggregation layer nodes. | Supports scale up (limited scale out, as a maximum of two aggregation layer nodes per mLAG can be used). |
| The more links to be added, the larger routing database. | The more links to be added, the larger the number of Address Resolution Protocol (ARP) entries. |
| ECMP flow-based load balancing. | Supports L3/L4 load-balancing hashing (load distribution) across the mLAG member links. |

Option 1, however, has several design limitations. For instance, without a careful interior gateway protocol (IGP) tuning, this design option can lead to a slow network convergence at the WAN edge, which can result (from a business point of view) in undesirable outcomes after a failure event. In addition, this design option has a potential of instability and lack of scalability when the network grows in terms of nodes connectivity and routing adjacencies (over a single shared LAN segment). In other words, this design option is the least resilient and scalable option among the other design options. Despite that, design option 1 can still meet some design requirements that do not need a tight convergence time or any scalability considerations, such as regional HQ WAN model with only a pair of WAN edge nodes and no future plan to increase the number of nodes or links. Therefore, the requirements always govern which design option is the best, factoring in whether any design constraint may influence the design choice.

## WAN Edge Connectivity Design Options

Enterprises can consider several WAN edge connectivity design options, such as single-homed or dual-homed, and using single or dual edge routers. The most important consideration here is to identify the business drivers and requirements that influence the selection of one option over others. Many variables will influence a business-driven design decision to select a specific WAN edge connectivity design option. The most common factors that drive this decision are as follows:

**Key Topic**

- Site type (for example, small branch versus data center versus regional office)

- Level of criticality (How much can the downtime cost? How critical is this site to the business if it goes offline for $x$ amount of time?)

- Traffic load (for example, the load on the HQ data center is more than that of the regional data center)

- Cost (Is cost-saving a top priority?)

Table 17-7 summarizes the various types of WAN edge connectivity design options depicted in Figure 17-6, along with the different considerations from a network design perspective.

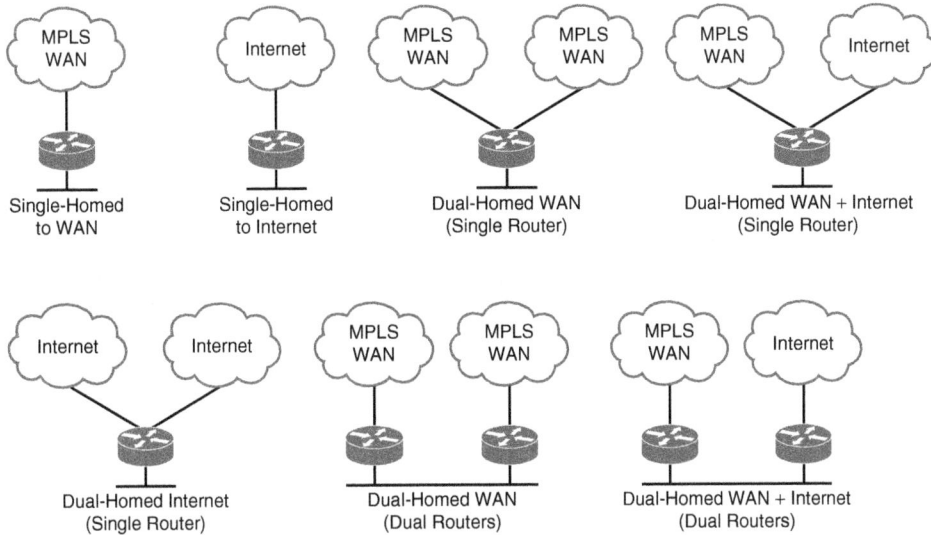

**Figure 17-6**  *WAN Edge Connectivity Options*

**Table 17-7**  WAN Edge Connectivity Design Options

| Connectivity Model | Redundancy | Reliability | Cost | Supported QoS model | Suitability |
|---|---|---|---|---|---|
| Single-homed to WAN | None | Moderate | Moderate | Consistent end to end | Small to medium-size branch sites with high traffic volume |
| Single-homed to Internet | None | Low | Low | Internet edge only | Small branch sites with low fault-tolerance requirements |
| Dual-homed WAN (single router) | Link redundancy only | Moderate | High | Consistent End to end | Medium-size to large, critical, or regional remote sites |
| Dual-homed WAN + Internet (single router) | Link redundancy only | Moderate (lower than MPLS) | Moderate | Consistent End to end over the MPLS path | Medium-size to large or regional remote sites |
| Dual-homed Internet (single router) | Link redundancy only | Moderate (lower than MPLS + Internet) | Moderate | Internet edge only | Small to medium-size remote site |
| Dual-homed WAN (dual routers) | Link and device redundancy | Very high (single provider versus dual providers) | Very high | Consistent End to end | Hub, HQ, DC, or large regional site |
| Dual-homed WAN + Internet (dual routers) | Link and device redundancy | High | Moderate to high | Consistent End to end over the MPLS path | Hub, large remote or regional sites |

17

### Single WAN Provider Versus Dual Providers

The previous section discussed the various WAN edge design options and the characteristics of each option from a design point of view. This section focuses on the dual WAN edge connectivity and takes it a step further to compare the impact of connecting a multihomed site to a single SP versus two different SPs, as summarized in Table 17-8.

**Table 17-8**   Single Provider Versus Dual Provider

|  | Single Service Provider | Dual (Different) Service Providers |
|---|---|---|
| Design simplicity (consistency, features) | Simple, consistent (for example, SLA, QoS design) | Can be inconsistent and more complex (for example, different SLAs, different QoS models, different routing protocols) |
| Cost | Fixed | May lead to a better competitive pricing |
| Availability | SP outage can lead to a WAN blackout | Offers a higher degree of WAN reliability and availability |
| Operational complexity | Simpler (consistent) | More complex (for example, dealing with different SLAs and maybe routing protocols) |

Large enterprises with large geographic distribution can mix between the connectivity options (single versus dual WAN) by using single and dual providers, based on the criticality of the site and business needs. For instance, regional hub sites and data centers can be dual-homed to two providers. In addition, this mixed connectivity design approach, where some remote sites are single-homed to a single provider while others are multihomed to dual providers (typically larger sites such as data centers or regional HQs), can offer a transit path during a link failure, as depicted in the scenario in Figure 17-7. Ideally the transit site should be located within the same geographic area or country (in the case of global organizations) to mitigate any latency or cost-related issues, when applicable, by reducing the number of international paths that traffic has to traverse. In addition, the second provider in Figure 17-7 can be an Internet-based transport such as DMVPN over the Internet.

## Remote Site (Branch) WAN Design Considerations

The WAN edge design options of a remote site can be based on any of the design options described in the previous section (see Table 17-7), where single or dual WAN edge routers can be used based on the requirements of each particular site. Most commonly, in large enterprises, remote sites are categorized based on different criteria, such as size, criticality, and location, and typically all the sites under the same categorization follow the same design standards.

**Europe-Primary Data Center**

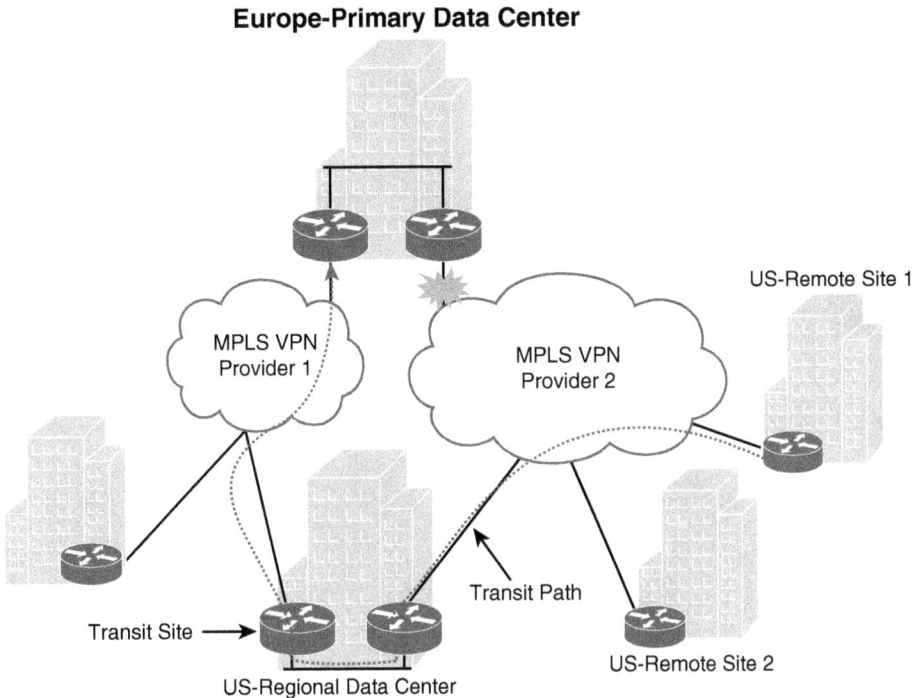

**Figure 17-7**  *Transit Path Scenario with Dual WAN Providers*

> **NOTE**   The edge node is usually either a CE node (for MPLS L3 or Layer 2 WAN) or a VPN spoke node. In some cases, a single WAN edge router can perform the role of both a CE router and VPN spoke router.

However, the level of availability is something that can be determined based on different variables, as discussed in the previous section, such as the level of criticality of the remote site. The rule of thumb for remote site availability is that the network ideally should tolerate single failure conditions, either the failure of any single WAN link or the failure of any single network device at the hub/HQ WAN site (by considering control plane or overlay failover techniques). However, as discussed earlier in this book, the different business drivers, constraints, and the level of site criticality can drive the level of availability of any given remote site. In other words, remote site availability is not always a requirement or a component that must be considered in the design. In general, remote sites with a single router and dual links must be able tolerate the loss of either of the WAN links. Remote sites with dual router, dual links can tolerate the loss of either a WAN edge router or a WAN link (multiple-failure scenarios).

In addition, from a design perspective, the selected WAN connectivity option has a significant influence on the LAN design of a remote site, as does the size of the site (in terms of the number of users and endpoints connected to the network). In general, the design models of a remote site fit into two primary models, as depicted in Figure 17-8 and Figure 17-9 (a and b) and compared in Table 17-9:

■ Single-tier design model

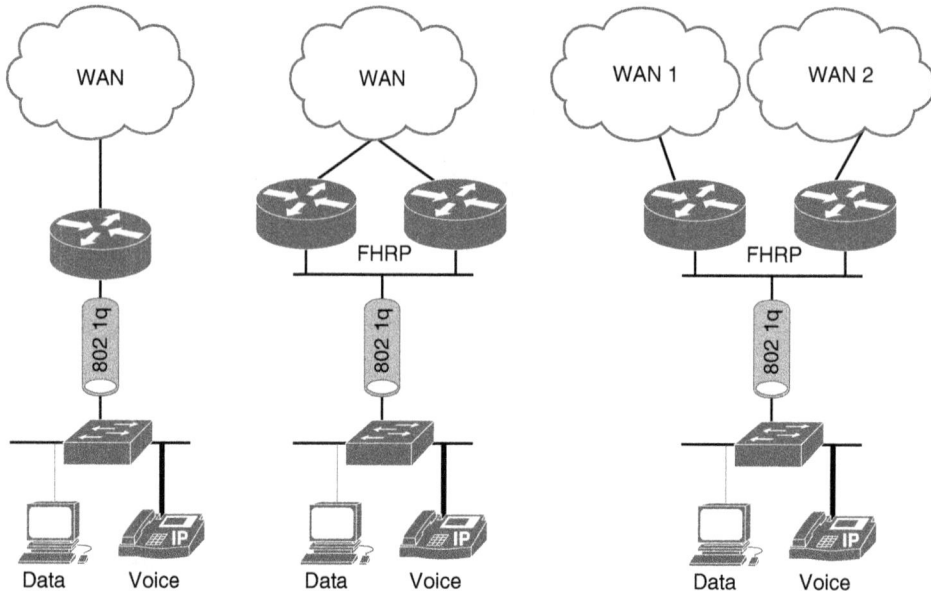

**Figure 17-8**  *Single-Tier Remote Site Design Model*

■ Multitier design model

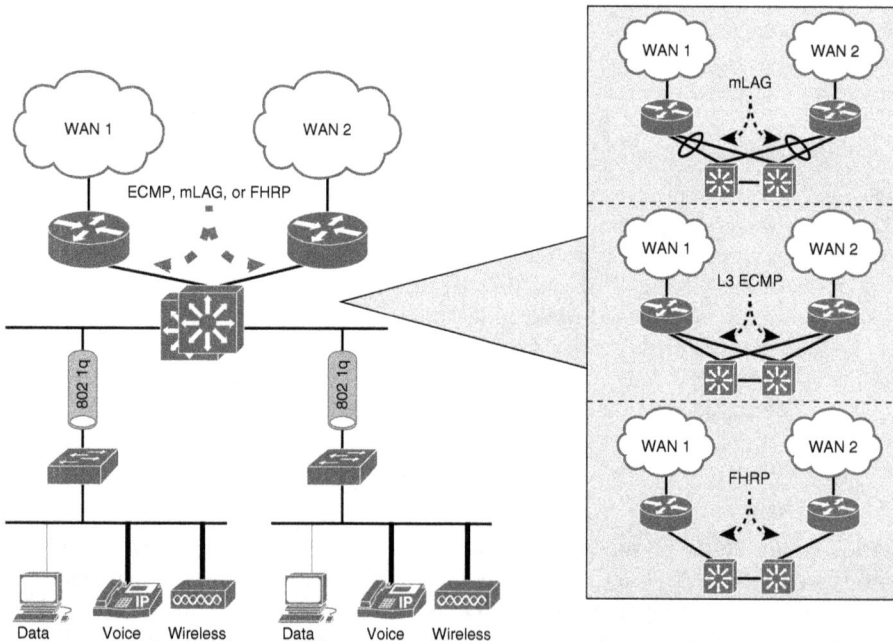

**Figure 17-9a**  *Multitier Remote Site: Dual WAN Router Design Model*

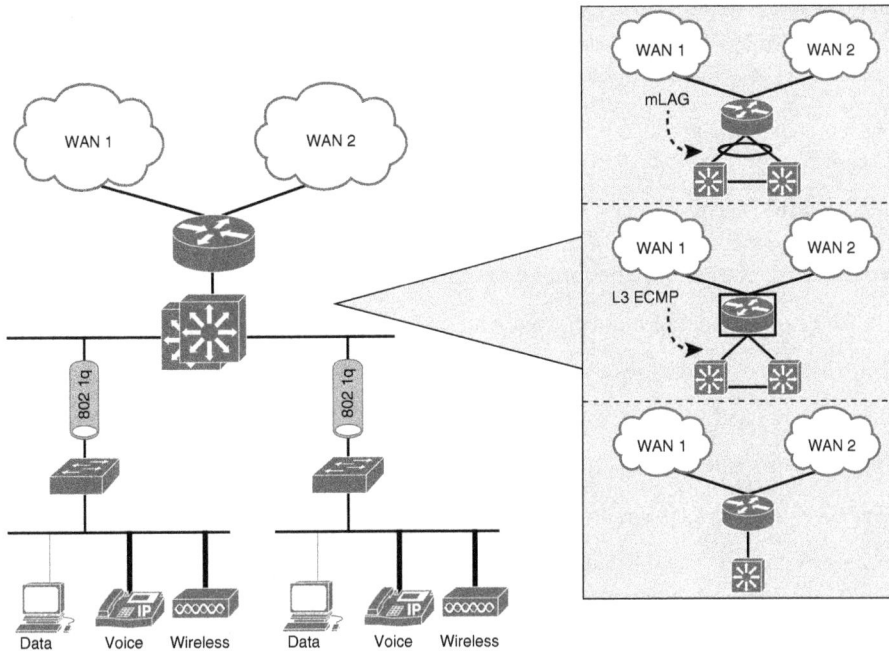

**Figure 17-9b**  *Multitier Remote Site: Single WAN Router Design Model*

**Table 17-9**  Remote Site Design Models

| | Single-Tier Design Model | Multitier Design Model |
|---|---|---|
| LAN scalability | Very limited | More scalable |
| First-hop Layer 3 service | WAN edge router | Distribution layer switches |
| LAN flexibility | Very limited | More flexible |
| LAN to WAN edge Layer 3 connectivity options | FHRP (First Hop Redundancy Protocol), in dual edge routers scenario | IGP over ECMP FHRP mLAG with FHRP mLAG with IGP |
| Supported number of endpoints | Small | Small to medium-sized |
| Supported WAN connectivity options | Single WAN, single router Dual WAN, single router Dual WAN, dual routers | Single WAN, single router Dual WAN, single router Dual WAN, dual routers |

**NOTE**  The WAN connectivity options in Table 17-9 apply for both private enterprise WAN and overlaid WAN over the Internet transport.

As discussed earlier in this section, using the Internet as a WAN transport in conjunction with DMVPN as the overlay transport offers several benefits to the business and enterprise WAN design, in particular for the remote site WAN connectivity, including the following:

- Offers a cost-effective and reliable (to a large extent) WAN connectivity over the Internet

- Reduces the time to add new remote sites over various media access types such as DMVPN over Internet over LTE, 4G/5G, or DSL, combined with the support of zero-touch configuration of hub routers when introducing new spokes to the network

- Provides automatic full-mesh connectivity with simple configuration of hub and spoke

- Supports (any-to-any) spoke-to-spoke direct connectivity fashion

- Supports dynamically addressed spokes

- Supports provisioning behind devices performing Network Address Translation (NAT)

- Supports features such as automatic IPsec triggering for building an IPsec tunnel

- Supports multiple flexible hub-and-spoke design options that serve different design goals, scales, and requirements, as illustrated in Figure 17-10

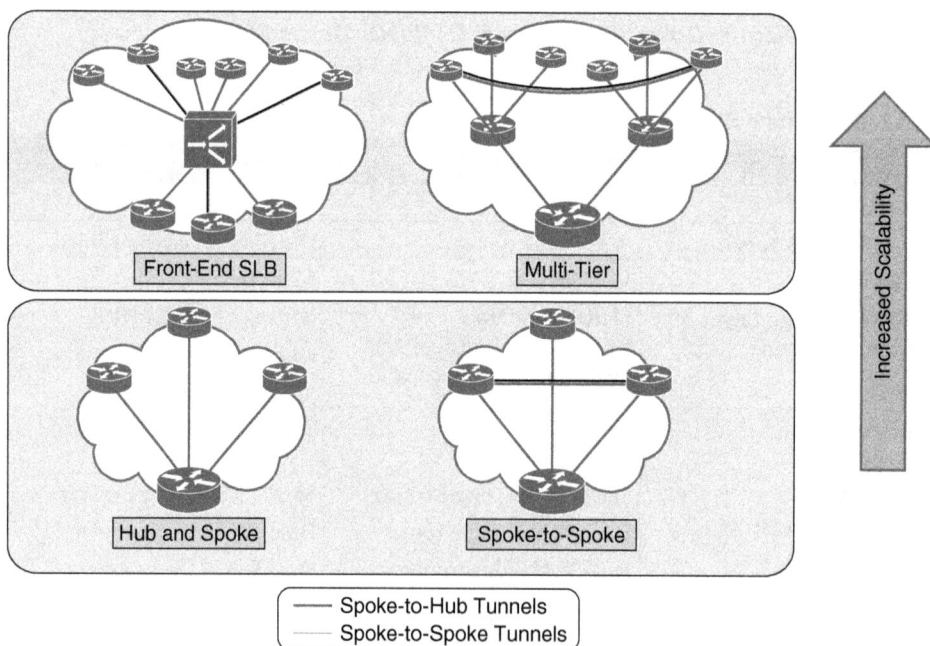

**Figure 17-10** *DMVPN Connectivity Models*

**NOTE**   Routing over GRE tunnels with large routing tables may require adjustments (normally lowering) to the maximum transmission unit (MTU) value of the tunnel interface.

As illustrated in Figure 17-11, three primary connectivity options comprise this design model (remote sites DMVPN-based WAN):

- Single router, single link

- Single router, dual links

- Dual routers, dual links

These connectivity options are primarily driven by the desired level of redundancy over the Internet. For instance, the dual routers, dual links connectivity option eliminates any single point of failure at the WAN/Internet edge, and with the single router, dual links option, the edge router poses a single point of failure to the design even though there are two redundant Internet links. Therefore, based on the different design requirements and constraints, the best "relevant" option can be selected. For example, if a retailer has a small remote site with a very small number of users that can perform transactions and save them locally in case of any WAN outage, adding redundancy may be considered overengineering in this particular scenario because it will not add significant value to the business, and it may be seen as additional unnecessary cost from the business point of view.

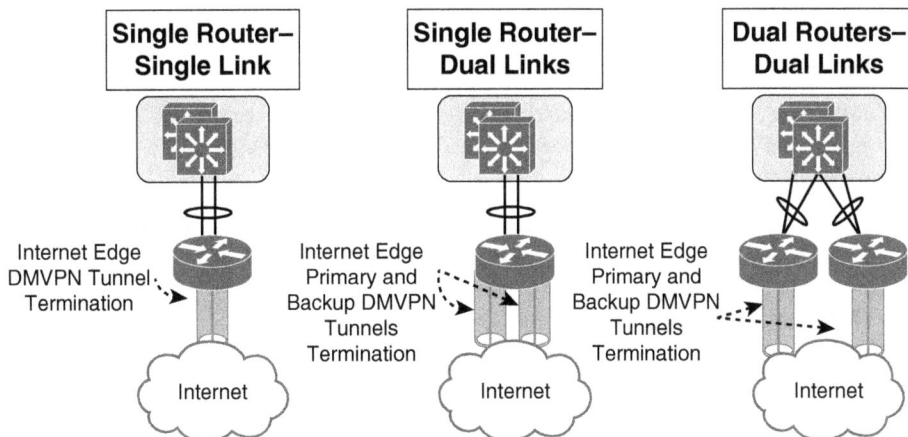

**Figure 17-11**   *Remote Site DMVPN-Based WAN Connectivity Options*

## Enterprise WAN Module Design Options

Previous sections in this chapter discussed the design elements and components of the enterprise WAN module. This section highlights the common proven design models you can use as a foundational reference for the enterprise WAN module design, based on the scale of the remote sites.

**NOTE** As stated, you can use these design models as foundation reference architecture and scale them based on the requirements. For instance, the design option 1 model can easily be migrated to design option 2 when the number of remote sites increases and requires a higher level of redundancy. Similarly, the number of edge access nodes (WAN/Internet edge routers) can be scaled out depending on the design requirements. For instance, an enterprise may consider design option 1 with an additional redundant edge router to a second MPLS WAN, while the Internet edge router is to be used only as a third level of redundancy with a tunneling mechanism.

**NOTE** The number of remote sites in the following categorization is a rough estimation only (based on the current Cisco Validated Design [CVD] at the time of this writing). Typically, this number varies based on several variables discussed earlier in this book, such as hardware limitations and routing design in terms of number of routes.

## Option 1: Small to Medium

This design option, illustrated in Figure 17-12, has the following characteristics:

- Dual redundant edge routers.

- Single WAN connectivity (primary path).

- Single Internet connectivity (backup path over VPN tunnel).

- Ideally, each of the WAN and Internet routers are dual-homed to the WAN module aggregation clustered switches using Layer 3 over mLAG (or Layer 3 ECMP in case of no switch clustering).

- This design model ideally can support a small to medium number of remote sites (ideally up to 100, taking into account hardware limitations as well).

## Option 2: Medium to Large

This design option, illustrated in Figure 17-13, has the following characteristics:

- Dual WAN connectivity and dual WAN routers.

- Dual Internet connectivity and dual Internet routers (typically backup path over VPN tunnel as well as primary for VPN-only remote sites).

- Each of the WAN and Internet routers is dual-homed to the WAN module aggregation clustered switches using Layer 3 over mLAG (or Layer 3 ECMP in case of no switch clustering).

- This design model supports a medium to large number of remote sites (can support up to a few thousand remote sites depending on the hardware capabilities of WAN routers and VPN termination routers).

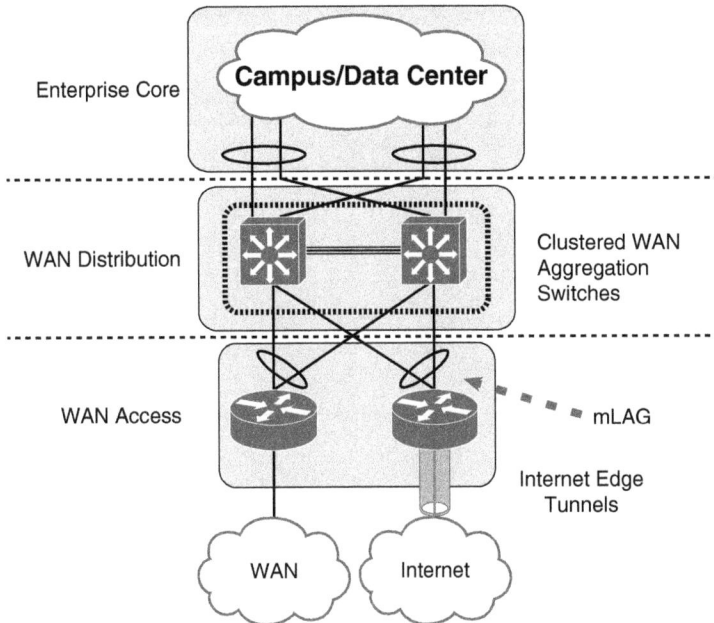

**Figure 17-12**  *Enterprise WAN Module Design Option 1*

17

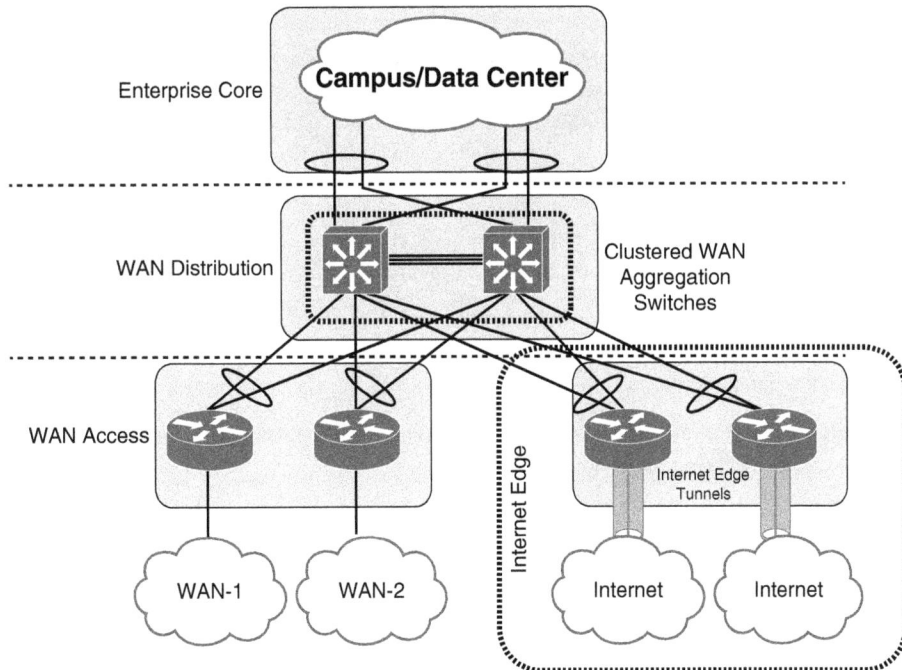

**Figure 17-13**  *Enterprise WAN Module Design Option 2*

Table 17-10 highlights the supported remote site WAN connectivity design options with regard to the two enterprise WAN design modules discussed earlier.

**Table 17-10** Supported Remote Site WAN Connectivity Design Options

| | WAN Module Design Option 1 | WAN Module Design Option 2 |
|---|---|---|
| Scalability with regard to the number of remote sites | Small to medium | High |
| Support remote sites with single WAN link (MPLS only) | Yes | Yes |
| Support remote sites with single WAN link (over Internet) | Yes (single hub, single point of failure) | Yes |
| Support remote sites with dual WAN links (MPLS only) | No | Yes |
| Support remote sites with dual WAN links (MPLS + Internet) | Yes | Yes |
| Support remote sites with dual WAN link (over Internet only) | Yes (single hub, single point of failure, which eliminates the benefit of the redundant Internet links of the remote site) | Yes |

### Option 3: Large to Very Large

This architecture, illustrated in Figure 17-14, targets very large-scale routed WAN deployments. This architecture encompasses branch, metro connectivity, and global core backbones.

This architecture consists of five primary modules:

- **Regional WAN:** Connects branch offices and aggregates remote locations

- **Regional MAN:** Connects remote offices and data centers across metro area transports

- **WAN core:** Interconnects regional networks and data centers within a country or theater or globally (provides connectivity between regional enterprises, interconnects within a theater and globally between theaters)

- **Enterprise edge:** Connects the enterprise network to other external networks and services (Internet service, mobile service)

- **Enterprise interconnect:** Used as an interconnection and aggregation point for all modules (provides connectivity between the regional WANs, MANs, data centers, enterprise edge, and campus networks)

This hierarchical structure offers flexibility for the design to be separated into different element tiers that are suitable to different environments. When a global footprint is required, all the elements of this architecture will likely apply. Whereas with a footprint that is solely within a single theater (region or country), it will not require the global core. However, it can be added when there is a requirement to expand into other regions. The design of each element, such as the regional WAN and remote sites, should follow the design options discussed earlier in this chapter. For example, the regional "enterprise WAN module" can be based on any of the WAN design options discussed earlier in this chapter.

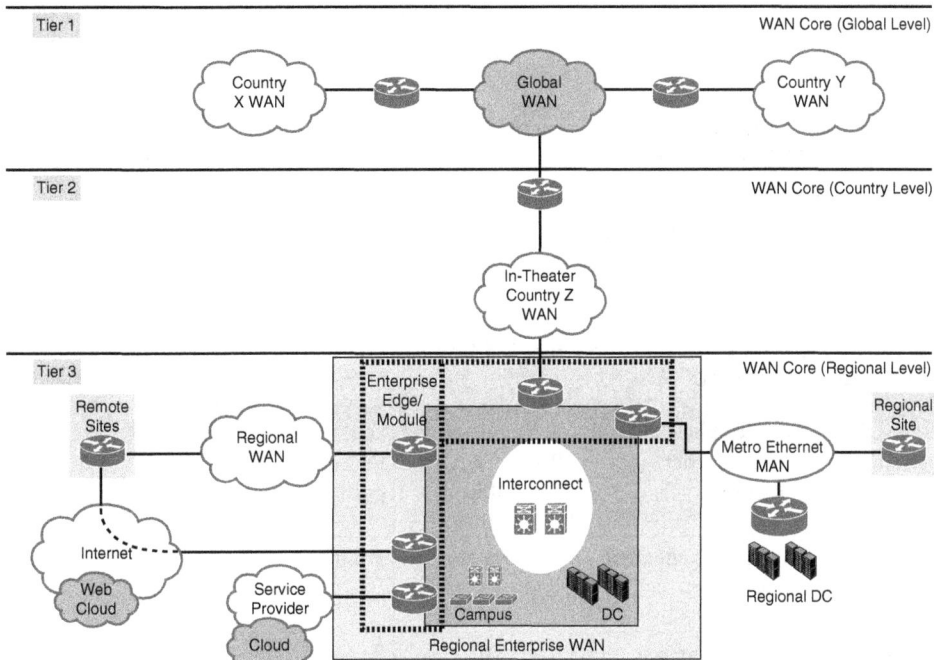

**Figure 17-14**   *Enterprise WAN Module Design Option 3*

# WAN Virtualization and Overlays Design Considerations and Techniques

Chapter 15, "Scalable Enterprise Campus Architecture Design," covered drivers and advantages of considering network virtualization techniques within the enterprise, in particular the enterprise campus network. This section focuses on how the enterprise can extend network virtualization across the WAN to maintain end-to-end path isolation for the different logical networks and groups. Network virtualization over the WAN (WAN virtualization) can be achieved using various approaches and techniques. This section highlights these various design options and the characteristics of each and suggests uses that can fit different business requirements. Overlay technologies are among the primary and common foundational technologies used by enterprises to facilitate achieving WAN virtualization. In fact, overlay (also referred to as *self-deployed VPN*) technologies are adopted by businesses to serve different purposes, from an enterprise network design perspective, such as the following:

- Build a cost-effective WAN model also known as Internet as WAN transport model (virtual private WAN). You can use this model either as a primary or backup WAN transport.

- Provide a mechanism to maintain end-to-end path isolation for logical groups or entities across the enterprise WAN.

- Secure IP communications over the private enterprise WAN or Internet.

- Provide a controlled remote access method for mobile users or third-party entities that might require access to certain internal resources.

- Provide overlaid transport for services and applications that are not supported by the underlay transport infrastructure, such as multicast-based applications and IPv6 applications over "unicast-only" IPv4 IP network.

Therefore, it is critical that network designers have a very good understanding of the different overlay (VPN) options in terms of supported design models, strengths, limitations, and suitable use cases.

Table 17-11 compares the different overlay (VPN) technologies that commonly are used to extend enterprise connectivity and facilitate network virtualization over IP transport networks.

**Key Topic**

**Table 17-11**  Network Overlay Technologies Comparison

|  | Remote Access VPN (Client Based) | DMVPN | IPsec | GRE |
|---|---|---|---|---|
| **Targeted transport network** | Public Internet transport | Private WAN and public Internet | Private WAN and public Internet | Private WAN and public Internet |
| **Supported network topology** | Hub-spoke (client to server) | Hub-spoke, spoke to spoke | Point-to-point | Point-to-point |
| **Routing technique** | Reverse-route injection | Static and dynamic routing | Reverse-route injection | Static and dynamic routing |
| **Encryption style** | Peer-to-peer protection | Peer-to-peer protection (with IPsec) | Peer-to-peer protection | Peer-to-peer protection (with IPsec) |
| **IP multicast** | Multicast replication at hub (if multicast is supported by the VPN client or may require GRE with IPsec) | Multicast replication at hub | Point-to-point (with GRE) | Point-to-point |
| **Scalability** | Moderate | High | Low | Low |
| **Design flexibility** | Limited to client/server communication style | High | Low | Moderate |
| **Operational complexity** | Moderate (only for remote-access users) | Moderate | High | High |
| **Network virtualization techniques** | Limited (VRF-aware remote access) | Flexible (end-to-end) | Flexible (VRF-aware IPsec) | Flexible (end-to-end) |

**NOTE**  The primary scalability limiting factor of any VPN solution is the supported number of sessions by the hardware platform that is used.

**NOTE**   GETVPN is an encryption mechanism that enables you to preserve IP header information that supports a true "any-to-any" encrypted IP connectivity model. Therefore, it is commonly used over private transport networks such as a private WAN instead of the other IP tunneling mechanisms. Having said that, GETVPN is not always the ideal or optimal overlay and encryption solution over the private WAN. For example, if the existing WAN platforms of an organization do not support GETVPN (and the business has no intention or plan to upgrade any network hardware/software), then you need to deal with the design constraints and consider other options here, such as IPsec with GRE or mGRE.

The different VPN technologies highlighted in Table 17-11 are the foundation of achieving WAN virtualization. However, modern large-scale enterprises can use other approaches to maintain end-to-end path separation, such as "self-deployed" MPLS L3VPN. The following section classifies, discusses, and compares all the different primary technologies and design options that can help you achieve WAN virtualization to suit different design requirements.

## WAN Virtualization

Introducing virtualization and path isolation over the WAN transport is commonly driven by the adoption of the network virtualization concept by the enterprise within the campus LAN, branches, or data center network. Therefore, to maintain end-to-end path isolation, network virtualization must be extended over the WAN transport in a manner that does not compromise path-isolation requirements. From a WAN design point of view, two primary WAN connectivity models drive the overall WAN virtualization design choices:

- **Customer- or enterprise-controlled WAN:** Also known as *self-deployed*, this model provides full control for the enterprise to use the desired core routing design and the type of virtualization techniques that meet their requirements, such as MPLS in the core or tunneling with multiple VRFs. Furthermore, all the techniques discussed in Chapter 15 "Scalable Enterprise Campus Architecture Design," section "Campus Network Virtualization Design Considerations" are applicable here. Typically, this model is based on the fact that the enterprise controls the WAN core infrastructure or transport, as depicted in Figure 17-15.

If the WAN SP in the middle provides L2 WAN transport, it can be categorized under the enterprise-controlled WAN model, because the enterprise will have the control and freedom, to a large extent, to deploy the desired end-to-end WAN virtualization techniques based on the business and technical requirements, such as MPLS-enabled virtualization or subinterfaces with VRFs, as illustrated in Figure 17-16.

- **SP-controlled WAN:** This model, compared to the previous model, provides the least control for the enterprise when it comes to routing an end-to-end network virtualization over an SP-controlled WAN transport, such as MPLS L3VPN, as depicted in Figure 17-17. Therefore, enterprises need to either extend the virtualization to the SP (to the PE node) or build an overlay over the SP managed network between their CE nodes to facilitate the formation of the required end-to-end network virtualization. This approach is commonly referred to as *over the top*.

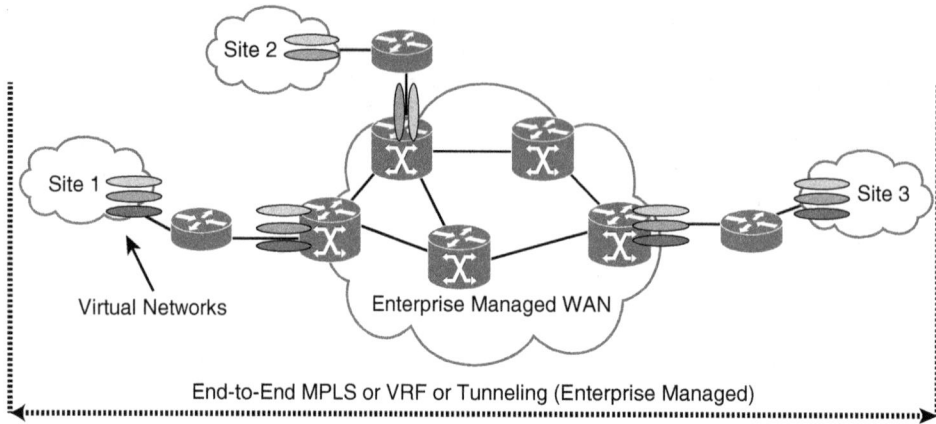

**Figure 17-15** *Enterprise-Managed WAN*

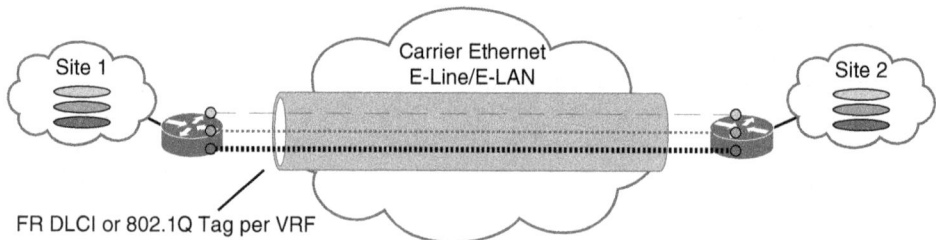

**Figure 17-16** *Enterprise-Managed WAN over L2VPN Cloud*

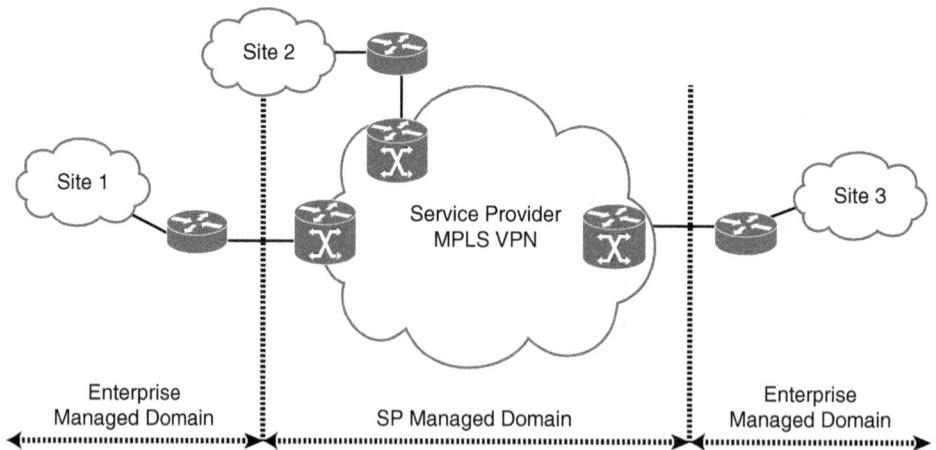

**Figure 17-17** *Unmanaged Enterprise WAN*

## Over-the-Top WAN Virtualization Design Options (Service Provider Coordinated/Dependent)

The following design options require coordination with the SP to support extending the enterprise network virtualization over the MPLS L3VPN provider network. The two common and proven approaches used to extend network virtualization of an enterprise over an unmanaged L3VPN SP network are as follows:

- **Back-to-back VRFs to provider PE:** This approach is based on using the concept of multi-VRF CE. This approach provides L3 path virtualization extension without exchanging labels over IP tunnels or physical interfaces (subinterfaces) with the provider PE, as illustrated in Figure 17-18. Typically, a routing instance (process) per VRF is required at each CE and PE node to exchange routing information per virtual network.

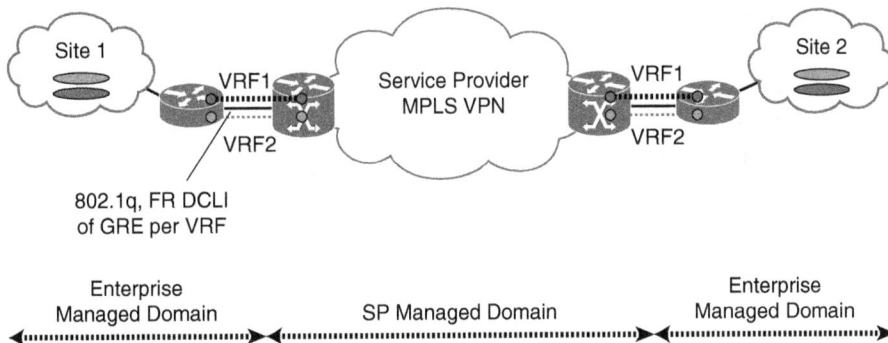

**Figure 17-18**  *Multi-VRF CE*

- **Enable MPLS (Label Distribution Protocol [LDP]) with provider PE:** This approach is based on the Carrier Supporting Carrier design model (CSC, RFC 8277), where the CE node can send packets along with MPLS label to the provider PE, which ultimately can facilitate for enterprises the formation of their own multiprotocol BGP (MP-BGP) peering across the SP MPLS L3VPN backbone, as illustrated in Figure 17-19.

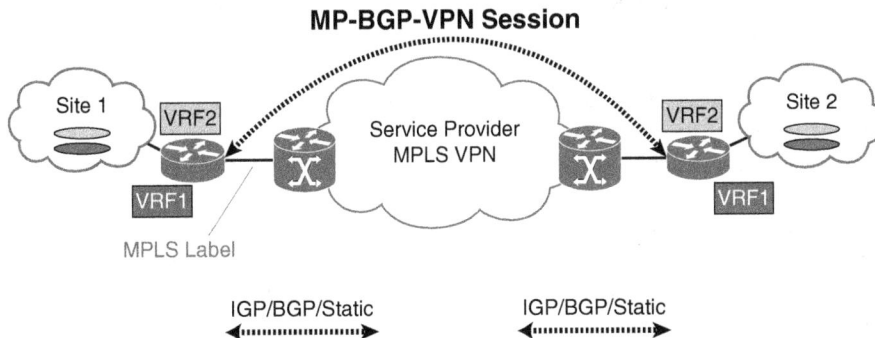

**Figure 17-19**  *CSC Model*

Table 17-12 compares these two design approaches from different design angles.

**Key Topic**

**Table 17-12**   Multi-VRF CE Versus CSC Model for Enterprise WAN Virtualization

| | CSC Model | Back-to-Back VRFs Model |
|---|---|---|
| Scalability | High | Low |
| Coordination with SP | Moderate | High |
| Design complexity | Moderate | The larger, the more complex |
| Dependencies on SP (for example, multicast support) | High | High |
| Extra cost | No | Yes (SP might charge per additional VRF) |
| Adding new virtual network/VRF requires coordination with SP | No | Yes |
| Requires label exchange with provider PE | Yes | No |
| Requires PE-CE routing instance per VRF | No | Yes |
| Control plane complexity | Moderate | High |
| Operational complexity | Moderate | High |
| Security and edge policy control | Moderate | High |
| QoS granularity | Moderate | High |

### Over-the-Top WAN Virtualization Design Options (Service Provider Independent)

This section discusses the different design options that use various overlay approaches, which can facilitate the extension of an enterprise network virtualization over an unmanaged L3 SP WAN. Unlike the approaches discussed in the previous section, the design options discussed in this section are end-to-end controlled and deployed by the customer or enterprise side without any coordination/dependencies with the WAN SP (simply because all the methods are based on the concept of using different tunneling [overlay] mechanisms that typically encapsulate and hide all the traffic and virtualization setup from the underlying SP transport network). The design options are as follows:

- **Point-to-point GRE tunnel per VRF:** This design option offers a simple private virtual network extension over GRE tunnels, where each GRE tunnel is assigned to a specific VRF per virtual network, without any need to coordinate with the WAN provider. However, this option can introduce operations and setup complexities in large deployments because of the large number of manual configurations of the point-to-point tunnels, each with its own control plane. In addition, this design option has the least scalability, because the number of tunnels can increase significantly when the number of sites and VRFs increases. For example, 60 sites with 3 VRFs each will require $(N - 1)$ tunnels per VRF, $(59 \times 3) = 177$ tunnels to create. Nevertheless, this design option can be a good choice for traffic isolation between a very small number of sites (ideally two or three sites only) with a very limited number of VRFs (ideally two or three), as illustrated in Figure 17-20. When point-to-point GRE tunnels have the same source and destination addresses, you must use a unique tunnel key as a demultiplexer in addition to having a unique VRF assigned.

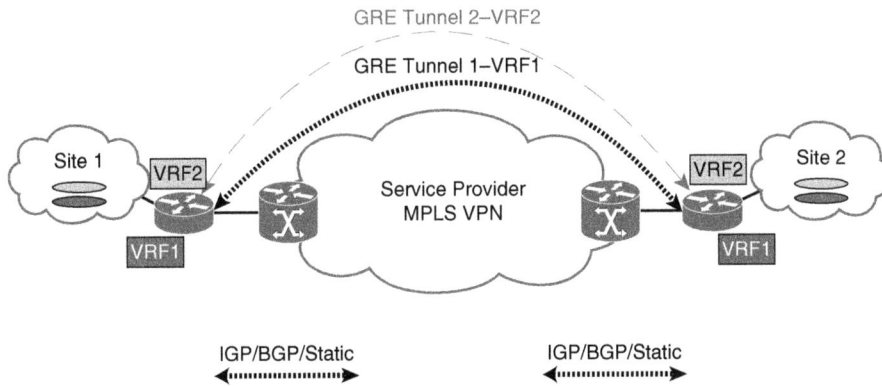

**Figure 17-20** *Point-to-Point GRE Tunnel per VRF*

**NOTE** For this design option and subsequent ones, it is hard to generalize and provide a specific recommended number of remote sites or VRFs, because the decision has to be made based on these two variables when measuring the scalability of the design option. For example, evaluating this design option for a network that requires path isolation between three sites, where each site has ten different virtual networks to transport, is different from when there are three sites with two virtual networks in each. In both cases, the number of sites is small; however, the number of VRFs (virtual networks) becomes the tiebreaker.

- **Dynamic multipoint GRE (DMVPN) per VRF:** This design option is typically based on using multipoint GRE tunnels (DMVPN) per virtual network, as illustrated in Figure 17-21, which helps to overcome some of the scalability issues of the previous option to some extent. This design option also supports direct spoke-to-spoke traffic forwarding (bypassing the hub) per VRF. Furthermore, it supports deployments of a larger scale than those of the point-to-point GRE tunnels. However, it still has scalability and operational limitations and complexities when the network grows, because there will be a DMVPN cloud per VRF. This means that the greater the number of VRFs required, the greater the number of DMVPN clouds that need to be created and operated, with a separate control plane for each. This design option ideally supports the following design combinations:

  - Large number of remote sites with very small number of VRFs (ideally two)

  - Small number of remote sites with small number of VRFs (ideally not more than three)

- **MPLS over point-to-point GRE tunnel:** This design option is based on the concept of encapsulating MPLS labels in a GRE tunnel, as described in RFC 4023, which helps to overcome some of the limitations of the point-to-point GRE tunnel per VRF design option, by using MPLS with an MP-BGP VPNv4/6 session over one GRE tunnel (RFC 4364 MP-BGP control plane style), as depicted in Figure 17-22. Consequently, there will be only one GRE tunnel required to carry LDP, IGP, and MP-BGP (VPNv4/6). Typically, there is no need to create a separate GRE tunnel per VRF with this design option. However, the number of remote sites is still a limiting factor in the scalability of this design option in the case where many remote sites need to be connected, either in a fully meshed manner or using hub-and-spoke overlay topology.

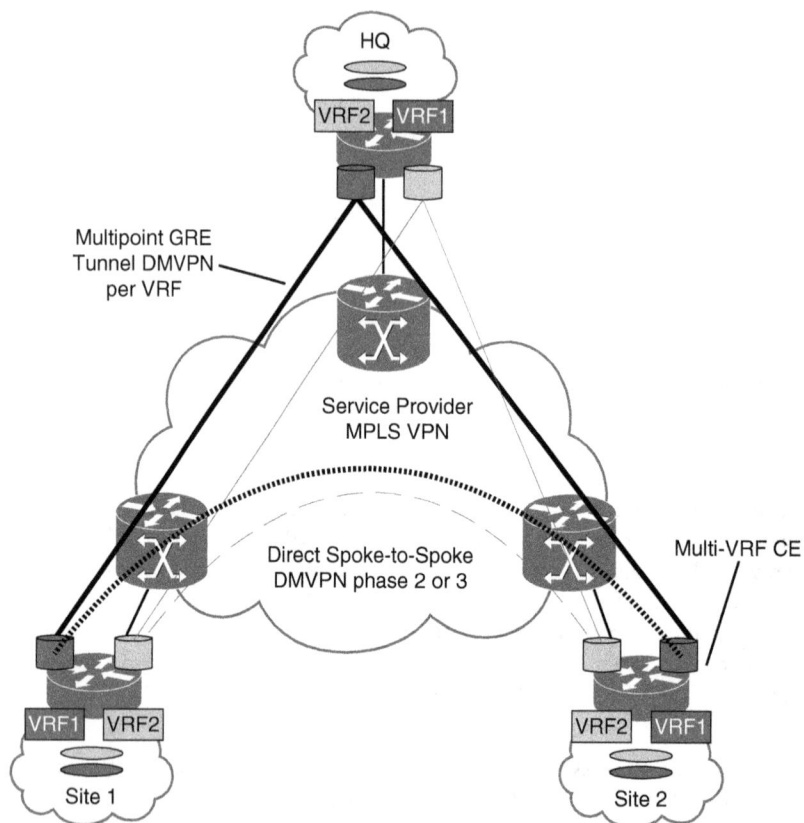

**Figure 17-21** *DMVPN per VRF*

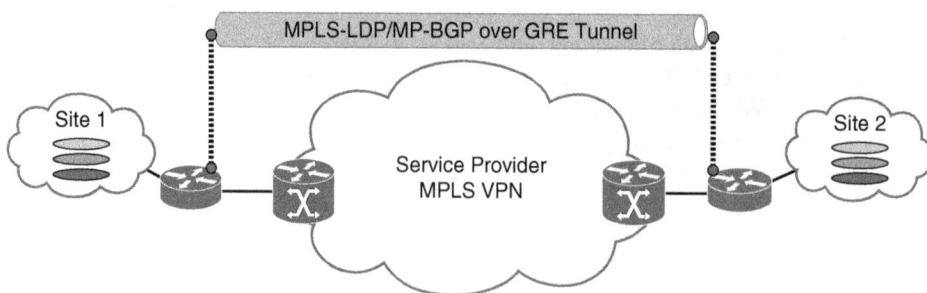

**Figure 17-22** *MPLS over Point-to-Point GRE Tunnel*

Furthermore, this design option can help simplify the interconnection of disjoint MPLS-enabled infrastructures over a native IP backbone. As illustrated in Figure 17-23, MPLS over GRE is used to extend the reachability between two MPLS-enabled islands over a non-MPLS backbone (native IP).

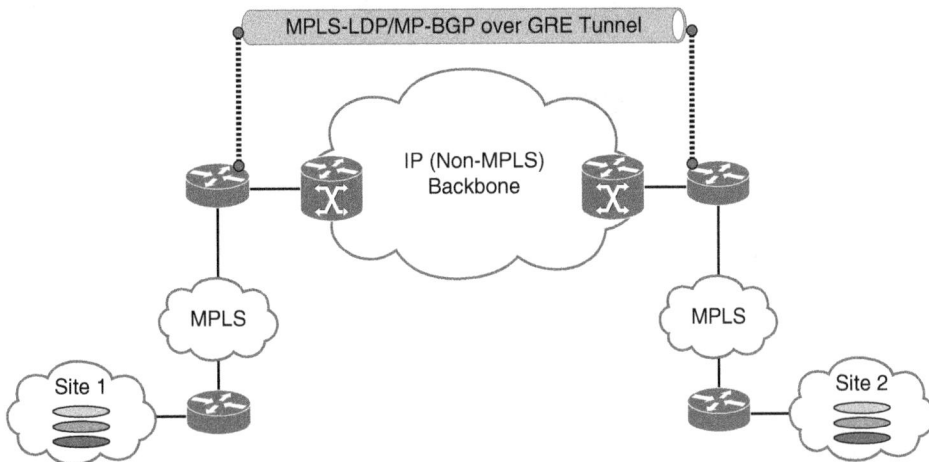

**Figure 17-23**   *Interconnecting MPLS-Enabled Islands over GRE*

■ **MPLS over dynamic multipoint GRE (DMVPN):** This design option, also known as
2547oDMVPN, is based on using MPLS over DMVPN tunnels (standard RFC 4364
MP-BGP control plane), which allows MPLS VPN to leverage the DMVPN framework
(Next Hop Resolution Protocol [NHRP] for dynamic endpoint discovery). Compared
to the DMVPN per VRF design option, using MPLS over the DMVPN will help to
avoid having a DMVPN cloud per VRF. In other words, there will be one DMVPN
cloud (carrying LDP, IGP, MP-BGP VPNv4) to transport all the VRFs between the different locations (sites) in a hub-and spoke-topology, as illustrated in Figure 17-24. This
makes it a very scalable solution for large hub-and-spoke deployments with multiple
distributed virtual networks. Also, this option supports direct spoke-to-spoke communication. Multicast traffic, however, must traverse the hub site if enabled.

■ **MPLS over multipoint GRE (using BGP for endpoint discovery):** MPLS over mGRE
simplifies the design and implementation of overlaid (self-deployed) MPLS VPN using
the standard RFC 4364 MP-BGP control plane, which offers dynamic tunnel endpoint
discovery using BGP as the control plane. This solution requires only one IP address
(typically a loopback address) of each of the enterprise CE routers to be advertised to
the interconnecting SP cloud network, as depicted in Figure 17-25. In addition, there is
no requirement to manually configure any GRE tunnel or enable LDP/RSVP (Resource
Reservation Protocol) on any interface. Instead, mGRE encapsulation is automatically
generated with the dynamic endpoint discovery capability. The VPNv4 label and VPN
payload are carried over the mGRE tunnel encapsulation. This solution offers a simplified and scalable any-to-any unicast (IPv4, IPv6 6VPE based) and multicast (MDT
based) MPLS VPN communication model.

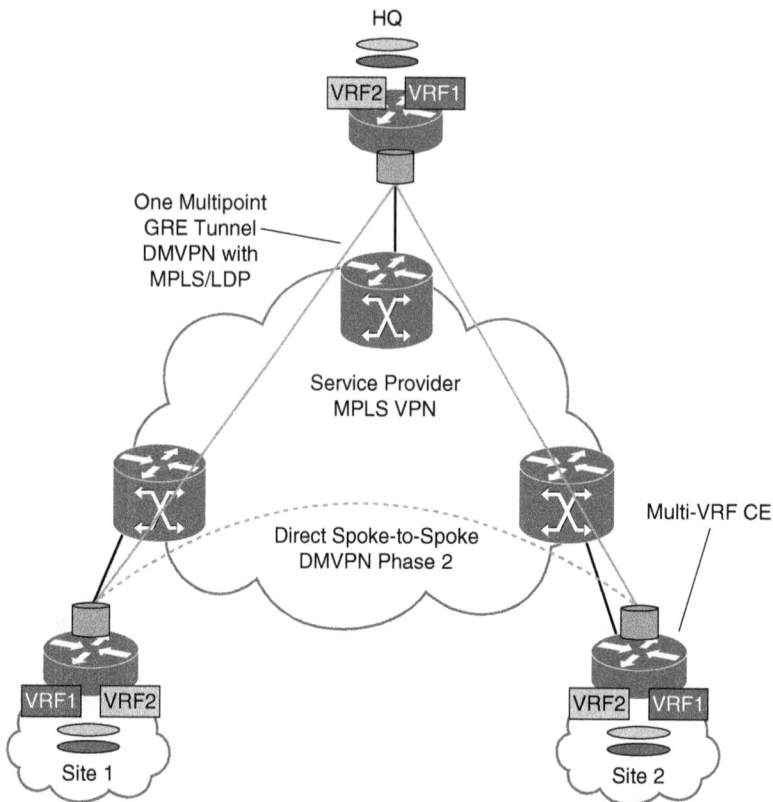

**Figure 17-24** *MPLS over DMVPN*

- **EIGRP Over the Top (OTP):** As the name implies, EIGRP OTP offers enterprise customers the opportunity to form EIGRP adjacencies across unmanaged WAN transport (typically over an L3VPN MPLS provider cloud) using unicast packets for peering and exchanging route prefixes without being injected into the provider's MP-BGP VPNv4/v6 routing table. With this approach, EIGRP OTP offers simplified dynamic multipoint encapsulation using Locator/Identifier Separation Protocol (LISP) to encapsulate its data traffic. EIGRP OTP relies on EIGRP routing tables rather than on the LISP mapping system to populate IP routing information. Furthermore, multiple instances of EIGRP can be deployed, along with other network virtualization techniques, to offer multiple routing instances.

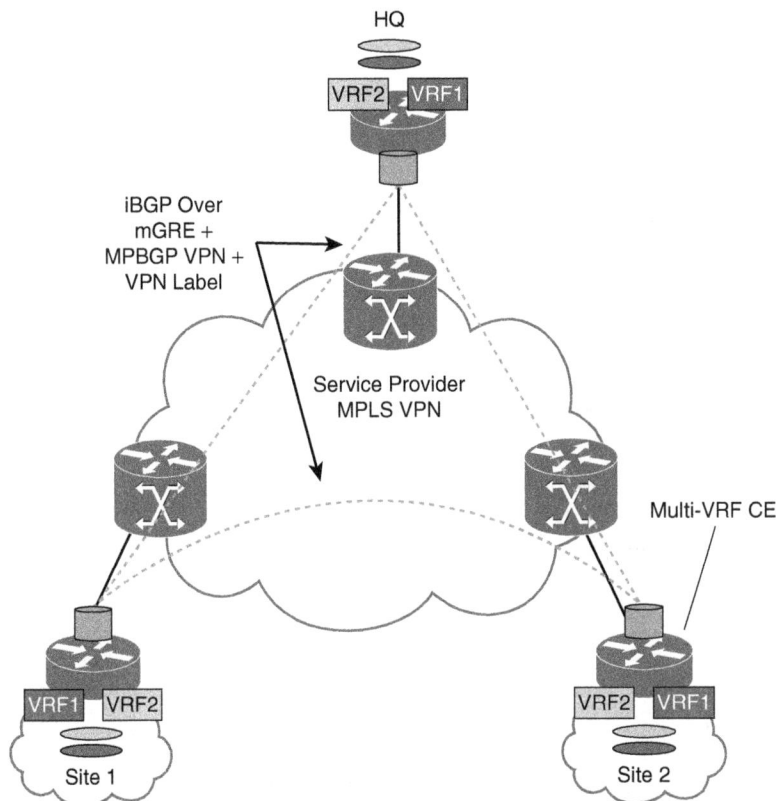

**Figure 17-25**   *MPLS over mGRE BGP Autodiscovery Based*

This design approach offers significant design flexibility for enterprise WAN connectivity because WAN networks will be seen as a virtual extension of the network, and enterprise customers can simply and transparently extend their infrastructure reachability over the provider's network using one unified control plane protocol, as shown in Figure 17-26.

## Comparison of Enterprise WAN Transport Virtualization Techniques

Table 17-13 provides a summarized comparison, from different design aspects, between the different WAN virtualization techniques discussed in this chapter.

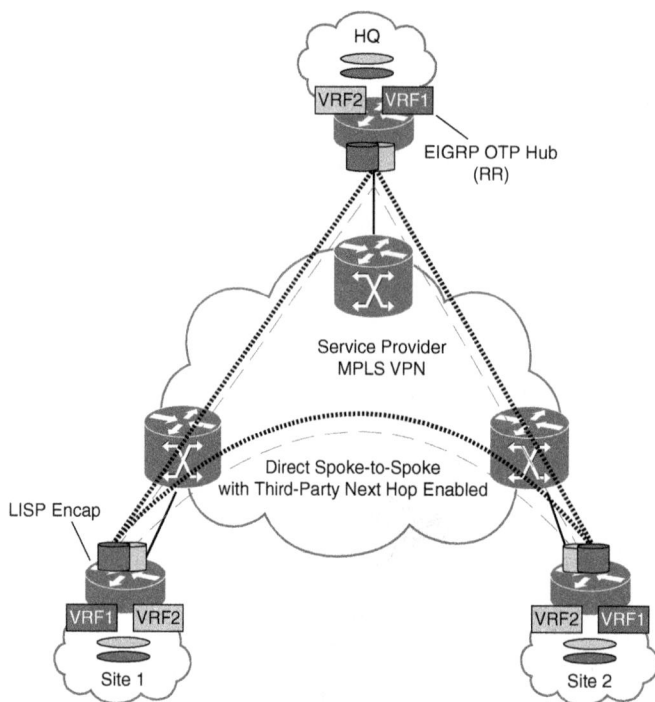

**Figure 17-26** *EIGRP OTP*

**Key Topic**

**Table 17-13** WAN Transport Virtualization Techniques Comparison

| | SP Dependent | Control Plane | Number of VNs Scalability | Number of Remote Sites Scalability | Direct CE-to-CE Forwarding | Multicast | IPv6 | Encryption |
|---|---|---|---|---|---|---|---|---|
| VRF Lite | Yes | IGP/BGP | Limited | Very Limited | Yes | SP dependent | SP dependent | GETVPN |
| CSC model | Yes | IGP + BGP/ MP-BGP | Scalable | Scalable | Yes | SP dependent | Yes | GETVPN |
| P2P GRE per VRF | No | IGP/BGP | Limited | Very Limited | Yes (full-mesh P2P tunnels) | Yes | Yes | IPsec |
| P2P GRE + MPLS | No | IGP + MP-BGP | Scalable | Very Limited | Yes (full-mesh P2P tunnels) | Yes | Yes | IPsec |
| DMVPN per VRF | No | IGP/BGP | Limited | Scalable | Yes | Yes (hub-and-spoke data path only) | Yes | IPsec |
| DMVPN + MPLS | No | IGP/ BGP + MP-BGP | Scalable | Scalable | Yes (DMVPN phase 2) | Yes (hub-and-spoke data path only) | Yes | IPsec |
| BGP mGRE + MPLS | No | MG-BGP | Scalable | Scalable | Yes | Yes | Yes | GETVPN |
| EIGRP OTP | No | EIGRP | Limited | Scalable | Yes (with third-party next hop enabled) | No | Yes | GETVPN |

**NOTE**   Operational complexity always increases when the network size increases and the WAN virtualization techniques used have limited scalability support, and vice versa.

## WAN Virtualization Design Options Decision Tree

Figure 17-27 is a summarized decision tree of the different design options for enterprise WAN virtualization.

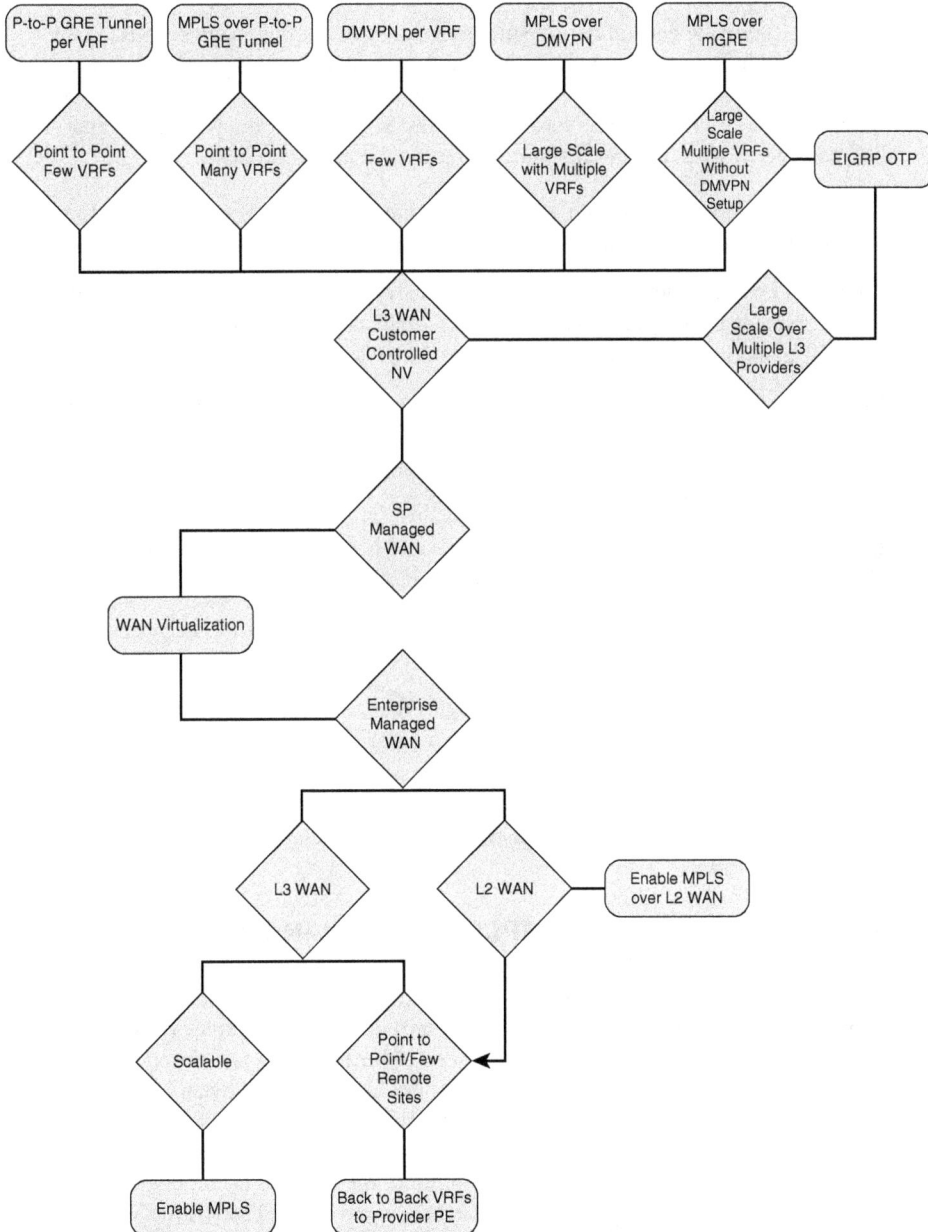

**Figure 17-27**   *Enterprise WAN Virtualization Decision Tree*

# Enterprise WAN Migration to MPLS VPN Considerations

Network migration is one of the most challenging and critical projects of network design. Network designers sometimes spend a large amount of time focusing on the "end state" of their network designs. However, a good and successful design must consider a migration strategy to move the network from its current state to the new state. Similarly, a good migration plan should address how network control plane protocols and applications will interact when the network is partially migrated.

It is impossible to provide general definite guidance for network migration because many variables drive and influence the migration plan and its strategy, such as business and application requirements, network size, and the technologies used. However, the following are some approaches and rules of thumb based on proven WAN migration experiences that should be considered when generating strategic or tactical migration plans:

- The phased migration approach is always recommended for large-scale networks with a large number of remote sites where multiple links or external exit points might exist.

- Logical and physical site architecture must be analyzed and taken into consideration during the planning phase, such as backdoor links, OSPF areas if the PE-CE protocol is OSPF, and BGP autonomous system (AS) numbering if BGP is used across the remote sites.

- In L3VPN, review the selected PE-CE routing protocol and consider how it can integrate with the existing routing setup and design without introducing any service interruption after moving to the new MPLS cloud. BGP is the most common choice and is often the only option a carrier has to provide.

- It is important to identify where the default route is generated and whether the remote sites will use a locally generated default route or over the MPLS cloud.

- Review the routing and identify the places that have summarization because, in some situations, this may lead to suboptimal routing or route summarization black holes.

- In L2VPN, there might be different topologies that can be provisioned by the provider (hub and spoke, full mesh, or partial mesh). A careful review of the topology and the effect of the path that traffic will take after the migration is very important, especially if you are moving from one topology to another. For example, you may be migrating the WAN from a hub-and-spoke topology over Frame Relay to a full-mesh L2 WAN (for example, VPLS).

## Migrating from Legacy WAN to MPLS L3VPN WAN Scenario

This section provides an example of a phased migration approach based on a migration scenario of an international media company that is currently using two separate Layer 2 (Frame Relay) clouds over hub-and-spoke topology across two different geographic regions (the United States and Europe). In addition, the hub sites of both the U.S. and Europe regions are interconnected over a Layer 2 virtual leased line (VLL) provisioned by an international Carrier Ethernet, as illustrated in Figure 17-28. After this media company finished a recent migration project of their video and broadcasting systems from a legacy time-division multiplexing (TDM)-based solution to an IP-based solution, they started facing degraded video quality because of the increased number of packets lost and delayed across the WAN. The obvious cause of this issue is the large number of video streams across the low-capacity

Frame Relay WAN links and because of the long trip each video stream has to take to reach its intended destination. For instance, traffic between two sites located within the same region has to traverse the hub site first. In addition, traffic between sites across different regions has to traverse the local and remote hub sites over the VLL link, as shown in Figure 17-28. Therefore, to overcome these limitations, a decision has been made to migrate the current hub-and-spoke Frame Relay WAN to an any-to-any topology over a single global MPLS L3VPN provider, to meet current traffic flow requirements with regard to the available bandwidth and delay and to achieve more consistent end-to-end QoS.

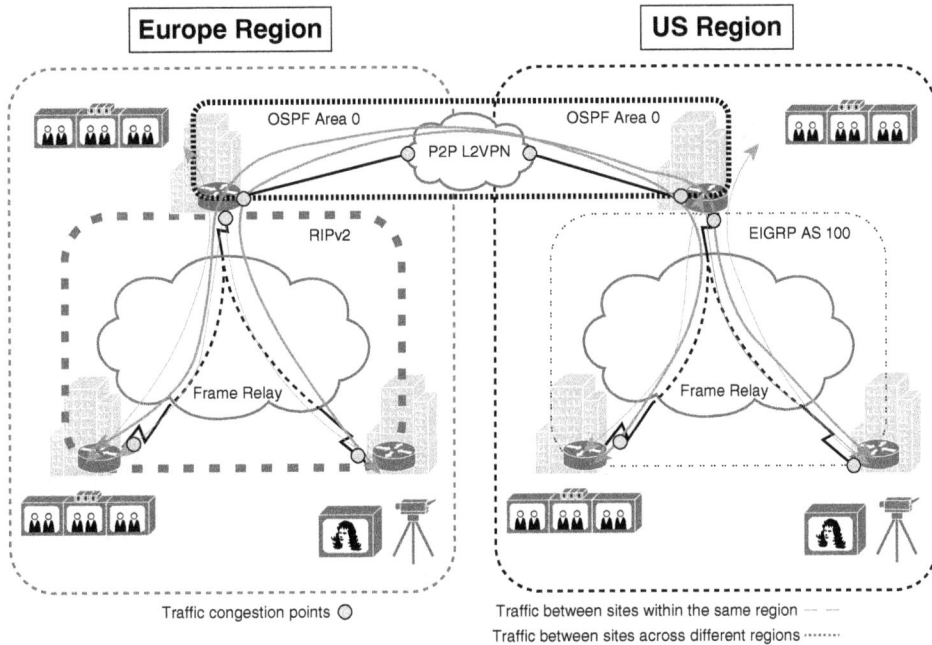

**Figure 17-28**  *Current Network Architecture and Traffic Flow*

## WAN Transport Virtualization Techniques Comparison

**Current state**

- WAN is based on hub-and-spoke topology over Frame Relay.

- EIGRP is the routing protocol used over the U.S. region WAN.

- RIPv2 is the routing protocol used over the Europe region WAN.

- OSPF is the routing protocol used by each hub site and between the hub sites over the VLL.

- OSPF area 0 is used over the VLL and each hub LAN network.

- External BGP (eBGP) is the proposed protocol to be used over the new MPLS L3VPN WAN.

**Assumption:** WAN IP addressing and the desired amount of bandwidth per site is organized with the MPLS VPN provider before the migration starts.

**Migration steps**

Step 1.   (Illustrated in Figure 17-29):

a.   Select the hub site to act as the transit site during this migration process (per region).

b.   Establish physical connectivity between the hub site (per region) and the new MPLS L3VPN SP.

c.   Establish an eBGP session between the hub (CE) and the SP (PE).

d.   Configure redistribution between OSPF and BGP at each hub site.

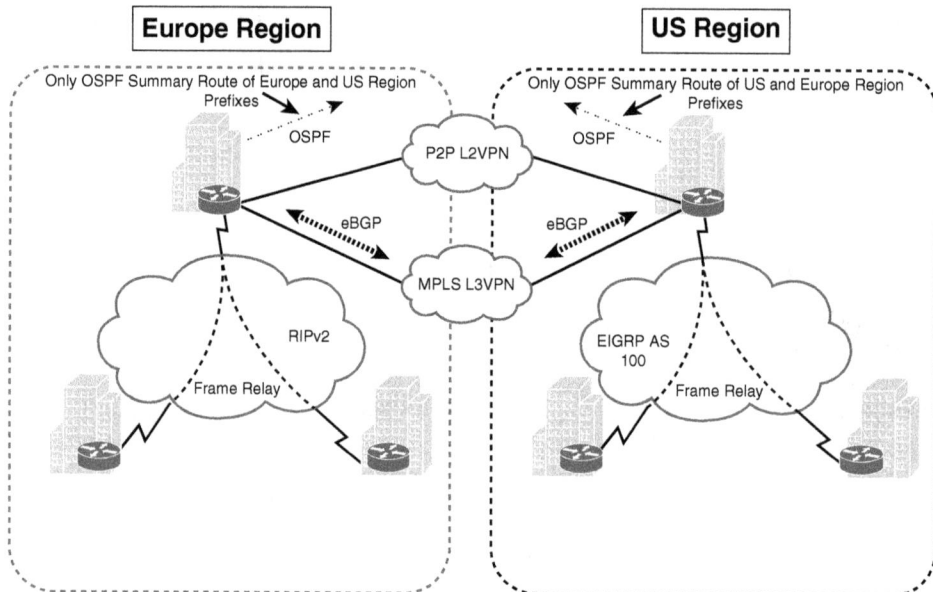

**Figure 17-29**   *WAN Migration Step 1*

**NOTE**   After the migration to the MPLS VPN, there will be two points of route redistribution with a backdoor link (over the L2VPN link between the hub sites). This scenario may introduce some route looping or suboptimal routing. Therefore, it is advised that over this L2VPN link each hub site (ASBR) advertises only the summary routes and not the specific to ensure that the MPLS WAN path is always preferred (more specific) without potential routing information looping. Alternatively, route filtering, as discussed in Chapter 8, "Layer 3 Technologies" (in the "Route Redistribution Design Considerations" section), must be considered.

Traffic between the LAN networks of each hub site will use the VLL (L2VPN) path as the primary path. Because the HQ LAN networks, along with the VLL link, are all part of the same area (area 0), the summarization at the hub

(Autonomous System Border Router [ASBR]) routers will not be applicable here (same area). By reducing OSPF cost over the VLL link, you can ensure that traffic between the HQ LANs will always use this path as the primary path.

**Step 2.**   (Illustrated in Figure 17-30):

a.   Connect one of the spoke routers intended to be migrated to the MPLS VPN.

b.   Establish an eBGP session with MPLS VPN SP and advertise the local subnet (LAN) using the BGP network statement (ideally without route retribution).

c.   Once the traffic starts flowing via the MPLS VPN (eBGP has a lower AD 20), disconnect the Frame Relay link.

d.   At this stage, traffic between migrated spokes and nonmigrated spokes will flow via the transit hub site.

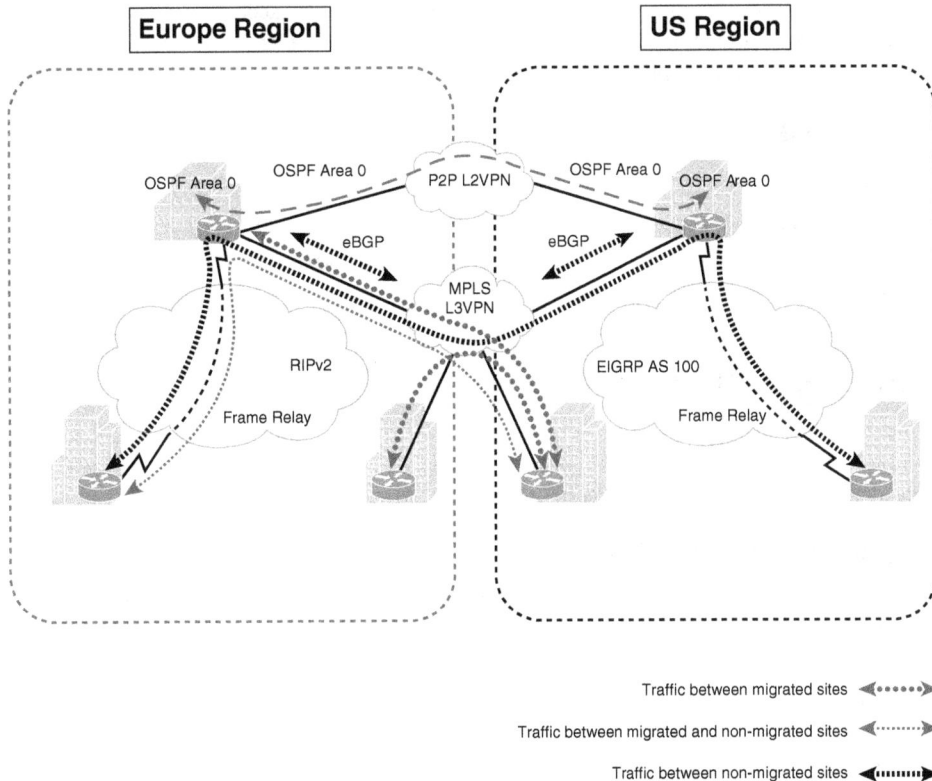

**Figure 17-30**   *WAN Migration Step 2*

**Step 3.**   Migrate the remaining spokes using the same phased approach.

> **NOTE**  With this approach, connectivity will be maintained between the migrated and nonmigrated sites without introducing any service interruption until the migration of the remaining remote sites is completed.

## Summary

Today's enterprise businesses are geographically dispersed, which makes them rely on the technology services of the WAN module to interconnect their disaggregated locations together. This is why the WAN module is one of the most vital and critical modules within the modern network architecture for businesses at the time of writing. Therefore, network designers must consider designs that can provide a common resource access experience to the remote sites and users, whether over the WAN or the Internet, without compromising any of the enterprise security requirements, such as end-to-end path separation between certain user groups. Last but not least, overlay integration at the WAN edge in today's networks can offer enterprises flexible and cost-effective WAN and remote-access connectivity, even considering the additional layer of control plane and design complexity that may be introduced into the overall enterprise architecture.

## References

- Al-shawi, Marwan, *CCDE Study Guide* (Cisco Press, 2015)

- Gartner Research, "Hybrid Will Be the New Normal for Next Generation Enterprise WAN" (Gartner, Inc., 2014; Reference Document ID: G00266397, https://www.gartner.com)

## Exam Preparation Tasks

As mentioned in the section "How to Use This Book" in the Introduction, you have a couple of choices for exam preparation: the exercises here, Chapter 18, "Final Preparation," and the exam simulation questions in the Pearson Test Prep Software Online.

## Review All Key Topics

Review the most important topics in this chapter, noted with the Key Topic icon in the outer margin of the page. Table 17-14 lists a reference of these key topics and the page numbers on which each is found.

**Table 17-14**  Key Topics for Chapter 17

| Key Topic Element | Description | Page Number |
|---|---|---|
| Table 17-2 | Layer 2 MPLS-Based WAN Advantages and Limitations | 452 |
| Table 17-3 | Layer 3 MPLS-Based WAN Advantages and Limitations | 454 |
| Paragraph | Internet as a primary transport solution | 455 |
| Table 17-4 | Internet as a WAN Transport Model Advantages and Limitations | 456 |
| Table 17-5 | Comparison of WAN Transport Models | 457 |

| Key Topic Element | Description | Page Number |
|---|---|---|
| List | WAN transport planning phase questions | 458 |
| List | Common factors that drive the decision for which WAN edge connectivity option to use | 460 |
| Table 17-9 | Remote Site Design Models | 465 |
| Table 17-11 | Network Overlay Technologies Comparison | 472 |
| Table 17-12 | Multi-VRF CE Versus CSC Model for Enterprise WAN Virtualization | 476 |
| Table 17-13 | WAN Transport Virtualization Techniques Comparison | 482 |
| List | WAN Migration example | 485 |

## Complete Tables and Lists from Memory

Print a copy of Appendix D, "Memory Tables" (found on the companion website), or at least the section for this chapter, and complete the tables and lists from memory. Appendix E, "Memory Tables Answer Key," also on the companion website, includes completed tables and lists to check your work.

## Define Key Terms

No new key terms were introduced in this chapter.

17

# CHAPTER 18

# Final Preparation

Part 1 of this book covered the different network design aspects that all network designers and CCDE candidates should know. Understanding network design fundamentals, principles, techniques, and pitfalls is highly critical to a network designer's success. All of these topics create a proper network design mindset. Part 2 of this book covered the technologies, protocols, and design considerations for them. Part 3 of this book focused on enterprise network architectures, merging the concepts presented in Part 1 and Part 2 within a specific network design use case.

Understanding all these topics is required to be prepared to pass the 400-007 CCDE Written Exam and eventually the CCDE Practical Exam to achieve the certification. While the previous 17 chapters supply the detailed information that you need to know for the exam, most people need more preparation than simply reading those chapters. Therefore, this chapter provides a set of tools and a study plan to help you complete your preparation for the exams.

This short chapter has three main sections. The first section helps you get ready to take the exam, and the second section lists the exam preparation tools useful at this point in the study process. The third section provides a suggested study plan you can use now that you have completed all the earlier chapters in this book.

## Getting Ready

Here are some important tips to keep in mind to ensure that you are ready for this rewarding exam:

- **Build and use a study tracker:** Consider using the exam objectives shown in this chapter to build a study tracker for yourself. Such a tracker can help ensure that you have not missed anything and that you are confident for your exam. As a matter of fact, this book offers a sample study planner as a website supplement.

- **Think about your time budget for questions on the exam:** When you do the math, you will see that, on average, you have one minute per question. While this does not sound like a lot of time, keep in mind that many of the questions will be very straightforward, and you will take 15 to 30 seconds on those. This leaves you extra time for other questions on the exam.

- **Watch the clock:** Check in on the time remaining periodically as you are taking the exam. You might even find that you can slow down pretty dramatically if you have built up a nice block of extra time.

- **Get some earplugs:** The testing center might provide earplugs, but get some just in case and bring them along. There might be other test takers in the center with you, and you do not want to be distracted by their moans and groans. I personally have no issue blocking out the sounds around me, so I never worry about this, but I know it is an issue for some.

- **Plan your travel time:** Give yourself extra time to find the center and get checked in. Be sure to arrive early. As you test more at a particular center, you can certainly start cutting it closer time-wise.

- **Online testing:** If participating in online testing, ensure that you have performed and passed the online system check. For more information, check here: https://www.cisco.com/c/en/us/training-events/training-certifications/online-exam-proctoring.html#~requirements

- **Get rest:** Most students report that getting plenty of rest the night before the exam boosts their success. All-night cram sessions are not typically successful.

- **Bring in valuables, but get ready to lock them up:** The testing center will take your phone, your smartwatch, your wallet, and other such items and will provide a secure place for them.

- **Take notes:** You will be given note-taking implements and should not be afraid to use them. I always jot down any questions I struggle with on the exam. I then memorize them at the end of the test by reading my notes over and over again. I always make sure I have a pen and paper in the car, and I write down the issues in the parking lot just after the exam. When I get home—with a pass or fail—I research those items!

## Tools for Final Preparation

This section lists some information about the available tools and how to access them.

### Pearson Cert Practice Test Engine and Questions on the Website

Register this book to get access to the Pearson IT Certification test engine (software that displays and grades a set of exam-realistic multiple-choice questions). Using the Pearson Cert Practice Test Engine, you can either study by going through the questions in Study mode or take a simulated (timed) 400-007 CCDE Written Exam.

The Pearson Test Prep practice test software comes with two full practice exams. These practice tests are available to you either online or as an offline Windows application. To access the practice exams that were developed with this book, please see the instructions in the introduction to the book.

**Step 1.** Go to www.PearsonTestPrep.com.

**Step 2.** Select Pearson IT Certification as your product group.

**Step 3.** Enter your email/password for your account. Use the same credentials you used to access this companion website.

**Step 4.** In the My Products tab, click the Activate New Product button

**Step 5.** Enter the unique access code found under the product listing in your account page under the Registered Products tab, and click the Activate button to activate your product.

**Step 6.** The product will now be listed in your My Products page. Click the Exams button to launch the exam settings screen and start your exam.

## Accessing the Pearson Test Prep Software Offline

If you wish to study offline, you can download and install the Windows version of the Pearson Test Prep software. You can find a download link for this software on the book's companion website, or you can just enter this link in your browser for a direct download: https://www.pearsonitcertification.com/content/downloads/pcpt/engine.zip

To access the book's companion website and the software, simply follow these steps:

**Step 1.** Click the Install Windows Version link under the Practice Exams section of the page to download the software.

**Step 2.** Once the software finishes downloading, unzip all the files on your computer.

**Step 3.** Double click the application file to start the installation, and follow the on-screen instructions to complete the registration.

**Step 4.** Once the installation is complete, launch the application and select the Activate Exam button on the My Products tab.

**Step 5.** Click the Activate a Product button in the Activate Product Wizard.

**Step 6.** Enter the unique access code found under the product listing in your account page under the Registered Products tab, and click the Activate button to activate your product.

**Step 7.** Click Next and then the Finish button to download the exam data to your application.

**Step 8.** You can now start using the practice exams by selecting the product and clicking the Open Exam button to open the exam settings screen.

Note that the offline and online versions sync together, so saved exams and grade results recorded on one version will be available to you in the other version as well.

## Customizing Your Exams

When you are in the exam settings screen, you can choose to take exams in one of three modes:

■ Study mode

■ Practice Exam mode

■ Flash Card mode

Study mode allows you to fully customize an exam and review answers as you are taking the exam. This is typically the mode you use first to assess your knowledge and identify information gaps. Practice Exam mode locks certain customization options in order to present a realistic exam experience. Use this mode when you are preparing to test your exam readiness. Flash Card mode strips out the answers and presents you with only the question stem. This mode is great for late-stage preparation, when you really want to challenge yourself to provide answers without the benefit of seeing multiple-choice options. This mode does not provide the detailed score reports that the other two modes provide, so it is not the best mode for helping you identify knowledge gaps.

In addition to these three modes, you will be able to select the source of your questions. You can choose to take exams that cover all of the chapters, or you can narrow your selection to just a single chapter or the chapters that make up specific parts in the book. All chapters are selected by default. If you want to narrow your focus to individual chapters, simply deselect all the chapters and then select only those on which you wish to focus in the Objectives area.

You can also select the exam banks on which to focus. Each exam bank comes complete with a full exam of questions that cover topics in every chapter. The two exams printed in the book are available to you, as are two additional exams of unique questions. You can have the test engine serve up exams from all four banks or just from one individual bank by selecting the desired banks in the exam bank area.

There are several other customizations you can make to your exam from the exam settings screen, such as the time allowed for taking the exam, the number of questions served up, whether to randomize questions and answers, whether to show the number of correct answers for multiple-answer questions, and whether to serve up only specific types of questions. You can also create custom test banks by selecting only questions that you have marked or questions on which you have added notes.

## Updating Your Exams

If you are using the online version of the Pearson Test Prep software, you should always have access to the latest version of the software as well as the exam data. If you are using the Windows desktop version, every time you launch the software, it will check to see if there are any updates to your exam data and automatically download any changes since the last time you used the software. This requires that you be connected to the Internet at the time you launch the software.

Sometimes, due to a number of factors, the exam data might not fully download when you activate your exam. If you find that figures or exhibits are missing, you might need to manually update your exams.

To update a particular exam you have already activated and downloaded, simply select the Tools tab and click the Update Products button. Again, this is only an issue with the desktop Windows application.

If you wish to check for updates to the Windows desktop version of the Pearson Test Prep exam engine software, simply select the Tools tab and click the Update Application button. Doing so allows you to ensure that you are running the latest version of the software engine.

### Premium Edition

In addition to the free practice exam provided on the website, you can purchase additional exams with expanded functionality directly from Pearson IT Certification. The Premium Edition of this title contains an additional two full practice exams and an eBook (in both PDF and ePub format). In addition, the Premium Edition title has remediation for each question to the specific part of the eBook that relates to that question.

To view the premium edition product page, go to https://www.informit.com/title/9780137601042.

### Chapter-Ending Review Tools

Chapters 1 through 17 each have several features in the "Exam Preparation Tasks" section at the end of the chapter. You might have already worked through these in each chapter. It can also be useful to use these tools again as you make your final preparations for the exam.

## Suggested Plan for Final Review/Study

This section lists a suggested study plan from the point at which you finish reading through Chapter 17 until you take the 400-007 CCDE Written Exam. You can ignore this plan, use it as is, or take suggestions from it.

The plan involves two steps:

**Step 1.**  Review key topics and "Do I Know This Already?" (DIKTA?) questions: You can use the table that lists the key topics in each chapter or just flip the pages, looking for key topics. Also, reviewing the DIKTA? questions from the beginning of the chapter can be helpful for review.

**Step 2.**  Use the Pearson Cert Practice Test engine to practice: The Pearson Cert Practice Test engine allows you to study using a bank of unique exam-realistic questions available only with this book.

## Summary

The tools and suggestions provided in this chapter have been designed with one goal in mind: to help you develop the skills required to pass the 400-007 CCDE Written Exam. This book has been developed from the beginning to not only tell you the facts but also help you learn how to apply the facts. No matter what your experience level leading up to taking the exam, it is my hope that the broad range of preparation tools, and even the structure of the book, will help you pass the exam with ease. I hope you do well on the exam.

# Answers to the "Do I Know This Already" Questions

## Chapter 1

1. **c and e.** Explanation: Constraints and mindset are the only two network design fundamentals listed. Security and scalability are both network design *principles*. The six network design fundamentals are mindset, design use cases, the business, constraints, requirements, and "why?"

2. **a, c, and d.** Explanation: Perimeter security, Zero Trust Architecture, and session- and transaction-based security are the security models that the industry has been shifting between over the last 20-plus years. Cybersecurity is not a security model but a component within these three models.

3. **c.** Explanation: Modularity allows for purpose-built building blocks to be leveraged. Failure isolation is a technique that creates boundaries within the network design to help contain problems from propagating. Shared failure state is where a device is performing multiple critical functions, and if it were to incur an outage from one of those functions, it would affect the other critical functions. Hierarchy is the process of creating layers within the architect for a specific purpose. The most common layers are Core, Distribution, Aggregation, and Access.

4. **a.** Explanation: Gold plating is the process of adding design elements that are excessive and do not meet any underlying requirement. Preconceived notions is when you as a network designer bring in outside information and assign that information as attributes to make design decisions, even though they do not apply to the situation in question. Best practices are the general list of configurations, features, and functions that should be deployed when there are no requirements governing, limiting, and restricting said item. Overthinking is a distractor option.

5. **d.** Explanation: The three categories of constraints are business, application, and technology. The only answer that has these three items is option d. The other options are incorrect.

6. **b.** Explanation: Resilience is the ability of the network to automatically fail over when an outage occurs. Reliability is best defined as a measure of how much of the network data gets from source to destination locations in the right amount of time to be leveraged correctly. Redundancy is the concept of having multiple resources performing the same function/role so that if one of them fails, the other takes over with limited to no impact on the production traffic. Routing failover is the ability of the routing protocol and configuration to fail traffic between different routing paths within the architecture.

7. **d.** Explanation: Failure isolation is a technique that creates boundaries within the network design to help contain problems from propagating. Hierarchy is the process of creating layers within the architecture for a specific purpose. The most common layers are core, distribution, aggregation, and access. Shared failure state is where a device is performing multiple critical functions and, if it were to incur an outage from one of those functions, would affect the other critical functions. Modularity allows for purpose-built building blocks to be leveraged.

8. **c.** Explanation: Hierarchy is the process of creating layers within the architecture for a specific purpose. The most common layers are core, distribution, aggregation, and access. Modularity allows for purpose-built building blocks to be leveraged. Failure isolation is a technique that creates boundaries within the network design to help contain problems from propagating. Shared failure state is where a device is performing multiple critical functions and, if it were to incur an outage from one of those functions, would affect the other critical functions.

9. **a.** Explanation: Reliability is best defined as how much of the network data gets from source to destination locations in the right amount of time to properly be leveraged correctly. Resilience is the ability of the network to automatically fail over when an outage occurs. Redundancy is the concept of having multiple resources performing the same function/role so that if one of them fails, the other would take over with limited to no impact on the production traffic. Routing Failover is the ability of the routing protocol and configuration to fail traffic between different routing paths within the architecture.

10. **b.** Explanation: Divest is where you as a network designer are splitting a business or company and the corresponding architecture into two or more functional independent architectures. Scaling is the network design use case that focuses on the scalability of the technology or holistic architecture in question. This could be as simple as a single flat area 0 OSPF design that just doesn't scale anymore to the business requirements, in which case you could leverage multiple areas, multiple area types, and LSA filtering techniques to help increase the scalability. In a design failure use case, there is a problem, and you have to resolve it. For example, not aligning the critical roles of Spanning Tree Protocol (STP) and First Hop Redundancy Protocol (FHRP). If your STP root bridge and your FHRP default gateways are not aligned to the correct devices, then you would have a suboptimal routing issue that would eventually lead to a design failure. A merger is when two or more different networks are integrated to include their own network design philosophies, to create one single end-to-end architecture.

# Chapter 2

1. **d.** Explanation: Each business has a set of priorities that are typically based on strategies adopted for the achievement of goals. These business priorities can influence the planning and design of IT network infrastructure. Therefore, network designers must be aware of these business priorities to align them with the design priorities, which ensures the success of the network they are designing by delivering business value.

2. c. Explanation: What does this business have to do and why? This is what we call a business driver. Business drivers are what organizations must follow. A business driver is usually the reason a business must achieve a specific outcome. It is the "why" the business is doing something. Required to maintain a specific compliance standard is a perfect example.

3. b. Explanation: A business outcome equates to the end result, such as saving money, diversifying the business, increasing revenue, or filling a specific need. Essentially, a business outcome is an underlying goal a business is trying to achieve. A business outcome will specifically map to a business driver.

4. a. Explanation: Business capabilities are not solutions. Business capabilities are what you get from a solution. A network access control solution provides a business the session- and transaction- based security capability, for example. Most solutions provide multiple capabilities. Some solutions provide parts of multiple capabilities, and when they are combined with other solutions, the business can get a number of capabilities that will make it successful.

5. c. Explanation: A decision matrix serves the same purpose as a decision tree, but with the matrix, network designers can add more dimensions to the decision-making process.

6. c. Explanation: Capital expenses (CAPEX), or expenditures, are the major purchases a business makes that are to be used over a long period of time. Examples of CAPEX include fixed assets, such as buildings and equipment.

7. a. Explanation: Operating expenses (OPEX), or expenditures, are the day-to-day costs incurred by a business to keep the business operational. Examples of OPEX include rent, utilities, payroll, and marketing.

8. b. Explanation: Return on investment (ROI) is the concept of identifying what the perceived potential benefit is going to be for the business if the business makes an investment in a new technology, solution, and/or capability. For example, completing that application upgrade is going to cost the business $5,000 in time and labor. The investment here is $5,000 but it doesn't have to be monetary. At some point, the business wants to recoup its investment, i.e., what is the business getting out of this investment? Included with this upgrade are new features and functionality that the business can charge a premium for. These premium services are how the business is going to recoup its initial $5,000 investment.

9. d. Explanation: The waterfall project management framework is linear, adverse to change, structured, and requires documentation at the end of each phase.

10. c. Explanation: The agile project management model is based on an incremental, iterative approach. Instead of in-depth planning at the beginning of the project, like that of the waterfall model, agile methodologies are open to changing requirements over time and encourage constant feedback from the different business leaders and stakeholders. Cross-functional teams work on iterations of a product over a period of time, and this work is organized into a backlog that is prioritized based on business or customer value. The goal of each iteration is to produce a working product. Agile welcomes change and, because of that, the end state is oftentimes not known at the beginning of the project. Throughout an agile project, feedback is welcomed and

requested on a reoccurring basis to the business leaders and stakeholders. Breaking down the project into an iterative process allows the project team to focus on higher-quality work that is collaborative and also promotes faster delivery.

# Chapter 3

1. **b.** Explanation: The single-server model is the simplest application model, and it is equivalent to running the application on a personal computer. All of the required components for an application to run are on a single application or server.

2. **c.** Explanation: The 2-tier application model is like a client/server architecture, where communication takes place between client and server. In this model, the presentation layer or user interface layer runs on the client side while the dataset layer gets executed and stored on the server side.

3. **a.** Explanation: The 3-tier application model has three tiers or layers called presentation (aka web), intermediate (aka application), and database.

4. **d.** Explanation: On-premises is the service model where a business owns and manages the infrastructure. A business will procure all of the infrastructure required to run the service and then fully manage, maintain, and operate it. In some situations, the management is outsourced but the infrastructure is procured and owned by the business.

5. **a.** Explanation: Platform as a Service (PaaS) is a service model where a vendor provides hardware and software tools, and customers use these tools to develop applications. PaaS users tend to be application developers.

6. **b.** Explanation: Infrastructure as a Service (IaaS) is a pay-as-you-go service model for storage, networking, and virtualization—all of a business's infrastructure needs. IaaS gives users cloud-based alternatives to on-premises infrastructure, so businesses can avoid investing in expensive onsite resources.

7. **c.** Explanation: Software as a Service (SaaS) is a service model where a vendor makes its software available to users, usually for a monthly or annual subscription service fee.

8. **c.** Explanation: Multi-cloud is the use of two or more cloud service providers (CSPs), with the ability to move workloads between the different cloud computing environments in real time as needed by the business.

9. **b.** Explanation: A private cloud consists of cloud computing resources used by one business. This cloud environment can be located within the business's data center footprint, or it can be hosted by a cloud service provider. In a private cloud, the resources, applications, services, data, and infrastructure are always maintained on a private network and all devices are dedicated to the business.

10. **a.** Explanation: A hybrid cloud is the use of both private and public clouds together to allow for a business to receive the benefits of both cloud environments while limiting their negative impacts on the business.

11. **b.** Explanation: Data governance is the planning of all aspects of data management. This includes availability, usability, consistency, integrity, and security of all data within the organization.

# Chapter 4

1.  **a, d. and f**. Explanation: Static factors are items that we know and therefore can preemptively base access and authorization on. The most common of these factors are credentials but could also include the level of confidence, device trust, network, physical location, biometrics, and device orientation. Threat intelligence, real-time data analytics, and GPS coordinates are all dynamic factors, sources of data that can be analyzed at the time of access to change what level of access and authorization (i.e., the trust score of the transaction in question) is being provided.

2.  **b, c, and e**. Explanation: Threat intelligence, real-time data analytics, and GPS coordinates are all dynamic factors, sources of data that can be analyzed at the time of access to change what level of access and authorization (i.e., the trust score of the transaction in question) is being provided. Static factors are items that we know and can preemptively base access and authorization on. The most common of these are credentials but could also include the level of confidence, device trust, network, physical location, biometrics, and device orientation.

3.  **b and d**. Explanation: A trust score is created by a combination of factors, both static and dynamic, and is used to continually provide identity assurance. Assets, applications, networks, and so forth—what we generally call resources—have levels of risk scores, which are thresholds that must be exceeded for access to be permitted. In general, the security plan categorization determines asset level or risk. Users have various roles and, based on those roles, are entitled to specific access to complete their job. A financial user would need a different level of access than a human resource user. These two users should not have the same access or authorization. They may have overlapping access to resources that they both need to complete their job functions.

4.  **a and d**. Explanation: For a resource such as a user or device to access another resource, such as an asset, application, or system, the requesting resource's authorization for access is determined by combining its entitlement level and trust score. Entitled to access (aka entitlement) means users have various roles and, based on those roles, are entitled to specific access to complete their job. A financial user would need a different level of access than a human resource user. These two users should not have the same access or authorization. They may have overlapping access to resources that they both need to complete their job functions. A trust score is created by a combination of factors, both static and dynamic, and is used to continually provide identity assurance. A trust score determines the level of access as required by the level of risk value of the asset being accessed. Although static factors contribute to the trust score, they do not directly comprise the authorization for access. Static factors are items that we know and can preemptively base access and authorization on. The most common of these are credentials but could also include the level of confidence, device trust, network, physical location, biometrics, and device orientation. Although dynamic factors contribute to the trust score, they do not directly comprise the authorization for access. Dynamic factors are sources of data that can be analyzed at the time of access to change what level of access and authorization (i.e., the trust score of the transaction in question) is being provided. The most

common is threat intelligence, but can also be geovelocity, GPS coordinates, and real-time data analytics around the transaction.

5.  **d.** Explanation: A policy engine is the location where policy is implemented, rules are matched, and associated access (authorization) is pushed to the policy enforcement points. This is also known as a policy administration point (PAP) and a policy decision point (PDP). A trust engine dynamically evaluates overall trust by continuously analyzing the state of devices, users, workloads, and applications (resources). It utilizes a trust score that is built from static and dynamic factors. This is also known as a policy information point (PIP). An endpoint device is any device an end user can leverage to access the enterprise network. Endpoint devices include business-owned assets and personally owned devices that are approved to access the enterprise network. Inventory is a single point of truth for all resources. This is an end-to-end inventory throughout the entire architecture/enterprise. This is also known as a policy information point.

6.  **a.** Explanation: A trust engine dynamically evaluates overall trust by continuously analyzing the state of devices, users, workloads, and applications (resources). It utilizes a trust score that is built from static and dynamic factors. This is also known as a policy information point (PIP). An endpoint device is any device an end user can leverage to access the enterprise network. Endpoint devices include business-owned assets and personally owned devices that are approved to access the enterprise network. Inventory is a single point of truth for all resources. This is an end-to-end inventory throughout the entire architecture/enterprise. This is also known as a policy information point. A policy engine is the location where policy is implemented, rules are matched, and associated access (authorization) is pushed to the policy enforcement points. This is also known as a policy administration point (PAP) and a policy decision point (PDP).

7.  **b.** Explanation: The CIA triad includes confidentiality, integrity, and availability. It does not include compliance. Confidentiality protects against unauthorized access to information to maintain the desired level of secrecy of the transmitted information across the internal network or public Internet. Integrity maintains accurate information end to end by ensuring that no alteration is performed by any unauthorized entity. Availability ensures that access to services and systems is always available and information is accessible by authorized users when required. Compliance is a distractor answer.

8.  **c.** Explanation: Integrity maintains accurate information end to end by ensuring that no alteration is performed by any unauthorized entity. Compliance is a distractor answer. Availability ensures that access to services and systems is always available and information is accessible by authorized users when required. Confidentiality protects against unauthorized access to information to maintain the desired level of secrecy of the transmitted information across the internal network or public Internet.

9.  **b.** Explanation: The Payment Card Industry Data Security Standard (PCI DSS) is a compliance standard focused on ensuring the security of credit card transactions. This standard specifies the technical and operational standards that businesses must follow to secure and protect credit card data provided by cardholders and

transmitted through card processing transactions. The Health Information Portability and Accountability Act (HIPAA) is a compliance standard focused on protecting health and patient information. Policy enforcement points are the locations where trust and policy are enforced. A policy engine is the location where policy is implemented, rules are matched, and associated access (authorization) is pushed to the policy enforcement points. This is also known as a policy administration point (PAP) and a policy decision point (PDP).

**10.** **a.** Explanation: The Health Information Portability and Accountability Act (HIPAA) is a compliance standard focused on protecting health and patient information. The Payment Card Industry Data Security Standard (PCI DSS) is a compliance standard focused on ensuring the security of credit card transactions. This standard refers to the technical and operational standards that businesses must follow to secure and protect credit card data provided by cardholders and transmitted through card processing transactions. Policy enforcement points are the locations where trust and policy are enforced. A policy engine is the location where policy is implemented, rules are matched, and associated access (authorization) is pushed to the policy enforcement points. This is also known as a policy administration point (PAP) and policy decision point (PDP).

# Chapter 5

**1.** **a.** Explanation: Business architecture enables everyone, from strategic planning teams to implementation teams, to get "on the same page" or to be synchronized, enabling them to address challenges and meet business objectives.

**2.** **b.** Explanation: Enterprise architecture is a process of organizing logic for business processes and IT infrastructure reflecting the integration and standardization requirements of the company's operating model.

**3.** **c.** Explanation: A business solution is a set of interacting business capabilities that delivers specific, or multiple, business outcomes.

**4.** **d.** Explanation: A business outcome is a specific measurable result of an activity, process, or event within the business.

**5.** **a.** Explanation: Technology Specific is a domain-specific architecture. Within this scope, the business is requiring help with finding and purchasing the right product or group of products in an architecture focus area. This might be data center, security, or enterprise networking focused but doesn't cross between the different architecture focus areas.

**6.** **b.** Explanation: Technology Architecture is a multi-domain architecture (MDA), also referred to as cross-architecture. In this scope, the business needs help understanding the benefits of multi-domain technology architecture and how to show the value it provides to the business. With this scope, two or more architecture focus areas are incorporated.

**7.** **c.** Explanation: Business Solutions is a partial business architecture scope for a business that requires expertise to help solve its business problems and determine how to measure the business impact of its technology investments (CAPEX, OPEX, ROI, TCO, etc.).

8. **d.** Explanation: Business Transformation is a business-led architecture scope for a business that requires help with transforming its business capabilities to facilitate innovation to accelerate the company's digitization.

9. **c.** Explanation: The Open Group Architecture Framework is an enterprise architecture methodology that incorporates a high-level framework for enterprises that focuses on designing, planning, implementing, and governing enterprise information technology architectures. TOGAF helps businesses organize their processes through an approach that reduces errors, decreases timelines, maintains budget requirements, and aligns technology with the business to produce business-impacting results.

10. **a.** Explanation: Information Technology Infrastructure Library is a set of best practice processes for delivering IT services to your organization's customers. ITIL focuses on ITSM and ITAM, and includes processes, procedures, tasks, and checklists that can be applied by any organization. The three focus areas of ITIL are Change Management, Incident Management, and Problem Management.

# Chapter 6

1. **a.** Explanation: The correct port-based Metro Ethernet transport mode for an E-Line service is Ethernet private line (EPL).

2. **d.** Explanation: The correct VLAN-based Metro Ethernet transport mode for an E-LAN service is Ethernet virtual private LAN (EVPLAN).

3. **b.** Explanation: The correct VLAN-based Metro Ethernet transport mode for an E-Line service is Ethernet virtual private line (EVPL).

4. **c.** Explanation: The correct port-based Metro Ethernet transport mode for an E-LAN service is Ethernet private LAN (EPLAN).

5. **c.** Explanation: The correct transport mode over pseudowire (PW) for a Frame Relay access connection is port-based per DLCI.

6. **a.** Explanation: The correct transport mode over PW for an ATM access connection is AAL5 protocol data units over PW cell relay over PW.

7. **b.** Explanation: The correct transport mode over PW for an Ethernet access connection is Protocol-based per VLAN.

8. **d and e.** Explanation: The Layer 2 transport options that are the best options for a very large enterprise Data Center Interconnect (DCI) solution are Provider Backbone Bridging with Ethernet VPN (PBB-EVPN) and Provider Backbone Bridging with Virtual Private LAN Service (PBB-VPLS). EVPN can support up to a large enterprise scenario, H-VPLS can support up to a medium enterprise scenario, and VPLS can only support up to a small scenario.

9. **b and d.** Explanation: The Layer 2 transport options that are best for MAC mobility are EVPN and PBB-EVPN because of the sequence number attribute being leveraged.

## Chapter 7

1. **b.** Explanation: PortFast is a feature that bypasses the listening and learning phases to transition directly to the forwarding state. STP edge port is another name for this but is less well-known.

2. **d.** Explanation: BPDU Filter is a feature that suppresses BPDUs on ports.

3. **a.** Explanation: Root Guard is a feature that prevents external switches—switches that are not part of your network or under your control—from becoming the root of the Spanning Tree Protocol tree.

4. **c.** Explanation: Multiple Spanning Tree (MST) is a protocol that is used to group multiple VLANs into a single STP instance. This also reduces the total number of spanning-tree instances that match the physical topology of the network, reducing the CPU load.

5. **c.** Explanation: Virtual Router Redundancy Protocol (VRRP) is a standards-based first-hop routing protocol that provides redundancy with a virtual router elected as the master.

6. **d.** Explanation: Link Aggregation Control Protocol (LACP) is a protocol defined in IEEE 802.3ad that provides a method to control the bundling of several physical ports to form a single logical channel.

7. **b.** Explanation: Virtual Switching System (VSS) is a Cisco technology that allows certain Cisco switches to bond together as a single virtual switch.

8. **d.** Explanation: The primary design concern for a Looped Triangle Topology is that STP limits the ability to utilize all the available uplinks within a VLAN or STP/MST instance.

9. **a.** Explanation: The primary design concern for a Loop-Free Inverted U Topology is that it introduces a single point of failure to the design if one distribution switch or uplink fails.

10. **b.** Explanation: The primary design concern for a Looped Square Topology is that a significant amount of access layer traffic might cross the interswitch link to reach the active FHRP.

11. **c.** Explanation: The primary design concern for Loop-Free U Topology is the inability to extend the same VLANs over more than a pair of access switches.

## Chapter 8

1. **c.** Explanation: Link-state advertisements (LSAs) are used by OSPF routers to exchange routing and topology information. When neighbors decide to exchange routes, they send a list of all LSAs in their respective topology database. Each router then checks its topology database and sends a Link State Request message requesting all LSAs that were not found in its topology table. Other routers respond with the Link State Update that contains all LSAs requested by the neighbor.

2. **a.** Explanation: Type 3 LSAs are generated by area border routers (ABRs) to advertise networks from one area to the rest of the areas in an autonomous system.

A

3.  **b.** Explanation: Point-to-multipoint indicates a topology where one interface can connect to multiple destinations. Each connection between a source and destination is treated as a point-to-point link. An example would be a Point-to-Multipoint Cisco Dynamic Multipoint VPN (DMVPN) topology. OSPF will not elect DRs and BDRs and all OSPF traffic is multicast to 224.0.0.5.

4.  **a, d, and e.** Explanation: Neighborship is discovered and maintained using hello packets. These packets are sent using multicast. Update messages are used to send routing information to neighbors. These packets are sent to either one neighbor via unicast or to multiple neighbors via multicast. They are sent using Reliable Transport Protocol. EIGRP uses query packets when a router loses a path to a network. The router sends a query packet to its neighbors, asking if they have information on that network. These packets are sent via multicast and using Reliable Transport Protocol.

5.  **b.** Explanation: An EIGRP stub router will inform neighbors via the hello packet that it's a stub; by doing so, neighbors will not send queries to the router. EIGRP stubs are typically used at spoke locations, as stubs cannot be used as transient routers.

6.  **b, d, and e.** Explanation: Update messages are used to send routing information to neighbors. These packets are sent to either one neighbor via unicast or to multiple neighbors via multicast. They are sent using Reliable Transport Protocol. EIGRP uses query packets when a router loses a path to a network. The router sends a query packet to its neighbors, asking if they have information on that network. These packets are sent via multicast and using Reliable Transport Protocol. Reply packets are used by routers that received the query packet to respond to the query. These are sent unicast to the router that sent the query and are sent using Reliable Transport Protocol.

7.  **a.** Explanation: IS-IS uses something like a designated router in OSPF, but in Intermediate System-to-Intermediate System (IS-IS) it's referred to as a Designated Intermediate System. A DIS is elected and is a pseudo node of the process. If you were to not have a DIS on a multiaccess environment, then all the LSPs would be flooded to other routers.

8.  **b.** Explanation: An Intermediate System-to-Intermediate System (IS-IS) level 2 router has the link-state information for the intra-area as well as inter-area routing. The L2 router sends only L2 hellos. IS-IS level 2 area is similar and often compared to OSPF backbone area 0.

9.  **c.** Explanation: The overload bit in Intermediate System-to-Intermediate System (IS-IS) is used to increase convergence and prevent black-holing of traffic in the environment. When the overload bit is set, it will gracefully redirect traffic around the device in which the bit is set, thus making it a non-transit router. By leveraging the overload bit, traffic will not be sent to routers where other processes (i.e., BGP) haven't converged yet and therefore due to the process not having all information would drop the traffic.

10. **d.** Explanation: To manipulate traffic inside your own AS, local preference can be used. Local preference is carried inside an AS (iBGP) so you can manipulate traffic at one node and the attribute is carried inside your AS.

11. **a and c.** Explanation: AS Path prepend is a very common way to influence traffic into your AS. If you want to prefer a router over another, then on the router that is less preferred, add additional AS to the path to make the route "look not as good." Another option, although less common, is the use of MED. MED may be valid when connected to the same neighboring AS with multiple connections versus connecting to different ASs.

# Chapter 9

1. **b.** Explanation: The unique route distinguisher (RD) per VPN allocation model is simple to design and manage, and requires lower hardware resource consumption compared to the other models.

2. **c.** Explanation: The unique RD per VPN per provider edge (PE) allocation model provides both active/active and active/standby connectivity design options for a multihomed remote site.

3. **b.** Explanation: EIGRP site of origin (SoO) is specifically used to help avoid or mitigate the impact of routing loops, and racing issues, in complex topologies leveraging EIGRP as a PE-CE routing protocol that contain both MPLS VPN and backdoor links.

4. **c.** Explanation: The OSPF DN bit is set when routes are redistributed from the BGP into OSPF. This bit is then checked when OSPF redistributes routes into BGP at another device. If the OSPF DN bit is set, those routes will not be redistributed back into the super backbone. BGP and EIGRP SoO are unrelated here. The OSPF domain ID is a mechanism to manually adjust the OSPF process ID and is normally used by the PE devices to ensure the OSPF neighbors are in the proper OSPF process.

5. **d.** Explanation: The route distinguisher is prepended per MP-BGP VPNv4/v6 prefix to seamlessly transport customer routes (overlapping and nonoverlapping) over one common infrastructure.

6. **a.** Explanation: The route target is specifically used to import and export routes from and to a VRF.

7. **b.** Explanation: The hub-and-spoke MPLS L3VPN topology sends all remote site traffic (spokes) to a specific hub location, in this case the data center, for a specific reason. In this case, the firewall in the data center must inspect the traffic before allowing spoke traffic to traverse to another spoke.

8. **d.** Explanation: The extranet and shared services MPLS L3VPN topology allows for resources to be properly shared between different VRFs and VPNs within the MPLS environment. In this case, the Internet service is the resource being shared. This is accomplished by setting the proper route target imports and exports.

9. **a.** Explanation: The underlay network is defined specifically by the physical switches and routers in the LAN.

10. **d.** Explanation: In the SD-WAN overlay, virtual networks (VNs) provide segmentation just like VRFs.

# Chapter 10

1. c. Explanation: The data plane is responsible for controlling fast-forwarding of traffic passing through a network device.

2. b. Explanation: The control plane is like the brain of the network node and usually controls and handles path selection functions.

3. a. Explanation: The management plane is the plane that is focused on the management traffic of the device, such as device access, configuration troubleshooting, and monitoring.

4. c. Explanation: Transit IP traffic is traffic for which a network device makes a typical routing and forwarding decision regarding whether to send the traffic over its interfaces to directly attached nodes.

5. d. Explanation: Exception IP traffic is any IP traffic carrying a nonstandard "exception" attribute, such as a transit IP packet with an expired TTL.

6. b. Explanation: Receive IP traffic is traffic destined to the network node itself, such as toward a router's IP address, and requires CPU processing.

7. a. Explanation: Non-IP traffic is typically related to non-IP packets and almost always is not forwarded, such as MPLS, IS-IS (CLNP), and Layer 2 keepalives.

8. d. Explanation: Application targeted attacks are mitigated by web proxy/filtering, e-mail proxy/filtering, and a web application firewall (WAF).

9. a and b. Explanation: A network direct access is mitigated by a Layer 2 iACL, Layer 3 iACL, and a firewall. Layer 2 attacks are mitigated by iACL, CoPP, and system and topological redundancy.

10. c. Explanation: A network DoS attack is mitigated by Layer 3 and Layer 2 network and device security considerations, and remotely triggered black hole (RTBH) anomaly-based IDS/IPS.

11. b. Explanation: EAP Transport Layer Security (EAP-TLS) authentication provides a certificate-based and mutual authentication of the client and the network. It relies on client-side certificates (identity certificates) and server-side certificates to perform authentication and can be used to dynamically generate user-based and session-based keys to secure future communications. One limitation of EAP-TLS is that certificates must be managed on both the client and server side. However, it is the most secure EAP type because of the mutual authentication of the client and network.

12. a. Explanation: EAP Message Digest Challenge (EAP-MD5) authentication provides a base level of EAP support and typically is not recommended for implementation because it may allow the user's password to be derived. However, it is the easiest of the EAP types to deploy because there is no requirement for certificate management on either the client or server sides.

# Chapter 11

1. **b.** Explanation: The receiver sensitivity is the most helpful because it defines the minimum usable signal strength a client can receive from an access point (AP). The AP cell size is determined by the distance a client can be located from the AP before the AP's signal falls below the receiver sensitivity.

2. **c.** Explanation: High density in a wireless design is determined by the number of clients per AP in an area. If the user population is high in a small area, all of the users might end up joining a single AP. The goal of a good wireless design would be to add additional APs and distribute the clients across them, maintaining an adequate level of performance for each AP. For a high-density design, the coverage area and cell size per AP is reduced to allow for higher performance for the clients leveraging each AP.

3. **a.** Explanation: The customer is wanting user authentication, so you could leverage RADIUS, AAA, or NAC (Cisco ISE) servers to meet that need.

4. **d.** Explanation: A data-only wireless deployment without any additional real-time applications being leveraged is usually used when clients use normal applications that have no specific performance requirements; thus, there is no need to account for jitter, latency, or packet loss.

5. **c.** Explanation: A voice deployment model is indicated because of the strict jitter requirement given. Jitter implies network performance that is necessary for real-time applications such as voice and video.

6. **b.** Explanation: If the AP is already at its lowest transmit power level setting, your next strategy should be to connect an external directional antenna to the AP. The patch antenna will focus the AP's RF energy into a smaller area and will help reduce the AP's cell size.

7. **a, b, and d.** Explanation: Wireless network designs focused on voice should use a minimum data rate of 12 Mbps. It is important to consider the number of simultaneous calls that each AP can support, based on the minimum data rate. As a general guideline, you should leverage the many 5-GHz channels, but carefully validate that you can use each DFS channel, only if radar D signals have not been detected on them.

8. **c.** Explanation: Of the options available, the floor plans will be most helpful, as you will leverage these directly into any wireless planning tool to help identify where to place APs within the wireless network design.

9. **d.** Explanation: The closer a client is located in relation to an AP, the stronger the AP's signal will be. With a stronger received signal, and constant or increasing SNR, the client will likely try to use a faster data rate.

# Chapter 12

1. **a and c.** Explanation: Zero-touch provisioning reduces the amount of time needed to deploy new infrastructure and eliminates the need to troubleshoot because of network outages caused by human error.

**2.** **a, c. and d.** Explanation: With ZTP we can have all of our cabling validated to ensure it's correct, ensure we have a saved configuration in a repository of our ZTP-enabled devices, and ensure all ZTP devices are running a specific code version.

**3.** **a and c.** Explanation: Infrastructure as code reduces the amount of time needed to deploy new infrastructure and eliminates the need to troubleshoot because of network outages caused by human error.

**4.** **a, c, and d.** Explanation: A CI/CD pipeline reduces the amount of time needed to deploy new infrastructure, reduces the need to troubleshoot because of network outages caused by human error, and increases time to market on new services.

**5.** **a.** Explanation: The proper steps in a CI/CD pipeline are source, build, test, and deploy.

# Chapter 13

**1.** **c.** Explanation: Internet Group Management Protocol (IGMP) snooping is a Layer 2 multicast protocol running on IPv4 networks that listens on multicast protocol packets between a Layer 3 multicast device and user hosts to maintain outbound interfaces of multicast packets. Multicasts may be filtered from the links that do not need them, conserving bandwidth on those links.

**2.** **a.** Explanation: Multicast Listener Discovery (MLD) snooping is a Layer 2 multicast protocol running on IPv6 networks that listens on multicast protocol packets between a Layer 3 multicast device and user hosts to maintain outbound interfaces of multicast packets. MLD snooping manages and controls multicast packet forwarding at the data link layer. Think of MLD snooping as IGMP snooping but for IPv6.

**3.** **b.** Explanation: Reverse path forwarding (RPF) is the mechanism used by Layer 3 nodes in the network to optimally forward multicast datagrams without loops.

**4.** **a.** Explanation: Protocol-Independent Multicast (PIM) Bidirectional (BIDIR) builds bidirectional shared trees connecting multicast sources and receivers. It never builds a shortest-path tree, so it scales well because it does not need a source-specific state. PIM-BIDIR is the best multicast routing protocol for many-to-many traffic pattern requirements.

**5.** **c.** Explanation: PIM Source-Specific Multicast (PIM-SSM) builds trees that are rooted in just one source. SSM eliminates the requirement for rendezvous points (RPs) and shared trees of sparse mode and only builds a shortest-path tree (SPT).

**6.** **d.** Explanation: PIM Bootstrap Router (PIM-BSR) is like Cisco's Auto-RP in that it is a protocol that is used to automatically find the rendezvous point (RP) in a multicast network. BSR is a standard and included in PIMv2, unlike Auto-RP, which is a Cisco-proprietary protocol. BSR sends messages on a hop-by-hop basis and does so by sending its packets to multicast address 224.0.0.13.

**7.** **a.** Explanation: With PIM-BIDIR, all traffic will follow the path through the RP, wherever in the network that RP is located. Because of this, a network designer will need to put the RP between the sources and receivers of the critical application.

8. **a, c, and d.** Explanation: There are four factors that influence the placement of a multicast RP: the multicast protocol that is used, the multicast tree model, the application multicast requirements, and a targeted network segment between the sources and receivers (LAN versus WAN).

9. **c.** Explanation: Anycast-RP is based on using two or more RPs configured with the same IP address on their loopback addresses. Typically, the Anycast-RP loopback address is configured as a host IP address (32-bit mask). From the downstream router's point of view, the Anycast-RP will be reachable via the unicast IGP routing. Multicast Source Discovery Protocol (MSDP) peering and information sharing is also required between the Anycast-RPs in this design because it is common for some sources to register with one RP and receivers to join a different RP.

10. **d.** Explanation: Phantom RP is a redundancy consideration for the RP in a PIM-BIDIR deployment. To create a phantom RP, two routers in a network segment will need to be configured with the same IP address but different subnet masks. Then the interior gateway protocol can control the preferred path for the root (phantom RP) of a multicast shared tree based on the longest match (longest subnet mask) where multicast traffic can flow through. The other router with the shorter mask can be used in the same manner if the primary router fails. This means the failover to the secondary shared tree path toward the phantom RP will rely on the unicast IGP convergence.

# Chapter 14

1. **d.** Explanation: Migrate the core to be in dual-stack mode first, and then other enterprise modules can be gradually migrated to IPv6-only or dual stack, depending on the goals and requirements of the business. Migrating to IPv6 this way ensures there is no service interruption.

2. **b.** Explanation: To provide IPv6 access either inbound or outbound at the enterprise Internet edge, a translation mechanism is required that is either based on a load balancer, pure DNS, or classical NAT64.

3. **a.** Explanation: Dual stack is when a device is running both IPv4 and IPv6 protocol stacks. When all of the devices in the network are running like this, it is called end-to-end dual stack.

4. **c.** Explanation: Generic Routing Encapsulation (GRE) is a protocol for encapsulating data packets that uses one routing protocol inside the packets of another protocol. GRE is one way to set up a direct point-to-point connection across a network. In this specific case, IPv6 would be running through the GRE tunnel that travels the IPv4 network.

5. **b.** Explanation: The WFQ algorithm offers a dynamic distribution among all traffic flows based on weights, like that of the DSCP values.

6. **d.** Explanation: First-in, first-out (FIFO) queuing is the default queuing when no other queuing is used. Although FIFO is considered suitable for large links that have a low delay with very minimal congestion, it has no priority or classes of traffic.

7. **c.** Explanation: Priority Queuing is typically four to six queues with different priority levels, and the higher priority queues are always serviced first.

8. **a.** Explanation: A LLQ supports real-time queuing and minimum bandwidth guarantee.

9. **b.** Explanation: Network Configuration Protocol (NETCONF) is a network management protocol defined by the IETF in RFC 6241. All NETCONF messages are encrypted with SSH and encoded with XML.

10. **c.** Explanation: RESTCONF, which is defined in RFC 8040, is an HTTP-based protocol that provides a programmatic interface for accessing YANG-modeled data. RESTCONF can be encoded in either XML or JavaScript Object Notation (JSON). In most design situations, it will be best to leverage NETCONF for router and switches and RESTCONF for controller north-bound communication.

# Chapter 15

1. **a and c.** Explanation: The routed access design model is easier to troubleshoot and has a faster convergence time than a classical Spanning Tree Protocol (STP)-based access design by eliminating the reliance on STP and First Hop Redundancy Protocol (FHRP), and relying on equal-cost multipath (ECMP) for traffic load sharing.

2. **b.** Explanation: One of the technical network design limitations of a routed access design model is the inability to span Layer 2 natively.

3. **d.** Explanation: A business network design limitation of the routed access design model is the higher monetary cost for the routing capabilities on the switch infrastructure for the corresponding licensed features required for routing protocols.

4. **b.** Explanation: Switch clustering is the best access-distribution connectivity model for flexibility.

5. **c.** Explanation: The routed access connectivity model supports both scale up and scale out.

6. **a.** Explanation: Enhanced Interior Gateway Routing Protocol (EIGRP) has the highest architecture flexibility without limitations to the number of tiers while including proper route summarization. Open Shortest Path First (OSPF) and Intermediate System-to-Intermediate System (IS-IS) have inherent limitations as the number of tiers increases that make them both not as flexible. The most flexible option is not listed, which is Border Gateway Protocol (BGP).

7. **b.** Explanation: Of the routing protocols present, OSPF is the only IGP that supports MPLS Traffic Engineering (MPLS-TE).

8. **d.** Explanation: The most scalable virtualization option available is MPLS with Multiprotocol Border Gateway Protocol (MP-BGP). With that said, this also comes with a high operational complexity and requires staff to have advanced routing experience.

9. **a.** Explanation: The design option that requires the least level of routing expertise is VLANs + 802.1Q + VRFs.

# Chapter 16

1. **a.** Explanation: LOCAL_PREFERENCE is a Border Gateway Protocol (BGP) attribute that is applied as routes come into the network to influence traffic outbound and is carried throughout the local autonomous system (AS).

2.   b. Explanation: AS-PATH prepend is a BGP attribute that is applied as routes leave the network (outbound via provider one) to influence traffic into the network (inbound via provider two) a different deterministic path back into the network.

3.   a. Explanation: Equal and unequal load sharing provides flexibility with limited monetary spending.

4.   b. Explanation: Active/standby is used in a situation where a backup link is needed but only as a requirement to survive an outage on the primary (active) link.

5.   c. Explanation: Equal and unequal load sharing with two edge routers is used when there are multiple data centers or multiple sites with a large campus environment hanging off of the inside of the Internet edge architecture with the need to leverage the internal campus routing information to determine the best Internet exit point.

# Chapter 17

1.   a. Explanation: An MPLS L2VPN WAN allows a customer to roll out new transport technologies and services, like IPv6 or multicast, rapidly without having to wait on the provider to make any changes.

2.   b. Explanation: An MPLS L3VPN WAN controls the number of routing neighborships deterministically, one or two BGP neighbors, while an MPLS L2VPN WAN can have hundreds of IGP neighbors across the same link. This second situation can be mitigated with proper configurations if needed.

3.   c. Explanation: Of the options provided, leveraging the Internet as a WAN is the cheapest solution.

4.   b. Explanation: The best WAN transport model for a very large number of remote sites is the MPLS L3VPN WAN option. The MPLS L2VPN WAN introduces some routing and adjacency issues and limitations with a large number of sites. The Internet as WAN option is limited to the VPN hardware supporting the number of concurrent VPN sessions required.

5.   b. Explanation: Dynamic multipoint VPN (DMVPN) is the most scalable option that also provides spoke registration and spoke-to-spoke communication dynamically.

6.   c and d. Explanation: IPsec and Generic Routing Encapsulation (GRE) both provide point-to-point network topologies, while remote access VPN and DMVPN provide hub-spoke network topologies.

7.   b. Explanation: Dynamic multipoint VPN (DMVPN) is the only option that inherently allows for spoke sites to base traffic directly between them, without going to a hub or data center location.

8.   b and d. Explanation: From the options provided, the best options are identifying where the default route is being generated (and how it will be propagated in the new WAN architecture) and knowing the logical and physical site architectures (like OSPF areas and area types, and if there are backdoor links).

# Cisco Certified Design Expert CCDE 400-007 Exam Updates

Over time, reader feedback allows Pearson to gauge which topics give our readers the most problems when taking the exams. To assist readers with those topics, the authors create new materials clarifying and expanding on those troublesome exam topics. As mentioned in the Introduction, the additional content about the exam is contained in a PDF on this book's companion website, at https://www.ciscopress.com/title/9780137601042.

This appendix is intended to provide you with updated information if Cisco makes minor modifications to the exam upon which this book is based. When Cisco releases an entirely new exam, the changes are usually too extensive to provide in a simple update appendix. In those cases, you might need to consult the new edition of the book for the updated content. This appendix attempts to fill the void that occurs with any print book. In particular, this appendix does the following:

- Mentions technical items that might not have been mentioned elsewhere in the book

- Covers new topics if Cisco adds new content to the exam over time

- Provides a way to get up-to-the-minute current information about content for the exam

## Always Get the Latest at the Book's Product Page

You are reading the version of this appendix that was available when your book was printed. However, given that the main purpose of this appendix is to be a living, changing document, it is important that you look for the latest version online at the book's companion website. To do so, follow these steps:

**Step 1.** Browse to https://www.ciscopress.com/title/9780137601042.

**Step 2.** Click the **Updates** tab.

**Step 3.** If there is a new Appendix B document on the page, download the latest Appendix B document.

**NOTE** The downloaded document has a version number. Comparing the version of the print Appendix B (Version 1.0) with the latest online version of this appendix, you should do the following:

- **Same version:** Ignore the PDF that you downloaded from the companion website.

- **Website has a later version:** Ignore this Appendix B in your book and read only the latest version that you downloaded from the companion website.

## Technical Content

The current Version 1.0 of this appendix does not contain additional technical coverage.

# GLOSSARY

# NUMERICS

**2-tier model**   An application model that is like the client/server architecture, where communication takes place between client and server. In this model, the presentation layer or user interface layer runs on the client side while the dataset layer gets executed and stored on the server side.

**3-tier model**   An application model that has three tiers (aka layers):   presentation (web), intermediate (application), and database.

# A

**add technology**   A network design use case that includes adding a new technology, functionality, or capability to an architecture.

**administrative distance (AD)**   A rating of the trustworthiness of a routing information source. A lower number is preferred.

**Anycast-RP**   Based on using two or more rendezvous points (RPs) configured with the same IP address on their loopback addresses. Typically, the Anycast-RP loopback address is configured as a host IP address (32-bit mask). From the downstream router's point of view, the Anycast-RP will be reachable via the unicast IGP routing. MSDP peering and information sharing is also required between the Anycast-RPs in this design, because it is common for some sources to register with one RP and receivers to join a different RP.

**AP cell**   The RF coverage area of a wireless access point; also called the basic service area (BSA).

**application requirements**   The items an application needs to properly function. For example, VoIP has specific requirements of latency, loss, and delay within the network to function properly.

**application tier**   The tier (aka layer) where all of the application's functions and logic occur. This layer processes tasks, functions, and commands; makes decisions and evaluations; and performs calculations. It also is how data is moved between the Web and database layers. This is often referred to as the Presentation or Logic layer of the application.

**area border router (ABR)**   An OSPF router that is connected to more than one area.

**authentication server**   A RADIUS server that contains an authentication database.

**authenticator**   A network device that the client is connecting to. In a wired deployment, this could be a network switch, and in a wireless deployment, this could be the local access point the client is connecting to or the wireless controller that manages the access points.

**authorization for access**   For a resource, in this case a user or device, to be authorized for access to another resource, in this case an asset, application, or system, the trust score and the entitlement level are combined to determine the authorization for access. Just because a trust score is high enough to access a resource, if the user or device doesn't have the correct entitlement, they will not have the appropriate authorization for access.

**autonomous system boundary router (ASBR)**   An OSPF router that injects external LSAs into the OSPF database.

**Auto-RP**   A Cisco-proprietary protocol that automatically communicates rendezvous points (RPs) within a PIM network. Candidate RPs send their announcements to the RP mapping agents with multicast address 224.0.1.39. The 224.0.1.40 address is used in Auto-RP discovery in the destination address for messages from the RP mapping agent to discover candidates. The RP mapping agents select the RP for a group based on the highest IP address of all candidate RPs.

**availability**   Ensures that access to services and systems is always available and information is accessible by authorized users when required.

# B

**Border Gateway Protocol (BGP)**   An interdomain routing protocol that allows BGP speakers residing in different autonomous systems to exchange routing information.

**bottom-up approach**   A design approach that focuses on selecting network technologies and design models first. This can impose a high potential for design failures, because the network will not meet the requirements of the business or its applications.

**BPDU Filter**   A feature that suppresses BPDUs on ports.

**BPDU Guard**   A feature that disables a PortFast-enabled port if a BPDU is received.

**brownfield**   A network design use case that already has an environment with production traffic running through it and now requires a network designer to modify the network architecture.

**business architecture (BA)**   Enables everyone, from strategic planning teams to implementation teams, to get "on the same page", enabling them to address challenges and meet business objectives.

**Business, Operations, Systems, and Technology (BOST)**   Framework that provides the structure for enterprise models, their elements, and relationships. Each of the four BOST elements has its own views. In this framework, the requirements flow downward through the four views. The capabilities flow upward in response to these requirements, creating a mapping between the requirements and capabilities. The success of this framework is based on the ability of a business to align its capabilities with the constantly changing requirements in all four views.

**business architecture scope**   Includes four levels of alignment, Technology Specific, Technology Architecture, Business Solutions, and Business Transformation.

**business capability**   A function provided by a solution that directly meets a business outcome. Business capabilities are not solutions. Business capabilities are what you get from a solution. A network access control solution provides a business the session- and transaction-based security capability, for example. Most solutions provide multiple capabilities. Some solutions provide parts of multiple capabilities; when they are combined with other solutions, the business can get a number of capabilities that will make them successful.

**business driver**   What a business has to do and why. Business drivers are what organizations must follow. A business driver is usually the reason a business must achieve a specific outcome. It is the "why" the business is doing something.

**business outcome**   An underlying goal a business is trying to achieve. A business outcome equates to the end result:   save money, diversify the business, make more money, or fill a specific need. A business outcome will specifically map to a business driver.

**business priority**   The top buckets the business is focused on that all other decisions must align with to ensure success. Each business has a set of priorities that are typically based on strategies adopted for the achievement of goals. These business priorities can influence the planning and design of IT network infrastructure. Therefore, network designers must be aware of these business priorities to align them with the design priorities. This ensures the success of the network they are designing by delivering business value.

**business solution**   A set of interacting business capabilities that delivers specific, or multiple, business outcomes.

**Business Solutions**   A partial business architecture scope for a business that requires expertise to help solve its business problems and determine how to measure the business impact of its technology investments (CAPEX, OPEX, ROI, TCO, etc.).

**Business Transformation**   A business-led architecture scope for a business that requires help with transforming its business capabilities to facilitate innovation to accelerate the company's digitization.

# C

**capital expenses (CAPEX)**   The major purchases a business makes that are to be used over a long period of time. Examples of CAPEX include fixed assets, such as buildings and equipment.

**carrier-neutral facility (CNF)**   A data center that is not owned by a network provider but rather is entirely independent of service providers.

**class-based weighted fair queuing (CBWFQ)**   Provides class-based queuing (user-defined classes) with a minimum bandwidth guarantee. It supports flow-based WFQ for undefined classes, such as class-default. It supports low-latency queuing (LLQ).

**cloud access point (CAP)**   A predefined location to get access to a cloud service provider (CSP) or multiple CSPs. Sometimes CAPs are located in a carrier-neutral facility (CNF).

**confidentiality**   Protects against unauthorized access to information to maintain the desired level of secrecy of the transmitted information across the internal network or public Internet.

**continuous delivery (CD)**    The practice of automatically preparing code changes for release into a production environment. With CD, all code changes are validated in a testing environment before being deployed into a production environment. When CD is properly implemented, the team will always have a deployment-ready build that has passed through the validation test process.

**continuous integration (CI)**    The practice of automating the integration of code changes from multiple team members into a single project.

**Control and Provisioning of Wireless Access Points (CAPWAP)**    A Control and Provisioning of Wireless Access Points (CAPWAP) is a logical network connection between access points and a wireless LAN controller. CAPWAP is used to manage the behavior of the APs as well as tunnel encapsulated 802.11 traffic back to the controller. CAPWAP sessions are established between the AP's logical IP address (gained through DHCP) and the controller's management interface.

**control plane**    Like the brain of the network node, it usually controls and handles all Layer 3 functions.

**controller**    Software-based component that is responsible for the centralized control plane of the SD-WAN fabric network. It establishes a secure connection to each edge router and distributes routes and policy information to it. It also orchestrates the secure data plane connectivity between the different edge routers by distributing crypto key information, allowing for a very scalable, IKE-less architecture.

# D

**data plane**    Responsible for controlling fast-forwarding of traffic passing through a network device.

**database tier**    The tier (aka layer) where information is stored and retrieved from a database. The information is then passed back to the intermediate (application) layer and then eventually back to the end user.

**decision matrix**    Serves the same purpose as a decision tree, but with the matrix, network designers can add more dimensions to the decision-making process.

**decision tree**    A helpful tool that a network designer can use to compare multiple design options, or perhaps protocols, based on specific criteria. A decision tree is a one-dimensional tool.

**Department of Defense Architecture Framework (DODAF)**    Defines a common approach for presenting, describing, and comparing DoD enterprise architectures across organizational, joint, or multinational boundaries. DODAF leverages common terminology, assumptions, and principles to allow for better integration between DoD elements. This framework is suited to large systems with complex integration and interoperability challenges. One element of this framework that is unique is its use of views. Each view offers an overview of a specific area or function and provides details for specific stakeholders within the different domains.

**design failure**   A network design use case that has a problem that needs to be resolved. A simple technical example of this is not aligning the critical roles of Spanning Tree Protocol (STP) and First Hop Redundancy Protocol (FHRP). If your STP root bridge and your FHRP default gateways are not aligned to correct devices, this would lead to a design failure situation.

**distance-vector routing protocol**   A routing protocol that advertises the entire routing table to its neighbors.

**divestment**   A network design use case that requires a network designer to split an architecture into two or more separate architectures that can each function as their own entity.

**dynamic factors**   Sources of data that can be analyzed at the time of access to change what level of access and authorization (i.e., the trust score of the transaction in question) is being provided. The most common dynamic factor is threat intelligence, but can also be geovelocity, GPS coordinates, and real-time data analytics around the transaction.

**Dynamic Frequency Selection (DFS)**   A mechanism that enables an AP to dynamically scan for RF channels and avoid those used by radar stations.

# E

**EAP Flexible Authentication via Secure Tunneling (EAP-FAST)**   Instead of using certificates to achieve mutual authentication, authenticates by means of a PAC (Protected Access Credential), which can be managed dynamically by the authentication server.

**EAP Message-Digest Challenge (EAP-MD5)**   Authentication that provides a base level of EAP support and is typically not recommended for implementation because it may allow the user's password to be derived.

**EAP Transport Layer Security (EAP-TLS)**   Authentication that provides a certificate-based and mutual authentication of the client and the network. It relies on client-side certificates (identity certificates) and server-side certificates to perform authentication and can be used to dynamically generate user-based and session-based keys to secure future communications. One limitation of EAP-TLS is that certificates must be managed on both the client side and server side.

**EAP Tunneled Transport Layer Security (EAP-TTLS)**   An extension of EAP-TLS that provides for certificate-based, mutual authentication of the client and network through an encrypted tunnel. Unlike EA-TLS, EAP-TTLS requires only server-side certificates, which makes it easier to deploy than EAP-TLS but less secure.

**edge control**   Network virtualization design element that represents the network access point. Typically, it is a host or end-user access (wired, wireless, or virtual private network [VPN]) to the network where the identification (authentication) for physical to logical network mapping can occur. For example, a contracting employee might be assigned to VLAN X, whereas internal staff are assigned to VLAN Y.

**edge router**   A device that sits at a physical site or in the cloud and provides secure data plane connectivity among the sites over one or more WAN transports. They are responsible for traffic forwarding, security, encryption, QoS, routing protocols such as BGP and OSPF, and more.

**Embedded-RP**   IPv6 Embedded-RP   Described in RFC 3306. Facilitates interdomain IPv6 multicast communication, in which the address of the rendezvous point (RP) is encoded in the IPv6 multicast group address, and specifies a PIM-SM group-to-RP mapping to use the encoding, leveraging, and extending unicast-prefix-based addressing. The IPv6 Embedded-RP technique offers network designers a simple solution to facilitate interdomain and intradomain communication for IPv6 Any-Source Multicast (ASM) applications without MSDP.

**endpoint device**   Any device an end user can leverage to access the enterprise network. Endpoint devices include business-owned assets and personally owned devices that are approved to access the enterprise network, potentially in a limited way like the common bring your own device (BYOD) deployment. In some vendor-specific implementations of Zero Trust, an endpoint device is also called a policy enforcement point (PEP).

**Enhanced Interior Gateway Routing Protocol (EIGRP)**   Cisco's proprietary enhanced distance-vector routing protocol.

**enterprise architecture (EA)**   A process of organizing logic for business processes and IT infrastructure reflecting the integration and standardization requirements of the company's operating model.

**entitled to access**   The resources a user is allowed to access based on the role they are in. Users have various roles and, based on those roles, are entitled to specific access to complete their job. A financial user would need a different level of access than a human resource user. These two users should not have the same access or authorization. They may have overlapping access to resources that they both need to complete their job functions.

**exception IP traffic**   Any IP traffic carrying a nonstandard "exception" attribute, such as a transit IP packet with an expired TTL.

**Extensible Authentication Protocol (EAP)**   Used to pass authentication information between the supplicant and the authentication server. There are a number of different types of EAP authentication options, with each handling the authentication differently.

# F

**failure isolation**   A technique that creates boundaries within the network design to help contain problems from propagating.

**Federal Enterprise Architecture Framework (FEAF)**   The industry standard framework for government enterprise architectures. Within this framework, the focus is on guiding the integration of strategic, business, and technology management architecture processes. One of the primary benefits of this framework is that it focuses on a common approach to technology acquisition within all U.S. federal agencies.

**feedback loop**    Continuous information sharing to allow for dynamic changes to policy based on constant analysis of new information via AI/ML, Big Data, and data lakes.

**First Hop Redundancy Protocol (FHRP)**    A protocol that deals with first-hop routing. Options are HSRP, VRRP, and GLBP.

**first-in, first-out (FIFO)**    The default queuing when no other queuing is used. Although FIFO is considered suitable for large links that have a low delay with very minimal congestion, it has no priority or classes of traffic.

**functional requirements**    Identify what the different technologies or systems will deliver to the business from a technological point of view. Specifically, functional requirements are the foundation of any system design because they define system and technology functions.

# G

**Gateway Load Balancing Protocol (GLBP)**    A Cisco-proprietary protocol that attempts to overcome the limitations of existing redundant router protocols by adding basic load-balancing functionality. In addition to being able to set priorities on different gateway routers, GLBP allows a weighting parameter to be set.

**General Data Protection Regulation (GDPR)**    A European Union (EU) regulation for data protection that sets guidelines for the collection and processing of personal information from individuals. It applies to the processing of personal data of people in the EU by businesses that operate in the EU. It's important to note that GDPR applies not only to firms based in the EU, but any organization providing a product or service to residents of the EU.

**gold plating**    Adding design elements that are excessive and do not meet any underlying requirement.

**greenfield**    A network design use case that is a clean slate or a clean canvas for a network designer to design an end-to-end architecture from beginning to end.

**gRPC Network Management Interface (gNMI)**    Developed by Google, provides the mechanism to install, manipulate, and delete the configuration of network devices and also to view operational data. The content provided through gNMI can be modeled using YANG. gRPC is a remote procedure call developed by Google for low-latency, scalable distributions with mobile clients communicating to a cloud server.

# H

**Health Information Portability and Accountability Act (HIPAA)**    A compliance standard focused on protecting health and patient information. Network designers must ensure that the network design for businesses that have to follow HIPAA leverage design options that meet the associated security controls specified by HIPAA.

**hierarchy**    The process of creating layers within the architecture for a specific purpose. The most common layers are core, distribution, aggregation, and access.

**Hot Standby Routing Protocol (HSRP)**   A Cisco-proprietary first-hop routing protocol that provides redundancy by creating a virtual router out of two or more routers.

**hybrid cloud**   The use of both private and public clouds together to allow for a business to receive the benefits of both cloud environments while limiting their negative impacts on the business.

# I

**Information Technology Infrastructure Library (ITIL)**   A set of best practice processes for delivering IT services to your organization's customers. ITIL focuses on ITSM and ITAM, and includes processes, procedures, tasks, and checklists that can be applied by any organization. The three focus areas of ITIL are change management, incident management, and problem management.

**Infrastructure as a Service (IaaS)**   A pay-as-you-go service model for storage, networking, and virtualization—all of a business's infrastructure needs. IaaS gives users cloud-based alternatives to on-premises infrastructure, so businesses can avoid investing in expensive onsite resources.

**Infrastructure as Code (IaC)**   The concept of managing and provisioning infrastructure through code instead of through manual processes. With IaC, configuration files are created that contain the network infrastructure specifications, which can then be edited and distributed depending on the need. It also ensures that the environments being provisioned are the same every time.

**infrastructure devices**   Networking devices within the enterprise, such as switches, routers, and firewalls. These devices are also called policy enforcement points (PEPs).

**integrity**   Maintains accurate information end to end by ensuring that no alteration is performed by any unauthorized entity.

**Intermediate System-to-Intermediate System (IS-IS)**   An interior gateway routing protocol with link-state characteristics that uses Dijkstra's shortest path algorithm to calculate paths to destinations.

**inventory**   Single point of truth for all resources. This is an end-to-end inventory throughout the entire architecture/enterprise. This is also known as a policy information point (PIP).

# L

**Lightweight Extensible Authentication Protocol (LEAP)**   An EAP authentication type that encrypts data transmissions using dynamically generated keys and supports mutual authentication. This is a Cisco proprietary protocol that Cisco licenses to other manufacturers and vendors to use.

**Link Aggregation Control Protocol (LACP)**   A protocol defined in IEEE 802.3ad that provides a method to control the bundling of several physical ports to form a single logical channel.

**link-state advertisement (LSA)**    A message that is used to communicate network information such as router links, interfaces, link states, and costs.

**link-state routing protocol**    A routing protocol that uses Dijkstra's shortest path algorithm to calculate the best path.

# M

**MAC Authentication Bypass (MAB)**    The authenticator sends the MAC address of the client device to the authentication server to check if it permits the MAC address. MAB is a very insecure option, but for devices that do not support 802.1X, it can be used.

**management plane**    Relates to the management traffic of the device, such as device access, configuration troubleshooting, and monitoring.

**many-to-one virtualization model**    Model in which multiple physical resources appear as a single logical unit. The classical example of many-to-one virtualization is the switch clustering concept. Other examples include firewall clustering, and FHRP with a single virtual IP (VIP) that front ends a pair of physical upstream network nodes (switches or routers).

**merger**    A network design use case that requires a network designer to combine two independent architectures into one holistic end-to-end architecture.

**modularity**    A concept that allows for purpose-built building blocks to be leveraged.

**Multicast Source Discovery Protocol (MSDP)**    Described in RFC 3618, used to interconnect multiple PIM-SM domains. MSDP reduces the complexity of interconnecting multiple PIM-SM domains by allowing the PIM-SM domains to use an interdomain source tree. With MSDP, the rendezvous points (RPs) exchange source information with RPs in other domains. Each PIM-SM domain uses its own RP and does not depend on the RPs in other domains. When an RP in a PIM-SM domain first learns of a new sender, it constructs a Source-Active (SA) message and sends it to its MSDP peers. All RPs that intend to originate or receive SA messages must establish MSDP peering with other RPs, either directly or via an intermediate MSDP peer.

**multi-cloud**    The use of two or more cloud service providers (CSPs), with the ability to move workloads between the different cloud computing environments in real time as needed by the business.

**Multiple Spanning Tree (MST)**    A protocol that is used to reduce the total number of spanning-tree instances that match the physical topology of the network, reducing the CPU load.

**multitier STP based**    Network design model that represents the classical or traditional way of connecting access to the distribution layer in the campus network. In this model, the access layer switches usually operate in Layer 2 mode only, and the distribution layer switches operate in Layer 2 and Layer 3 modes. The primary limitation of this design model is the reliance on Spanning Tree Protocol (STP) and First Hop Redundancy Protocol (FHRP).

# N

**Network Configuration Protocol (NETCONF)**   A network management protocol, defined in RFC 6241, that provides rich functionality for managing configuration and state data. The protocol operations are defined as remote procedure calls (RPCs) for requests and replies in XML-based representation. NETCONF supports running, candidate, and startup configuration datastores. The NETCONF capabilities are exchanged during session initiation. Transaction support is also a key NETCONF feature. NETCONF is a client/server protocol and is connection-oriented over TCP. All NETCONF messages are encrypted with SSH and encoded with XML. A NETCONF manager is a client, and a NETCONF device is a server. The initial contents of the <hello> message define the NETCONF capabilities that each side supports. The YANG data model defines capabilities for the supported devices.

**network manager**   Centralized network management system that provides a GUI to easily monitor, configure, and maintain all SD-WAN devices and links in the underlay and overlay network.

**non-IP traffic**   Typically related to non-IP packets and almost always is not forwarded, such as MPLS, IS-IS (CLNP), and Layer 2 keepalives.

# O

**one-to-many virtualization model**   Model in which a single physical resource can appear as many logical units, such as virtualizing an x86 server, where the software (hypervisor) hosts multiple virtual machines (VMs) to run on the same physical server. The concept of network function virtualization (NFV) can also be considered as a one-to-many system virtualization model.

**on-premises**   The service model where a business owns and manages the infrastructure. A business will procure all of the infrastructure required to run the service and then fully manage, maintain, and operate it. In some situations, the management is outsourced but the infrastructure is procured and owned by the business.

**Open Short Path First (OSPF)**   A link-state routing protocol that uses Dijkstra's shortest path algorithm to calculate paths to destinations.

**operating expenses (OPEX)**   The day-to-day costs incurred by a business to keep the business operational. Examples of OPEX include rent, utilities, payroll, and marketing.

**orchestrator**   Software-based component that performs the initial authentication of edge devices and orchestrates controller and edge device connectivity. It also has an important role in enabling the communication of devices that sit behind NAT.

**overlay network**   Runs over the underlay network to create a virtual network. Virtual networks isolate both data plane traffic and control plane behavior among the physical networks of the underlay. Virtualization is achieved inside SD-LAN by encapsulating user traffic over IP tunnels that are sourced and terminated at the boundaries of SD-LAN. Network virtualization extending outside of the SD-LAN is preserved using traditional virtualization technologies such as virtual routing and forwarding (VRF)-Lite, MPLS VPN, or SD-WAN. Overlay networks can run across all or a subset of the underlay network devices. Multiple overlay networks can run across the same underlay network to support multitenancy through virtualization.

# P

**Payment Card Industry Data Security Standard (PCI DSS)**    A compliance standard focused on ensuring the security of credit card transactions. This standard refers to the technical and operational standards that businesses must follow to secure and protect credit card data provided by cardholders and transmitted through card processing transactions. Network designers have to ensure that the network design for businesses that have to follow PCI DSS leverages design options that meet the associated security controls specified by PCI DSS.

**perimeter security (aka Turtle Shell)**    Legacy security model that leverages a security device at the edge or perimeter that is the gatekeeper into the network. This security device has a bunch of security capabilities that limit what traffic can get into the network and what can leave it. Inside the network, behind the security device, there are no other security devices. In this model, there is full east–west (lateral movement) traffic between users and resources.

**phantom RP**    A redundancy consideration for the rendezvous point (RP) in a PIM-BIDIR deployment. To create a phantom RP, two routers in a network segment will need to be configured with the same IP address but different subnet masks. Then IGP can control the preferred path for the root (phantom RP) of a multicast shared tree based on the longest match (longest subnet mask) where multicast traffic can flow through. The other router with the shorter mask can be used in the same manner if the primary router fails. This means the failover to the secondary shared tree path toward the phantom RP will rely on the unicast IGP convergence.

**PIM Bidirectional (BIDIR)**    Defined in RFC 5015, a variant of PIM-SM that builds bidirectional shared trees connecting multicast sources and receivers. It never builds a shortest-path tree, so it scales well because it does not need a source-specific state. PIM-BIDIR eliminates the need for a first-hop route to encapsulate data packets being sent to the RP. PIM-BIDIR dispenses with both encapsulation and source state by allowing packets to be natively forwarded from a source to the RP using the shared tree state.

**PIM Source-Specific Multicast (PIM-SSM)**    A variant of PIM-SM that builds trees that are rooted in just one source. SSM, defined in RFC 3569, eliminates the requirement for rendezvous points (RPs) and shared trees of sparse mode and only builds a shortest-path tree (SPT). SSM trees are built directly based on the receipt of group membership reports that request a given source. SSM is suitable for when well-known sources exist within the local PIM domain and for broadcast applications.

**Platform as a Service (PaaS)**    A service model where a vendor provides hardware and software tools, and customers use these tools to develop applications. PaaS users tend to be application developers.

**policy enforcement point (PEP)**    Location where trust and policy are enforced.

**policy engine**    Implements policy, matches rules, and pushes associated access to the policy enforcement points. This is also known as a policy administration point (PAP) and a policy decision point (PDP).

**Priority Queuing (PQ)**    Typically supports four queues with different priority levels, and the higher-priority queues are always serviced first.

**private cloud**    Consists of cloud computing resources used by one business. This cloud environment can be located within the business's data center footprint, or it can be hosted by a cloud service provider (CSP). In a private cloud, the resources, applications, services, data, and infrastructure are always maintained on a private network and all devices are dedicated to the business.

**Protected Extensible Authentication Protocol (PEAP)**    Provides a method to transport securely authentication data, including legacy password-based protocols. PEAP accomplishes this by using tunneling between PEAP clients and an authentication server.

**public cloud**    The most common type of cloud computing, the cloud computing resources are owned and operated by a cloud service provider (CSP). All infrastructure components are owned and maintained by the CSP. In a public cloud environment, a business shares the same hardware, storage, virtualization, and network devices with other businesses.

# R

**radio frequency (RF)**    A wireless electromagnetic signal used as a form of data communication.

**receive IP traffic**    Traffic destined to the network node itself, such as toward a router's IP address, and requires CPU processing

**recovery point objective (RPO)**    The amount of data that can be lost during an outage at peak business demand before harm occurs to the business. The amount of data that can be lost from an RPO perspective is given a specific time value, which is measured against the last backup that took place.

**recovery time objective (RTO)**    The time an application, system, network, or resource can be offline without causing significant business damage as well as the time it takes to restore the service in question. RTO is focused on the time to recover a failing system or network outage.

**redundancy**    The concept of having multiple resources performing the same function/role so that if one of them fails, the other takes over with limited to no impact on the production traffic.

**reliability**    How much of the network data gets from source to destination locations in the right amount of time to be leveraged correctly.

**replace technology**    A network design use case that includes replacing a technology, function, or capability to an architecture with another one. For example, replacing OSPF with EIGRP as a routing protocol.

**resilience (aka resiliency)**    The ability of the network to automatically fail over when an outage occurs.

**RESTCONF**    Defined in RFC 8040, an HTTP-based protocol that provides a programmatic interface for accessing YANG modeled data. RESTCONF uses HTTP operations to provide create, retrieve, update, and delete (CRUD) operations on a NETCONF datastore containing YANG data. RESTCONF is tightly coupled to the YANG data model definitions. It supports HTTP-based tools and programming libraries. RESTCONF can be encoded in either XML or JSON.

**return on investment (ROI)**    The concept of identifying what the perceived potential benefit is going to be for the business if the business does an action (i.e., is the investment going to be profitable for the business?).

**reverse path forwarding (RPF)**    The mechanism used by Layer 3 nodes in the network to optimally forward multicast datagrams. The RPF algorithm uses the following rules:

- **Receiving:**  If the multicast packet has arrived on the RPF interface and the router receives it on an interface used to send unicast packets to the source subnet.

- **Forwarding:**  If the packet arrives on the RPF interface, the router forwards it out the interfaces that are present in the outgoing interface list of a multicast routing table entry.

- **Dropping:**  If the packet does not arrive on the RPF interface, the packet is silently discarded to avoid multicast loops.

**Risk Management Framework (RMF)**    Focuses on the integration of security, privacy, and cyber supply chain processes into the system development life cycle. This risk-based approach to control the selection process considers effectiveness, efficiency, and legal constraints and directives. Managing organizational risk is critical to the security, safety, and privacy of information systems. RMF can be applied to new and legacy systems, leveraging any technology, and within any specific market or sector.

**risk score**    The minimum threshold level to access a specific resource. Assets, applications, networks, and so forth—what we generally call resources—have levels of risk scores, which are thresholds that must be exceeded for access to be permitted. In general, the security plan categorization determines asset level or risk.

**Root Guard**    A feature that prevents external switches from becoming the root of the Spanning Tree Protocol tree.

**route distinguisher (RD)**    For an MPLS L3VPN to support having multiple customer VPNs with overlapping addresses and to maintain the control plane separation, the PE router must be capable of using processes that enable overlapping address spaces of multiple customer VPNs. In addition, the PE router must also learn these routes from directly connected customer networks and propagate this information using the shared backbone. This is accomplished by using an RD per VPN or per VRF. As a result, the MPLS core can seamlessly transport customers' routes (overlapped and nonoverlapped) over one common infrastructure and control plane protocol to take advantage of the RD prepended per MP-BGP VPNv4/v6 prefix.

**route target (RT)**    Route targets (RTs) are an additional identifier and are considered part of the primary control plane elements of a typical MPLS L3VPN architecture because they facilitate the identification of which VRF can install which VPN routes. In fact, RTs represent the policies that govern the connectivity between customer sites. This is achieved via controlling the import and export RTs. Technically, in an MPLS VPN environment, the export RT is to identify a VPN membership with regard to the existing VRFs on other PEs, whereas the import RT is associated with each PE local VRF. The import RT recognizes and maps the VPN routes (received from remote PEs or leaked on the local PE from other VRFs) to be imported into the relevant VRF of any given customer. In other words, RTs can offer network designers a powerful full capability to control what MP-BGP VPN route is to be installed, in any given VRF/customer routing instance. In addition, it provides flexibility to create various logical L3VPN (WAN) topologies for the enterprise customer, such as any to any, hub and spoke, and partially meshed, to meet different connectivity requirements.

**routed access**    Network design model in which access layer switches act as Layer 3 routing nodes, providing both Layer 2 and Layer 3 forwarding. In other words, the demarcation point between Layer 2 and Layer 3 is moved from the distribution layer to the access layer. Based on that, the Layer 2 trunk links from access to distribution are replaced with Layer 3 point-to-point routed links.

# S

**scaling**    A network design use case that requires a network designer to address design limitations with a current production design and suggest modification to increase its scalability. This can be as simple as a single flat area 0 OSPF design that doesn't scale to the business requirements, in which case a network designer can leverage multiple areas, multiple area types, and LSA filtering techniques to increase the scalability of the network design.

**service set identifier (SSID)**    A unique ID that is used as a wireless network name and can be made up of case-sensitive letters, numbers, and special characters. When designing wireless networks, we give each wireless network a name which is the SSID. This allows end users to distinguish from one wireless network to another.

**services virtualization**    Network virtualization design element that represents the extension of the network virtualization concept to the services edge, which can be shared services among different logically isolated groups, such as an Internet link or a file server located in the data center that must be accessed by only one logical group (business unit).

**session- and transaction-based security**    A security model in which users and devices are locked down. Resources like printers, applications, security cameras, and so forth can be locked down to have access only to what they need access to. In this model, east–west traffic is secured dynamically.

**shared failure state (aka fate sharing)**    A state in which a device is performing multiple critical functions and if it were to incur an outage from one of those functions, it would affect the other critical functions.

**shared tree**    The multicast tree roots somewhere between the network's source and receivers. The root is called the rendezvous point (RP). The tree is created from the RP throughout the network with no loops. Sources will first send their multicast traffic to the RP, which then forwards data to the member of the group in the shared tree.

**shortest-path tree (SPT)**   Also called source trees, the multicast tree roots from the source of the multicast group and then expands throughout the network to the destination hosts. These paths are created without having to go through a rendezvous point (RP).

**signal-to-noise ratio (SNR)**   The difference between a received signal's strength and the noise floor.

**single-server model**   The simplest application model, equivalent to running the application on a personal computer. All of the required components for an application to run are on a single application or server.

**Software as a Service (SaaS)**   A service model where a vendor makes its software available to users, usually for a monthly or annual subscription service fee.

**Spanning Tree Protocol (STP)**   A protocol that prevents loops from being formed when switches are interconnected via multiple paths.

**static factors**   Items that we know and can preemptively base access and authorization on. The most common static factors are credentials but could also include the level of confidence, device trust, network, physical location, biometrics, and device orientation.

**strategic planning approach**   Typically targets planning to long-term business outcomes and strategies.

**supplicant**   A software client running on the end device that passes authentication information to the authentication server.

**switch clustering**   Network design model that provides the simplest and most flexible design compared to the other design models. By introducing the switch clustering concept across the different functional modules of the enterprise campus architecture, network designers can simplify and enhance the design to a large degree. This offers a higher level of node and path resiliency, along with significantly optimized network convergence time.

# T

**tactical planning approach**   Typically targets planning to overcome an issue or to achieve a short-term goal.

**technical requirements**   The technical aspects that a network infrastructure must provide in terms of security, availability, and integration. These requirements are often called *nonfunctional requirements*.

**technology architecture**   A multi-domain architecture (MDA), also referred to as cross-architecture. In this scope, the business needs help understanding the benefits of multi-domain technology architecture and how to show the value it provides to the business. With this scope, two or more architecture focus areas are incorporated.

**technology specific**   A domain-specific architecture. Within this scope, the business is requiring help with finding and purchasing the right product or group of products in an architecture focus area. This might be data center, security, or enterprise networking focused, but it doesn't cross between the different architecture focus areas.

**The Open Group Architecture Framework (TOGAF)**   An enterprise architecture methodology that incorporates a high-level framework for enterprises that focuses on designing, planning, implementing, and governing enterprise information technology architectures. TOGAF helps businesses organize their processes through an approach that reduces errors, decreases timelines, maintains budget requirements, and aligns technology with the business to produce business-impacting results.

**three-tier model**   Network design model that is typically used in large enterprise campus networks, which are constructed of multiple functional distribution layer blocks. This model has dedicated core, distribution, and access layers.

**top-down approach**   A design approach that simplifies the design process by splitting the design tasks to make it more focused on the design scope and performed in a more controlled manner, which can ultimately help network designers to view network design solutions from a business perspective.

**transit IP traffic**   Traffic for which a network device makes a typical routing and forwarding decision regarding whether to send the traffic over its interfaces.

**transport virtualization**   Network virtualization design element that represents the transport path that will carry different virtualized networks over one common physical infrastructure, such as an overlay technology like a generic routing encapsulation (GRE) tunnel. The terms *path isolation* and *path separation* are commonly used to refer to transport virtualization.

**trust engine**   Dynamically evaluates overall trust by continuously analyzing the state of devices, users, workloads, and applications (resources). Utilizes a trust score that is built from static and dynamic factors. This is also known as a policy information point (PIP).

**trust score**   A combination of factors, both static and dynamic, this is used to continually provide identity assurance. A trust score determines the level of access as required by the level of risk value of the asset being accessed.

**two-tier model**   Network design model that is more suitable (than a three-tier model) for small to medium-size campus networks (ideally not more than three functional disruption blocks to be interconnected), where the core and distribution functions can be combined into one layer, also known as *collapsed core-distribution architecture*. This model has two layers, a collapsed core-distribution layer and an access layer.

# U

**underlay network**   Defined by the physical switches and routers that are part of the LAN. All network elements of the underlay must establish IP connectivity via the use of a routing protocol. Theoretically, any topology and routing protocol can be used, but the implementation of a well-designed Layer 3 foundation to the LAN edge is highly recommended to ensure performance, scalability, and high availability of the network. In the SD-LAN, end-user subnets are not part of the underlay network but instead are part of the overlay network.

**unstated requirements**   The concept that customers do not articulate their specific requirements explicitly. Customers assume requirements, which leave the network designer to figure out what requirements are important to the design.

# V

**virtual local-area network (VLAN)**    A broadcast domain that is isolated within Layer 2 and defined logically. Ports in a LAN switch are assigned to different VLAN numbers.

**virtual network (VN)**    Provides segmentation, much like Virtual Routing and Forwarding (VRF) instances. Each VN is isolated from other VNs and each has its own forwarding table. An interface or subinterface is explicitly configured under a single VN and cannot be part of more than one VN. Labels are used in the management protocol route attributes and in the packet encapsulation, which identifies the VN a packet belongs to. The VN number is a 4-byte integer with a value from 0 to 65530.

**Virtual Router Redundancy Protocol (VRRP)**    A standards-based first-hop routing protocol that provides redundancy with a virtual router elected as the master.

**Virtual Routing and Forwarding (VRF)**    One of the primary mechanisms used in today's modern networks to maintain routing isolation on a Layer 3 device level. In MPLS architecture, each PE holds a separate routing and forwarding instance per VRF per customer. Typically, each customer's VPN is associated with at least one VRF. Maintaining multiple VRFs on the same PE is similar to maintaining multiple dedicated routers for customers connecting to the provider network.

**VPN label**    Typically, VPN traffic is assigned to a VPN label at the egress PE (LER) that can be used by the remote ingress PEs (LER), where the egress PE demultiplexes the traffic to the correct VPN customer egress interface based on the assigned VPN label. In other words, the VPN label is generated and assigned to every VPN route by the egress PE router, then advertised to the ingress PE routers over an MP-BGP update. Therefore, it is only understood by the egress PE node that performs demultiplexing to forward traffic to the respective VPN customer egress interface/CE based on its VPN label.

# W

**web tier**    The front end of the application that all end users access. This is how an end user sees and interacts with the application. This is often called the web or GUI tier (aka layer) of the application. The main function of this tier is to translate tasks and results to something the end user can understand.

**weighted fair queuing (WFQ)**    Algorithm that offers a dynamic fair distribution among all traffic flows based on weight.

# Y

**Yet Another Next Generation (YANG)**    An IETF standard (RFC 6020) data modeling language used to describe the data for network configuration protocols such as NETCONF and RESTCONF.

# Z

**Zachman Framework**   Provides a means to classify a business's architecture in a structured manner. This is a proactive business tool that is used to model a business's functions, elements, and processes to help the business manage change throughout the organization.

**Zero Trust Architecture**   A security model that adds real-time capture and analytics tools to the mix to allow for real-time AI/ML decision making. Every device, user, application, server, service, and resource (even data itself) is assigned a trust score. This trust score changes based on what the analytics engine sees happening.

**zero-touch provisioning (ZTP)**   Capability used to automatically configure a new device once it is plugged into the network. When configured, the new device will pull a DHCP address from a preconfigured DHCP server that tells the new device where to pull its initial configuration and OS image. If the image downloaded is not the same as the image that is running on the new device, the new device will complete an upgrade to the downloaded image. Once the upgrade is completed, the new device will apply the downloaded configuration, which is commonly called the day zero configuration. At this point, the device is ready for future day one configuration.

# Index

# C

web (presentation) layer, 3-tier application model, 69, 70

WFQ (Weighted Fair Queuing), 377

"Why?" identifying requirements with, 15–16

wireless networks

CAPWAP, 306–309

device capabilities, 303–305

enterprise wireless network design, 309–313

high density wireless network design, 309–312

IEEE 802.11 wireless standard, 303–305

LAN controller services ports, 309

security, 305–306

traffic flows, 306–309

video design, 312–313

voice design, 312–313

wireless security, 291–292

# X - Y

YANG (Yet Another Next Generation), 396–397, 398

# Z

Zachman Framework, 112, 115

zero successor routes, 175

Zero Trust Architectures, 17, 91–92, 94

authorization for access, 93

dynamic factors, 92

endpoint devices, 93

entitled to access, 93

feedback loops, 94

infrastructure devices, 94

inventories, 93

migrating, 94–95

PEP, 94

policy engines, 93

risk scores, 93

static factors, 92

trust engines, 93

trust scores, 93

ZTP (Zero-Touch Provisioning), 318–319, 324